BAYESIAN INFERENCE AND MAXIMUM ENTROPY METHODS IN SCIENCE AND ENGINEERING

Proceedings in the Series of Workshops on Bayesian Inference and Maximum Entropy Methods in Science and Engineering

Year		Held in	Publisher	ISBN
2007	27th	Saratoga Springs, NY, U.S.A.	AIP Conf. Proceedings Vol. 954	978-0-7354-0468-7
2006	26th	Paris, France	AIP Conf. Proceedings Vol. 872	978-0-7354-0371-0
2005	25th	San José, CA, U.S.A.	AIP Conf. Proceedings Vol. 803	0-7354-0292-2
2004	24th	Garching, Germany	AIP Conf. Proceedings Vol. 735	0-7354-0217-5
2003	23rd	Jackson Hole, WY, U.S.A.	AIP Conf. Proceedings Vol. 707	0-7354-0182-9
2002	22nd	Moscow, ID, U.S.A.	AIP Conf. Proceedings Vol. 659	0-7354-0119-5
2001	21st	Baltimore, MD, U.S.A.	AIP Conf. Proceedings Vol. 617	0-7354-0063-6
2000	20th	Gif-sur-Yvette, France	AIP Conf. Proceedings Vol. 568	0-7354-0004-0
1999	19th	Boise, ID, U.S.A.	AIP Conf. Proceedings Vol. 567	0-7354-0003-2

To learn more about these titles, or the AIP Conference Proceedings Series, please visit the webpage **http://proceedings.aip.org/proceedings**

BAYESIAN INFERENCE AND MAXIMUM ENTROPY METHODS IN SCIENCE AND ENGINEERING

27th International Workshop on Bayesian Inference and Maximum Entropy Methods in Science and Engineering

Saratoga Springs, New York   8 – 13 July 2007

EDITORS

Kevin H. Knuth
Ariel Caticha
Adom Giffin
Carlos C. Rodríguez
University at Albany, New York

Julian L. Center, Jr.
Creative Research Corp.
Andover, Massachusetts

All papers have been peer reviewed

SPONSORING ORGANIZATIONS
The Edwin T. Jaynes International Center for Bayesian Methods
   and Maximum Entropy, Boise State University, Idaho
University at Albany, New York
MaxEnt Workshops, Inc.

Melville, New York, 2007
AIP CONFERENCE PROCEEDINGS ■ VOLUME 954

Editors:

Kevin H. Knuth
Ariel Caticha
Adom Giffin

Department of Physics
1400 Washington Avenue
University at Albany
Albany, NY 12222
USA

E-mail: kknuth@albany.edu
ariel@albany.edu
physics101@gmail.com

Carlos C. Rodríguez
Department of Mathematics
1400 Washington Avenue
University at Albany
Albany, NY 12222
USA

E-mail: carlos@math.albany.edu

Julian L. Center, Jr.
Creative Research Corp.
Andover, MA 01810
USA

E-mail: jcenter@ieee.org

Authorization to photocopy items for internal or personal use, beyond the free copying permitted under the 1978 U.S. Copyright Law (see statement below), is granted by the American Institute of Physics for users registered with the Copyright Clearance Center (CCC) Transactional Reporting Service, provided that the base fee of $23.00 per copy is paid directly to CCC, 222 Rosewood Drive, Danvers, MA 01923, USA. For those organizations that have been granted a photocopy license by CCC, a separate system of payment has been arranged. The fee code for users of the Transactional Reporting Services is: 978-0-7354-0468-7/07/$23.00.

© 2007 American Institute of Physics

Permission is granted to quote from the AIP Conference Proceedings with the customary acknowledgment of the source. Republication of an article or portions thereof (e.g., extensive excerpts, figures, tables, etc.) in original form or in translation, as well as other types of reuse (e.g., in course packs) require formal permission from AIP and may be subject to fees. As a courtesy, the author of the original proceedings article should be informed of any request for republication/reuse. Permission may be obtained online using Rightslink. Locate the article online at http://proceedings.aip.org, then simply click on the Rightslink icon/"Permission for Reuse" link found in the article abstract. You may also address requests to: AIP Office of Rights and Permissions, Suite 1NO1, 2 Huntington Quadrangle, Melville, NY 11747-4502, USA; Fax: 516-576-2450; Tel.: 516-576-2268; E-mail: rights@aip.org.

L.C. Catalog Card No. 2007938734
ISBN 978-0-7354-0468-7
ISSN 0094-243X
Printed in the United States of America

Contents

Preface ix
Dedication xi

TUTORIALS

Foundations? 3
        J. Skilling
Information and Entropy 11
        A. Caticha
Lattice Theory, Measures and Probability 23
        K. H. Knuth

FOUNDATIONS

Probability and Geometry 39
        J. Skilling
Wrong Priors 47
        C. C. Rodríguez
Towards a Cross-Level Theory of Neural Learning 56
        A. J. Bell
Updating Probabilities with Data and Moments 74
        A. Giffin and A. Caticha
Bayesians *Can* Learn from Old Data 85
        W. H. Jefferys
The Marginalization Paradox and the Formal Bayes' Law 93
        T. C. Wallstrom
Positive Evidence for Non-Arbitrary Assignments of Probability 101
        W. M. Briggs
Cellular Automata Generalized to an Inferential System 109
        D. J. Blower
From Maxent to Machine Learning and Back 117
        T. D. Sears
Induced Semantics for Undirected Graphs: Another Look at the Hammersley-Clifford Theorem 125
        T. D. Sears and P. Sunehag
Origins of the Combinatorial Basis of Entropy 133
        R. K. Niven
On Shannon-Jaynes Entropy and Fisher Information 143
        V. I. Dimitrov
The Role of Information in the Probabilistic Reconstruction of Quantum Theory 153
        P. Goyal
From Information Geometry to Newtonian Dynamics 165
        A. Caticha and C. Cafaro

Information Geometry and Chaos on Negatively Curved Statistical
Manifolds... 175
    C. Cafaro
Gravity from a Probabilistic Point of View 185
    P. Martin

METHODS

The Concept of Integrated Data Analysis of Complementary
Experiments... 195
    R. Fischer and A. Dinklage
Designing Intelligent Instruments...................................... 203
    K. H. Knuth, P. M. Erner, and S. Frasso
Deconvolution Using Thin-Plate Splines 212
    U. v. Toussaint and S. Gori
Comparison of Numerical Methods for Evidence Calculation................ 221
    R. Preuss and U. v. Toussaint
Regression for Proportion Data 229
    J. L. Center, Jr.
Inverse Covariance Simplification for Efficient Uncertainty
Management 237
    A. Jalobeanu and J. A. Gutiérrez
Propagation of Statistical Information through Non-Linear Feature
Extractions for Robust Speech Recognition 245
    R. F. Astudillo, D. Kolossa, and R. Orglmeister
Maximum a Posteriori Maximum Entropy Signal Denoising.................. 253
    A.-K. Seghouane and L. Knockaert
FBST for Unit Root Problems 260
    M. Diniz, C. A. B. Pereira, and J. M. Stern
The Problem of Separate Hypotheses via Mixture Models.................. 268
    M. de Souza Lauretto, S. R. de Faria, Jr., B. B. Pereira, C. A. B. Pereira,
    and J. M. Stern
Assigning Priors for Ordered and Bounded Parameters 276
    P. M. Goggans and C.-Y. Chan
Bayesian Inference Featuring Entropic Priors........................... 283
    T. Neumann
Strong Nonlinear Correlations, Conditional Entropy and Perfect
Estimation 293
    C. S. Jones, J. M. Finn, and N. Hengartner

APPLICATIONS

A Bayesian Re-Analysis of HD 11964: Evidence for Three Planets.......... 307
    P. C. Gregory
Bayesian Analysis of RR Lyrae Luminosities and Kinematics................ 315
    T. R. Jefferys, T. G. Barnes, A. Dambis, and W. H. Jefferys

Estimating Background Spectra 322
      M. K. Tse, J. Choinsky, D. F. Carbon, and K. H. Knuth
Parameter Estimation in Ultrasonic Measurements on Trabecular
Bone 329
      K. R. Marutyan, C. C. Anderson, K. A. Wear, M. R. Holland, J. G. Miller,
      and G. L. Bretthorst
Model Selection in Ultrasonic Measurements on Trabecular Bone 337
      C. C. Anderson, K. R. Marutyan, K. A. Wear, M. R. Holland, J. G. Miller,
      and G. L. Bretthorst
Computing the Probability of Local Brain Connectivity Using
Diffusion Tensor Imaging 346
      J. S. Shimony, A. A. Epstein, and G. L. Bretthorst
Generalised Skilling-Bryan Minimisation for Micro-Rotation Imaging
in Light Microscopy 354
      D. Laksameethanasan and S. S. Brandt
Bayesian Estimator for Angle Recovery: Event Classification and
Reconstruction in Positron Emission Tomography 362
      A. M. K. Foudray and C. S. Levin
Bayesian Tomographic Reconstruction of Microsystems 372
      S. Fekih Salem, A. Vabre, and A. Mohammad-Djafari
A Nonparametric Bayesian Approach for Emission Tomography
Reconstruction 381
      É. Barat and T. Dautremer
Filter Out High Frequency Noise in EEG Data Using the Method of
Maximum Entropy 386
      C.-Y. Tseng and H. C. Lee
Bayesian Estimation of the Learning Effects of Repeated Pointing
Tasks 394
      K. Kyo
Classification of Maize and Weeds by Bayesian Networks 402
      M. Chapron, A. Oprea, B. Sultana, and L. Assemat
Separation of Stochastic and Deterministic Information from
Seismological Time Series with Nonlinear Dynamics and Maximum
Entropy Methods 410
      R. M. Gutiérrez, G. M. Useche, and E. Buitrago
Using Prior Information in Bayesian Inference - with Application to
Fault Diagnosis 418
      A. Pernestål and M. Nyberg
Application of the Maximum Entropy Method to Estimating
Parameter Distributions for Sonar Signal Processing 427
      R. L. Culver, H. J. Camin, J. A. Ballard, C. W. Jemmott, and L. H. Sibul
Modeling the Multiple-Antenna Wireless Channel Using Maximum
Entropy Methods 435
      M. Guillaud, M. Debbah, and A. L. Moustakas
Analysis of Intrusion Detection and Attack Proliferation in Computer
Networks 443
      P. Rangan and K. H. Knuth

Bayesian Extraction of Deep UV Resonance Raman Signature of Fibrillar Cross-β Sheet Core based on H-D Exchange Data 450
    V. K. Shashilov and I. K. Lednev

Lessons about Likelihood Functions from Nuclear Physics 458
    K. M. Hanson

Assessment of Electron Energy Distributions in Discharges by Optical Emission Spectroscopy 468
    D. Dodt, A. Dinklage, R. Fischer, and R. Preuss

Group Photo 476
Yearbook 479
Photos 483
Author Index 491

PREFACE

The Twenty-Seventh International Workshop on Bayesian Inference and Maximum Entropy Methods in Science and Engineering (MaxEnt 2007) was held in Saratoga Springs, New York, USA from July 8-13, 2007. In celebration of the 50th anniversary of Ed Jaynes' 1957 paper, "Information Theory and Statistical Mechanics", we focused the tutorial sessions on the foundations of probability theory and entropy and their relationships to other areas of study. The workshop featured 5 tutorials, 6 invited speakers, 38 oral presentations, and 23 poster presentations, and was attended by 89 participants from 17 countries.

Like the previous MaxEnt Workshops, this meeting encompassed all aspects of probabilistic inference such as foundations, techniques and applications. The topics represented included probability theory and information theory, models and model testing, sampling techniques, astronomy and astrophysics, biology and medicine, geophysics and earth science, physics and quantum mechanics, image analysis, text analysis, source separation, classification, and communication.

The workshop began with the tutorial sessions where John Skilling presented "Foundations?," Jose M. Bernardo presented "Bayesian Statistics as Applied Probability," Ariel Caticha presented "Information and Entropy," Carlos Rodríguez presented "Geometry and Prior Information," and Kevin Knuth presented "Lattice Theory, Measures and Probability".

We were extremely fortunate to have a distinguished set of invited speakers, which included Professor Shun-ichi Amari (RIKEN, Japan), Dr. Tony Bell (University of California Berkeley), Professor Jose M. Bernardo (Universitat de Valencia, Spain), Dr. Philip Goyal (Cambridge University, United Kingdom), Professor Phil Gregory (University of British Columbia, Canada), and Professor Stephen Roberts (University of Oxford, United Kingdom).

For the first time in the history of this workshop, the meeting's talks were videotaped. With the permission granted by the speakers, these videotaped lectures will be made permanently available on the internet at http://maxent2007.org. We would like to thank Carlos Rodríguez for suggesting and engineering this new way of preserving the ideas and atmosphere of MaxEnt 2007.

The MaxEnt workshops are unique in several respects, and MaxEnt 2007 was no different. The strong attendance of this workshop by both masters and novices alike and the ease with which they mix and intermingle is critical in the transmission of ideas across generations. The organizers would like to extend a special thank you to the experienced professionals who throughout the years have spent time with and have inspired the new eager scientists at this workshop. This is perhaps one of the greatest strengths of this workshop, and has led to a coherence and exchange of ideas that is unparalleled in the scientific community. MaxEnt also exhibits a great diversity of techniques and applications all presented in a single-track format. In this sense, these meetings are truly interdisciplinary, which leads to cross-fertilization that again is unparalleled.

The aim of the MaxEnt workshops is to push at the forefront of our understanding. We contend that this is impossible unless diverse and possibly controversial points of view are deliberately encouraged. Just as the impressionist painters found their work rejected at the annual exhibitions of the Salon de l'Académie, we feel that some of the papers collected here could very well be rejected by established scientific journals. And just as the impressionists did not compromise quality as they explored the esthetic frontier, we hold that these papers do not compromise on quality as they explore the scientific frontier. The ideas they contain may be controversial, and some of them will eventually be proven wrong, but at this early stage they need to be nurtured rather than prematurely dismissed. Let MaxEnt 2007 play the essential role of a Salon des Refusés.

This year, several confirmed participants found that they were unable to attend the workshop due to difficulties obtaining visas. As fellow scientists, many of us at the workshop were extremely disappointed by this news. Science unifies humanity in ways that no other human endeavor can. It unites across race, religion, age, and nationality. Our hopes are that we can move away from the differences that divide us, and look forward to increasing unity across the human race.

This meeting would not have been possible without the support of several groups and individuals. We would like to thank Boise State University and the E.T. Jaynes Foundation for their continued support. We would like to acknowledge the University at Albany, which supported this workshop at many levels. Specifically, we would like to thank Dean Peter Bloniarz (Dean of the College of Computing and Information), Dean Joanne Wick-Pelletier (Dean of the College of Arts and Sciences), John Kimball (Chair of Physics), and Donna Collins (Physics Departmental Secretary) for their support and efforts. We would like to acknowledge the International Society for Bayesian Analysis and the Valencia International Meetings on Bayesian Statistics for their support in advertising this workshop to an ever-growing community.

We the editors, would also like to acknowledge and thank the local organizers Ning Xiang, Deniz Gençağa, and Nabin Malakar for their time and effort, as well as the advisory committee and the reviewers who worked to improve the quality of the work presented in this volume. We would also like to acknowledge the efforts of the student assistants: Newshaw Bahreyni, Tilman Birnstiel, Philip Erner, Haley Maunu, Muhammad Mubeen, and Man Kit Tse. Last, we especially acknowledge and thank Emily Knuth, Marta Gomez, Lee Anne Center, Tracy Giffin, and Monica Rodríguez for their help and support when it was most needed.

Once again, we thank John Burg for his work in establishing the E. T. Jaynes Foundation, which has supported this meeting for the last seven years. Last we extend a special acknowledgement to Gary Erickson, who has worked with the E.T. Jaynes Foundation, and through his efforts greatly facilitated the organization and execution of these workshops.

Commemorating 50 years since the publication of E. T. Jaynes' pioneering paper "Information Theory and Statistical Mechanics," Physical Review **106**, 106 (1957).

We dedicate this volume to the memory of E. T. Jaynes who showed us the way. He would have been pleased and (we hope) astonished at the multiplicity of roads that we have followed.

TUTORIALS

Foundations?

John Skilling

Maximum Entropy Data Consultants Ltd, Killaha East, Kenmare, Kerry, Ireland

Abstract. *Measure* is a valuation on elements of a lattice: it obeys the sum rule. *Probability* is a bi-valuation on elements of a lattice within a context: it obeys the sum and product rules. *Maximum entropy* is the variational principle which assigns measures and probabilities. These results derive from simple and general symmetries. In inference, it is the propositions being investigated that form the appropriate lattice on which measures and probabilities are defined, and there is no rational alternative to the Bayesian approach of using standard probability calculus. This philosophy permeates science and beyond.

Keywords: Probability, entropy, foundations, epistemology, religion.
PACS: 02.50.Cw, 02.50.Tt.

SCIENCE

Science (Figure 1) is an intricate web of connected ideas about the world. It used to be called *natural philosophy* and it gives us the predictive power that enables technology.

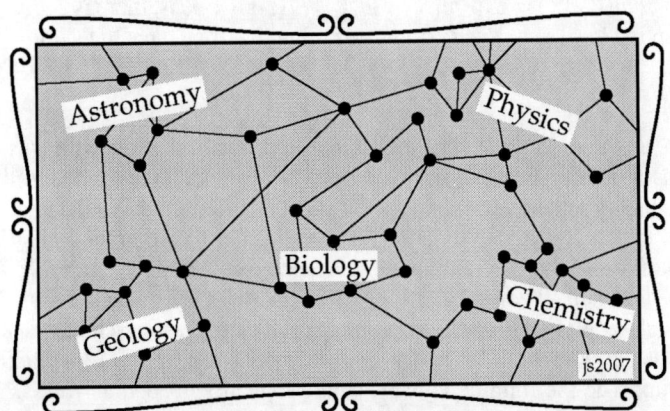

FIGURE 1. Science: The web of understanding.

The range of the connections is far-reaching. An early example from Newtonian gravity is the equivalence of the fall of an apple to the Moon's orbit. This "gravity" model brings a huge range of observations together, with correspondingly great predictive power. For example, we can construct useful tide tables. A *"gravity is bogus"* sceptic cannot be logically contradicted — but he won't produce a good tide table unless he has a model at least as good as the scientist's. Probability calculus lets us quantify the quality of our models. It is fundamental to inference, and hence to our technology. To use it, we need not claim our model to be true. Indeed we could not recognise the truth as such even if some believer told us about it. How could we tell?

PROBABILITY

We seek a calculus for the quality of connections:

If we accept X, then is Y plausible?

There are several styles of approach to this question, from detailed mathematical axiomatics (measure theory [1], set theory [2]) to plausible analogies such as betting [3] through to the plain inadequate (fuzzy logic [4]). The difficulty I have with detailed axioms is that they fail to rule out alternatives.

For example, measure theory shares features with probability calculus. Measure theory starts with the "+" operator, basically

$$\text{Definition of measure: } m(X \cup Y) = m(X) + m(Y) \text{ for } X \cap Y = \emptyset, \qquad (1)$$

in order to quantify some system with join operator $\cup$ and meet $\cap$. Actually, the definition is usually written "for generality" with potentially infinite combinations $\sum_{i=1}^{\infty}$ of elements $X_1, X_2, \ldots$, and the subject immediately gets technical with discussion of the limits involved. Measure theory is successful, yet why should we not adopt some alternative, such as $m(X \cup Y) = \exp(m(X) + m(Y))$? As it happens, there is a reason. Measure theory is applied to *associative* systems, meaning that they have the symmetry

$$(X \cup Y) \cup Z = X \cup (Y \cup Z) \qquad (2)$$

It is then a *theorem* [5] that the *only* numerical quantification with this associative symmetry uses ordinary arithmetical addition "+" (or can be re-graded to this standard form). Thus the hypothetical generalisation above is prohibited. A related generalisation $\mu(X \cup Y) = \exp(\log(\mu(X)) + \log(\mu(Y)))$ is valid, but the re-graded $m = \log \mu$ obeys ordinary addition, so there is nothing new. Associative symmetry underlies and justifies the standard measure-theory definition, so is (in my view) a better foundation. It is natural to start applying it to finite problems with only a few elements X, Y, Z, only passing to the infinite limit as and when an application appears to require it. Finite definitions underlie the infinite anyway, so are logically deeper, as well as being more convincing to the mathematically unsophisticated.

We now use this "symmetry of finite systems" outlook to define our calculus of inference, in a style originally due to Cox [6] and popularised by Jaynes [7] (chapter 2). Probability calculus is more subtle than measure theory because the propositions $x, y, z, \ldots$ it deals with are always in the context t of some model, which may iteslf be variable. We seek $p(x \mid t)$, and not just $m(X)$. Hence we will need *two* requirements, not just one.

- Requirement 1: **Associativity of content.**
  For disjoint x, y, z within t, for example

$$x = \boxed{\cdot} \qquad y = \boxed{\cdot\cdot} \qquad z = \boxed{\cdot} \qquad t = \{\boxed{\cdot},\boxed{\cdot\cdot},\boxed{\cdot\cdot},\boxed{\cdot\cdot\cdot},\boxed{\cdot\cdot\cdot},\boxed{\cdot\cdot\cdot}\}$$

we have $(x \vee y) \vee z = x \vee (y \vee z)$, where "$\vee$" is the join operator OR (and "$\wedge$" is the meet AND). According to the associativity theorem, our original p can be re-graded through some function ψ to yield addition:

$$\psi(p(x\,|\,t)) + \psi(p(y\,|\,t)) = \psi(p(x \vee y\,|\,t)) \qquad (3)$$

- Requirement 2: **Associativity of context.**
  Let x, y, z, t be nested propositions $x \leq y \leq z \leq t$, for example

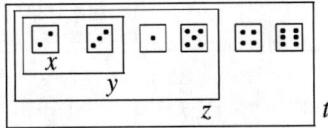

The corresponding implications can be learned in either order, because $(x \Rightarrow y) \Rightarrow z$ and $x \Rightarrow (y \Rightarrow z)$ are both equivalent to $x \Rightarrow z$, within any encompassing context t. Again according to the associativity theorem, our original p can be re-graded through some other function ϕ to yield addition:

$$\phi(p(x\,|\,y)) + \phi(p(y\,|\,z)) = \phi(p(x\,|\,z)) \qquad (4)$$

Some straightforward algebra proves from (3) and (4) that the original p can be re-graded onto a standard *probability* scale P on which

$$\left.\begin{array}{ll} 0 = P(\emptyset\,|\,t) \leq P(x\,|\,t) \leq P(t\,|\,t) = 1 & \text{(Range)},\\[4pt] P(x\,|\,t) + P(y\,|\,t) = P(x \vee y\,|\,t) + P(x \wedge y\,|\,t) & \text{(Sum rule)},\\[4pt] P(x\,|\,y)P(y\,|\,z) = P(x\,|\,z) & \text{(Product rule)}. \end{array}\right\} \qquad (5)$$

These are the standard equations of probability calculus. There is no alternative (other than trivial re-grading into percentages, logarithms, or such), if the associativity symmetries are to be preserved. Conversely, any supposed alternative must contradict one of these requirements, thereby exposing it to simple and convincing counter-example.

An immediate corollary of these equations is that probability values can be written in the form of a ratio

$$P(x\,|\,t) = \frac{m(x \wedge t)}{m(t)} \qquad (6)$$

of measures, reminiscent of the old frequentist definition [8] as a ratio of numbers of occurence. Since (6) implies the original standard equations (5), it could logically be used as a plausible definition of probability. However, the derivation of (5) from associative symmetries shows that those equations are *required*, and not merely plausible.

Interestingly, our derivation of probability calculus holds for *any* lattice. In particular, negation "NOT" is not required, so that applications include non-Boolean lattices. However, it happens that probability values become over-determined and trivial if the lattice is not distributive, so that the domain of application is to systems modelled by a distributive lattice.

PROBABILITY ASSIGNMENT

The calculus, whether of measure or probability, allows arbitrary numerical assignments on each join-irreducible element of the lattice — in inference, these elements are the elementary "atomic" propositions. Such assignment of "prior" values of measure or probability can be a rather subjective matter of judgment. However, if there is no reason to favour one atom over any other, then that equivalence requires all the assignments to be the same (a "uniform" prior). Simple examples are $P_i = \frac{1}{6}$ for the six faces of a die, $P(x) = (2\pi)^{-1}$ on the circle $[0, 2\pi)$, and $P(x,y) = 1$ on the unit toroid $[0,1)^2$, and so on.

The latter examples are distributions with respect to continuous parameters. If such parameters are transformed, say from x to $\xi(x)$ (where different ξ are no longer equivalent), then the sum rule requires the distribution to transform by the standard rule $P(\xi) = P(x)|\frac{dx}{d\xi}|$. Similarly in two dimensions, an initially uniform distribution over x and y would transform to the factorised form $P(\xi, \eta) = f(\xi)g(\eta)$ if x were transformed to ξ, and y to η independently. More generally, consider the direct product $Z = X \times Y$

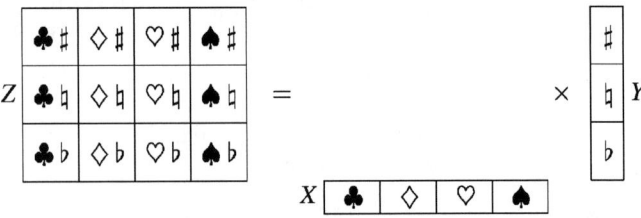

of two lattices, in which elements $x \in X$ and $y \in Y$ are independent (as cards X and music Y might well be). In such case, we require measures to combine by multiplication, so that $m(z) = m(x)m(y)$ for element $z = x \times y$. Otherwise, choices of card suit and music key would interfere in subsequent analysis.

ENTROPY

Here, we seek a variational principle to underlie numerical valuations on lattices. Can we get values for measure m or probability P by varying some scalar potential S?

Just as for probabilities, associativity of content $(x \vee y) \vee z = x \vee (y \vee z)$ requires any valuation to be additive over the atoms

$$S(m\text{'s}) = \sum_i \sigma(m_i) \qquad (7)$$

for some scalar function σ of a single real argument (or can be re-graded to such form without loss of generality). Although this additive form is what one would naturally propose, it's not arbitrary because it's been justified.

- Requirement 3: **Independence of models.**

Any general formula should apply to special cases. One special case is the direct product $Z = X \times Y$ illustrated above. In the variational approach, we must be able to build the measure $\zeta = m(z)$ on the composite lattice by varying S subject to the only constraints available, which are the factor variables $\xi = m(x)$ and $\eta = m(y)$. Then the variational equation (for a single element z) is

$$\sigma'(\zeta) = \lambda_1(\xi) + \lambda_2(\eta) \tag{8}$$

where λ_1 and λ_2 are the Lagrange multipliers for the x and y constraints. This must be satisfied by the known, factorised, solution

$$\zeta = \xi\eta \tag{9}$$

A little algebra then shows that the only valuation of additive form (7) having variational solution (9) is the *entropy* [9]

$$S(m\text{'s} \mid \mu\text{'s}) = \sum_i \left(m_i - \mu_i - m_i \log \frac{m_i}{\mu_i} \right) \quad \text{of } measure. \tag{10}$$

Freely varying (in fact maximising) the entropy lets each m revert to an original value μ. Injecting extra constraints modifies m appropriately.

When applying this formalism to probabilistic inference within some context, the probabilities form a measure which sums to 1. Embedding this hard constraint reduces (10) to the common form of entropy [10]

$$S(P\text{'s} \mid \pi\text{'s}) = -\sum_i P_i \log \frac{P_i}{\pi_i} \quad \text{of } probability \text{ (within context)} \tag{11}$$

where π is an original probability distribution and P is its modification under extra constraints.

Just as before, the domain of application is any distributive lattice.

MAXIMUM ENTROPY

The "maximum entropy" variational principle, in which S is maximised under constraints, is used for assignment. It is a *selection* procedure. It gives a single answer, and not a probabilistic or otherwise plausible range of results.

A classic application is the assignment of a distribution of prior probability (prior, that is, to subsequent use) under a "testable" constraint $\sum c_i P_i = \texttt{constant}$. If there is an original measure π_i (perhaps uniform), then maximisation of (11) under the new constraint, and subject to normalisation, yields the required distribution

$$P_i \propto \pi_i e^{\lambda c_i} \tag{12}$$

where λ is the Lagrange multiplier needed to fit the constraint value.

Positive and additive distributions are often found in data analysis, in such fields as imagery and spectroscopy, where they are modelled as measures (not necessarily summing to 1). When seeking a "best" measure m in such work, it is often useful to acquire it by maximising the general entropy (10) under a likelihood constraint $L(m) \geq L_{\text{plausible}}$ derived from the data.

A special case of this is where a two-dimensional array m_{ij} is assigned by maximising

$$S(m) = -\sum_{ij} m_{ij} \log m_{ij} \qquad (13)$$

(with uniform original measure μ) under marginal constraints

$$\sum_j m_{ij} = \xi_i, \qquad \sum_i m_{ij} = \eta_j \qquad (14)$$

(with common sum Z). The maximum-entropy "best" choice is the factorised form

$$m_{ij} = \xi_i \eta_j / Z \qquad (15)$$

In fact, this is basically how S was derived in the previous section.

SUMMARY

The picture is both simple and clear. The only numerical valuation that can be placed on the elements x of a lattice is an additive *measure* $m(x)$, obeying the sum rule. That result is forced by associativity. The lattice should also be distributive, otherwise the assigned numbers become over-determined and not useful. However, the lattice need not be Boolean.

When a *bi*-valuation is needed, involving both an element x and its context t, we need a *probability* $P(x \mid t)$, obeying the product rule as well as the sum rule. Again, it is simply the basic symmetry of associativity that requires this form. "Probability" is now a defined term, manipulated with a defined, unique calculus. It gives meaning to the more amorphous ideas of plausibility, belief, credibility and so on, which commonly motivate investigation of the foundations of inference. In inference, the lattice elements x are propositions, each accompanied by a negation NOT x, so that the lattice is Boolean. However, probability calculus remains the only acceptable calculus for any distributive lattice, even if non-Boolean — such as the lattice of questions about propositions [11].

The "maximum entropy" variational principle is used to assign measures and probabilities. Its "$p \log p$" formula (equations (10) and (11)) is the only one which allows independent models to stay independent when analysed together. Alternative forms, such as the $\sum p_i^\alpha$ formula associated with Rényi [12] and Tsallis [13], would break that symmetry and introduce mutual interference if used in maximisation, and it is confusing to describe such indices of diversity as "entropies".

It is no accident that applications as diverse as the Kullback-Leibler deviation of one distribution from another, Shannon information, and thermodynamic disorder, all share the "$p \log p$" form. They must, because they are all applicable to independent models, and thereby require the associative symmetry that forces that formula. Indeed, the variational derivation is what makes "entropy" a defined term, which gives meaning to originally more amorphous ideas such as deviation and information.

These results about measure, probability and entropy are generated "from below", by considering small, finite problems. It is also possible (actually, common) to axiomatise measure theory and probability theory from the infinite domain. However, that makes the subject look difficult, distracts from the real job of understanding a finite world, and suggests that there might be nooks and crannies where paradox could arise. A finite derivation seems preferable, and is consonant with the obvious fact that our thoughts and computations in the real world are always limited and finite.

DISCUSSION

Science is about connections. These can be far-reaching, and critically important to our civilisation. Inference lets us investigate them rationally.

Practical inference extends throughout science — and permeates ordinary life too. Using probability calculus leads to the usual Bayesian structure of prior-likelihood-evidence-posterior. Through the "evidence" (also called the prior predictive), the calculus quantifies the value of models, as well as making "posterior" predictions from them. It is, of course, not required to use probabilistic (Bayesian) terminology when using the calculus. Specifically, "fuzzy" would be a convenient shorthand for "probabilistic" if the term had not been hi-jacked to suggest counter-Bayesian manipulations. All that matters is that the calculus is used correctly.

There is no way that probability calculus can break away from the Bayesian structure. We always process models of the truth. We do not see "the truth", and even if one of our models happened to be true, we would never know it. We may motivate ourselves by aiming to search out the truth, but it's a myth. The best that we can actually do is to provide models that do a good job of collating and cross-predicting data over some reasonably wide field of endeavour. Ontology (the study of what is "really" out there) is dead. Epistemology (the study of what we individually *think* might be there) lives. It is, incidentally, remarkable that such philosophical depth can result from simple attention to elementary thoughts about the basis of inference.

Any dogmatic, absolute statement — to which classical religions are particularly prone — denies all contradictory ideas, and hence is liable to fracture the network of possible connections (Figure 2). For example, I once read a claim that the dinosaurs died out because Noah did not have room for them on the Ark. Maybe the author believed it. However, that hypothesis would badly fracture the web of connections that comprise scientific geology. If anybody bothered to do the sums, they would find a huge reduction in the coherence of the geological record, as measured by the Bayesian evidence, when comparing the Noah hypothesis with the conventional account.

The danger lies not so much in the intellectual poverty following such disconnection, but in the consequential damage to predictive power and technical ability. Would you really want an evolution-denying creationist to be in charge of our defences against bird flu? Rationality matters. It matters a lot, yet dogma denies the free thinking that we need in order to understand, predict, and take action.

Science is commonly, but wrongly, perceived as truth-seeking. Actually, it's a quest for predictive connections, and those connections are of practical value. In science's war with the irrational, we should fight from the firm ground of practicality, and not on the weak ground of some supposedly-authoritative knowledge of mythical truth.

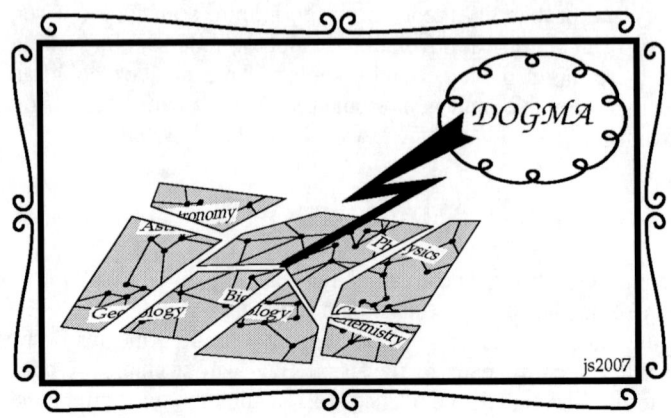

FIGURE 2. Dogma: The breaking of understanding.

ACKNOWLEDGMENTS

It is a pleasure to acknowledge help and encouragement from Kevin Knuth, Ariel Caticha and Carlos Rodríguez at SUNY, whose work has led to renewed development and interest in foundational issues. This work was supported by Maximum Entropy Data Consultants Ltd of Cambridge (England).

REFERENCES

1. A. N. Kolmogorov, *Foundations of the theory of probability*, Chelsea Publishing Co, New York (1950) (original in German, 1933).
2. T. L. Fine, *Theories of probability*, Academic Press, New York (1973).
3. B. de Finetti, "Prevision, its logical laws, its subjective sources", in Kyburg, H.E, and Smokler, H.E. (Eds.) *Studies in subjective probability*, 93–158, Wiley, New York (1964).
4. L. A. Zadeh, "Fuzzy algorithms", *Information and Control* **5**, 94–102 (1968).
5. J. Aczél, "The associativity equation re-visited", in *Bayesian inference and maximum entropy methods*, ed. G. J.Erickson and Y. Zhai, AIP **707**, 195–203 (2003).
6. R. T. Cox, "Probability, frequency and reasonable expectation", *Amer. J. Phys.* **14**, 1–13 (1946).
7. E. T. Jaynes, *Probability theory: the logic of science*, Cambridge Univ. Press (2003).
8. J. Neyman, "Outline of a theory of statistical estimation based on the classical theory of probability", *Phil. Trans. Roy. Soc. A* **236**, 333–380 (1937).
9. J. Skilling, "The axioms of maximum entropy," in *Maximum entropy and Bayesian methods*, ed. G. J.Erickson and C. R. Smith, Kluwer, Dordrecht (1988).
10. C. E. Shannon, "A mathematical theory of communication", *Bell System Tech. J.* **27**, 379–423 and 623–656 (1948).
11. K. H. Knuth, "Deriving laws from ordering relations", in *Bayesian inference and maximum entropy methods*, ed. G. J.Erickson and Y. Zhai, AIP **707**, 204–235 (2003).
12. A. Rényi, "On measures of information and entropy", *Proc. 4th Berkeley symposium on mathematics, statistics and probability*, 547–561 (1961).
13. C. Tsallis, "Possible generalization of Boltzmann-Gibbs statistics", *J. Statistical Physics*, **52**, 479–487 (1988).

Information and Entropy

Ariel Caticha

Department of Physics, University at Albany–SUNY, Albany, NY 12222, USA

Abstract. What is information? Is it physical? We argue that in a Bayesian theory the notion of information must be defined in terms of its effects on the beliefs of rational agents. Information is whatever constrains rational beliefs and therefore it is the force that induces us to change our minds. This problem of updating from a prior to a posterior probability distribution is tackled through an eliminative induction process that singles out the logarithmic relative entropy as the unique tool for inference. The resulting method of Maximum relative Entropy (ME), which is designed for updating from *arbitrary* priors given information in the form of *arbitrary* constraints, includes as special cases both MaxEnt (which allows arbitrary constraints) and Bayes' rule (which allows arbitrary priors). Thus, ME unifies the two themes of these workshops – the Maximum Entropy and the Bayesian methods – into a single general inference scheme that allows us to handle problems that lie beyond the reach of either of the two methods separately. I conclude with a couple of simple illustrative examples.

Keywords: Information, Entropy, Bayesian Inference, Method of Maximum Entropy
PACS: 04.20.Cv,02.50.Tt,02.40.Ky

INTRODUCTION

The general problem of inductive inference is to update from a prior probability distribution to a posterior distribution when new information becomes available. This raises several basic questions which are the subject of this paper. First, what is information? It is clear that data "contains" or "conveys" information, but what does this precisely mean? Is information some sort of physical fluid that can be contained or transported? Is information *physical*? Can we measure amounts of information? Do we need to? What is entropy?

A second set of questions revolves around our methods to process information. We know that Bayes' rule is the natural way to update probabilities when the new information is in the form of data and we know that Jaynes' method of maximum entropy, MaxEnt, is designed to handle information in the form of constraints [1]. At first sight these two methods appear unrelated. Are they compatible with each other? Are there other methods? Moreover, the range of applicability of either method is somewhat limited: Bayes' rule can handle arbitrary priors and data, and it can even handle some constraints, but not arbitrary constraints. On the other hand, MaxEnt can handle arbitrary constraints even data, but not arbitrary priors. Can we extend these methods?

As discussed in [2] the Shannon-Jaynes interpretation of entropy as a measure of uncertainty or of amount of information is somewhat problematic. The issue is not purely academic because the way equations are set up to solve a problem and even the kind of problems that we are willing to consider are affected by the particular meaning attributed to quantities such as entropy or probability. The Shannon-Jaynes interpretation was fairly adequate for their purposes, namely, communication theory and statistical mechanics, but it is not at all clear that their entropy with its attendant interpretation was the appropriate tool for the very different problem of updating probabilities.

The important contribution of Shore and Johnson [3] was the realization that any confusion surrounding the meaning of entropy could be, if not resolved, at least evaded by directly axiomatizing the procedure for updating probabilities instead of seeking dubious measures for a vaguely defined notion of information. Their argument, which is based on demanding consistency – if a problem can be solved in two different ways the two solutions must agree – is fundamentally sound. However, the detailed assumptions in their derivation have been criticized in [4, 5].

Another approach to entropy was proposed by Skilling [6]. Although his axioms were clearly inspired by Shore and Johnson, the method was very different in two respects. First, Skilling was not directly concerned with the problem of updating probabilities; his method was designed for the determination of positive-additive functions such as intensities in an image. In retrospect we see that the application to this particular problem was quite unfortunate because when the method failed to produce good image reconstructions the natural reaction was a widespread loss of confidence about entropy methods in general.

The second difference, which I think is a truly significant contribution, is that Skilling's approach is a systematic method for induction. He spelled out in full detail how to construct a general theory from known special cases. The fundamental inductive principle is deceptively trivial: *'If a general theory exists it must apply to special cases'*. The basic idea is that when there exists a special case that happens to be known all candidate theories that fail to reproduce it must be discarded. Thus, the known special cases – called the axioms of the theory – constrain the form of the general theory, and the idea is that a sufficient number of such constraints will determine the general theory completely. Of course, there is always the unfortunate possibility that the desired general theory does not exist, but if it does, then the search can be conducted in a systematic and orderly way.

Philosophers already had a name for such a method: they called it *eliminative induction* [7]. On the negative side, eliminative induction, like any other form of induction, is not guaranteed to work. It failed, for example, in Skilling's image reconstruction problem. On the positive side, eliminative induction adds an interesting twist to Popper's scientific methodology. According to Popper scientific theories can never be proved right, they can only be proved false; a theory is corroborated only to the extent that all attempts at falsifying it have failed. Eliminative induction is fully compatible with Popper's notions but the point of view is just the opposite. Instead of focusing on *failure* to falsify one focuses on *success*: it is the successful falsification of all rival theories that corroborates the surviving one. The advantage is that one acquires a more explicit understanding of why competing theories are eliminated.

The present paper is the third in a sequence devoted to clarifying the use of relative entropy as a tool for processing information and updating probabilities [2, 8]. In [2] we applied Skilling's method to the problem of Shore and Johnson. The answer to the question 'What is entropy?' turns out to be trivial and somewhat surprising: *entropy needs no interpretation*. We do not need to know what 'entropy' means, we only need to know how to use it. This explains why the "correct" interpretation had been so elusive – there is none. In [2] and then again in [8] the special cases, the axioms, were increasingly polished to clarify how alternative entropies are ruled out. Furthermore, in [2] we also discussed the question, central to any general method of updating, of the extent to which the distribution of maximum entropy is to be preferred over all others, the extent to which distributions with entropies less than the maximum are to be ruled out.

In this paper we review how eliminative induction leads to a unique candidate for a general theory of inference, the method of Maximum relative Entropy (ME), which is designed for updating from *arbitrary* priors given information in the form of *arbitrary* constraints. The three axioms used in [8] – locality, coordinate invariance, and consistency for independent subsystems – are sufficient to single out the logarithmic relative entropy as the unique tool for updating. In particular, we wish to elaborate further on the use of the third axiom – consistency for independent subsystems – to eliminate alternative entropies [12].

The idea is rather simple. The known special cases covered under axiom 3 also include situations in which we have a large number N of independent identical systems where all sorts of inferences can be reliably carried out using various asymptotic techniques (laws of large numbers, large deviation theory, etc.). The close connection with the method of maximum entropy has been repeatedly emphasized by several authors [9]-[11]. We conclude that the logarithmic relative entropy is the only candidate for a general method for updating probabilities. Alternative entropies can be useful for other purposes – for example, when studying the information geometry of statistical manifolds – but not for a general theory of updating.

In [8] we showed that the ME method includes both MaxEnt and Bayes' rule as special cases and therefore it unifies the two dominant themes of these workshops – the Maximum Entropy and Bayesian methods – into a single general inference scheme that allows us to handle problems that lie beyond the reach of either of the two methods separately. I conclude with a couple of simple illustrative examples.

In a companion paper [13] we discuss the problem of multiple constraints. Should the constraints be processed simultaneously or sequentially and, if so, in what order? There we also give an explicit example in which ME is used to simultaneously process information in the form of data and moment constraints.

WHAT IS INFORMATION?

It is not unusual these days to hear that systems "carry" or "contain" information and that "information is physical". This mode of expression can perhaps be traced to the origins of information theory in Shannon's theory of communication. We say that we have received information when among the vast variety of messages that could conceivably have been generated by a distant source, we discover which particular message was actually sent. It is thus that the message "carries" information. The analogy with physics is straightforward: the set of all possible states of a physical system can be likened to the set of all possible messages, and the actual state of the system corresponds to the message that was actually sent. Thus, the system "conveys" a message: the system "carries" information about its own state. Sometimes the message might be difficult to read, but it is there nonetheless.

This language – information is physical – useful as it has turned out to be, does not exhaust the meaning of the word 'information'. The goal of information theory, or better, communication theory, is to characterize the sources of information, to measure the capacity of the communication channels, and to learn how to control the degrading effects of noise. It is somewhat ironic but nevertheless true that this "information" theory is unconcerned with the central Bayesian issue of how the message affects the beliefs of a rational agent. A fully Bayesian information theory demands an explicit account of the relation between information and beliefs.

Our desire to update from one state of belief to another is driven by the conviction that not all probability assignments are equally good. One can argue that what makes one probability assignment better than another is that it better reflects some objective feature of the world, that it provides a better guide to the "truth" – whatever this might mean. The updating mechanism is supposed to allow us to incorporate information about the world into our beliefs.

The implication is that when confronted with new information our choices as to what we are honestly and rationally allowed to believe should become correspondingly restricted. This, I propose, is the defining characteristic of information: *Information is whatever constrains rational beliefs.* An important aspect of this notion is that for a rational agent the updating is not optional; it is a moral imperative. *Information is whatever forces a change of rational beliefs.*

Our definition captures an idea of information that is directly related to changing our minds: information is the driving force behind the process of learning. Note also that although there is no need to talk about amounts of information, whether measured in units of bits or otherwise, our notion of information allows precise quantitative calculations. Indeed, by information in its most general form, we mean the set of constraints on the family of acceptable posterior distributions and this is precisely the kind of information the method of maximum entropy has been designed to handle.

It may be worthwhile to point out an analogy with Newtonian dynamics. The state of motion of a system is described in terms of momentum – the "quantity" of motion – while the change from one state to another is explained in terms of an applied force. Similarly, in Bayesian inference a state of belief is described in terms of probabilities – the "quantity" of belief – and the change from one state to another is due to information. Just as a force is defined as that which induces a change in motion, so information is that which induces a change of beliefs.

UPDATING PROBABILITIES: THE ME METHOD

Consider a variable x which can be discrete or continuous, in one or several dimensions. The uncertainty about x is described by a probability distribution $q(x)$. Our goal is to update from the prior distribution $q(x)$ to a posterior distribution $P(x)$ when new information – that is, constraints – becomes available. The constraints could be given in terms of expected values but this is not necessary. The question is: of all those distributions $p(x)$ within the family defined by the constraints, which do we select?

As suggested by Skilling [6] to select the posterior it seems reasonable to rank the candidate distributions in *order of increasing preference*. It is clear that to accomplish this goal the ranking must be transitive: if distribution p_1 is preferred over distribution

p_2, and p_2 is preferred over p_3, then p_1 is preferred over p_3. Such transitive rankings are represented by assigning to each $p(x)$ a real number $S[p]$, which we will henceforth call entropy, in such a way that if p_1 is preferred over p_2, then $S[p_1] > S[p_2]$. The selected distribution P (one or possibly many, for on the basis of the available information there may be several equally preferred distributions) will be that which maximizes the entropy $S[p]$. We are thus led to a method of Maximum Entropy (ME) that is a variational method involving entropies which are real numbers. These features are imposed on purpose; they are dictated by the function that the ME method is *designed* to perform.

Next, to define the ranking scheme, we must decide on the functional form of $S[p]$. First, the purpose of the method is to update from priors to posteriors. The ranking scheme must depend on the particular prior q and therefore the entropy S must be a functional of both p and q. Thus the entropy $S[p,q]$ produces a ranking of the distributions p *relative* to the given prior q: $S[p,q]$ is the entropy of p *relative* to q. Accordingly $S[p,q]$ is commonly called *relative entropy*. Since all entropies are relative, even when relative to a uniform distribution, the modifier 'relative' is redundant and will be dropped.

Second, since we deal with incomplete information the method, by its very nature, cannot be deductive: *the method must be inductive*. The best we can do is use those special cases where we know what the preferred distribution should be to eliminate those entropy functionals $S[p,q]$ that fail to provide the right update. The known special cases will be called (perhaps inappropriately) the *axioms* of the theory. They play a crucial role: they define what makes one distribution preferable over another.

The three axioms below are chosen to reflect the conviction that information collected in the past and codified into the prior distribution is very valuable and should not be frivolously discarded. This attitude is maximally conservative: the only aspects of one's beliefs that should be updated are those for which new evidence has been supplied. Furthermore, since the axioms do not tell us what and how to update, they merely tell us what not to update, they have the added bonus of maximizing objectivity – there are many ways to change something but only one way to keep it the same. Thus, we adopt the

Principle of Minimal Updating (PMU): *Beliefs should be updated only to the extent required by the new information.*

The three axioms, a brief motivation for them, and their consequences for the functional form of the entropy are listed below; more details and proofs are given in [2] and [8]. As will become immediately apparent the axioms do not refer to merely three cases; any induction from such a weak foundation would hardly be reliable. The reason the axioms are convincing and so constraining is that they refer to three infinitely large classes of known special cases.

Axiom 1: Locality. *Local information has local effects.*

Suppose the information to be processed does not refer to a particular subdomain $\mathcal{D}$ of the space $\mathcal{X}$ of x's. In the absence of any new information about $\mathcal{D}$ the PMU demands we do not change our minds about $\mathcal{D}$. Thus, we design the inference method so that $q(x|\mathcal{D})$, the prior probability of x conditional on $x \in \mathcal{D}$, is not updated. The selected conditional posterior is $P(x|\mathcal{D}) = q(x|\mathcal{D})$. The consequence of axiom 1 is that non-overlapping domains of x contribute additively to the entropy. Dropping additive terms and multiplicative factors that do not affect the overall ranking, the entropy functional can be simplified to the form

$$S[p,q] = \int dx\, F(p(x), q(x), x),\tag{1}$$

where F is some unknown function.

Axiom 2: Coordinate invariance. *The system of coordinates carries no information.* The points x can be labeled using any of a variety of coordinate systems. One can *always* change coordinates but this should not affect the ranking of the distributions. The consequence of axiom 2 is that $S[p,q]$ can be written in terms of coordinate invariants such as $dx\, m(x)$ and $p(x)/m(x)$, and $q(x)/m(x)$:

$$S[p,q] = \int dx\, m(x) \Phi\left(\frac{p(x)}{m(x)}, \frac{q(x)}{m(x)}\right).\tag{2}$$

(Again, additive terms and multiplicative factors that do not affect the overall ranking have been dropped.) Thus the unknown function F which had three arguments has been replaced by two unknown functions, one is a density $m(x)$, and the other is a function Φ with two arguments. Next we determine the density $m(x)$ by invoking the locality axiom 1 once again.

Axiom 1 (special case): *When there is no new information there is no reason to change one's mind.*

When no new information is available the domain $\mathcal{D}$ in axiom 1 coincides with the whole space $\mathcal{X}$. The conditional probabilities $q(x|\mathcal{D}) = q(x|\mathcal{X}) = q(x)$ should not be updated and the selected posterior distribution coincides with the prior, $P(x) = q(x)$. The consequence is that up to normalization $m(x)$ must be the prior distribution $q(x)$, which restricts the entropy to functionals of the form

$$S[p,q] = \int dx\, q(x) \Phi\left(\frac{p(x)}{q(x)}\right).\tag{3}$$

Axiom 3: Consistency for independent subsystems. *When a system is composed of subsystems that are known to be independent it should not matter whether the inference procedure treats them separately or jointly.*

Suppose the information on two independent subsystems 1 and 2 is such that the prior distributions $q_1(x_1)$ and $q_2(x_2)$ are respectively updated to $P_1(x_1)$ and $P_2(x_2)$ when they are treated separately. When treated as a single system the joint prior is $q_1(x_1)q_2(x_2)$ and the family of potential posteriors is $p(x_1, x_2) = p_1(x_1)p_2(x_2)$. The entropy functional must be such that the selected posterior is $P_1(x_1)P_2(x_2)$. The consequence of axiom 3 for this particular case of just two subsystems is that entropies are restricted to the one-parameter family given by

$$S_\eta[p,q] = \frac{1}{\eta(\eta+1)}\left[1 - \int dx\, p(x)\left(\frac{p(x)}{q(x)}\right)^\eta\right].\tag{4}$$

Once again, additive terms and multiplicative factors that do not affect the overall ranking scheme can be freely chosen. The $\eta = 0$ case reproduces the usual logarithmic

relative entropy,
$$S[p,q] = -\int dx\, p(x) \log \frac{p(x)}{q(x)} \tag{5}$$

[Use $y^\eta = \exp \eta \log y \approx 1 + \eta \log y$ in eq.(4) and let $\eta \to 0$ to get eq.(5).]

In [8] we argued that the index η has to be the same for all systems. To see why consider any two independent systems characterized by η_1 and η_2. Consistency between the joint and separate updates requires that $\eta_1 = \eta_2$ therefore η must be a universal constant. From the success of statistical mechanics as a theory of inference we inferred that the value of this constant must be $\eta = 0$ leading to the logarithmic entropy, eq.(5). Here we offer a different argument also based on a broader application of axiom 3:

Axiom 3 (special case): Consistency for large numbers of independent identical subsystems.

The known special cases covered under axiom 3 include situations in which we have a large number N of independent identical systems. In such cases either the weak law of large numbers or large deviation theory in the form of Sanov's theorem are sufficient to make the desired inferences. Entropy considerations are not needed.

Let the x variables be discrete x_i with $i = 1\ldots m$. The identical priors for the individual systems are q_i and the available information is that the potential posteriors p_i are subject, for example, to an expectation value constraint such as $\langle a \rangle = A$, where A is some specified value and $\langle a \rangle = \sum a_i p_i$.

Consider the set of N systems treated jointly. Let the number of systems found in state i be n_i, and let $f_i = n_i/N$ be the corresponding frequency. In the limit of large N the frequencies f_i converge (in probability) to the desired posterior P_i while the sample average $\bar{a} = \sum a_i f_i$ converges (also in probability) to the expected value $\langle a \rangle = A$. The probability of a particular frequency distribution $f = \{f_1 \ldots f_n\}$ generated by the prior q is multinomial,

$$Q_N(f|q) = \frac{N!}{n_1! \ldots n_m!} q_1^{n_1} \ldots q_m^{n_m} \quad \text{with} \quad \sum_{i=1}^m n_i = N, \tag{6}$$

and for large N we have

$$Q_N(f|q) \approx \exp N(S[f,q] + r_N), \tag{7}$$

where $S[f,q]$ given by eq.(5), and where r_N is a correction that vanishes as $N \to \infty$. To find the most probable frequency distribution satisfying the constraint $\bar{a} = A$ one maximizes $Q_N(f|q)$ subject to $\bar{a} = A$, which is equivalent to maximizing the entropy $S[f,q]$ subject to $\bar{a} = A$. The corresponding problem for the individual systems is that of maximizing $S_\eta[p,q]$ subject to $\langle a \rangle = A$. The two procedures agree only when we choose $\eta = 0$. Therefore, entropies S_η with $\eta \neq 0$ are not consistent with the laws of large numbers and must be discarded.

Csiszar [10] and Grendar [11] have argued that the asymptotic argument above provides a valid justification for the ME method of updating. An agent whose prior is q receives the information $\langle a \rangle = A$ which can be reasonably interpreted as a sample average $\bar{a} = A$ over a large ensemble of N trials. The agent's beliefs are updated so that the posterior P coincides with the most probable f distribution. This is quite compelling but, of course, as a justification of the ME method it is restricted to situations where it is natural to think in terms of ensembles with large N. This justification is not nearly as compelling for singular events for which large ensembles either do not exist or are too unnatural and contrived. From our point of view the asymptotic argument above does not by itself provide a fully convincing justification for the universal validity of the ME method but it does provide considerable inductive support. It serves as a valuable consistency check that must be passed by any inductive inference procedure that claims to be of *general* applicability.

The results are summarized as follows:

The ME method: *The objective is to update from a prior distribution q to a posterior distribution given the information that the posterior lies within a certain family of distributions p. The selected posterior $P(x)$ is that which maximizes the entropy $S[p,q]$. Since prior information is valuable the functional $S[p,q]$ has been chosen so that beliefs are updated only to the extent required by the new information. No interpretation for $S[p,q]$ is given and none is needed.*

BAYES' RULE AND ITS GENERALIZATIONS

The problem is to update our beliefs about $\theta \in \Theta$ (θ represents one or many parameters) on the basis of three pieces of information: (1) the prior information codified into a prior distribution $q(\theta)$; (2) the data $x \in \mathcal{X}$ (obtained in one or many experiments); and (3) the known relation between θ and x given by the model as defined by the sampling distribution or likelihood, $q(x|\theta)$. The updating consists of replacing the *prior* probability distribution $q(\theta)$ by a *posterior* distribution $P(\theta)$ that applies after the data has been processed.

The crucial element that will allow Bayes' rule to be smoothly incorporated into the ME scheme is the realization that before the data information is available not only we do not know θ, we do not know x either. Thus, the relevant space for inference is not Θ but the product space $\Theta \times \mathcal{X}$ and the relevant joint prior is $q(x,\theta) = q(\theta)q(x|\theta)$. We should emphasize that the information about how x is related to θ is contained in the *functional form* of the distribution $q(x|\theta)$ – for example, whether it is a Gaussian or a Cauchy distribution – and not in the actual values of the arguments x and θ which are, at this point, still unknown.

Next we collect data and the observed values turn out to be x'. We must update to a posterior that lies within the family of distributions $p(x,\theta)$ that reflect the fact that x is known,

$$p(x) = \int d\theta\, p(\theta, x) = \delta(x - x') \,. \tag{8}$$

This data information constrains but is not sufficient to determine the joint distribution

$$p(x,\theta) = p(x)p(\theta|x) = \delta(x - x')p(\theta|x') \,. \tag{9}$$

Any choice of $p(\theta|x')$ is in principle possible. Additional input is needed and it is at this point that we invoke the Principle of Minimal Updating: beliefs need to be revised only to the extent required by the data. Accordingly the conditional prior $q(\theta|x')$ requires no revision and the selected posterior $P(x,\theta)$ is such that $P(\theta|x') = q(\theta|x')$, or

$$P(x,\theta) = \delta(x-x')q(\theta|x') \, . \tag{10}$$

The corresponding marginal posterior probability $P(\theta)$ is

$$P(\theta) = \int dx \, P(\theta,x) = q(\theta|x') = q(\theta)\frac{q(x'|\theta)}{q(x')} \, , \tag{11}$$

which is recognized as Bayes' rule. This is extremely reasonable: we *maintain* those beliefs about θ that are consistent with the data values x' that turned out to be true. Data values that were not observed are discarded because they are now known to be false. 'Maintain' is the key word: it reflects the PMU in action.

Remark: Bayes' rule is usually written in the form

$$q(\theta|x') = q(\theta)\frac{q(x'|\theta)}{q(x')} \, , \tag{12}$$

and called Bayes' theorem. This formula is very simple; perhaps it is too simple. It is just a restatement of the product rule – valid for any x' whether observed or not – and therefore it is a simple consequence of the *internal* consistency of the *prior* beliefs. The drawback of this formula is that the left hand side is not a *posterior* but rather a *prior* (conditional) probability; it obscures the fact that an additional principle – the PMU – was needed for updating.

Next we show that Bayes' rule is consistent with, and indeed, is a special case of the ME method [8]. This is not too surprising given that the ME is also based on the PMU. According to the ME method the selected joint posterior $P(x,\theta)$ is that which maximizes the entropy,

$$S[p,q] = -\int dx d\theta \, p(x,\theta) \log \frac{p(x,\theta)}{q(x,\theta)} \, , \tag{13}$$

subject to the appropriate constraints. Note that the information in the data, eq.(8), represents an *infinite* number of constraints on the family $p(x,\theta)$: for each value of x there is one constraint and one Lagrange multiplier $\lambda(x)$. Maximizing S, (13), subject to (8) and normalization,

$$\delta\{S + \alpha\left[\int dx d\theta \, p(x,\theta) - 1\right] + \int dx \, \lambda(x)\left[\int d\theta \, p(x,\theta) - \delta(x-x')\right]\} = 0 \, , \tag{14}$$

yields the joint posterior,

$$P(x,\theta) = q(x,\theta)\frac{e^{\lambda(x)}}{Z} \, , \tag{15}$$

where Z is a normalization constant, and $\lambda(x)$ is determined from (8),

$$\int d\theta \, q(x,\theta)\frac{e^{\lambda(x)}}{Z} = q(x)\frac{e^{\lambda(x)}}{Z} = \delta(x-x') \, , \tag{16}$$

so that the joint posterior is

$$P(x,\theta) = q(x,\theta)\frac{\delta(x-x')}{q(x)} = \delta(x-x')q(\theta|x) , \qquad (17)$$

from which we recover Bayes' rule, eq.(11).

I conclude with a couple of very simple examples that show how the ME allows generalizations of Bayes' rule. The background for these generalized Bayes problems is the familiar one: We want to make inferences about some variables θ on the basis of information about other variables x. As before, the prior information consists of our prior knowledge about θ given by the distribution $q(\theta)$ and the relation between x and θ is given by the likelihood $q(x|\theta)$; thus, the prior joint distribution $q(x,\theta)$ is known. But now the information about x is much more limited.

Example 1.– The data is uncertain: x is not known. The marginal posterior $p(x)$ is no longer a sharp delta function but some other known distribution, $p(x) = P_D(x)$. This is still an infinite number of constraints

$$p(x) = \int d\theta \, p(\theta, x) = P_D(x) , \qquad (18)$$

that are easily handled by ME. Maximizing S, (13), subject to (18) and normalization, leads to

$$P(x,\theta) = P_D(x)q(\theta|x) . \qquad (19)$$

The corresponding marginal posterior,

$$P(\theta) = \int dx \, P_D(x)q(\theta|x) = q(\theta) \int dx \, P_D(x)\frac{q(x|\theta)}{q(x)} , \qquad (20)$$

is known as Jeffrey's rule.

Example 2.– Now we have even less information: $p(x)$ is not known. All we know about $p(x)$ is an expected value

$$\langle f \rangle = \int dx \, p(x) f(x) = F . \qquad (21)$$

Maximizing S, (13), subject to (21) and normalization,

$$\delta\{S + \alpha[\int dxd\theta \, p(x,\theta) - 1] + \lambda \int dxd\theta \, p(x,\theta)f(x) - F\} = 0 , \qquad (22)$$

yields the joint posterior,

$$P(x,\theta) = q(x,\theta)\frac{e^{\lambda f(x)}}{Z} , \qquad (23)$$

where the normalization constant Z and the multiplier λ are obtained from

$$Z = \int dx \, q(x)e^{\lambda f(x)} \quad \text{and} \quad \frac{d\log Z}{d\lambda} = F . \qquad (24)$$

The corresponding marginal posterior is

$$P(\theta) = q(\theta) \int dx \, \frac{e^{\lambda f(x)}}{Z} q(x|\theta) . \qquad (25)$$

The two posteriors (20) and (25) are sufficiently intuitive that one could have written them down directly without deploying the full machinery of the ME method, but they do serve to illustrate the essential compatibility of Bayesian and Maximum Entropy methods. A less trivial example is given in [13].

CONCLUSIONS

Any Bayesian account of the notion of information cannot ignore the fact that Bayesians are concerned with the beliefs of rational agents. The relation between information and beliefs must be clearly spelled out. The definition we have proposed – that information is that which constrains rational beliefs and therefore forces the agent to change its mind – is convenient for two reasons. First, the information/belief relation very explicit, and second, the definition is ideally suited for quantitative manipulation using the ME method.

The other main conclusion is that the logarithmic relative entropy is the only candidate for a general method for updating probabilities – the ME method – which includes MaxEnt and Bayes' rule as special cases; it unifies them into a single theory of inductive inference.

It is true that there exist many different ways to define measures of separation, or divergence between distributions and that these "entropies" can be useful in a wide variety of ways. In fact, it was precisely this wealth of possibilities that Shore and Johnson intended to avoid. These other "entropies" can be useful for other purposes but not for updating; at least not for an updating theory that strives to achieve universal applicability. Let us emphasize that the reason the ME method uses the logarithmic entropy as the tool for updating is not that this entropy has been shown to provide the *correct* measure of distance – there are many other such measures. We do not even claim that inferences on the basis of the ME method are guaranteed to be *correct* – this is induction; there are no guarantees. It is just that all alternative entropies are much worse because in known cases they give answers that are demonstrably wrong.

Acknowledgements: I would like to acknowledge valuable discussions with C. Cafaro, N. Caticha, A. Giffin, K. Knuth, and C. Rodríguez.

REFERENCES

1. E. T. Jaynes, Phys. Rev. **106**, 620 and **108**, 171 (1957); R. D. Rosenkrantz (ed.), *E. T. Jaynes: Papers on Probability, Statistics and Statistical Physics* (Reidel, Dordrecht, 1983); E. T. Jaynes, *Probability Theory: The Logic of Science* (Cambridge University Press, Cambridge, 2003).
2. A. Caticha, "Relative Entropy and Inductive Inference," in *Bayesian Inference and Maximum Entropy Methods in Science and Engineering*, ed. by G. Erickson and Y. Zhai, AIP Conf. Proc. **707**, 75 (2004) (arXiv.org/abs/physics/0311093).
3. J. E. Shore and R. W. Johnson, IEEE Trans. Inf. Theory **IT-26**, 26 (1980); IEEE Trans. Inf. Theory **IT-27**, 26 (1981).
4. S. N. Karbelkar, Pramana – J. Phys. **26**, 301 (1986).
5. J. Uffink, Stud. Hist. Phil. Mod. Phys. **26B**, 223 (1995).
6. J. Skilling, "The Axioms of Maximum Entropy" in *Maximum-Entropy and Bayesian Methods in Science and Engineering*, G. J. Erickson and C. R. Smith (eds.) (Kluwer, Dordrecht, 1988).

7. J. Earman, *Bayes or Bust?: A Critical Examination of Bayesian Confirmation Theory* (MIT Press, Cambridge, 1992).
8. A. Caticha and A. Giffin, "Updating Probabilities", *Bayesian Inference and Maximum Entropy Methods in Science and Engineering*, Ali Mohammad-Djafari (ed.), AIP Conf. Proc. **872**, 31 (2006) (arxiv.org/abs/physics/0608185).
9. J. M. van Campenhout and T. M. Cover, IEEE Trans. Inform. Theory, **IT-27** 483 (1981).
10. I. Csiszar, "An extended maximum entropy principle and a Bayesian justification," in *Bayesian Statistics 2*, J. M. Bernardo et al. (eds.) (North Holland, Amsterdam, 1985); I. Csiszar, "MaxEnt, Mathematics, and Information Theory," *Maximum Entropy and Bayesian Methods in Science and Engineering*, K. M. Hanson and R. N. Silver (ed.) (Kluwer, 1996).
11. M. Grendar, Jr. and M. Grendar, "What is the question that MaxEnt answers? A probabilistic interpretation," *Bayesian Inference and Maximum Entropy Methods in Science and Engineering*, Ali Mohammad-Djafari (ed.), AIP Conf. Proc. **568**, 83 (2001) (arxiv.org/abs/math-ph/0009020); "Maximum Entropy and Maximum Entropy methods: Bayesian interpretation" (arxiv.org/abs/physics/0308005).
12. A. Renyi, "On measures of entropy and information," *Proc. 4th Berkeley Symposium on Mathematical Statistics and Probability*, Vol. **1**, 547-461 (U. of California Press, 1961); C. Tsallis, J. Stat. Phys. **52**, 479 (1988).
13. A. Giffin and A. Caticha, "Updating Probabilities with Data and Moments", in these proceedings (arxiv:0708.1593).

Lattice Theory, Measures and Probability

Kevin H. Knuth

Department of Physics
Department of Informatics
University at Albany (SUNY)
Albany NY, 12222, USA
(e-mail: kknuth@albany.edu, http://knuthlab.rit.albany.edu)

Abstract. In this tutorial, I will discuss the concepts behind generalizing ordering to measuring and apply these ideas to the derivation of probability theory. The fundamental concept is that anything that can be ordered can be measured. Since we are in the business of making statements about the world around us, we focus on ordering logical statements according to implication. This results in a Boolean lattice, which is related to the fact that the corresponding logical operations form a Boolean algebra.

The concept of logical implication can be generalized to degrees of implication by generalizing the zeta function of the lattice. The rules of probability theory arise naturally as a set of constraint equations. Through this construction we are able to neatly connect the concepts of order, structure, algebra, and calculus. The meaning of probability is inherited from the meaning of the ordering relation, implication, rather than being imposed in an ad hoc manner at the start.

Keywords: lattice, order, poset, measure, probability
PACS: 01.70.+w, 02.10.-v, 02.50.-r

INTRODUCTION

We begin by considering one of the most basic of concepts: *ordering*. Two objects are ordered by simply comparing them to one another. This basis of comparison relies on some sort of rule that decides which item is 'bigger'. This rule for comparing two objects is called a *binary ordering relation*. We can then extend this activity to a set of objects by considering all pairwise comparisons. These pairwise comparisons will result in a *partially-ordered set*. We say that the set is partially-ordered since it may be the case that two objects may be incomparable according to the binary ordering relation we have chosen. Whatever the case may be, given a set or elements and a binary ordering relation, we can generate a partially-ordered set, or *poset* for short. This abstraction is important because the concept of ordering is fundamental. In hindsight, we will not be surprised at the deep connections that we will uncover here.

Measuring is a more advanced concept as it takes ordering one step further. Rather than just saying that one element is bigger than another, measuring specifies how much bigger. Measuring is based on ordering, and we shall see that the rules for manipulating measures originate as constraints imposed by the partial order. Another way to think about it is that the partial order dictates the algebra and the measure dictates the calculus.

At this point we have kept this discussion purposefully abstract. The reason is to highlight the fundamental nature of the concepts. In this paper, we will use ordering relations to derive probability theory. This methodology is far more fundamental than that introduced by Richard Cox [1, 2], who was the original inspiration for this work. We will leave no room for arguing about poorly-defined concepts such as belief. Instead, we will work with the set of logical statements and apply the straightforward binary ordering relation known as implication.

What we will discover is applicable to measures on *all* partially-ordered sets isomorphic to our resulting poset of logical statements ordered by implication. We will also see relationships shared by other partially-ordered sets, and find that the commonalities are due to the underlying ordering relations.

MATHEMATICAL PRELIMINARIES

My intention is to keep this tutorial as basic as possible without delving deep into the mathematics. However, some introduction to order theory is necessary to properly visualize these partially-ordered sets. I encourage the reader to work through this section as the results are extremely fruitful. Remember, we are merely formalizing the extremely familiar notion of ordering.

Consider a set of elements S, and a binary ordering relation denoted $\leq$. Note that $\leq$ may mean the usual 'less than or equal to', but here we use the symbol to represent any binary ordering relation. We say that z *includes* x if $x \leq z$. If it is true that x is not equal to z, then we can write $x < z$. Last, if $x < z$, and there exists no element y in the set S such that $x < y$ and $y < z$ then we say that z *covers* x, and write $x \prec z$. In this case we can view z as succeeding x in a hierarchy. We will use this concept of covering to construct diagrams of partially-ordered sets (Figure 1). It is important to stress that inclusion, as described above, is another way of looking at ordering.

Once we have a partially-ordered set, we can consider an element of the set and note where it lies in the ordering. In this sense, the perspective of a partially-ordered set leads to the concept of structure. The set of elements that include a given element x is called the upper bound of x. The set of elements included by x is called the lower bound of x. We can also talk about upper and lower bounds of a subset of S, so that the set of all elements of S that include a subset U of S is the upper bound of U. Likewise for the lower bound.

When it happens that both the upper and lower bound exist for all pairs of elements in the partially-ordered set, then we say that the set is a lattice. For two elements x and y in the lattice $\mathcal{S}$, composed of the set S and a binary-ordering relation $\leq$, one can denote the upper bound of x and y as $x \vee y$ and the lower bound of x and y as $x \wedge y$. The symbols $\vee$ and $\wedge$ can be seen to act as operators that take a pair of elements and map them to a third element. These operators are called the *join* and *meet*, respectively. This is the algebraic perspective. Thus a lattice can be viewed as either a structure or an algebra.

Some lattices have special elements called the top $\top$ and bottom $\bot$. The top is the unique greatest upper bound, whereas the bottom is the unique least lower bound. Note that not all posets have tops and bottoms. In Figure 1 neither poset (a) nor (b) has a top element.

All lattices share similar structural features that can be expressed as algebraic rules. We will find later that these rules impose constraints on the measures that we assign, and appear as constraint equations that are found universally throughout the sciences.

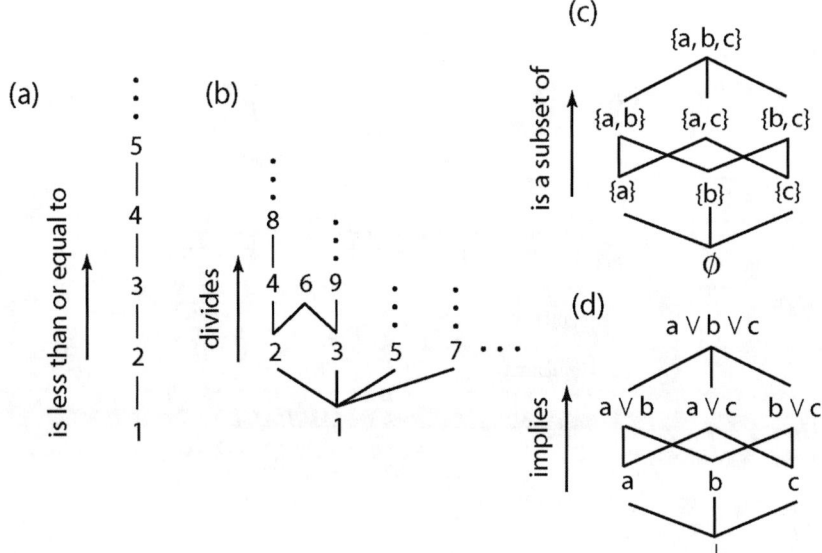

FIGURE 1. Four examples of posets. (a) The poset of integers ordered according to the usual '*less than or equal to*'. (b) The poset of integers ordered according to '*divides*'. (c) The poset of of subsets of $\{a,b,c\}$ ordered according to '*is a subset of*'. (d) The poset of logical statements ordered according to '*implies*'. Note that even though posets (a) and (b) both involve integers, the ordering relation results in distinct structures. Conversely, posets (c) and (d) are isomorphic despite the fact that both the sets and ordering relations are different.

We will find that some elements cannot be expressed as the join of any two other elements in the poset. Such elements are called *join-irreducible*. As an example, the elements covering the bottom in all four posets of Figure 1 are join-irreducible. In the lattice of logical statements, these are the exhaustive set of mutually-exclusive statements. In the set of integers ordered by '*divides*', these are the prime numbers. Already we are seeing strange connections. The poset of integers ordered by '*less than or equal to*' is composed entirely of join-irreducible elements. Incidentally, this poset is called a *chain*.

By formalizing the notion of ordering, we will see that there are general features that are shared by a wide array of partially ordered sets. These general features are the prize that awaits those who are willing to back away from the specifics and to generalize the operations involved in ordering. In this sense we are unifying all things ordered.

MEASURING

By focusing on the concepts underlying the process of ordering, we can now see measuring in a new light. Measuring is a generalization of ordering. Since ordering is represented by inclusion, one way to generalize ordering is to generalize the binary concept

TABLE 1. Features of Lattices

	ORDERING
Order	$\leq$
Reflexivity	For all A, $A \leq A$
Antisymmetry	If $A \leq B$ and $B \leq A$ then $A = B$
Transitivity	If $A \leq B$ and $B \leq C$ then $A \leq C$

	OPERATIONS
Join	$\vee$
Meet	$\wedge$
Idempotency	$A \vee A = A$
	$A \wedge A = A$
Commutativity	$A \vee B = B \vee A$
	$A \wedge B = B \wedge A$
Associativity	$A \vee (B \vee C) = (A \vee B) \vee C$
	$A \wedge (B \wedge C) = (A \wedge B) \wedge C$
Absorption	$A \vee (A \wedge B) = A \wedge (A \vee B) = A$
Distributivity	$A \wedge (B \vee C) = (A \wedge B) \vee (A \wedge C)$
	$A \vee (B \wedge C) = (A \vee B) \wedge (A \vee C)$

CONSISTENCY
$A \leq B \quad \Leftrightarrow \quad A \wedge B = A \quad \Leftrightarrow \quad A \vee B = B$

of inclusion to *degrees of inclusion*. Since we are required to preserve *transitivity*, we can use Real Numbers to quantify degrees of inclusion. There are other mathematical quantities that we could use as well, but for now we will focus on real numbers.

Inclusion can be encoded in terms of the zeta function, which quantifies whether the lattice element y includes the lattice element x, that is

$$\zeta(x,y) = \begin{cases} 1 & \text{if } x \leq y \\ 0 & \text{if } x \not\leq y. \end{cases} \quad \text{(zeta function)} \quad (1)$$

Here I define the dual of the zeta function $\zeta^\partial(x,y)$, which quantifies whether x includes y, that is

$$\zeta^\partial(x,y) = \begin{cases} 1 & \text{if } x \geq y \\ 0 & \text{if } x \not\geq y. \end{cases} \quad \text{(dual of the zeta function)} \quad (2)$$

Note that if x does not include y, we group the ordered pair into an equivalence class where the answer is zero.

We now introduce a generalization of the function above where

$$z(x,y) = \begin{cases} 1 & \text{if } x \geq y \\ 0 & \text{if } x \wedge y = \bot \\ z & \text{otherwise, where } 0 < z < 1. \end{cases} \quad \text{(degrees of inclusion)} \quad (3)$$

The motivation here is that if x does not include y, we would like to consider the degree to which x includes y. It is here that the generalization takes place. It should be noted that

TABLE 2. The function $z(x,y)$ in equation (3). The x-values are aligned in the rows, and the y-values are in the columns. Note that this is a generalization of the dual of the zeta function (2) where a subset of the zero values are now unknown values between zero and one. These are indicated by the question marks. This function describes the degree to which y implies x. Clearly, we need more information to assign these values.

	$\perp$	a	b	c	$a \vee b$	$a \vee c$	$b \vee c$	$\top$
$\perp$	1	0	0	0	0	0	0	0
a	1	1	0	0	?	?	0	?
b	1	0	1	0	?	0	?	?
c	1	0	0	1	0	?	?	?
$a \vee b$	1	1	1	0	1	?	?	?
$c \vee a$	1	1	0	1	?	1	?	?
$b \vee c$	1	0	1	1	?	?	1	?
$\top$	1	1	1	1	1	1	1	1

this generalization is not a measure per se, which takes one argument to a real number, but rather a bi-valuation, which takes two lattice elements to a real number. The notation we are using here $z(x,y)$ emphasizes this fact, and we shall be careful not to lose sight of this as we adopt other notations specific to certain lattices.

Now it is important to note that this generalization does not tell us which particular values our measure should take for each pair of lattice elements. Some of the values of this bi-valuation are inherited directly from the zeta function. These are the situations when $x \geq y$, so that $z(x,y) = \zeta^\partial(x,y) = 1$, or when $x \wedge y = \perp$, so that $z(x,y) = \zeta^\partial(x,y) = 0$. However, we now need to determine the values of the function $z(x,y)$ when $x \not\geq y$ and $x \wedge y \neq \perp$. This is similar to the idea of analytic continuation when extending real-valued functions to complex-valued functions. To accomplish this, we need another principle.

CONSISTENCY

If our bi-valuation is to represent the lattice structure, the value we assign to any given pair of lattice elements must not be independent of the assignments we make to all the other pairs. If the assignments were independent of the lattice structure, then the measure cannot possibly represent the ordering. Dependence is necessary.

To derive these dependencies among the bi-valuation assignments, we turn to the lattice algebra. For element, $a \vee b$, one would expect that $z(a \vee b, c)$ must somehow depend on $z(a,c)$ and $z(b,c)$. Our goal is to determine the functional relationship by considering the lattice structure. Skilling's Principle of Generality [3] suggests that if a function is to be generally applicable, then it must work in special cases. If such a function exists, and we can constrain its form with a sufficient number of special cases, then we will be able to determine the functional form by relying on these constraints.

The most effective way to impose a special case is to use the algebra to write an expression two different ways. Consistency requires that two distinct expressions of the same quantity must give the same result when used as an argument in the bi-valuation. If this were not the case, then we would have two different measures assigned to the same pair of lattice elements, and we would not have a function. Let us see where this leads.

Constraints Imposed by the Lattice

Associativity of the Join

We first consider associativity of the join operation. The idea behind this derivation is that we consider an expression such as $u \vee (v \vee w)$ and use associativity to write this expression a second way $(u \vee v) \vee w$. Since these two expressions represent the same element, the degree to which $u \vee (v \vee w)$ includes a fourth statement t must equal the degree to which $(u \vee v) \vee w$ includes t.

We now consider the special case where $u \wedge v = v \wedge w = u \wedge w = \bot$. This is helpful since $z(x, \bot) = 0$ for all elements x, and we can neglect any dependence on the various meets, such as $u \wedge v$. If the assignments of the bi-valuation are to be consistent with one another and if $u \wedge v = \bot$ then $z(u \vee v, t)$ must be a related to $z(u, t)$ and $z(v, t)$. We express this idea by introducing a function $S(\cdot, \cdot)$ such that

$$z(u \vee v, t) = S[z(u,t), z(v,t)]. \tag{4}$$

The next step is to rewrite the two bi-valuations $z(u \vee (v \vee w), t)$ and $z((u \vee v) \vee w, t)$ in terms of the function $S[\cdot, \cdot]$

$$z(u \vee (v \vee w), t) = z((u \vee v) \vee w, t), \tag{5}$$

which results in

$$S[z(u,t), z((v \vee w), t)] = S[z((u \vee v), t), z(w,t)]. \tag{6}$$

Using the relation (4) again, we find that

$$S[z(u,t), S[z(v,t), z(w,t)]] = S[S[z(u,t), z(v,t)], t), z(w,t)]. \tag{7}$$

Substituting $a = z(u,t)$, $b = z(v,t)$ and $c = z(w,t)$ we obtain

$$S[a, S[b,c]] = S[S[a,b], c], \tag{8}$$

which is a functional equation known as the *associativity equation*. The general solution [4] is

$$S(a,b) = f(f^{-1}(a) + f^{-1}(b)), \tag{9}$$

where f is an arbitrary function. This is simplified by letting $g = f^{-1}$

$$g(S(a,b)) = g(a) + g(b). \tag{10}$$

Writing this in terms of the original expressions we find that,

$$g(z(u \vee v, t)) = g(z(u,t)) + g(z(v,t)), \tag{11}$$

which reveals that there exists a function $g: \mathbb{R} \to \mathbb{R}$ that re-maps these numbers to a more convenient representation [5, 6]. Defining $p(u,t) \equiv g(z(u,t))$ we get

$$p(u \vee v, t) = p(u,t) + p(v,t), \qquad (12)$$

which is the sum rule for the join of two atomic elements.

In general, for any pair of lattice elements x and y one can show [6] that the constraint equation becomes the familiar *sum rule*

$$p(u \vee v, t) = p(u,t) + p(v,t) - p(u \wedge v, t). \qquad (13)$$

This 'rule' is a constraint equation that ensures that associativity of the join is preserved by the bi-valuation assignments across the lattice.

This result is ubiquitous throughout mathematics and the sciences; so much so, that the relations often appear obvious [7]. For instance, the sum rule appears in many contexts, including probability theory, measure theory, geometry, quantum mechanics, information theory, etc. Here are some familiar examples:

$$p(x \vee y | \top) = p(x|\top) + p(y|\top) - p(x \wedge y | \top) \qquad (14)$$

$$I(X;Y) = H(X) + H(Y) - H(X,Y) \qquad (15)$$

$$max(A,B) = A + B - min(A,B) \qquad (16)$$

where the fact that these functions are bi-valuations has been suppressed by the standard notation. In the last example, A and B are Real numbers. Keep in mind that this relation is a constraint equation that enforces associativity. The constraint of associativity appears everywhere, once you know to look for it.

Associativity of Order and Distributivity

Given a chain of lattice elements $u \leq v \leq w \leq t$, we can consider the degree to which u includes t, $z(u,t)$. For the lattice of logical statements ordered by implication this is the degree to which t implies u. Since it is clear that t includes v, consistency requires that $z(u,t)$ be related to the degree to which u includes v, $z(u,v)$, and the degree to which v includes t, $z(v,t)$. We express this relationship as an unknown function $P(\cdot, \cdot)$

$$z(u,t) = P[z(u,v), z(v,t)]. \qquad (17)$$

Using the four elements in the chain, we can use associativity to write this two ways

$$z(u,t) = P[z(u,v), z(v,t)] = P[z(u,w), z(w,t)]. \qquad (18)$$

We can now write $z(v,t)$ and $z(u,w)$ using the same function. This results in

$$P[z(u,v), P[z(v,w), z(w,z)]] = P[P[z(u,v), z(v,w)], z(w,t)]. \qquad (19)$$

Substituting $a = z(u,v)$, $b = z(v,w)$ and $c = z(w,z)$ we again obtain the associativity equation

$$P[a,P[b,c]] = P[P[a,b],c]. \qquad (20)$$

The solution is of the form given in (9). The fact that there is an arbitrary function involved means that we can obtain a solution to this functional equation, which is commensurate with our previous sum rule. By exponentiating, we obtain the *product rule*

$$p(u,t) = p(u,v)p(v,t). \qquad (21)$$

This may look a bit more familiar if we use the fact that $u \leq v$ to write $u = u \wedge v$, and the fact that $v \leq t$ to write $v = v \wedge t$.

$$p(u \wedge v, t) = p(u, v \wedge t)p(v,t). \qquad (22)$$

This constraint equation was derived in the special case of a chain. If a constraint equation is to hold in general, it must hold in special cases. It is straightforward to show that this form does indeed hold in general when the lattice elements do not lie in a chain. This can be done by considering the constraint of distributivity of $\wedge$ over $\vee$ [5, 6].

Commutativity

The fact that $u \wedge v = v \wedge u$ allows us to write the product rule two ways. This gives us our final constraint equation, which is known as *Bayes' Theorem*

$$p(v,u)p(u,t) = p(u,v)p(v,t). \qquad (23)$$

In the case of our chain, the first term $p(v,u) = 1$ and this reduces to our original equation (21) as a triviality. The product rule and Bayes' Theorem are much more interesting when the elements do not lie in a chain

$$p(v, u \wedge t)p(u,t) = p(u, v \wedge t)p(v,t). \qquad (24)$$

Bayes' Theorem is so easily derived that one might be tempted to dismiss the constraint of commutativity as being a triviality. However, one can imagine ordered structures that are not commutative, and hence not lattices. In such structures associativity would give the usual sum and product rules, but the lack of commutativity would deny us a Bayes' Theorem. This is suggestive of the structure of quantum mechanics.

PROBABILITY THEORY

The derivations of the previous sections are applicable to distributive lattices in general. One such lattice is the Boolean lattice of logical statements. The bi-valuation $p(x,t)$, which describes the degree to which the statement x includes the statement t, or equivalently the degree to which t implies x. The *meaning* of the function $p(\cdot,\cdot)$ is inherited from the *meaning* of the ordering relation. Meaning cannot be assigned arbitrarily. For this reason, we will not talk about degrees of belief. Nor shall we require the notion of truth. The game we are playing involves only computations regarding the degree to which one statement implies another.

To highlight how these results relate to probability theory, we simply change notation

$$p(x|y) \doteq p(x,y). \tag{25}$$

Note that $p(x|y)$ is a bi-valuation that takes two logical statements as arguments. I will call this function *probability*, since it performs exactly the way we have come to expect probabilities to perform.

These logical statements are not some ill-defined prior information I that Jaynes introduces. Instead they are logical statements that belong to the lattice. While this function is traditionally read as 'the probability of x given y', we read 'the degree to which y implies x'. The prior information I that acts to externally constrain the values of the bi-valuation is not an argument of this function. Instead, this prior information I acts as an additional constraint. This may be an excellent place to adopt a new notation introduced by Skilling to represent the external information that goes into constraining the probability assignments

$$p(x|y) \qquad\qquad || \; I. \tag{26}$$

This careful analysis highlights the fact that probability is a bi-valuation, and not a measure in the usual sense. Expressions like $p(x)$ have no meaning in this context. Of course one is free to define a new function that takes one argument by fixing the value of the second argument, such as

$$p_t(x) \doteq p(x|t). \tag{27}$$

This is usually done for $t = \top$ and written by overloading the function p

$$p(x) \doteq p(x|\top). \tag{28}$$

This notation is unfortunate as it obscures the fact that this function is a bi-valuation.

Product Spaces

We rarely work in a single space. Instead we form product spaces, and it is essential to look at this explicitly to clear up some confusion that has persisted. We begin with the *Model Space*, $\mathcal{M}$, (also called the *state space*) which consists of a set of atomic statements that describe our physical system. If we are describing fruit, the atomic statements might be:

'a = It is an apple!'
'b = It is a banana!'
'c = It is a citrus fruit!'.

One may choose other statements, perhaps some that refine the notion of a citrus fruit to oranges, grapefruit, lemons, limes and kumquats. That is your model; you construct the space! The only requirement is that the atoms be mutually exclusive and exhaustive. If they are not, then the 'garbage in, garbage out' rule applies.

A lattice can be formed from the atomic statements by considering statements that consist of all possible disjunctions of the atomic statements. The resulting space is known as the *hypothesis space*, $\mathcal{H}$. The hypothesis space entertains statements such as 'It is an apple or a banana!' Clearly the fruit can never be in such a state. However an observer can make such a statement and thus convey the information that he or she possesses about the system. All of our calculations are performed in the hypothesis space. In this sense our results do not describe the universe so much as they describe the predictive power of one who might make statements about the universe. Since statistical mechanics is based on inferences in the hypothesis space, statistical mechanics is a theory describing the statements one might make about a thermodynamic system rather than being a theory about the thermodynamic system itself. It is an important, but subtle distinction. One might go as far as suggesting that the laws of physics are actually laws of inference [8].

For the remainder of this section, I will consider a more limited model space by considering just apples and bananas: a and b. The lattice describing the hypothesis space is shown in the upper left-hand corner of Figure 2 and is isomorphic to 2^M where M is the number of atomic statements in the model space $\mathcal{M}$. In the text that follows, I will make reference to probabilities defined in the hypothesis space, and will denote this function as $p_{\mathcal{H}}(\cdot|\cdot)$ to differentiate it from probabilities defined in the other spaces we will consider.

Other statements can be made about our fruit. Imagine that we have a color sensor that can generate two statements:

$$\begin{aligned} `r &= \text{It is red!'} \\ `y &= \text{It is yellow!'} \end{aligned}$$

These atomic statements form a second Boolean lattice, isomorphic to 2^N, with N being the number of atomic statements about the measurements. I will call this space the *Data Space* $\mathcal{D}$. Probabilities defined in this space will be denoted by $p_{\mathcal{D}}(\cdot|\cdot)$.

If we choose to consider these statements that the color sensor produces along with statements in the hypothesis space, we can create the product space $\mathcal{H} \times \mathcal{D}$, which results in a more complicated lattice isomorphic to $2^{(M+N)}$. This is the space in which the Bayesian does his or her work. Probabilities defined in this joint space will be denoted $p_{\mathcal{H} \times \mathcal{D}}(\cdot|\cdot)$, or just $p(\cdot|\cdot)$ for short.

The structure of the joint lattice imposes constraints on the values of the probability. These constraints are given by the three constraint equations: the sum rule, the product rule, and Bayes theorem. Our job is to assign the unknown values of $p(\cdot|\cdot)$, (for example, refer to Table 2). To do this, we need to find the values for all combinations of arguments, which amounts to $(2^{(M+N)})^2$ values. Many of these are trivial, and several are constrained to be related to one another by our three constraint equations, but a few remain that are left unconstrained. This is where external constraints in the form of prior information come into the problem. If one possesses prior information about how these statements imply one another, then this information acts as an external constraint and must be used to guide the assignments of the values for our function $p(\cdot|\cdot)$. There

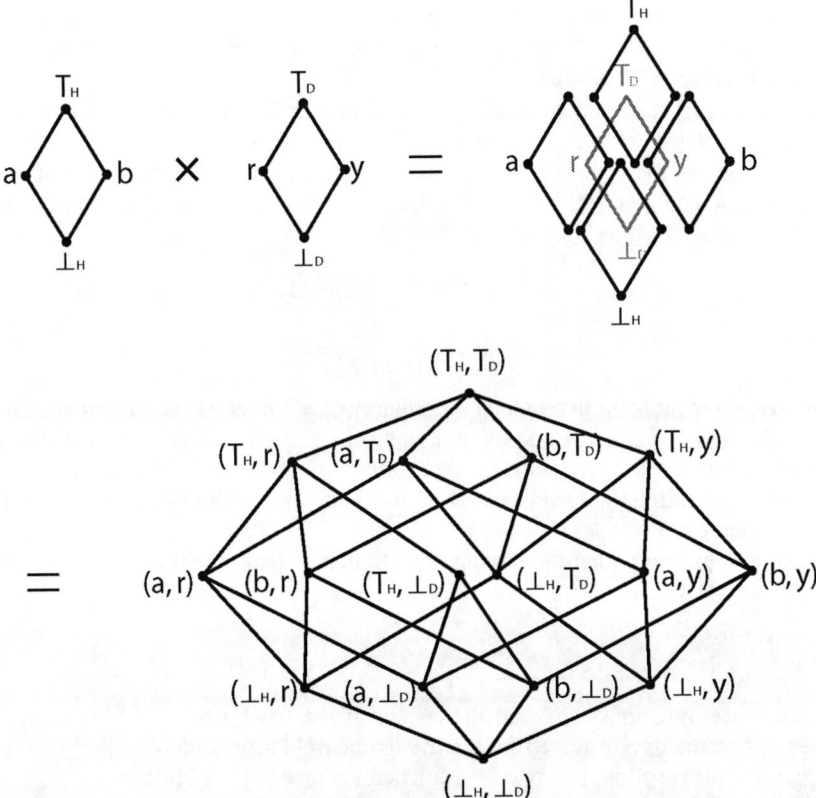

FIGURE 2. This figure illustrates the generation of a product space from a hypothesis space $\mathcal{H}$ and a data space $\mathcal{D}$. The lattice product is formed from the Cartesian product of their respective elements. Our task is to assign bi-valuations to each pair of elements in this joint space. Refer to the text for a complete description of the spaces.

are many ways to do this. One could assign the degree to which the joint atomic statements, such as (a,r), are implied by the joint top $\top \equiv (\top_H, \top_D)$. This degree is written $p((a,r)|\top)$.

One way to do this is by defining all of the prior probabilities in the hypothesis space. In this case, we would assign $p_\mathcal{H}(a|\top_H)$ and $p_\mathcal{H}(b|\top_H)$ in the hypothesis space. For consistency, these bi-valuations must be equal to their counterparts in the joint space where all possible data statements are considered. That is:

$$p_{\mathcal{H}\times\mathcal{D}}((a,r) \vee (a,y)|\top) \doteq p_\mathcal{H}(a|\top). \tag{29}$$

The probability on the left can be written in a more compact notation by defining $(a, \top_D) \doteq (a,r) \vee (a,y)$, and by dropping the messy joint subscript on p

$$p((a,\top_D)|\top) \doteq p_\mathcal{H}(a|\top). \tag{30}$$

The values of the prior probabilities in the hypothesis space $p_\mathcal{H}(a|\top)$ carry over into the joint space $p((a,\top_D)|\top)$ when we consider the top element in the data space, $\top_D$, which in this case is $\top_D = r \vee y$.

The next assignments to be made are commonly called likelihoods. These are the bi-valuations of the form $p((\top_H, r)|(a, \top_D))$. These probabilities are only defined in the joint space as they consider the degree to which a statement about the hypothesis implies a statement about the data. Note that $(\top_H, r)$ is a statement in the joint space that considers all the possible statements, so that

$$(\top_H, r) \equiv (a, r) \vee (b, r) \tag{31}$$

and that

$$(a, \top_D) \equiv (a, r) \vee (a, y). \tag{32}$$

The posterior probabilities, which are again just a subset of the possible values of the bi-valuation $p(\cdot|\cdot)$, are found by applying the constraint equation known as Bayes Theorem. These are the probabilities, such as $p((a, \top_D)|(\top_M, r))$. The evidence, or the marginal likelihood $p((\top_M, r)|\top)$ can be found using the constraint that all probabilities should sum to one.

An important, and perhaps surprising, fact to note is that the prior probability Π, the likelihood L, the evidence E, and the posterior P in Bayes Theorem

$$P = \Pi \frac{L}{E} \tag{33}$$

are not separate functions! They are all the generalized zeta function $p(\cdot|\cdot)$ with different classes of statements considered in its two arguments. There is no need to use different symbols for this function. It is one function defined on one joint lattice.

That these four terms are each manifestations of the same probability function defined on the joint space, highlights the fact that the likelihood is indeed a probability. It is the degree to which a statement describing a hypothesis implies a statement about the data. It does not matter if one data statement is realized, or even if anyone ever performs the experiment. The likelihood is not just a probability, but it is the same function as the posterior, the prior, and the evidence. This is important since the prior information that we use to constrain the prior probability assignments can also be used to constrain the likelihood assignments and vice versa. They are the same function, and the same information can be used to constraint the values of that function. One can even conceive that one can examine the likelihood assignment, identify the constraints that were applied and then apply them to the prior probability assignments. There is absolutely no problem with this since the entire game is about using information to constrain the probability assignments. All that is required is that this information be applied consistently. This is essentially the approach taken by Rodríguez [9] and Caticha and Preuss [10] to assign entropic priors.

The job of the Bayesian is to extend the definition of this bi-valuation across the joint lattice. From this perspective, there is no updating—only assignment. This is a 'timeless' perspective roughly analogous to the four-dimensional view of space-time in a many-worlds picture. All possible models and all possible data are represented, along with all possible combinations of their joins. The hypothesis space is potentially vast.

When a data value is observed, a logical statement about the data is made. The set of all other complementary data statements were not made and no longer need to be part of the lattice. They are the worlds that did not happen. These unspoken statements are grouped into the same equivalence class with the bottom element, and the lattice collapses. Furthermore, if a hypothesis is ruled out with certainty, that statement too can be grouped into an equivalence class with the bottom, and the lattice experiences further collapse. In both cases the information provided by these statements has constrained a specific class of bi-valuations to be zero. We can carry these values on in a large lattice, or simplify the lattice by adopting an equivalence class as described above and effectively adopt a smaller, yet functionally equivalent, lattice.

CONCLUSION

The most basic binary comparisons lead to the concept of ordering. The particular set of elements being compared, and the nature of the comparison give rise to a partially-ordered set (poset). We can then generalize inclusion on this poset to degrees of inclusion, which is implemented by generalizing the zeta function which encodes inclusion on the lattice. This process is analogous to the procedure of analytic continuation of real-valued functions into the complex plane. The end result is that there are a host of undefined values of this function. However, these undefined values are related to one another via a set of constraint equations that encode the structural, or algebraic, properties of the lattice. Three constraint equations arise in distributive lattices: the sum rule, the product rule and Bayes' Theorem. The universal nature of these lattices explains why these three constraint equations appear throughout mathematics and the sciences.

The game we Bayesians are playing is based on assigning the undefined values of this new function, which encodes degrees of inclusion. In most cases, we work with product spaces where one set of statements refer to what we call the hypothesis space, and the other set/s of statements refer to what we call the data space. We artificially divide this function into classes based on which space the statements forming the arguments originate. This gives rise to the notion of the prior, the likelihood, the evidence, and the posterior. However, each of these 'functions' are merely manifestations of this generalized function defined across the joint lattice. They are one and the same function—one and the same probability. The process of updating is simply that of assignment. Furthermore, if one can examine the likelihood assignments and extract the information being used to further constrain their values, one can use this information to constrain the values of the priors. This is just another example of the applications of consistency and honesty, which are fundamental to these theories.

This new perspective is timeless with respect to data acquisition. It is fundamentally distinct from Cox's approach, which directly relied on controversial and poorly-defined concepts such as belief. Furthermore, this perspective is unifying as it applies to measuring in general.

ACKNOWLEDGMENTS

I am deeply indebted to Ariel Caticha, John Skilling, Carlos Rodríguez, Janos Aczél, Deniz Gençağa and Jeffrey Jewell for insightful and inspiring discussions, and many invaluable remarks and comments.

REFERENCES

1. R. T. Cox, *Am. J. Physics* **14**, 1–13 (1946).
2. R. T. Cox, *The Algebra of Probable Inference*, Johns Hopkins Press, Baltimore, 1961.
3. J. Skilling, "The Axioms of Maximum Entropy," in *Maximum-Entropy and Bayesian Methods in Science and Engineering*, edited by G. J. Erickson, and C. R. Smith, Kluwer, Dordrecht, 1988, pp. 173–187.
4. J. Aczél, *Lectures on Functional Equations and Their Applications*, Academic Press, New York, 1966.
5. A. Caticha, *Phys. Rev. A* **57**, 1572–1582 (1998).
6. K. H. Knuth, *Neurocomputing* **67**, 245–274 (2005).
7. K. H. Knuth, "Deriving laws from ordering relations.," in *Bayesian Inference and Maximum Entropy Methods in Science and Engineering, Jackson Hole WY, USA, August 2003*, edited by G. J. Erickson, and Y. Zhai, AIP Conference Proceedings 707, American Institute of Physics, New York, 2004.
8. A. Caticha, and C. Cafaro, "From Information Geometry to Newtonian Dynamics," in *Bayesian Inference and Maximum Entropy Methods in Science and Engineering, Saratoga Springs NY, USA, July 2007*, edited by K. H. Knuth, A. Caticha, J. Center, J. L., G. A., and C. C. Rodríguez, AIP Conference Proceedings, ????
9. C. C. Rodríguez, Entropic priors, Tech. rep., Department of Mathematics, University at Albany, Albany NY (1991).
10. A. Caticha, and R. Preuss, *Phys. Rev. E* **70**, 046127 (2004).

FOUNDATIONS

Probability and Geometry

John Skilling

Maximum Entropy Data Consultants Ltd., Killaha East, Kenmare, County Kerry, Ireland

Abstract. Probability calculus is understood, and uniquely defined as the only rational tool for consistent inference. Yet two problems remain. One is a matter of principle: *how do we assign the prior distribution that expresses the question we wish to ask?* The other is a matter of practice: *how do we navigate the parameter space in order to compute the posterior inference?* Probability distributions have a natural geometry, which can be used to help in both these. But, like any other professional tool, geometry should be used with intelligence and care.

Keywords: prior, probability, geometry, metric, Fisher.
PACS: 02.50.Cw, 02.50.Tt

INTRODUCTION

We always see probability distributions through mean values induced by linear filters. For example, suppose we assign probabilities $p^1, p^2, \ldots, p^6$ (summing to 1) to the faces of a 6-faced die. From a set of tosses, we might accumulate the mean number of dots, involving $p^1 + 2p^2 + \ldots + 6p^6$. We might acquire prizes from valuations of the various outcomes, such as $100 for tossing a five, involving $100p^5$. Rational decisions would be based on loss functions, which are valuations viewed negatively by pessimists.

All such connections with practical consequences are linear, of the form $\sum_i Q_i p^i$, where **Q** is an "observable" vector from the space dual to that of **p**. Probabilities can be chained, as in the product rule of probability calculus, but the connection with reality is always linear. This suggests that there may be a geometry (which the superscript/subscript notation anticipated) in which an observation is identified as a scalar product

$$\langle \mathbf{Q} \rangle \equiv \mathbf{Q} \cdot \mathbf{p} = \sum_i Q_i p^i \qquad (1)$$

In Riemannian geometry [1], with "contravariant" components x^i denoting the position of a vector **x**, small increments **u** and **v** have scalar product $\mathbf{u} \cdot \mathbf{v} = \sum_{ij} g_{ij} u^i v^j$, where $g_{..}$ is the symmetric "metric" tensor, which usually depends on position. "Covariant" components of **x** are defined by $x_i = \sum_j g_{ij} x^j$, this transformation being reversed through the inverse metric g^{ij}. The length of a small increment is defined as $\ell(\delta \mathbf{x}) = \sqrt{\delta \mathbf{x} \cdot \delta \mathbf{x}}$, and the macroscopic "geodesic" distance $\ell(\mathbf{x}, \mathbf{y})$ is defined as the shortest sum of small lengths along paths joining **x** and **y**. Meanwhile, points within a given small distance of **x** define an ellipsoid of volume proportional to $1/\prod \sqrt{\lambda_k} = 1/\sqrt{\det g_{..}}$ where λ_k are the eigenvalues of $g_{..}$. Hence it is natural to assign a local density $\rho \propto \sqrt{\det g_{..}}$ in the neighbourhood of points of the Riemannian "manifold".

Can ℓ help with navigation? Can ρ help assign priors? To investigate these questions, we need to discover the appropriate metric.

THE RIEMANNIAN SIMPLEX

Working at first in the $(n-1)$-dimensional simplex of arbitrary probability distributions $\mathbf{p}$ (normalised to $\sum_{i=1}^{n} p^i = 1$), we already have the scalar product (1) between covariant $\mathbf{Q}$ and contravariant $\mathbf{p}$, but we need the metric g to get $\mathbf{A} \cdot \mathbf{B}$ and $\mathbf{p} \cdot \mathbf{q}$. On the observable (covariant) side, the expectation of the product

$$\langle \mathbf{AB} \rangle = \sum_i A_i B_i p^i \tag{2}$$

is [2] the only plausible candidate for the scalar product

$$\mathbf{A} \cdot \mathbf{B} = \sum_{ij} A_i B_j g^{ij} \tag{3}$$

Hence we assign the inverse metric as $g^{ij} = p^i$ if $i = j$ (0 otherwise), yielding the metric

$$g_{ij} = \begin{cases} 1/p^i & \text{if } i = j \\ 0 & \text{otherwise.} \end{cases} \tag{4}$$

This expression (4) is the "Fisher metric" [3]. It can alternatively be derived (apart from normalisation) by defining the length-squared separation of neighbouring distributions to be their cross-entropy. Under this metric, the length element and local density (normalised to unity) are (Fig. 1)

$$\ell(\delta \mathbf{p}) = \left(\sum_i \frac{(\delta p^i)^2}{p^i} \right)^{1/2} \qquad \rho(\mathbf{p}) = \frac{\Gamma(\frac{n}{2})}{\pi^{n/2}} \frac{\delta(\sum p^i - 1)}{\sqrt{p^1 p^2 \cdots p^n}} \tag{5}$$

Macroscopic distances can be derived (without Christoffel symbols) by transforming to square-roots $r^i = \sqrt{p^i}$, in terms of which the simplex $\sum p^i = 1$ becomes the positive orthant of the unit sphere $\sum (r^i)^2 = 1$, with length element $\ell = 2\sqrt{\sum (\delta r^i)^2}$. This latter geometry is Euclidean (scaled by 2), so geodesics are arcs of great circles around the surface, with

$$\ell(\mathbf{p}, \mathbf{q}) = 2\alpha, \quad \text{where } \cos \alpha = \sum_i \sqrt{p^i q^i}, \tag{6}$$

α being the angle between $\mathbf{p}$ and $\mathbf{q}$. (The related straight-line distance through the sphere interior is known as the Hellinger distance.) It follows that distances lie between 0 (for which $\mathbf{p} = \mathbf{q}$) and π (for which $\mathbf{p}$ and $\mathbf{q}$ are disjoint with no mutual support).

Meanwhile, the local density ρ transforms to being constant over the sphere orthant.

Is density useful as a prior on the simplex?

The density ρ in (5), if used as a prior for an unknown distribution $\mathbf{p}$ of proportions (which might or might not represent probabilities), is Dirichlet: $\Pr(\mathbf{p}) = \mathscr{D}(\frac{1}{2}, \frac{1}{2}, \ldots, \frac{1}{2})$. Thus, with $n = 2$, the geometrical prior on a proportion $\mathbf{p} = (\theta, 1-\theta)$ is

$$\Pr(\theta) = \pi^{-1}/\sqrt{\theta(1-\theta)} \tag{7}$$

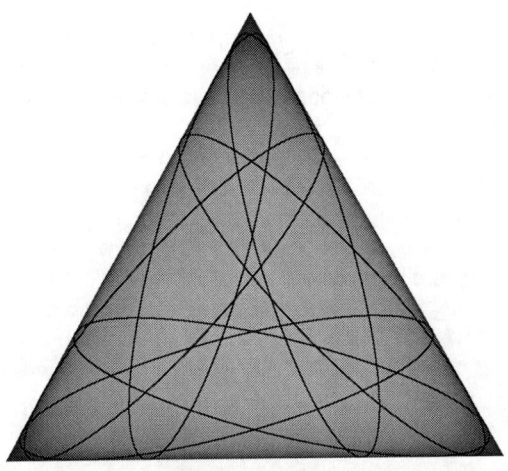

FIGURE 1. The simplex $p^1 + p^2 + p^3 = 1$, with a selection of geodesic paths and shaded by density ρ.

For many purposes, this may be quite reasonable. It's normalised and behaves sensibly near $\theta = 0$ and $\theta = 1$. Transformed to log-odds ($= \log(\theta/(1-\theta))$), the prior is inverse cosh, favouring log-odds that are $O(1)$ rather than wildly unbalanced. On putting $\theta = \sin^2 \phi$, prior mass is seen to be distributed uniformly over angle $0 \leq \phi \leq \pi/2$. On the other hand, (7) may not be reasonable. Background prior information could easily favour unbalanced log-odds, modelled better by $\Pr(\theta) \propto (\theta(1-\theta))^{-1+\varepsilon}$ for small (but not zero!) ε. The geometric proposal would not then be appropriate.

Another mis-application of (7) would be to estimate proportions $p(x)$ along a range of continuous coordinate x. Here, we would want to be able to compute on an arbitrarily fine grid, in order to approach the continuum limit properly. Yet, on a very fine grid, the Dirichlet distribution $\mathscr{D}(\frac{1}{2}, \frac{1}{2}, \frac{1}{2}, \ldots, \frac{1}{2})$ says that $p(x)$ is almost certainly almost flat as regards any macroscopic observable. For a sensible continuum limit, the Dirichlet indices α_i in $\mathscr{D}(\alpha_1, \alpha_2, \ldots)$ have to be a *measure* over x, summing to a finite total. They cannot all be $\frac{1}{2}$.

In short, the geometric prior on the simplex can be useful, but it's not automatically appropriate. Use with care.

Is distance useful for navigating the simplex?

Again, the message is mixed. When maximising some function $f(\mathbf{p})$, practical experience shows that the inverse metric *is* useful in converting covariant derivatives $\partial f/\partial p^i$

into contravariant search directions $p^i \partial f/\partial p^i$ (to be used within fairly small distances $\ell \lesssim 1$), which are less prone to break through the simplex boundaries into prohibited domains of negative p-values. Geodesics may be less useful, though, because their curvature tends to break the concave property of f that underlies most maximisation procedures. Instead of using geodesics, it seems better to seek provisional maximisation in the tangent space, holding the metric fixed (even though this too is capable of over-shooting into negative p-values). So, in algorithm design, geometry is useful but not dominant.

When navigating large distances (approaching π) between distributions that are nearly disjoint, the geodesic path between them is little more than linear combinations. Indeed, in $\sqrt{p}$ coordinates, the geodesic great circle *is* just linear combination. Thus the left column of Fig. 2 shows the geodesic path, through the simplex, between two separated normal distributions $N(-2,1)$ and $N(2,1)$, where $N(\mu, \sigma)$ stands for the usual Gaussian $p(x|\mu, \sigma) = \frac{1}{\sqrt{2\pi}\sigma} \exp \frac{-(x-\mu)^2}{2\sigma^2}$. In general, the distance between $N(\mu_1, \sigma_1)$ and $N(\mu_2, \sigma_2)$ evaluates to

$$\ell(\mu_1, \sigma_1; \mu_2, \sigma_2) = 2\cos^{-1}\left(\sqrt{\frac{2\sigma_1\sigma_2}{\sigma_1^2 + \sigma_2^2}} \exp \frac{-(\mu_1 - \mu_2)^2}{4(\sigma_1^2 + \sigma_2^2)}\right) \tag{8}$$

It would be more sophisticated to consider paths restricted to stay within the manifold of normal distributions. Geometry is available here too.

THE RIEMANNIAN MANIFOLD

Let parameters θ describe a m-dimensional manifold $\mathbf{p}(\theta)$ within the simplex. Assuming continuity and differentiability, a small change $\delta\theta$ induces a change in $\mathbf{p}$ with length-squared

$$(\delta L)^2 = \sum_k \frac{1}{p^k}(\delta p^k)^2 = \sum_k \frac{1}{p^k}\left(\sum_i \frac{\partial p^k}{\partial \theta^i}\delta\theta^i\right)\left(\sum_j \frac{\partial p^k}{\partial \theta^j}\delta\theta^j\right) \equiv \sum_{ij} G_{ij}\delta\theta^i\delta\theta^j, \tag{9}$$

$$G_{ij} = \sum_k \frac{1}{p^k}\frac{\partial p^k}{\partial \theta^i}\frac{\partial p^k}{\partial \theta^j} = \text{Fisher metric on manifold}. \tag{10}$$

For example, normal distributions have $m = 2$ (from the μ and σ components of θ), and their Fisher metric evaluates to

$$G = \begin{pmatrix} \sigma^{-2} & 0 \\ 0 & 2\sigma^{-2} \end{pmatrix} \tag{11}$$

The length element and local density are

$$dL = \sqrt{\frac{(d\mu)^2}{\sigma^2} + 2\frac{(d\sigma)^2}{\sigma^2}}, \qquad \rho \propto \frac{1}{\sigma^2}. \tag{12}$$

and the geodesic distance through the manifold evaluates to

$$L(\mu_1, \sigma_1; \mu_2, \sigma_2) = \sqrt{2}\cosh^{-1}\left(\frac{(\mu_1 - \mu_2)^2 + 2(\sigma_1^2 + \sigma_2^2)}{4\sigma_1\sigma_2}\right) \tag{13}$$

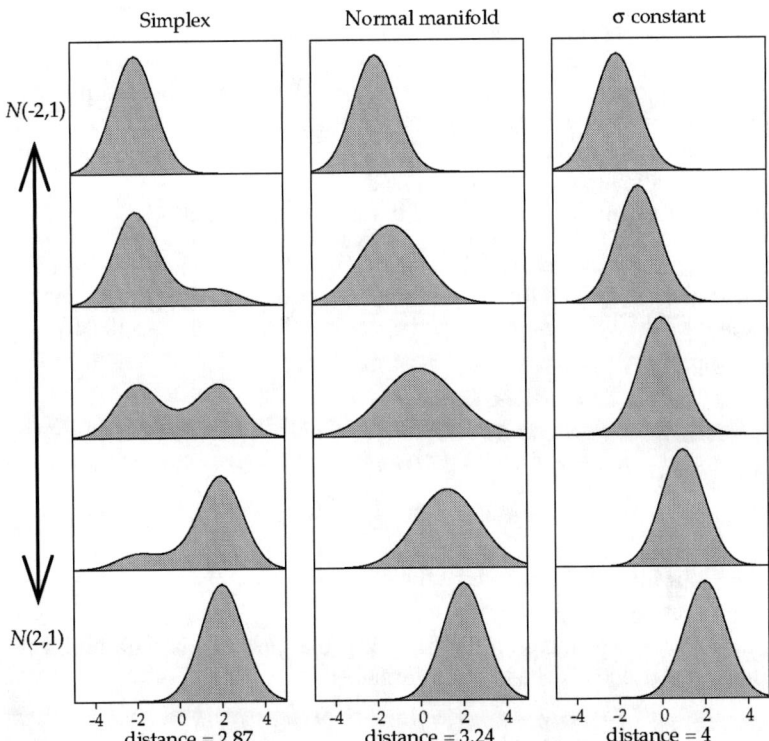

FIGURE 2. Different routes between $N(-2,1)$ and $N(2,1)$.

As it must be, this is rather larger than the simplex distance because of the constraint on the path. The second column of Fig. 2 shows the geodesic path through the manifold, with μ and σ both allowed to change. For comparison, the third column shows the path if μ alone is allowed to vary: this path is longer again.

Is density useful as a prior on the manifold?

An immediate danger is that the Fisher density is sensitive to parameterisation. For example, suppose that normal distributions were inadvertently coded with three parameters as $p(x|\mu_1,\mu_2,\sigma) = \frac{1}{\sqrt{2\pi}\sigma} \exp\frac{-(x-(\mu_1+\mu_2))^2}{2\sigma^2}$ because the user was unaware that only two are needed. The Fisher metric

$$G = \begin{pmatrix} \sigma^{-2} & \sigma^{-2} & 0 \\ \sigma^{-2} & \sigma^{-2} & 0 \\ 0 & 0 & 2\sigma^{-2} \end{pmatrix} \quad (14)$$

would be degenerate with two identical rows, so $\det G$ would vanish, destroying the $\sqrt{\det G}$ density being proposed for a prior $\Pr(\mu_1,\mu_2,\sigma)$. This particular behaviour, with two of the parameters coalescing into a single $\mu = \mu_1 + \mu_2$ would be easy to notice, and to correct.

Near-degeneracy would be less visible, so more dangerous. For example, let **p** depend strongly on θ^1 but only weakly on θ^2. As the latter dependence shrinks towards zero, it can be tweaked to make the density $\sqrt{g_{11}g_{22} - g_{12}g_{12}}$ tend to any desired shape, which need bear no relation to the single-variable density $\sqrt{g_{11}}$.

Yet the density can *sometimes* be useful. Consider the bitnet [4]

$$① \longrightarrow ② \longrightarrow ③$$

This is graphical shorthand for a 5-parameter manifold $\theta = (p,q,r,s,t)$ on the 8-vertex simplex of probability distributions on the 8 points 000 to 111. Specifically,

$$\begin{aligned}
p^1 &= \Pr(111) = pqs \\
p^2 &= \Pr(110) = pq(1-s) \\
p^3 &= \Pr(101) = p(1-q)t \\
p^4 &= \Pr(100) = p(1-q)(1-t) \\
p^5 &= \Pr(011) = (1-p)rs \\
p^6 &= \Pr(010) = (1-p)r(1-s) \\
p^7 &= \Pr(001) = (1-p)(1-r)t \\
p^8 &= \Pr(000) = (1-p)(1-r)(1-t)
\end{aligned}
\left.\begin{matrix} \\ \\ \\ \\ \\ \\ \\ \\ \end{matrix}\right\}
\begin{matrix} pq \\ p(1-q) \\ (1-p)r \\ (1-p)(1-r) \end{matrix}
\left.\begin{matrix} \\ \\ \\ \\ \end{matrix}\right\}
\begin{matrix} p \\ 1-p \end{matrix} \quad (15)$$

In this model, the first bit is ON with probability p, then the second is ON with probability q or r depending on the first, then the third is ON with probability s or t depending on the second. The Fisher metric is

$$G = \operatorname{diag}\left(\frac{1}{p(1-p)}, \frac{p}{q(1-q)}, \frac{1-p}{r(1-r)}, \frac{pq+(1-p)r}{s(1-s)}, \frac{p(1-q)+(1-p)(1-r)}{t(1-t)}\right)$$

whence the density (normalised to 1) is

$$\rho = 8\pi^{-5}\sqrt{\frac{(pq+(1-p)r)(p(1-q)+(1-p)(1-t))}{q(1-q)r(1-r)s(1-s)t(1-t)}} \quad (16)$$

This is helpfully objective, and looks perfectly reasonable since the required probability values are presumably $O(1)$. The geometric density is, of course, invariant under change of coordinates, so does not rely on the perhaps-arbitrary (p,q,r,s,t) labelling of the manifold.

Is distance useful for navigating the manifold?

Possibly, if the manifold is smooth. For example, in the manifold of normal distributions, geodesics are semicircles in the $(\mu, \sqrt{2}\sigma)$ plane, centred on the μ axis. This is convenient, and sympathetic to the nature of the problem.

A less forgiving example is the manifold of exponential distributions $E(a,w)$ of location a and width w, defined by

$$p(x|a,w) = \begin{cases} w^{-1}e^{(a-x)/w} & \text{for } x \geq a, \\ 0 & \text{otherwise.} \end{cases} \quad (17)$$

The simplex distance between two such distributions (ordered with $a_1 \leq a_2$) is

$$\ell(a_1, w_1; a_2, w_2) = 2\cos^{-1}\left(\frac{\sqrt{4w_1 w_2}}{w_1 + w_2} \exp\frac{a_1 - a_2}{2w_1}\right) \tag{18}$$

as plotted in Fig. 3. Moving within unit distance $\ell \leq 1$ allows the width to change by nearly a factor 3 (the scale of w in the figure is logarithmic), and the location by up to $\frac{1}{4}$ of the width: distributions within this domain overlap reasonably well, so tend to be mutually accessible within navigation algorithms. At smaller distances, though, the accessible domains shrink to needles, and not the smooth ellipses usually envisaged.

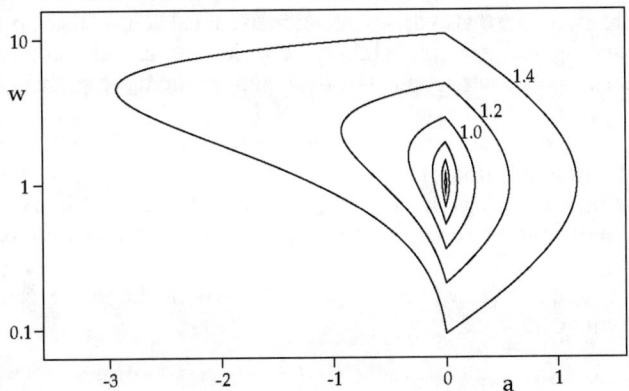

FIGURE 3. Contours of simplex distance $\ell = 0.4, 0.6, 0.8, 1.0, 1.2, 1.4$ around $E(0,1)$.

The technical reason for this is that the Fisher metric is singular

$$G = \begin{pmatrix} \infty & w^{-2} \\ w^{-2} & w^{-2} \end{pmatrix} \tag{19}$$

so the manifold is infinitely curved in a. Accordingly, geodesics are vertical lines unable to change location, so staying within the manifold works against useful navigation. Of course, this is a consequence of the sharp leading edge of the exponential which (counter to assumption) is not differentiable. On the other hand, smoothing the edge at some small scale ε still leaves geodesic lengths $O(\varepsilon^{-1/2})$ large, which is uncomfortable.

What has gone wrong here is basic to the geometry of manifolds. The Fisher metric (10) involves *two* derivatives $\partial/\partial\theta$, allowing *micro*scopic structure to dominate *macro*scopic behaviour, which is opposite to what is usually preferred. Add a function like $\varepsilon\sin(\theta/\varepsilon^2)$, which is everywhere small but has large and rapidly-varying slope, to a probability distribution $\mathbf{p}(\theta)$ and Bayesian inferences of macroscopic properties are almost unaffected because the perturbation is small. Do the same to a manifold, and the density and geodesic lengths that we want as tools become dominated by the fussy detail of locally-large slope.

When a continuous parameter x is divided into, say, a million cells for computation, the probability distributions $p(x)$ occupy a million-dimensional simplex. A path in this simplex can turn through a million right-angles (with length $10^6 \pi$) before any backtracking, even though no simplex distance can exceed π. To avoid this trap, manifolds need to be smooth. Geometry offers no automatic escape from the curse of dimensionality.

CONCLUSIONS

Geometry offers an interesting and potentially useful adjunct to probability calculus. It draws attention to a definition of geodesic distance between distributions which can be helpful when designing navigation algorithms — perhaps particularly useful in time-evolving applications [5]. It also draws attention to a local density that can be interpreted as a natural setting for prior probability. Both density and distance are attractively invariant under re-parameterisation, and they address the basic problems encountered in practical Bayesian inference.

However, geometry needs to be applied in a benign environment, because its proposals appear dubious in manifolds that are highly curved, or of high dimension. Geometry seems to have little to say about such problems. Equally, the applications of geometry range beyond probability calculus, to non-normalised distributions and beyond. So geometry cannot be identified with probability. Nevertheless, there is an overlap where both probability and geometry can be applied. When doing inference, geometry *may* be useful. Use it with care.

ACKNOWLEDGMENTS

I express particular thanks to Carlos Rodríguez for helpful, informative and enthusiastic discussions of geometry.

REFERENCES

1. S.-i. Amari, "Differential geometry methods in statistics", Springer-Verlag, Berlin (1985)
2. W. K. Wootters, "Statistical distance and Hilbert space", *Phys. Rev.* D**23**, 357-362 (1981)
3. R. A. Fisher, "On the mathematical foundations of theoretial statistics", *Phil. Trans. Roy. Soc. London* A**222**, 309-368 (1922)
4. C. C. Rodríguez, "The volume of bitnets", in *Maximum Entropy and Bayesian Methods* edited by R. Fischer, R. Preuss and U. von Toussaint, AIP Conf. Proc. **735**, 555-564 (2004)
5. H. Snoussi and A. Mohammad-Djafari, "Particle filtering on Riemannian manifolds", in *Maximum Entropy and Bayesian Methods* edited by A. Mohammad-Djafari, AIP Conf. Proc. **872**, 219-226 (2006)

Wrong Priors

Carlos C. Rodríguez

http://omega.albany.edu:8008/
Department of Mathematics and Statistics
The University at Albany, SUNY
Albany, NY 12222
USA

Abstract. All priors are not created equal. There are right and there are wrong priors. That is the main conclusion of this contribution. I use, a cooked-up example designed to create drama, and a typical textbook example to show the pervasiveness of wrong priors in standard statistical practice.

Keywords: Information Geometry, Volume Prior, Bayesian Inference, Bayesian Information Geometry, Ignorance Priors
PACS: 02.40.-k,02.50.Tt

INTRODUCTION

The information geometry available in regular statistical models can be used to build objectively meaningful prior distributions. When the information volume of the model is finite, the uniform distribution over the model manifold coincides with Jeffreys invariant rule. A simple example in two dimensions shows that the popular naive diffuse prior over the parameters of this model is in fact a *wrong* prior, requiring more than ten thousand observations to match Jeffreys rule with only 100 samples. Bayesian inference is suffering from an epidemic of wrong priors and to prove it I consider standard simple logistic regression with naive diffuse priors and with uniform priors over the manifold. The results are obviously less dramatic but similar to the previous cooked-up example. When the information volume of the model is infinite, the uniform distribution over the model does not exist. However, the available geometry can still be exploited and it provides a semiparametric family of invariant objectively ignorant priors.

A SIMPLE EXAMPLE

Consider bivariate normals with unit covariance matrix and mean vector restricted to a region of the euclidean plane. Specifically, for given values a and b the experiment consists of choosing (x,y) at random on the euclidean plane with,

$$x = \exp\left(-50(a^2+b^2)\right) + \varepsilon_1$$
$$y = \exp\left(-50((a-c)^2+(b-c)^2)\right) + \varepsilon_2.$$

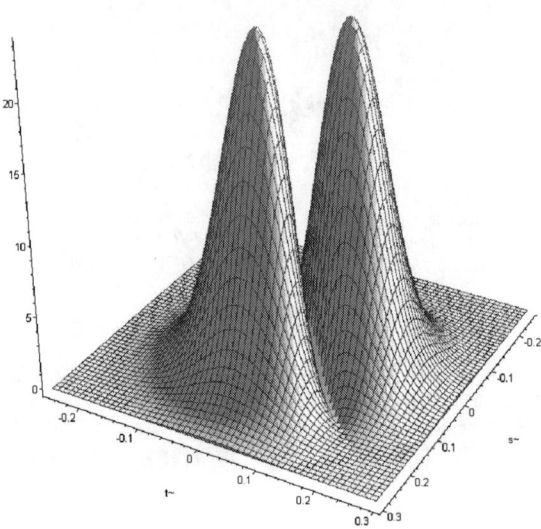

FIGURE 1. The prior π_1 is the true uniform over the model. Notice that the picture is not drawn at scale. The actual peaks should be more than 40 times taller than the ones displayed.

The unknown parameters are $a,b \in R$ but $c = 0.1$ is assumed known, and ε_1 and ε_2 are independent standard normals. The problem consists of learning the parameters $\theta = (a,b)$ from n independent observations $(x_1,y_1),\ldots,(x_n,y_n)$. We want to compare the performance of two priors on (a,b). The naive "'ignorant'" prior π_0 that takes a and b independently from $N(0,100)$ and the uniform prior over the manifold model, π_1 given by,

$$\pi_1(a,b) = \frac{|a-b|}{Z} \exp\left(-100(a^2+b^2) + 10(a+b)\right) \quad (1)$$

where Z is a finite normalization constant. Equation (1) is just the normalized volume form of the model computed trivially as $\sqrt{\det g}/Z$ with g as the information matrix (minus the expected values of the second derivatives of the log likelihood). This prior puts positive mass on the entire (a,b) plane (except on the line $a = b$ but that region has measure 0) but it is very far from uniform as it is shown in figure 1. Notice also that there are two peaks because the likelihood is invariant under the exchange of a with b. The volume prior respects this symmetry.

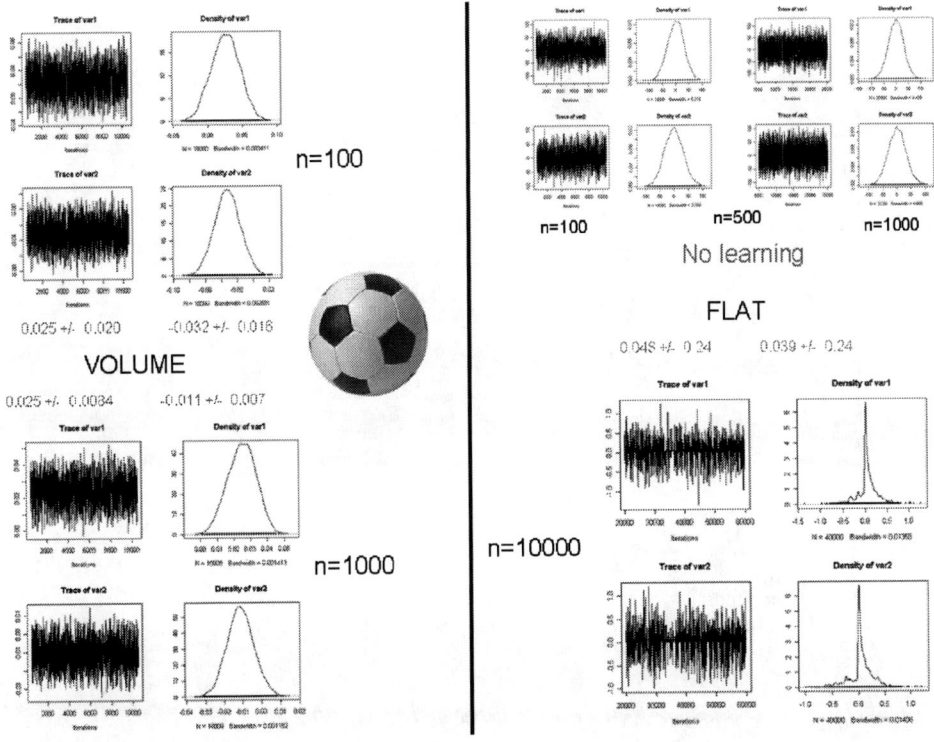

FIGURE 2. The right side shows the posteriors for the naive flat prior. The left side shows the posterior computed with the true uniform over the model.

Posterior Inference

With the help of the free MCMC package [1, 2] it only takes a few lines of code to realize the inadequacy of the naive prior for this example. The results of the MCMC simulations are summarized in figure 2. The true parameters where fixed at $a = 0.025$ and $b = -0.01$ and independent samples were chosen from the distribution with those parameters. With the naive flat prior the posteriors after observing 100, 500 and 1000 samples were essentially identical to the priors $N(0, 100)$, i.e. nothing was learned from the data. With 10000 observations the program was able to learn the values $(0.048 \pm 0.24, 0.039 \pm 0.24)$ for the true parameters. In contrast, just after 100 observations the posterior with the true uniform prior estimates the parameters very precisely as $(0.025 \pm 0.020, -0.032 \pm 0.016)$, still one order of magnitude of extra accuracy over the posterior with the flat prior with two orders of magnitude of extra data!

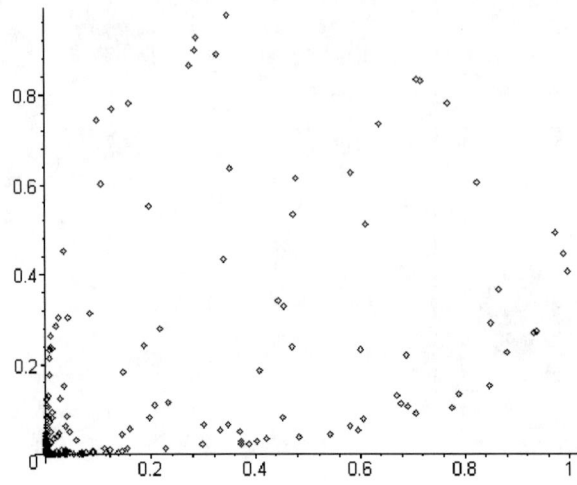

FIGURE 3. Ten thousand (u,v) points from (a,b) points chosen uniformly inside a ball of radius 3 centered at 0. Notice that more than 9000 points disappear into the origin.

Why is the volume prior so good?

To understand why the naive flat prior is so bad and the volume prior so good let's identify the transformed region of means (u,v) given by,

$$u = \exp(-50(a^2+b^2))$$
$$v = \exp(-50((a-c)^2+(b-c)^2))$$

as (a,b) range over the entire plane. An easy way to find the shape of this region is to pick points (a,b) at random on the plane and plot the corresponding (u,v) points. Figure 3 shows $10000(u,v)$ points obtained from $10000(a,b)$ points uniformly distributed inside a circle centered at the origin of radius 3. Notice that lots of points disappear into the origin!. Now take another $10000(a,b)$ points but now distributed according to π_1 with density given in (1). Figure 4 shows these (a,b) points. Notice that they are all highly concentrated about two points close to the origin. The corresponding (u,v) points are shown in figure 5. Got it?

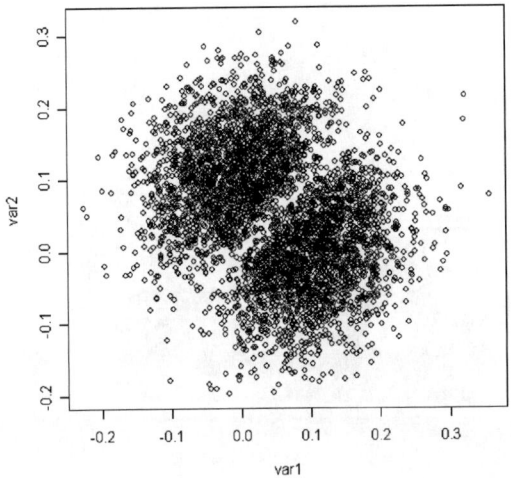

FIGURE 4. Ten thousand (a,b) points from the true uniform over the model

TABLE 1. Bioassay data from [3, p.88]

Dose, x_i (log g/ml)	Number of animals, n_i	Number of deaths, y_i
-0.863	5	0
-0.296	5	1
-0.053	5	3
0.727	5	5

The equation of the boundary of the leaf of (u,v) points

The computation of the exact equation of the leaf boundary in figure 5 is a nice exercise in simple optimization: Find max and min of v subject to the constraint that $u = t$. The max is given by the Red (dark) $R(t)$ curve in figure 6, with

$$R(t) = \exp(-(\sqrt{-\log t} - 1)^2). \qquad (2)$$

The min is given by the Green (light) curve,

$$G(t) = \exp(-(\sqrt{-\log t} + 1)^2) \qquad (3)$$

with $0 < t < 1$ in both cases.

Notice that there is a non-removable corner singularity at $t = 0$ but it is a piece of euclidean space so the curvature is zero at every point.

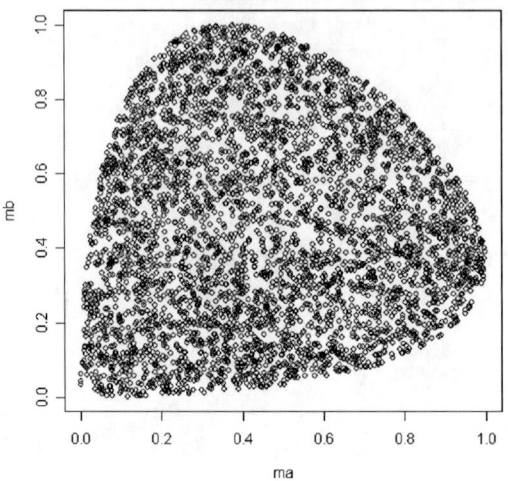

FIGURE 5. Ten thousand (u,v) points from (a,b) chosen from the density (1).

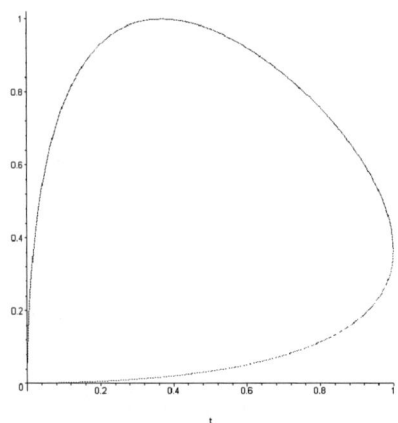

FIGURE 6. The exact boundary of the (u,v) points. Upper dark (red) part is $R(t)$. Lower light (green) part is $G(t)$.

TEXTBOOK EXAMPLE

Perhaps the first non-trivial example of a multiparameter bayesian model is simple logistic regression (see [3, p.88]). Twenty animals were tested, five at each of four dose levels (see Table 1). The standard model for this kind of data is,

$$(x_i, n_i, y_i); \; i = 1, \ldots, k,$$

assumed independent with,
$$y_i | \theta_i \sim \text{Bin}(n_i, \theta_i),$$

where θ_i is the probability of death for animals given dose x_i. The standard logistic dose-response relation is:

$$\log \frac{\theta_i}{1 - \theta_i} = a + bx_i \qquad (4)$$

The joint distribution of $(y_1, \ldots, y_k)$ is a function of the unknown parameters (a, b) and straight (but tedious) calculations give the volume element $dV = \sqrt{\det g} \, dadb$ in the (a, b) parameterization as

$$\begin{aligned}
dV &= T \sigma \, dadb \\
T &= \sum_j w_j = \sum_j n_j \theta_j (1 - \theta_j) \\
\sigma &= \text{stand. dev. of } X \text{ defined as,} \\
& P\{X = x_j | \theta_j\} \propto w_j \\
\theta_j &= \frac{1}{1 + \exp(-a - bx_j)}
\end{aligned}$$

This is a strange looking density (see figure 7). In particular this prior is proper and it assigns correlation of about 0.5 between a and b. This correlation is known a priori from the underlying geometry. In fact, the volume prior provides a better fit to the data than the standard diffuse naive prior that models a and b as independent variables with large variances. Figure 8 shows the results of the posterior simulations with both priors. Left panel with naive prior, right panel with volume prior. The red (dark) middle curves represent the logistic curves associated to the mean posterior values for (a, b) (100 thousand of them). The pictures also show 500 logistic curves obtained by sampling 500 (a, b) pairs from the available posterior samples. There is clearly more spread of logistic curves on the right than on the left panel. This is compatible with the fact that the volume prior samples uniformly over the manifold. Just like in the cooked-up example the over-spread (a, b) points cover only a small region of the manifold.

BEYOND FINITE VOLUMES

When the information volume $V(M)$ of the model M,

$$V(M) = \int_M dV = \int_\Theta \sqrt{\det g} \, d\theta$$

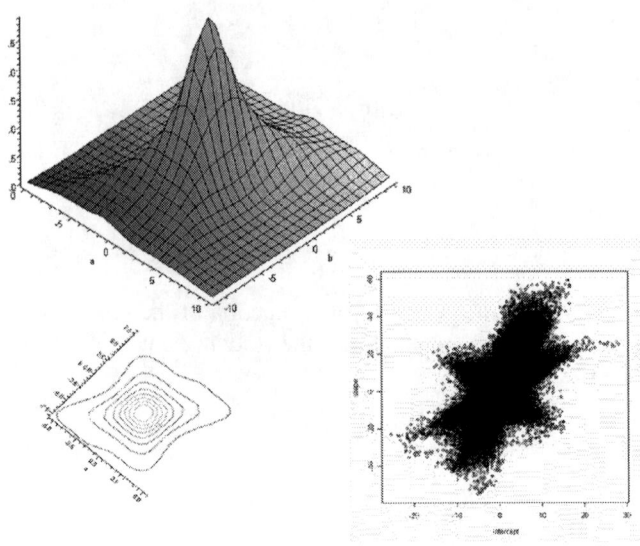

FIGURE 7. Uniform prior for logistic regression in the (a,b) parameterization. Bottom left: Contours. Bottom right: 2.5×10^5 samples from this prior.

is infinite; there is no uniform distribution over M. However, the underlying information geometry provides the following class of priors given as scalar density fields defined invariantly on M by,

$$\pi(p|t,v,\delta,\alpha) = \frac{1}{Z}[1+\alpha v I_\delta(p:t)]^{-\frac{1}{v}} \quad (5)$$

where $p \in M$, t is a probability distribution guessing the actual distribution of the data, δ, v are scalar parameters in $[0,1]$, $\alpha > 0$ large enough so that $Z < \infty$ and $I_\delta(p:t)$ is the δ-information deviation between (unnormalized) distributions p and t given by,

$$I_\delta(p:t) = \frac{1}{\delta(1-\delta)} \int [\delta p + (1-\delta)t - p^\delta t^{1-\delta}]$$

where the integral is over the whole data space manifold. This family of priors exists for any regular model and it has many remarkable properties. In particular this family maximizes a simple and objective notion of ignorance. For details see my *A geometric theory of ignorance*. The hyper parameters can be estimated with priors of the same kind or with a nonparametric prior of the Dirichlet Process type (which could itself be

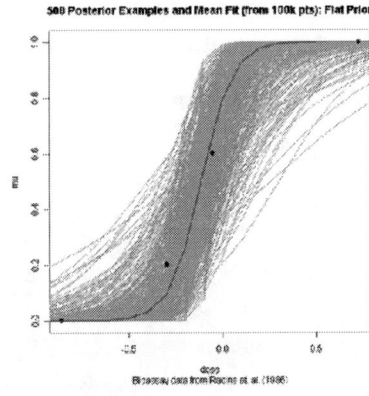 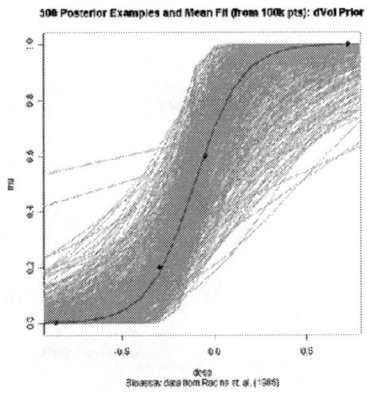

FIGURE 8. Posterior logistic curves. Left panel with naive prior. Right panel with volume prior.

seen as part of this family if we allow M to be infinite dimensional). There are still many open problems but the road ahead seems clear: More geometry.

ACKNOWLEDGMENTS

I am in debt to Phil Dawid whose invitation to talk at UCL prompted the finding of the example in the paper and to Ariel Caticha, Kevin Knuth, and John Skilling for many interesting discusions.

REFERENCES

1. A. D. Martin, , and K. M. Quinn, *MCMCpack: Markov chain Monte Carlo (MCMC) Package* (2007), URL http://mcmcpack.wustl.edu, r package version 0.8-2.
2. R Development Core Team, *R: A Language and Environment for Statistical Computing*, R Foundation for Statistical Computing, Vienna, Austria (2007), URL http://www.R-project.org, ISBN 3-900051-07-0.
3. A. Gelman, J. Carlin, H. Stern, and D. Rubin, *Bayesian Data Analysis, second edition*, London: CRC press, 2004.

Towards a Cross-Level Theory of Neural Learning

Anthony J. Bell

Redwood Center for Theoretical Neuroscience
University of California at Berkeley
3210F Tolman Hall, MC# 3192
Berkeley, CA 94720-3192

Abstract. This paper reviews ideas and results from unsupervised learning theory that have given the best explanation yet of how neural firing rates self-organise to code natural images in area V1 of visual cortex. It then discusses the generalisation of these ideas to self-organising spike-coding networks. A mismatch between the resulting spike-learning algorithm and the known physiological processes of synaptic plasticity is then used as a motivation to introduce the rather obvious idea that neurons are not sending their information to other neurons, but to synapses – more microscopic structures. This prompts a survey of other inter-level communications in the brain and inside cells. It is proposed on the basis of this that information flows all the way up and down the reductionist hierarchy – an idea that transforms many of our ideas about machine learning and neuroscience. What it transforms them into is not yet clear, but the remainder of the paper discusses this.

Keywords: Machine learning, Neuroscience, Spiking neurons, Reductionist hierarchy
PACS: 87.19.lo, 87.19.L-, 87.19.lv

THE NEED FOR PROGRESS

Despite many technical advances in probabilistic machine learning no-one has been able to connect its ideas convincingly with the learning processes occurring in the brain. At the same time, efforts to understand biological self-organisation with ideas from physics have not yielded as much as might be hoped. And to complete the triangle, the project to connect physical law to principles of information and computation is still a marginal activity, despite some fascinating results (for example [13] in this volume).

It is nonetheless anticipated that these 3 lines of enquiry (physics, biology, inference) will converge before long and a new science of complex systems will be invented. The mother-to-be of this invention is necessity. We face enormous challenges in climate, ecology, health and education – in the organisation of our societies and in their relationships to the biological systems that they contain and that contain them. At the same time, our communications and biotechnologies are transitioning to a new level of sophistication. It is hard to believe that we will be able to use our new technologies responsibly and find solutions to our problems without a better understanding of what life is, what learning is – for what characterises life perhaps above everything else, is its ability to adapt to and create new circumstances. We need to understand what gives biological systems their amazing adaptive abilities. This paper makes a serious attempt to propose a new line of thinking about this, using results and controversies in neuroscience and statistical learning theory as a guide.

MACHINE LEARNING AND THE BRAIN

The modern theories of statistical machine learning and probabilistic inference, although they emerged largely from the neural networks community, have little to say to an experimental neuroscientist. A comfortable and unexamined consensus seems to exist along the lines of "optimal perceptual inference to build representations, and optimal decision theory to choose actions". This view is so common that it may be called 'the Standard Model'. Its first component (inference) imagines the cortex utilising Bayesian procedures to estimate the values of variables in a scene (like depth). The theory is supported by the finding that humans can combine relevant information in Bayesianly optimal ways (for example visual and touch information [19]). But the question of which of the combinatorial number of possible 'hidden variables in the world' one should estimate is not answered by the Bayesian framework. The theory only works if it is already known what should be estimated. Our brains cannot always be estimating all variables of potential relevance. Ideas about attention, involving guiding feedback from higher cortical areas, have not yet matured into an accepted theory, and require the brain to have a motivation, to which we now turn.

Somewhere in the middle of the brain, the problem of representing turns into the problem of choosing and executing actions. At this point, the second stage of the Standard Model imagines cortex computing actions that maximise an abstract utility function, sometimes called 'reward', based on motivational information supplied by sub-cortical structures (the basal ganglia). The problem with this theory is that the circuits carrying this information themselves need to learn how to convert sensory input into motivational signals. There is no *given* reward function. The (rather elegant) mathematics of reinforcement learning, in which the reward signal comes in on a 'special wire' from outside unfortunately does not map onto the real situation where the neuromodulatory connections flooding cortex from subcortical structures are themselves plastic (or else they could not be altered by addictive substances). Much of what is rewarding is in constant flux as the needs of the physical organism change from moment to moment. The fact that reward is necessarily a plastic function within the system, and not a value judgement mystically arriving from outside is never more apparent than when one visits a robotics lab where reinforcement learning is being used. Like the undefined fitness functions of an Artificial Life simulation, the undefined reward functions of reinforcement learning signal the inadequacy of the theory underlying them to apply to the neurophysiological situation.

Even if we were to somehow blend the decision theoretic stage into the perceptual inference stage, so the Standard Model looked less like a two-stage homuncular hangover from a Cartesian worldview, we would still be stuck with the reward concept, defined when the rat is in the box, but much more elusive *in vivo*. Perhaps it is time to dispense with this concept and its attendant goal of explaining the complex social behaviours of N creatures as being that of the optimisation of N separate undefined scalar reward functions.

Aside from concerns about attention and reward, the Standard Model, with its focus on sensory and motor processes, has nothing to say about what the brain is doing when it is just thinking. Why for that matter do we need to sleep? All creatures with nervous systems above a certain complexity need to sleep or their nervous systems become epileptic, causing death. This leads many to believe that sleep is a neural requirement. It is unlikely that sleep exists just to conserve energy since we use as much when we are asleep as when we are awake and resting.

If reward is at best a learned function, then the brain's learning must be unsupervised. Unsupervised learning attacks a host of problems, including clustering, but in the absence of an *a priori* need to cluster, it is perhaps best viewed as simply density estimation. The line that I will follow in this paper is that density estimation is the correct way to think about learning in the brain. To summarise the theory in one sentence: the brain is saying 'how likely was that?' and adjusting itself accordingly. The density estimation occurs not between some external world input and some internal brain representation, but rather *across levels of the reductionist hierarchy*. But we are getting ahead of our story. First we must explain what density estimation is and why anyone would think it adequate to explain something so unsensory as behaviour.

SENSORY-MOTOR DENSITY ESTIMATION

The goal of density estimation is to fit a model probability density function (pdf) to data, as when we find the mean and variance that best fits a normal curve to a histogram of data values of, for example, peoples' heights. The objective that is minimised is called the Kullback-Leibler divergence between the data pdf p and the model pdf q, defined as $D[p|q] = \langle \log(p/q) \rangle_p$ where $\langle \cdot \rangle_p$ means the average over the pdf p.

Now if we had some arbitrary parameter w that we wanted to learn (say the strength of a synapse in the brain), we could learn it by computing the gradient of the KL divergence and running down it till we reach a minimum. A few lines of calculus shows this gradient to be:

$$\partial_w D[p|q] = \left\langle \left(1 + \log \frac{p}{q}\right) \partial_w \log p - \partial_w \log q \right\rangle_p \tag{1}$$

The second term is the gradient of the log likelihood of the data under the model, which (before we take the Bayesian path and talk about prior distributions over models) is the gradient normally followed in density estimation algorithms. This term is the *sensory* term, corresponding to changing ones model to fit the data. The first term, on the other hand, is a *motor* term: it corresponds to changing ones data to fit the model. While this would be a disreputable activity for a statistician, it is nonetheless a part of life, if we consider that a synaptic weight change may change the probability of future data.

Unless the dynamics of the world's data-generation process is deterministic and known, it seems impossible to evaluate this first term. But its very existence wakes us up to the fact that density estimation need not be a sensory game alone: the gradient of the KL divergence and the log likelihood are different. Furthermore, the tables are turned on a common complaint against unsupervised or Shannon information based learning models: that they do not distinguish between meaningful information and noise. The appearance of a second term dependent on the motor-influence of w suffices to make some data more relevant than other data. This term has the potential to actually provide a foundation for the signal/noise distinction, containing, as it does, the subjectivity of how ones brain-state effects the world.

But how can this term do so in a way that is meaningful to a creature - things like finding food and keeping warm? We do not have an answer to this question. But if we can find a cross-level theory of learning such that even cells and molecules can be seen as modeling and changing their local environments and contributing to a global model, then when they are too cold or lack energy, they will display dynamics that, just like an agent, will give rise to emergent properties that cause macroscopic behaviour changes, such as eating or finding shelter.

Such a theory may be out of reach at the moment, but we have to start somewhere. We will start by exploring the technicalities of statistical density estimation in order to set the stage for the cross-level ideas introduced later on. This will necessitate a dip into the mathematics, but for the unitiated, hold on - the story gets better later.

If the sensory-motor problem does have the unsupervised structure argued for by eq.(1), then the introduction of the motor term, for all its analytic intractability, *cannot* make things worse. It is *easier* to find the hidden degrees of freedom of the world when we can manipulate objects than when we are just looking at pictures. Children learn by doing, not just observing. It is easier to understand a world that we are in the process of creating than one that is given to us. (Of course, we can always go too far in this, leading to solipsism. A question for later sections is: what force tends to keep an adaptive agent away from a solipsistic solution?)

Knuth et al ([32] and references therein) describe a different unsupervised approach to the sensory-motor problem. It uses the combined calculus of inference and inquiry to design instruments that perform actions to maximise expected information gain. This is eminently sensible, but biological systems do not appear to function this way. Rather, organisms seem to converge on *autopoetically* stable reverberations with partially self-constructed environments [36], and this is closer to the idea we are trying to reach. A robot maximising information gain would never stop to sing a song.

Another interesting unsupervised approach maximises the information flow in the perception-action loop [30], though it is subject to the this same criticism.

SENSORY DENSITY ESTIMATION

Density estimation learning methods have already provided our best computational model to date for how the brain might self-organise from experience. One can show tiny images or movies to a neural network and train it to have receptive fields similar to those measured by single-neuron recordings from area V1 of visual cortex of cats and monkeys. Figure 1a viewable on the internet, shows a movie of an experiment done by Hubel and Wiesel on a cat in the 1960s, demonstrating how the notion of a receptive field arose.

Theoretical results of this kind were first obtained by Olshausen and Field [38] using the idea that neurons should try be sparse (fire rarely) and decorrelated. But the receptive fields can be obtained using density estimation alone. The results in Figure 1b and 1c were obtained by an Independent Component Analysis (ICA) network [5] and a kind of *Dependent* Component Analysis (DCA) network related to Hyvärinen and Hoyer's 'Topographic ICA' [27, 39, 49]). These receptive fields are static. Dynamic (spatio-temporal) receptive fields were obtained by van Hateren and Ruderman [48],

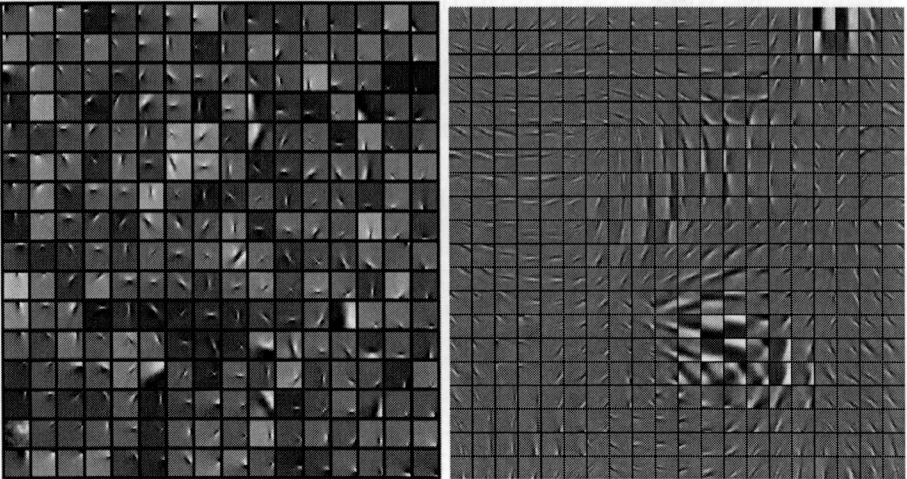

FIGURE 1. Receptive fields learned from natural image data. The full version of this figure is on the internet at www.snl.salk.edu/ tony/RecFields.html. (a) On web: A movie showing Hubel and Wiesel's discovery of visual receptive fields in cat. (b) Left above: ICA-learned image bases. Each picture is a learned axis in image space, corresponding to a column of $\mathbf{W}^{-1}$ (see text). (c) Right above: a typical basis set obtained with a model closely related to Topographic-ICA [49] [Thank you to Simon Osindero for permission to reprint his figure]. (d) On web: spatio-temporal receptive fields learned from natural movies by van Hateren and Ruderman [48]

and are shown on the web (Figure 1d). In both cases shown here, receptive fields are 'Gabor-like': localised in space, orientation, spatial frequency and phase, like Hubel and Wiesel's simple cells of V1. (Phase-invariant complex cells may also be learned [27]). In the topographic ICA case, the neurons are also spatially ordered in a 2D grid, or 'map', very much as V1 cells are arranged across the sheet of cortex, ie: in a orientation 'column' where position, orientation and spatially frequency vary continuously across the map, except at discontinuities called pinwheels visible in Figure 1c.

Both ICA and DCA are simple density estimation networks. They take a multivariate data distribution and find a new set of axes in it (just as a Fourier transform or Principal Component Analysis does). Unlike PCA, the new coordinate system is chosen entirely on the basis of the statistics of the data (PCA and Fourier bases impose the additional constraint that the axes be orthogonal in the original space). [It is important here not to confuse the orthogonality of the transform with the decorrelation of the resulting output variables.] The axes are found by training a complete set of filters (ie: a square matrix $\mathbf{W}$) to transform the data by $\mathbf{u} = \mathbf{W}\mathbf{x}$ into a new vector space $\mathbf{u}$ where the elements u_i are either as statistically independent as possible, or statistically dependent in some specified way. The training is done by presenting the images one at a time and changing the filter matrix according to one of the following equations:

$$\text{ICA}: \quad \Delta \mathbf{W} \quad \propto \quad \left(\mathbf{I} - \langle \mathbf{f}(\mathbf{u})\mathbf{u}^T \rangle_p \right) \mathbf{W} \qquad (2)$$

$$\text{DCA}: \quad \Delta \mathbf{W} \propto \left(\langle \mathbf{f(u)u}^T \rangle_q - \langle \mathbf{f(u)u}^T \rangle_p \right) \mathbf{W} \tag{3}$$

where $\mathbf{I}$ is the identity matrix. The learned axes (or *basis functions*) actually correspond to the columns of the inverse of the filter matrix $\mathbf{W}^{-1}$.

Both algorithms linearly transform the data into a **u**-space where a certain statistical model, $q(\mathbf{u})$, (a 'shaping density') is imposed. The optimisation is to fit the transformed data to this model by gradient ascent in the log likelihood of the data under this model via $\Delta \mathbf{W} \propto \langle \partial_\mathbf{W} \log q(\mathbf{x}) \rangle_p$, just as in eq.(1) without the motor term. The models on the input and output neurons are related by $q(\mathbf{x}) = q(\mathbf{u})|\mathbf{W}|$, where $|\cdot|$ means the absolute determinant.

In ICA, the model factorises: $q(\mathbf{u}) = \prod_i q(u_i)$, and the details of the univariate marginals, $q(u_i)$, may also be learned (though it is often un-necessary to do so). The vector of functions $f(\mathbf{u})$ has entries $f_i(\mathbf{u}) = -\partial_{u_i} \log q(\mathbf{u})$, and these are called the score functions. If these were linear, a condition satisfied by having a gaussian models on the $q(u_i)$, then ICA can be seen to stabilise on average ($\langle \Delta \mathbf{W} \rangle_p = 0$) when $\mathbf{I} = \langle \mathbf{uu}^T \rangle_p$, in other words when the outputs are unit variance decorrelated. To make non-gaussian signals independent we need statistics higher than second-order and these are provided by the Taylor-expansion of the score functions.

Much more could (and has) been said about this, but the main point to make here is that DCA is the completely general form, turning into ICA when the model we impose is that of independence, ie: $\langle \mathbf{f(u)u}^T \rangle_q = \mathbf{I}$. The DCA form can be derived by writing the model density in the completely general Gibbs' form:

$$q(\mathbf{u}) = \frac{1}{Z} e^{-E(\mathbf{u})} \tag{4}$$

involving an 'energy' $E(\mathbf{u})$ and a normaliser called the partition function Z. The two averages over the model and data densities in eq.(3) are then seen to arise from the gradients with respect to $\mathbf{W}$ of the log partition function and the energy respectively. The learning equation for a single weight has exactly a Boltzmann machine structure [24] consisting of a Hebbian (correlational) term sampled over the data density and an anti-Hebbian (anti-correlational) term sampled over the model density. This is said by some to be accomplished by alternately learning from data in an awake phase, then unlearning from the model in an asleep phase. This idea, while intriguing, has yet to condense into a serious neurobiological theory of sleep, but it is one to which we will return. There are few applications of DCA, for the same reason that there are few for the Boltzmann Machine, namely that the training (sampling from q) is just too slow.

The topographic ICA results in Figure 1c [27, 39] are actually obtained by a very simple DCA model[1], but more complicated models run into this need to integrate over the model density (the first term of eq.(3)). This integration, which is a universal bother in machine learning and statistical physics is only tractable in simple cases (like Gaussian or ICA models), and otherwise, as mentioned above, we must resort to

---

[1] overlapping neighbourhoods on a map: $q(\mathbf{u}) \propto \prod_K q(\mathbf{u}_K)$ where $\mathbf{u}_K$ means neighbourhood K of the map, and radially symmetric laplacian marginals, $q(\mathbf{u}_K) \propto \exp(\|\mathbf{u}_K\|)$

sampling from the model density using one of many schemes (Monte-Carlo Markov Chain (MCMC) or Contrastive Divergence [26] being two such schemes).

Were we able to solve the model-selection problem (the choice of $q(\mathbf{u})$) and the gradient of the partition function, we would be in good shape to attempt the Holy Grail problem of building hierarchical representations just from data, as we could use the resulting groupings of variables (like the neighbourhoods of the topographic map) to non-linearly recoordinatise the data at each layer and then look for new structure 'unwrapped' by the non-linear recoordinatisation. An example would be to re-express data fitting a radially symmetric laplacian model in spherical coordinates (phases and an amplitude) and input this to a higher network.

Many have travelled this road (my attempt is in [7]) and few have emerged unbloodied and with meaningful results (the few are [29, 27, 39]). It is not a problem I would recommend to a graduate student unless he had a good new idea. Rather I would recommend stepping back to look at the problem afresh, and biology can be a great inspiration in redesigning ones question until the answer looks right. In other words, if one is struggling with the problems of model selection and partition function gradients then perhaps it is a good idea to ask how on earth these problems map onto the tissue inside our skulls. That is the track that we will follow in the next part of the story.

To conclude, this section has been a quite dense summary of much technical work by many people just to arrive at two equations. eq.(3) is Amari et al's Natural Gradient [2] transformation of Hinton et al's view [25] of the original Infomax-ICA algorithm [4], which is identical to the maximum likelihood approach (see [12] for an explanation). The Natural Gradient concept (optimisation in the metric space of matrices) is explained in detail (without reference to Amari's Information Geometry) in [47, 37]. The ICA method in eq.(2) is also in the natural gradient form proposed by Amari et al where the weight space is given a Fisher metric based on reasons of information geometry [3]. A review of related sparse-coding techniques is found in [45].

SPIKING DENSITY ESTIMATION

Looking at biology, there is quite a variety of phenomena to draw inspiration from. Since it was not at all clear what model selection and the partition function gradient might mean in neural tissue, I decided to focus on a problem that had disturbed me for a long time: the issue of learning with neurons that spike. My earlier attempts on this had floundered (actually giving the Infomax-ICA algorithm as a by-product). The reason to tell the story is that it moves us close enough to biology that we can derive the *reductio ad absurdem* which sends us in a completely new direction. The story is interesting and I hope the reader will indulge me.

The problem was as follows: most real neurons communicate with each other not by sending real numbers (as in neural network models) but by sending pulses called spikes which last about 1ms. You can hear them crackle away in the Hubel & Wiesel movie in Figure 1a (on the internet). Unresolved controversy has raged in the neuroscience community for decades about whether or not the timings of these spikes is meaningful since they sound so much like Geiger counters popping randomly. Cortical pyramidal cells in area V1 which are *not* driven by their preferred visual stimulus (so-called

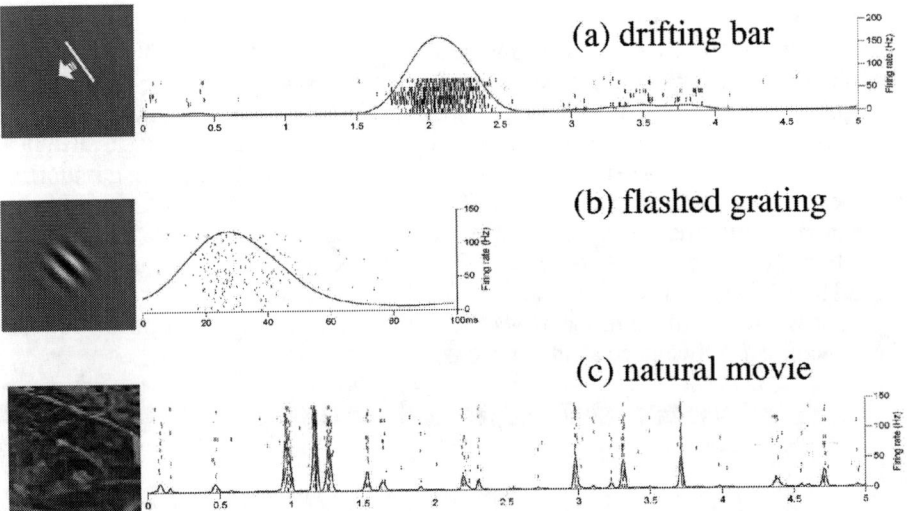

FIGURE 2. Recordings from an excitatory pyramidal cell in area V1 of an anaesthetised cat's visual cortex when shown (a) a drifting bar (see also the movie in Figure 1a), (b) a flashed grating, and (c) a natural movie. In each case, each row of dots represents a single trial, and each dot is a neuron spiking. Data from Blanche et al [10] with permission.

spontaneously firing cells) have roughly Poisson firing statistics: that is - they look like completely random point processes. And when we repeatedly give a neuron its preferred stimulus, its rate repeatably goes up while the detailed structure of its spike timings are different on each trial, as can be seen in Figure 2a and 2b.

There are two interpretations of this seemingly noisy Poisson-like firing. The dominant one has been that neurons are 'noisy rate coders' of their preferred stimuli. But a radically different explanation emerges if we consider that Poisson firing could also be the consequence of a neuron trying to maximise its information transmission rate. When we compress signals to maximise their information rate (as in image or video compression), the elements of the code become statistically independent (minimally redundant). If such an optimisation were to occur in spiking neurons, we would expect to see neurons firing with Poisson-like statistics.

The noisy rate coding idea is diametrically opposed to the idea that spike timings look noisy because they are highly informative. If one of these ideas is correct, the other one is wrong. Evidence for spike timing codes has built up over the years in studies of sensory neurons, but it is harder to demonstrate in cortical neurons because they are further from the sensory input and receive many unknown inputs from higher in the brain. The crucial breakthrough in studying this came when researchers started to record from cortical neurons while the animal was exposed to naturalistic stimuli instead of drifting gratings and bars designed to elicit maximum rate responses. An example of this is shown in Figure 2c. In multiple presentations of a natural movie to an anaesthetised cat while recording from an area V1 cortical pyramidal cell, the spiking pattern was quite repeatable from trial to trial, and when a neuron fired, it fired 1-3 spikes reliably usually within a 50ms time window. Such responses to natural stimuli (which the neuron is presumably more used to) are not consistent with the noisy rate coding hypothesis, but they are consistent with a picture where individual spikes signal the precise timing of the perceptual events they encode (see also [17] for an example from rat auditory cortex).

Although this debate is by no means settled, it does stimulate the theorist to attempt a proof-in-concept that spike-timing codes can self-organise. I embarked upon this project, together with Lucas Parra and Jeff Beck [8, 40]. Our idea was to use the same density estimation learning described earlier, but where the elements of the neural code are spike-timings, not real numbers representing rates, as in the simpler neural networks trained by ICA or DCA.

The principles are the same, but the network is an integrate-and-fire network [22]. For the ith neuron, the time-dependent voltage is:

$$u_i(t) = \sum_j \mathbf{W}_{ij} \sum_k R_{ij}(t - t_k) \qquad (5)$$

It sums over synaptic inputs j and spikes k arriving at that synapse at times t_k. The functions R_{ij} are the shapes of the potentials caused by the spikes, except R_{ii} which is the shape of the voltage reset after $u_i(t)$ reaches a threshold value and neuron i itself fires a spike. Our learning algorithm works by maximising the sensitivity of all output spike timings to input spike timings in a single-layer feedforward network. Without going into too much detail, there is a density model $q(\mathbf{t}_{in})$ ($\mathbf{t}_{in}$ being the vector of all input timings) and it is a function of the weight matrix $\mathbf{W}$ and the output timings $\mathbf{t}_{out}$. For every input spike l that helps cause an output spike k, the relevant synaptic weight $\mathbf{W}_{ij}$ changes according to:

$$\Delta \mathbf{W}_{ij} \propto \frac{\mathbf{T}_{kl}}{\mathbf{W}_{ij}} \left(\left[\mathbf{T}^{T\#} \right]_{kl} - \left[\mathbf{TT}^{T\#} \right]_{kk} \right) - f(r_i) r_j \qquad (6)$$

in which the matrix $\mathbf{T}$ is the spike-timing Jacobian (or sensitivity) matrix, having entries $\partial t_k / \partial t_l$ and the last term is a non-linear Hebbian term in the input (r_j) and output (r_i) spike rates, appropriately defined. As in eq.(2), f is again a score function.

We were very disappointed with this rule. It was a lot of work to find it, a lot of work to simulate it, and it is utterly biologically implausible. The simplicities of the ICA/DCA algorithms were not replicated in the spiking situation. The notation $\left[\mathbf{T}^{T\#} \right]_{kl}$, for example, represents the kl-th entry of the pseudoinverse of the transpose of the matrix representing the sensitivity of all output spike timings to all input spike timings, defined over all time and all neurons. The learning algorithm is horrendously non-local in space and time (meaning synaptic weight changes cannot be made using time-local pre- and post-synaptic information). Furthermore, the algorithm only works if there are more output spikes than input spikes (an *overcomplete* mapping being required to make a non-lossy map more probable). Were it not for Lucas Parra's persistence, this network learning rule would never have been derived, proven correct in simulations or published.

The answer was so complicated that the question had to be wrong. Referring again to the neurophysiology (always a good idea) soon revealed why. In focusing all our attention in the mapping between input spikes and output spikes, we had treated the dendrites of a neuron as if they were simple feedforward functions, designed to get information to the next neuron. In reality, there is also feedback from the output of the cell back to the synapses (called the back-propagating action potential), as well as electrical communication between synapses in the dendrites. In addition, the growing experimental literature on spike-based synaptic learning (called Spike Timing-Dependent Plasticity, or STDP), clearly showed that this information fed back to the synapse was implicated in synaptic plasticity. The physiology of synaptic plasticity is inordinately complex and controversial [15, 44], but one common theme emerges from the literature: calcium converts electrical signals into the molecular changes required to alter synapses. There are at least two distinct calcium currents operating in and around synapses to do this. The first enters through ion channels opened by neurotransmitter (the various kinds of NMDA receptor). The second enters through ion channels which are opened by changes in voltage internal to the cell (the NMDA receptor also has a voltage dependency). The other kind of receptor at excitatory synapses, the AMPA receptor does not let in calcium and thus cannot drive plasticity directly. Details aside, what this means is that the synapse integrates activity external and internal to the cell to determine how it should change.

We had used the mapping from input spikes to output spikes as our trainable density model, ignoring the backward and sideways information pathways in the dendrites. Our learning rule was clearly unbiological, also in the way it required the feedforward neural mapping to be non-lossy.

The conclusion was obvious: when we added the other information pathways in, the mapping relevant for the purpose of learning was not from input spike to output spike, *but from input spike to synaptic readout*. This readout was done by calcium currents local to the synapse, not a thresholding mechanism at the axon hillock. And what was read out were three kinds of spiking activity: spikes arriving at that synapse, and the influence of spikes arriving at other synapses and propagating back from the cell body, the latter two signalling to the synapse through graded potentials in the dendrites.

This may not sound startling to a physiologist, but from the direction we were approaching, the implications were startling indeed. Firstly, since there are roughly 1000 times as many synapses in the brain as neurons, the mapping was 1000 times overcomplete, easily solving the problem that an invertable mapping was required for the density estimation maths to work. In fact the state variables at the neural level (ie: spikes) could now be as lossy as they liked, because they no longer had to model the statistics of other spikes - this job could be done by new state variables (driven by calcium) operating at a different level of the system: the synaptic level. Suddenly neural information was preserved in the map to *synaptic* readout, while (in all likelihood) thrown away in the map to *neural* readout. This bypassed the main criticism of information theoretic neural learning algorithms: that they could not throw away information, as neurons clearly did. It also made sense to have the informational readout at the site of learning, rather than the output of the cell, thus decoupling the circuit's statistical model from its computation, two essentially different tasks which were conflated in the ICA/DCA case which had no synaptic state variables.

To make the model concrete, it is proposed that synaptic plasticity (at excitatory glutamatergic synapses anyway) operates roughly within the following framework. The neuron is a network of protein complexes (post-synaptic densities and the axon hillock), communicating similarly to eq.(5):

$$u_a(t) = \sum_b w_b \sum_k R_{ab}(t - t_k, u_b) \qquad (7)$$

except that now the indices a and b refer to these sites on the membrane. Each site has a learnable synaptic weight w_b and the transfer functions R_{ab} represent the effects that spikes k at site b can have on the voltage at site a (R_{aa} is the local synaptic response). This is essentially just the cable equation for linear electrical communication in dendrites, with a non-linear voltage-dependence added to account for the NMDA receptor voltage-dependency and conductance effects. At each site a, there are two calcium readouts, the first being synaptic (NMDA receptor) calcium c_a^+ and the second being intrinsic (voltage-dependent) calcium c_a^- carrying information from the rest of the cell:

$$c_a^+(t) = \lambda_a^+ w_a \sum_k R_{aa}(t - t_k, u_a) \qquad (8)$$

$$c_a^-(t) = \lambda_a^- \sum_{b \neq a} w_b \sum_k R_{ab}(t - t_k, u_b) \qquad (9)$$

The new 'plasticity parameters' λ_a^+ and λ_a^- represent the fraction of the local synaptic and intrinsic ionic currents which are calcium-carrying and thus available to drive molecular change. (Hippocampal excitatory synapses, for example, are much more plastic than cortical synapses, having much higher NMDA receptor counts. So-called 'silent synapses', common in developing nervous systems, and largely lacking AMPA receptors, would have λ_a^+ close to 1.)

These two kinds of calcium drive a first-order kinetic scheme involving a phenomenological variable y_a which is the 'readout':

$$\dot{y}_a = e^{c_a^+}(1 - y_a) - e^{-c_a^-} y_a \qquad (10)$$

You may ask: where did these equations come from? The answer is that they are pure guesses based on intuition, a reading of the literature on the physiology of synaptic plasticity and a desire to simplify things. They are included here merely to illustrate what may be the essential features of a calcium-based synaptic readout: a dynamic computation that compares external input with internal activity to determine how a weight should change. The real situation is much more complex [15], and varies greatly with synapse-type. However the kinetic scheme is not a complete fantasy: the push-pull of c_a^+ and c_a^- is meant to represent the actions of calcium-driven kinase and phosphotase proteins (like CAM-K2 and calcineurin) which activate opponent processes controlling the delivery to and recycling from the membrane of AMPA receptors, or the alteration of their sensitivity through phosphorylation.

It is useful to try to make a concrete model that shows information flowing from the neural to the synaptic level. But we have no learning rule here, just some equations for synaptic readout that suggest that macroscopic (neural) activity might be statistically modeled by a more microscopic set of dynamic variables located at synapses. What is missing is an understanding of this kind of inter-level communication in general. It is to this that we now turn.

LEVELS IN BIOLOGY

As scientists, we usually like to believe that the level at which we work is the important level for understanding more macroscopic phenomena, more microscopic phenomena being irrelevant. This is understandable because science is the search for lawful behaviour – this search involves adjusting experimental conditions (macro-variables) until the things which are measured (meso-variables) behave deterministically. Micro-variables are then not needed to explain these cases, and are then often regarded as merely implementation detail for the observed laws, or if they interfere with the lawful behaviour, they are "noise". In other words, the scientific method creates a series of self-reinforcing parochialisms, each centred on a certain level of description, each behaving deterministically largely only under the experimental conditions imposed.

No-one is specially to blame here. A molecular biologist who regards the quantum level as irrelevant cannot criticise a social psychologist for whom the skull is a reflecting barrier. These points may seem obvious, but think how often we read phrases like "the genetic basis of behaviour" or "the social basis of religion". The word 'basis' betrays a fundamentalism that seek to diminish the importance ordered emergence from the microsphere. And the notion that higher laws (ie: more compact determinisms) in the macrosphere are not much better ways of talking is also implicit here.

We have already seen two examples of problematic thinking in neuroscience that can arise from this: the rate-coding neuron which disappears when we show the cat a natural movie, and reward-maximising behaviours that become elusive when the rat escapes from the box. Rewards and rates are not defined under more natural conditions.

The other error mentioned above is to assume that deviations from deterministic behaviour are a consequence of 'noise'. The inability of an experimenter to predict a phenomenon does not mean that there are not meaningful hidden variables producing it. (You may not expect to read this sentence but that doesn't mean I am noisy!) Thus it is a mistake to talk about synaptic transmission as unreliable just because an experimenter cannot predict if a spike arriving at a presynaptic bouton will cause a vesicle release or not. Presynaptic filtering of spikes based on a bouton's internal state is probably an intelligent process.

Similar mistakes are made at the cellular level. Those studying bulk ion channel kinetics regard the motions of individual channels as noisy. Yet for molecules in the neighbourhood of single channels, these motions are a signal, particularly if any calcium flows through the channel. Calcium concentrations vary meaningfully over distances of 10nm – the width of a membrane protein. Calcium ions address individual molecules. Calcium may be to the post-synaptic density (PSD) what voltage is to the neuron as a whole: a spatially varying field communicating between proteins the way voltage communicates between protein complexes called synapses. ([44] is an up-to-date review of the amazing structure of the PSD.)

Turning to the cytoplasm, for the classical biochemist, enzyme reactions occur as they do *en masse* in a thermal aqueous medium, molecules bumping into each other randomly. But a more modern picture of the cytoplasm reveals that there is 'macromolecular crowding' (in which up to 50% of the cell's volume may be taken up by immobilised proteins and polynucleotides). This has opened the fascinating possibility that the remaining space (water) is *completely ordered* into a 3-dimensional protein-gated switching network, lined by charged hydration shells, continuously semi-conducting ions and small molecules differentially based on their size, valence and shape [46, 18]. The cellular metabolism is "vectorial". The same seems true of the membrane, a packed and organised 2-dimensional fluid in which cholesterol-based structures of all sizes (called lipid rafts) control the positions and motions of every integral protein [23, 28]. The quiet revolution in cell biology that has produced this new picture of ordered and meaningful multiscale micro-organisation mirrors the developments in systems neuroscience away from preoccupation with rate-coding feature detectors and towards an understanding of cross-level interactions involving spikes, and correlated activity on all scales. We turn now to the latter.

At the level above that of the neuron, many new lawful relations are coming to light. Global brain oscillations in the delta range (1-4 Hz) may well constrain less global oscillations in the theta range (4-8 Hz) [34], just as theta oscillations seem to constrain oscillations in the even more spatio-temporally local gamma range (40-200 Hz) [14], and just as gamma oscillations seem to determine when spikes are most likely to occur [20, 21]. In each case, it is at a particular phase of the lower frequency, more global, oscillation that the higher frequency, more local, effect is likely to increase its amplitude (or probability of occurence, in the case of gamma-to-spike coupling). The timing of spikes relative to the gamma oscillation is emerging as an important information-carrying factor in visual coding [33] and elsewhere [21]. See also [11] for information on this and other related topics, including important findings about how theta-to-gamma coupling in the hippocampus codes information about whereabouts a rat is in its environment (place-coding). A mini-review on these topics appears in [43]. It is noteworthy that the structured relations between spike timings and oscillation phases can be disrupted for several minutes by zapping the local tissue with a magnetic field [1].

All of this indicates a very structured organisation of the brain's oscillatory activity across space and time. Some neuroscientists have argued that these oscillations are 'epiphenomena', since they are just the result of many synchronised membrane currents. The computation, goes the argument, is occuring at the synapses – the junction points of the spiking computer. Others disagree, arguing that these oscillating electrical fields *can* influence spike timings directly [41], so-called ephaptic interactions. But even if this is not the case, describing these fields as epiphenomena is like saying that the US government is an epiphenomenon as it is just the result of the activity of many people talking to each other. The law-like influences of higher-order structures on lower is not an indicator of mystical downward causality but an epistemic statement about which groupings of variables are sufficient to summarise causal dependency. When we converse, we hear each others words, not each others spikes filtered through the air.

PUTTING IT TOGETHER?

We have mused on the problem of self-organising sensory-motor systems, explained some statistical learning theory, dived into the physiology of synaptic plasticity and outlined the flows of information across scales in the nervous system and inside cells. But we have been dancing around our main theme: that of uniting learning theory with

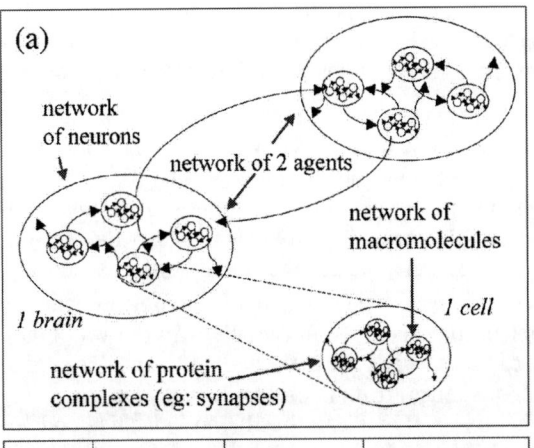
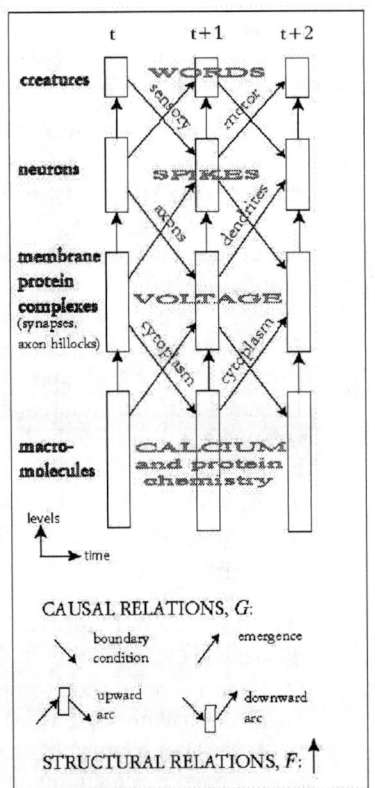

FIGURE 3. Three schematic views of the multi-level organisation found in biology. (a) Nervous systems viewed as networks of networks. (b) The units at each level are the networks at the level beneath. Adaptivity is given a different name at each level, but it is really a single unitary process, viewed at different scales, like an image viewed at different resolutions in space and time. Natural selection and self-organisation are thus not different processes, or competing explanations for biological adaptivity, but different spatio-temporal accounts of the same underlying thing. (c) A discrete-time 'cartoon' (for didactic purposes only) of the temporal evolution of a multiresolution state vector. Structural relations reduce the state vector as we ascend the reductionist hierarchy. Causal relations are thus similarly transformed. Objects at a given level do not communicate directly with each other, but rather through downward and upward arcs. For example, spikes can influence other spikes through dendrites and axons or through agents communicating.

this multi-level picture. This is because when there is no good theory (note the first word of this paper's title!), it is important to to survey all clues, empirical and theoretical, as well as to identify the shortcomings in existing ideas.

But there is no shying away any more. The central observation so far, that neurons talk to synapses not neurons, may be trivial. But if we repeat it at different scales, it naturally leads to to a view of the nervous system which is at odds with most of the main strands of computational and experimental thinking, yet this view is consistent with all the empirical data and throws the concepts of machine learning into new unexplored inter-level scenarios. This view is presented in Figure 3. It has the following characteristics:

- Biological organisation consists of networks within networks (Figure 3a). Microscopic networks are inside the nodes of macroscopic networks, like the network of synapses (eq.(7)) inside a neuron. The outputs and inputs of a network are signals to and from the network above in the hierarchy. Phases of groups of neurons, timings of individual neurons, flows of voltage and flows of calcium carry the network information at the cell assembly, neural, synaptic and molecular levels respectively.
- There is thus no 'functionalist cut-off level' [2] anywhere in the biological hierarchy [6]. Nature does not seem to shield the macro from the micro in the way that a computer does. A single photon can save a cat's life in the dark [42], and this kind of structured amplification (called emergence) from microscopic matter is happening continually all through biological tissue. There can thus be no "machine code of the brain". The ordered flow of information from the microscopic is represented in Figure 3c by the upward diagonal arrows.
- This information flow is bidirectional. As we communicate with each other using words, we change each other's gene expression, through the downward diagonal arrows in Figure 3c. Words (for example) cause spikes, spikes cause calcium flows and these send a message to the nucleus to make different proteins. See [9] for a mechanism by which this can be accomplished.[3]
- The sensory-motor loop, properly considered, is an inter-level interaction. Going upward, behaviours like words emerge from spikes, and going downward, they are the (social) boundary condition for a listener's neural activity, just as spikes are the neural boundary condition on the membrane for voltage flows in the dendrites.
- The sensory-motor loop is just one stage in a multi-resolution hierarchy of nested similar inter-level dynamics. It is not special, just as "agents" are not a special stage in this hierarchy, but merely a level of description.
- Spikes (for example) can communicate with each other through two different mechanisms: an upward arc via the social network, and a downward arc via the dendritic network (see Figure 3c). Viewed this way, there are no horizontal arrows, just flows of information up and down. Even messages between synapses are differently processed by different voltage or calcium-sensitive macromolecules in the post-synaptic density of the receiving synapse (the lowest downward arc in Figure 3c).
- These processes are *all the same thing expressed at different resolutions*. The state vectors are transformed by dimensionality-reducing structural relations in 3c, and the causal relations are transformed with them.

---

[2] Functionalism is the idea that you could build a computer out of beer cans and string.

[3] The mechanism is the Endoplasmic Reticulum. Quoting Berridge: "the ER and plasma membrane form a binary membrane system that functions to regulate a variety of neuronal processes including excitability, associativity, neurotransmitter release, synaptic plasticity and gene transcription".

A cross-level theory of learning would be a theory taking place in a causal structure like that of Figure 3c. Normally in (Bayesian) learning theory, we separate the data $\mathbf{x}$ from the model (the learned weights $\mathbf{W}$), and define quantities like the likelihood $q(\mathbf{x}|\mathbf{W})$, the prior $q(\mathbf{W})$, and the posterior $q(\mathbf{W}|\mathbf{x})$. In a cross-level theory, the data (for example spike timings) is just the network traffic which flows through connections which are nodes in the network below, as descibed. Thus quantities like the likelihood and the posterior are actually conditional distributions of groupings of variables across scale. It is a tantalising task to connect the framework known as 'hierarchical Bayes' to the hierarchy of matter observed in experiments.

The key to this may lie in generalising to the learning-scenario a framework in physics known as the Renormalisation Group (RG) [50]. RG has been called the most important new mathematical idea discovered in the 20th century. It is what enabled the creation of quantum field theory from generic quantum mechanics. It is also the idea used to understand equilibrium phase transitions in certain physical systems (like the 2D Ising model). It allows one to investigate the changes of a physical system as one views it at different spatial scales. It applies to scale-invariant systems, the correlation functions of which change in ordered ways as the resolution is altered. It does not apply in biology because biology has special scales, permitting the separation of nested networks that we have described.

As mentioned, the statistics of the activity in scale-invariant systems transform lawfully as the scale is varied. RG provides equations that capture these constraints. These equations are used to solve for the partition function of the system, leading to the accurate prediction of important physical quantities. This partition function, Z, is exactly the same mathematical quantity we encountered before in eq.(4).

In machine learning, as indicated earlier, it is not Z but its *gradient* which matters (remember the first term in eq.(3)). Could it be that something like RG could enable the calculation of an appropriate learning gradient for the multiresolution state vectors shown in Figure 3c?

The empirical relations between statistics of activity at different scales of the brain's material hierarchy (like the dendritic currents, spikes and local field potentials we have discussed) are only beginning to unveiled by an array of sophisticated new multiresolution probes. The principles underlying the brain's multi-level statistical model of its sensory-motor world may even be identified by experimenters before the theorists get their act together.

Physics, biology, inference – the strings are not yet tied, the questions not even really formalised. In physics, even reductionism is not necessarily on solid ground [35]. Noise, signal, control, reward, agency, the brain as a logic device – the ideas that lay behind the cybernetics movement in the 1950s, and have so heavily influenced our thinking today, are starting to look like they may not be the primitives of a future emergent understanding. What kind of statistical self-modelling is occuring in matter? How can we draw these loose strands together? The game is open. There's everything to play for.

ACKNOWLEDGMENTS

Thanks to Kilian Koepsell and Lucas Parra for many discussions on this, to Tim Blanche and Simon Osindero for permission to use figures from their work (in Tim's case of data not yet published in a journal [10]), to all at the Redwood Center, to Geoff Hinton (and CIAR) and Terry Sejnowski for ideas and support, and to MaxEnt '07, CosNet/Complex '07 in Australia, and Okinawa Institute of Science and Technology in Japan for chances to air these issues The work was funded by the Swartz Foundation and the NSF Science of Learning Center (TDLC). As well as providing the primary funding, Jerry Swartz's encouragement, enthusiasm and perception has been a crucial help.

REFERENCES

1. Allen E.A et al. 2007. Transcranial magnetic stimulation elicits coupled neural and hemodynmaic consequences. *Science*, **317**, 1918-21
2. Amari S.-I., Cichocki A. and Yang H.H. 1996. A New Learning Algorithm for Blind Signal Separation. *Advances in Neural Information Processing Systems 8*, 757-763, MIT Press, Cambridge MA
3. Amari S-I. 1997. Natural gradient works efficiently in learning. *Neural Computation*, **10**, 251-276
4. Bell A.J. and Sejnowski T.J. 1995. An information-maximization approach to blind separation and blind deconvolution. *Neural Computation*, 7, 6, 1129-1159
5. Bell A.J. and Sejnowski T.J. 1997. The 'independent' components of natural images are edge-filters. *Vision Research*, **37**, 3327-3338
6. Bell A.J. Levels and loops: the future of Artificial Intelligence and Neuroscience. 1999. *Phil. Trans. R. Soc. Lond. B* **354**, 2013-2020
7. Bell A.J. 2002. The co-information lattice. *4th International Symposium on Independent Component Analysis and Blind Source Separation*, Nara, Japan
8. Bell A.J. and Parra L.C. 2005. Maximising sensitivity in a spiking network. *Advances in Neural Information Processing Systems 17*, MIT Press
9. Berridge M.J. 1998. Neuronal calcium signaling. *Neuron*, **21**, 13-26
10. Blanche T.J., Freiwald W.A. and Swindale N.V. 2006. Neural sparseness in cat and monkey visual cortex studied with silicon polytrode arrays. *Society of Neuroscience Abstracts*, Atlanta, Georgia
11. Buzsaki G. 2006. *Rhythms of the brain*. Oxford Univ. Press
12. Cardoso J.-F. 1997. Infomax and maximum likelihood for blind source separation. *IEEE Signal Processing Letters*, **4**, 4, 112-114
13. Caticha A. and Cafaro C. 2008. From information geometry to Newtonian dynamics. *In [31] (this volume)*
14. Canolty R.T. et al. 2006. High Gamma Power Is Phase-Locked to Theta Oscillations in Human Neocortex. *Science*, **313**, 5793, 1626-28
15. Dan, Y. & Poo, M.-m. 2004. Spike timing-dependent plasticity of neural circuits. *Neuron* **44**, 23-30
16. Dewar R.C. 2005. Maximum entropy production and the fluctuation theorem. *J. Phys. A: Math. Gen.*, **38**, L371-381
17. DeWeese M.R., Wehr M. and Zador A.M. 2003. Binary spiking in auditory cortex. *J. Neurosci.*, **23**, 7940
18. Ellis R. J. 2001. Macromolecular crowding: obvious but underappreciated. *Trends in Biochem. Sci.*, **26**, 10, 597-604
19. Ernst M.O. and Banks M.S. 2002. Humans integrate visual and haptic information in a statistically optimal fashion. *Nature*, **415**, 6870, 429-33
20. Fries P. 2005. A mechanism for cognitive dynamics: neuronal communication through neuronal coherence. *Trends Cogn. Sci.*, **9**, 474
21. Fries P., Nicolic D. and Singer W. 2007. The gamma cycle. *Trends Neurosci.*, **30**, 7, 309-16
22. Gerstner W. and Kistner W.M. 2002. *Spiking neuron models*. Cambridge University Press

23. Hancock J.F. Lipid rafts: contentious only from simplistic standpoints. 2006. *Nature Rev. Molec. Cell Biol.*, **7**, 456-462
24. Hinton G.E. and Sejnowski T.J. 1986. Learning and relearning in Boltzmann machines. In Rumelhart D.E. and McClelland J.L. (eds) *Parallel Distributed Processing: Explorations in the Microstructure of Cognition. Volume 1: Foundations*, MIT Press
25. Hinton G.E. et al. 2001. A New View of ICA, *Proceedings of ICA 2001*, San Diego, CA.
26. Hinton, G.E. 2002. Training Products of Experts by Minimizing Contrastive Divergence. *Neural Computation*, **14**, 1771-1800
27. Hyvärinen A. and Hoyer P. 2001. A Two-Layer Sparse Coding Model Learns Simple and Complex Cell Receptive Fields and Topography from Natural Images. *Vision Research*, **41**, 18, 2413-23
28. Jacobson K., Mouritsen O.G. and Anderson R.G.W. 2007. Lipid rafts: at a crossroads between cell biology and physics. *Nature Cell Biol.*, **9**, 1, 7-14
29. Karklin Y. and Lewicki M.S. 2005. A hierarchical Bayesian model for learning non-linear statistical regularities in non-stationary natural signals. *Neural Computation*, **17**, 2, 397-423
30. Klyubin A.S., Polani D. and Nehaniv C.L. 2007. Representations of space and time in the maximization of information flow in the perception-action loop. *Neural Computation*, **19**, 2387-2432
31. Knuth K.H. et al. 2007. *Bayesian Inference and Maximum Entropy Methods in Science and Engineering, Saratoga Springs, NY, USA*, AIP Conf. Proc., Melville NY:AIP
32. Knuth K.H., Erner P.M. and Frasso S. 2007. Designing intelligent instruments. *[31] (this volume)*
33. Koepsell K. et al. 2008. Retinal oscillations carry visual information to cortex. *under review*
34. Lakatos P. et al. 2005. An oscillatory hierarchy controlling neuronal excitability and stimulus processing in the auditory cortex. *J. Neurophysiol.*, **94**, 1904-11
35. Laughlin R.B. 2006. *A different universe: reinventing physics from the bottom down.* Basic Books, NY
36. Maturana H. and Varela F. 1973. *Autopoiesis and cognition: the realization of the living.* Dordecht: D. Reidel
37. Moon T.K. and Gunther J.H. 2002. Contravariant adaptation on structured matrix spaces. *Signal Processing*, **82**, 10, 1389-1410
38. Olshausen B. and Field D.F. 1997. Emergence of simple-sell receptive field properties by learning a sparse code for natural images. *Nature*, **381**, 607-609
39. Osindero S., Welling M. and Hinton, G.E. 2006. Topographic Product Models Applied To Natural Scene Statistics. *Neural Computation*, **18**, 2
40. Parra L.C., Beck J. and Bell A.J. 2008. On the maximisation of information flow between spiking neurons *under review*
41. Radman T. et al. 2007. Spike timing amplifies the effect of electric fields on neurons: implications for endogenous field effects, *J. Neurosci.*, **27**, 3030-3036
42. Rieke F. and Baylor D.A. 1998. Single-photon detection by rod cells of the retina. *Rev. Mod. Phys.* **70**, 1027-36
43. Sejnowski T.J. and Paulsen O. 2006. Network oscillations: emerging computational principles. *J. Neurosci.*, **26**, 6, 1673-76
44. Sheng M. and Hoogenraad C.C. 2007. The postsynaptic architecture of excitatory synapses: a more quantitative view, *Annu. Rev. Biochem.*, **76**, 823-47
45. Simoncelli E.P. and Olshausen B.A. 2001. Natural image statistics and neural representation, *Annu Rev Neurosci.* **24**, 1193-216
46. Spitzer J.J. and Poolman B. 2005. Electrochemical structure of the crowded cytoplasm. *Trends in Biochem. Sci.*, **30**, 10, 536-541
47. Theis F.J. 2005. Gradients on matrix manifolds and their chain rule. *Neural Information Processing Letters*, **9**, 1, 1-13
48. van Hateren J.H. and Ruderman D.L. 1998. Independent component analysis of natural image sequences yields spatio-temporal filters similar to simple cells in primary visual cortex. *Proc. R. Soc. Lond. B*, **265**, 2315-2320
49. Welling, M., Hinton, G.E. and Osindero, S. 2003. Learning Sparse Topographic Representations with Products of Student-t Distributions. *Advances in Neural Information Processing Systems 15*, MIT Press, Cambridge, MA
50. Zinn-Justin J. 2002. *Quantum field theory and critical phenomena.* Oxford Univ. Press

Updating Probabilities with Data and Moments

Adom Giffin and Ariel Caticha

Department of Physics, University at Albany–SUNY, Albany, NY 12222, USA

Abstract. We use the method of Maximum (relative) Entropy to process information in the form of observed data and moment constraints. The generic "canonical" form of the posterior distribution for the problem of simultaneous updating with data and moments is obtained. We discuss the general problem of non-commuting constraints, when they should be processed sequentially and when simultaneously. As an illustration, the multinomial example of die tosses is solved in detail for two superficially similar but actually very different problems.

INTRODUCTION

The original method of Maximum Entropy, MaxEnt [1], was designed to assign probabilities on the basis of information in the form of constraints. It gradually evolved into a more general method, the method of Maximum relative Entropy (abbreviated ME) [2]-[6], which allows one to update probabilities from arbitrary priors unlike the original MaxEnt which is restricted to updates from a uniform background measure.

The realization [5] that ME includes not just MaxEnt but also Bayes' rule as special cases is highly significant. First, it implies that ME is *capable of reproducing every aspect of orthodox Bayesian inference* and proves the complete compatibility of Bayesian and entropy methods. Second, it opens the door to tackling problems that could not be addressed by either the MaxEnt or orthodox Bayesian methods individually. The main goal of this paper is to explore this latter possibility: the problem of processing data plus additional information in the form of expected values.[1]

When using Bayes' rule it is quite common to impose constraints on the prior distribution. In some cases these constraints are also satisfied by the posterior distribution, but these are special cases. In general, constraints imposed on priors do not "propagate" to the posteriors. Although Bayes' rule can handle *some* constraints, we seek a procedure capable of enforcing *any* constraint on the posterior distributions.

After a brief review of how ME processes data and reproduces Bayes' rule, we derive our main result, the general "canonical" form of the posterior distribution for the problem of simultaneous updating with data and moment constraints. The final result is deceivingly simple: Bayes' rule is modified by a "canonical" exponential factor. Although this result is very simple, it should be handled with caution: once we consider several sources of information such as multiple constraints we must confront the problem of non-commuting constraints. We discuss the question of whether they should be

---

[1] For simplicity we will refer to these expected values as *moments* although they can be considerably more general.

processed simultaneously, or sequentially, and in what order. Our general conclusion is that these different alternatives correspond to different states of information and accordingly we expect that they will lead to different inferences.

As an illustration, the multinomial example of die tosses is solved in some detail for two problems. They appear superficially similar but are in fact very different. The first die problem requires that the constraints be processed sequentially. This corresponds to the familiar situation of using MaxEnt to derive a prior and then using Bayes to process data. The second die problem, which requires that the constraints be processed simultaneously, provides a clear example that lies beyond the reach of Bayes' rule.

UPDATING WITH DATA USING THE ME METHOD

Our first concern when using the ME method to update from a prior to a posterior distribution is to define the space in which the search for the posterior will be conducted. We wish to infer something about the value of a quantity $\theta \in \Theta$ on the basis of three pieces of information: prior information about θ (the prior), the known relationship between x and θ (the model), and the observed values of the data $x \in \mathcal{X}$.[2] Since we are concerned with both x and θ, the relevant space is neither $\mathcal{X}$ nor Θ but the product $\mathcal{X} \times \Theta$ and our attention must be focused on the joint distribution $P(x, \theta)$. The selected joint posterior $P_{new}(x, \theta)$ is that which maximizes the entropy,

$$S[P, P_{old}] = -\int dx d\theta\, P(x, \theta) \log \frac{P(x, \theta)}{P_{old}(x, \theta)}, \tag{1}$$

subject to the appropriate constraints. All prior information is codified into the *joint prior* $P_{old}(x, \theta) = P_{old}(\theta) P_{old}(x|\theta)$. Both $P_{old}(\theta)$ (the familiar Bayesian prior distribution) and $P_{old}(x|\theta)$ (the likelihood) contain prior information.[3] The new information is the observed data x', which in the ME framework must be expressed in the form of a constraint on the allowed posteriors. The family of posteriors $P(x, \theta)$ that reflects the fact that x is now known to be x' is such that

$$P(x) = \int d\theta\, P(x, \theta) = \delta(x - x'). \tag{2}$$

This amounts to an *infinite* number of constraints on $P(x, \theta)$: for each value of x there is one constraint and one Lagrange multiplier $\lambda(x)$.

Maximizing S, (1), subject to the constraints (2) plus normalization,

$$\delta\left\{S + \alpha\left[\int dx d\theta\, P(x, \theta) - 1\right] + \int dx\, \lambda(x)\left[\int d\theta\, P(x, \theta) - \delta(x - x')\right]\right\} = 0, \tag{3}$$

---

[2] We use the concise notation θ and x to represent one or many unknown variables, $\theta = (\theta_1, \theta_2 \ldots)$, and one or multiple experiments, $x = (x_1, x_2 \ldots)$.

[3] The notion that the likelihood function contains prior information may sound unfamiliar from the point of view of standard Bayesian practice. It should be clear that the likelihood is *prior* information in the sense that its functional form is known *before* the actual data is known, or at least before it can be processed.

yields the joint posterior,

$$P_{new}(x,\theta) = P_{old}(x,\theta)\frac{e^{\lambda(x)}}{z}, \qquad (4)$$

where z is a normalization constant, and $\lambda(x)$ is determined from (2),

$$\int d\theta\, P_{old}(x,\theta)\frac{e^{\lambda(x)}}{z} = P_{old}(x)\frac{e^{\lambda(x)}}{z} = \delta(x-x'). \qquad (5)$$

The final expression for the joint posterior is

$$P_{new}(x,\theta) = \frac{P_{old}(x,\theta)\,\delta(x-x')}{P_{old}(x)} = \delta(x-x')P_{old}(\theta|x), \qquad (6)$$

and the marginal posterior distribution for θ is

$$P_{new}(\theta) = \int dx P_{new}(x,\theta) = P_{old}(\theta|x'), \qquad (7)$$

which is the familiar Bayes' conditionalization rule.

To summarize: $P_{old}(x,\theta) = P_{old}(x)P_{old}(\theta|x)$ is updated to $P_{new}(x,\theta) = P_{new}(x)P_{new}(\theta|x)$ with $P_{new}(x) = \delta(x-x')$ fixed by the observed data while $P_{new}(\theta|x) = P_{old}(\theta|x)$ remains unchanged. We see that in accordance with the minimal updating philosophy that drives the ME method *one only updates those aspects of one's beliefs for which corrective new evidence (in this case, the data) has been supplied.*

SIMULTANEOUS UPDATING WITH MOMENTS AND DATA

Here we generalize the previous section to include additional information about θ in the form of a constraint on the expected value of some function $f(\theta)$,

$$\int dx d\theta\, P(x,\theta)f(\theta) = \langle f(\theta)\rangle = F. \qquad (8)$$

We emphasize that constraints imposed at the level of the prior need not be satisfied by the posterior. What we do here differs from the standard Bayesian practice in that we *require* the constraint to be satisfied by the posterior distribution.

Maximizing the entropy (1) subject to normalization, the data constraint (2), and the moment constraint (8) yields the joint posterior,

$$P_{new}(x,\theta) = P_{old}(x,\theta)\frac{e^{\lambda(x)+\beta f(\theta)}}{z}, \qquad (9)$$

where z is a normalization constant,

$$z = \int dx d\theta\, e^{\lambda(x)+\beta f(\theta)} P_{old}(x,\theta). \qquad (10)$$

The Lagrange multipliers $\lambda(x)$ are determined from the data constraint, (2),

$$\frac{e^{\lambda(x)}}{z} = \frac{\delta(x-x')}{ZP_{\text{old}}(x')} \quad \text{where} \quad Z(\beta,x') = \int d\theta\, e^{\beta f(\theta)} P_{\text{old}}(\theta|x'), \tag{11}$$

so that the joint posterior becomes

$$P_{\text{new}}(x,\theta) = \delta(x-x') P_{\text{old}}(\theta|x') \frac{e^{\beta f(\theta)}}{Z}. \tag{12}$$

The remaining Lagrange multiplier β is determined by imposing that the posterior $P_{\text{new}}(x,\theta)$ satisfy (8). This yields an implicit equation for β,

$$\frac{\partial \log Z}{\partial \beta} = F. \tag{13}$$

Note that since $Z = Z(\beta,x')$ the resultant β will depend on the observed data x'. Finally, the new marginal distribution for θ is

$$P_{\text{new}}(\theta) = P_{\text{old}}(\theta|x') \frac{e^{\beta f(\theta)}}{Z} = P_{\text{old}}(\theta) \frac{P_{\text{old}}(x'|\theta)}{P_{\text{old}}(x')} \frac{e^{\beta f(\theta)}}{Z}. \tag{14}$$

For $\beta = 0$ (no moment constraint) we recover Bayes' rule. For $\beta \neq 0$ Bayes' rule is modified by a "canonical" exponential factor.

COMMUTING AND NON-COMMUTING CONSTRAINTS

The ME method allows one to process information in the form of constraints. When we are confronted with several constraints we must be particularly cautious. In what order should they be processed? Or should they be processed at the same time? The answer depends on the nature of the constraints and the question being asked.

We refer to constraints as *commuting* when it makes no difference whether they are handled simultaneously or sequentially. The most common example is that of Bayesian updating on the basis of data collected in multiple experiments: for the purpose of inferring θ it is well-known that the order in which the observed data $x' = \{x'_1, x'_2, \ldots\}$ is processed does not matter. The proof that ME is completely compatible with Bayes' rule implies that data constraints implemented through δ functions, as in (2), commute. It is useful to see how this comes about.

When an experiment is repeated it is common to refer to the value of x in the first experiment and the value of x in the second experiment. This is a dangerous practice because it obscures the fact that we are actually talking about *two* separate variables. We do not deal with a single x but with a composite $x = (x_1, x_2)$ and the relevant space is $\mathscr{X}_1 \times \mathscr{X}_2 \times \Theta$. After the first experiment yields the value x'_1, represented by the constraint $c_1 : P(x_1) = \delta(x_1 - x'_1)$, we can perform a second experiment that yields x'_2 and is represented by a second constraint $c_2 : P(x_2) = \delta(x_2 - x'_2)$. These constraints

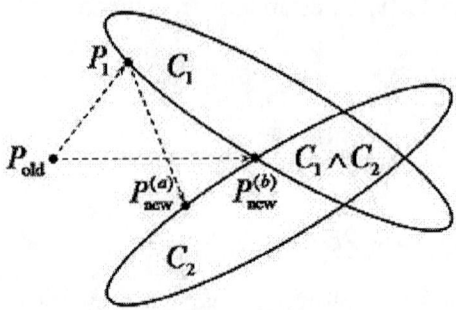

FIGURE 1. Illustrating the difference between processing two constraints C_1 and C_2 sequentially ($P_{\text{old}} \to P_1 \to P_{\text{new}}^{(a)}$) and simultaneously ($P_{\text{old}} \to P_{\text{new}}^{(b)}$ or $P_{\text{old}} \to P_1 \to P_{\text{new}}^{(b)}$).

c_1 and c_2 commute because they refer to *different* variables x_1 and x_2. An experiment, once performed and its outcome observed, cannot be *un-performed* and its result cannot be *un-observed* by a second experiment. Thus, imposing one constraint does not imply a revision of the other.

In general constraints need not commute and when this is the case the order in which they are processed is critical. For example, suppose the prior is P_{old} and we receive information in the form of a constraint, C_1. To update we maximize the entropy $S[P, P_{\text{old}}]$ subject to C_1 leading to the posterior P_1 as shown in Figure 1. Next we receive a second piece of information described by the constraint C_2. At this point we can proceed in essentially two different ways:

(a) Sequential updating. Having processed C_1, we use P_1 as the current prior and maximize $S[P, P_1]$ subject to the new constraint C_2. This leads us to the posterior $P_{\text{new}}^{(a)}$.

(b) Simultaneous updating. Use the original prior P_{old} and maximize $S[P, P_{\text{old}}]$ subject to both constraints C_1 and C_2 simultaneously. This leads to the posterior $P_{\text{new}}^{(b)}$.[4]

To decide which path (a) or (b) is appropriate, we must be clear about how the ME method treats constraints. The ME machinery interprets a constraint such as C_1 in a very mechanical way: all distributions satisfying C_1 are in principle allowed and all distributions violating C_1 are ruled out.

Updating to a posterior P_1 consists precisely in revising those aspects of the prior P_{old} that disagree with the new constraint C_1. However, there is nothing final about the distribution P_1. It is just the best we can do in our current state of knowledge and we fully expect that future information may require us to revise it further. Indeed, when new information C_2 is received we must reconsider whether the original C_1 remains valid or not. Are *all* distributions satisfying the new C_2 really allowed, even those that violate C_1? If this is the case then the new C_2 takes over and we update from P_1 to $P_{\text{new}}^{(a)}$.

---

[4] At first sight it might appear that there exists a third possibility of simultaneous updating: (c) use P_1 as the current prior and maximize $S[P, P_1]$ subject to both constraints C_1 and C_2 simultaneously. Fortunately, and this is a valuable check for the consistency of the ME method, it is easy to show that case (c) is equivalent to case (b). Whether we update from P_{old} or from P_1 the selected posterior is $P_{\text{new}}^{(b)}$.

The constraint C_1 may still retain some lingering effect on the posterior $P_{\text{new}}^{(a)}$ through P_1, but in general C_1 has now become obsolete.

Alternatively, we may decide that the old constraint C_1 retains its validity. The new C_2 is not meant to revise C_1 but to provide an additional refinement of the family of allowed posteriors. In this case the constraint that correctly reflects the new information is not C_2 but the more restrictive $C_1 \wedge C_2$. The two constraints should be processed simultaneously to arrive at the correct posterior $P_{\text{new}}^{(b)}$.

To summarize: sequential updating is appropriate when old constraints become obsolete and are superseded by new information; simultaneous updating is appropriate when old constraints remain valid. The two cases refer to different states of information and therefore *we expect* that they will result in different inferences. These comments are meant to underscore the importance of understanding what information is being processed; failure to do so will lead to errors that do not reflect a shortcoming of the ME method but rather a misapplication of it.

SEQUENTIAL UPDATING: A LOADED DIE EXAMPLE

This is a loaded die example illustrating the appropriateness of sequential updating. The background information is the following: A certain factory makes loaded dice. Unfortunately because of poor quality control, the dice are not identical and it is not known how each die is loaded. It is known, however, that the dice produced by this factory are such that face 2 is on the average twice as likely to come up as face number 5.

The mathematical representation of this situation is as follows. The fact that we deal with dice is modelled in terms of multinomial distributions. The probability that casting a k-sided die n times yields m_i instances for the ith face is

$$P_{\text{old}}(m|\theta) = P_{\text{old}}(m_1...m_k|\theta_1...\theta_k, n) = \frac{n!}{m_1!...m_k!} \theta_1^{m_1}...\theta_k^{m_k}, \quad (15)$$

where $m = (m_1, \ldots, m_k)$ with $\sum_{i=1}^{k} m_i = n$, and $\theta = (\theta_1, \ldots, \theta_k)$ with $\sum_{i=1}^{k} \theta_i = 1$. The generic problem is to infer the parameters θ on the basis of information about moments of θ and data m'. The additional information about how the dice are loaded is represented by the constraint $\langle \theta_2 \rangle = 2 \langle \theta_5 \rangle$. Note that this piece of information refers to the factory as a whole and not to any individual die. The constraint is of the general form of (8)

$$C_1 : \langle f(\theta) \rangle = F \quad \text{where} \quad f(\theta) = \sum_i^k f_i \theta_i. \quad (16)$$

For this particular factory $F = 0$, and all $f_i = 0$ except for $f_2 = 1$ and $f_5 = -2$. Now that the background information has been given, here is our first example.

We purchase a die. On the basis of our general knowledge of dice we are led to write down a joint prior

$$P_{\text{old}}(m, \theta) = P_{\text{old}}(\theta) P_{\text{old}}(m|\theta). \quad (17)$$

(The particular form of $P_{\text{old}}(\theta)$ is not important for our current purpose so for the sake of definiteness we can choose it flat.) At this point the only information we have is that we have a die and it came from a factory described by C_1. Accordingly, we use ME to update to a new joint distribution. This is shown as P_1 in Figure 1. The relevant entropy is

$$S[P,P_{\text{old}}] = -\sum_m \int d\theta \, P(x,\theta) \log \frac{P(x,\theta)}{P_{\text{old}}(x,\theta)}, \qquad (18)$$

where

$$\sum_m = \sum_{m_1\ldots m_k=0}^{n} \delta(\textstyle\sum_{i=1}^{k} m_i - n) \quad \text{and} \quad \int d\theta = \int d\theta_1 \ldots d\theta_k \, \delta(\textstyle\sum_{i=1}^{k} \theta_i - 1),$$

Maximizing S subject to normalization and C_1 gives the P_1 posterior

$$P_1(m,\theta) = \frac{e^{\lambda f(\theta)}}{Z_1} P_{\text{old}}(m,\theta), \qquad (19)$$

where the normalization constant Z_1 and the Lagrange multiplier λ are determined from

$$Z_1 = \int d\theta \, e^{\lambda f(\theta)} P_{\text{old}}(\theta) \quad \text{and} \quad \frac{\partial \log Z_1}{\partial \lambda} = F. \qquad (20)$$

The joint distribution $P_1(m,\theta) = P_1(\theta) P_1(m|\theta)$ can be rewritten as

$$P_1(m,\theta) = P_1(\theta) P_{\text{old}}(m|\theta) \quad \text{where} \quad P_1(\theta) = P_{\text{old}}(\theta) \frac{e^{\lambda f(\theta)}}{Z_1}. \qquad (21)$$

To find out more about this particular die we toss it n times and obtain data $m' = (m'_1,\ldots,m'_k)$ which we represent as a new constraint

$$C_2 : P(m) = \delta(m - m'). \qquad (22)$$

Our goal is to infer the θ that apply to our particular die. The original constraint C_1 applies to the whole factory while the new constraint C_2 refers to the actual die of interest and thus takes precedence over C_1. As $n \to \infty$ we expect C_1 to become less and less relevant. Therefore the two constraints should be processed sequentially.

Using ME, that is (6), we impose C_2 and update from $P_1(m,\theta)$ to a new joint distribution (shown as $P_{\text{new}}^{(a)}$ in Figure 1)

$$P_{\text{new}}^{(a)}(m,\theta) = \delta(m - m') P_1(\theta|m). \qquad (23)$$

Marginalizing over m and using (21) the final posterior for θ is

$$P_{\text{new}}^{(a)}(\theta) = P_1(\theta|m') = P_1(\theta) \frac{P_1(m'|\theta)}{P_1(m')} = \frac{1}{Z_2} e^{\lambda f(\theta)} P_{\text{old}}(\theta) P_{\text{old}}(m'|\theta). \qquad (24)$$

where

$$Z_2 = \int d\theta \, e^{\lambda f(\theta)} P_{\text{old}}(\theta) P_{\text{old}}(m'|\theta). \qquad (25)$$

The readers will undoubtedly recognize that (24) is precisely the result obtained by using MaxEnt to obtain a prior, in this case $P_1(\theta)$ given in (21), and then using Bayes' theorem to take the data into account. This familiar result has been derived in some detail for two reasons: first, to reassure the readers that ME does reproduce the standard solutions to standard problems and second, to establish a contrast with the example discussed next.

SIMULTANEOUS UPDATING: A LOADED DIE EXAMPLE

Here is a different problem illustrating the appropriateness of simultaneous updating. The background information is the same as in the previous example. The difference is that the factory now hires a quality control engineer who wants to learn as much as he can about the factory. His initial knowledge is described by the same prior $P_{old}(m, \theta)$, (17). After some inquiries he is told that the only available information is $C_1 : \langle \theta_2 \rangle = 2 \langle \theta_5 \rangle$. Not satisfied with this limited information he decides to collect data that reflect the production of the whole factory. Randomly chosen dice are tossed n times yielding data $m' = (m'_1, \ldots, m'_k)$ which is represented as a constraint,

$$C_2 : P(m) = \delta(m - m') . \tag{26}$$

The apparent resemblance with (22) may be misleading: (22) refers to a single die, while (26) now refers to the whole factory. The goal here is to infer the distribution of θ that describes the overall population of dice produced by the factory. The new constraint C_2 is information in addition to, rather than instead of, the old C_1: the two constraints should be processed simultaneously. From (12) the joint posterior is [5]

$$P_{new}^{(b)}(m, \theta) = \delta(m - m') P_{old}(\theta | m') \frac{e^{\beta f(\theta)}}{Z} . \tag{27}$$

Marginalizing over m the posterior for θ is

$$P_{new}^{(b)}(\theta) = P_{old}(\theta | m') \frac{e^{\beta f(\theta)}}{Z} = \frac{1}{\zeta} e^{\beta f(\theta)} P_{old}(\theta) P_{old}(m' | \theta) , \tag{28}$$

where the new normalization constant is

$$\zeta = \int d\theta \, e^{\beta f(\theta)} P_{old}(\theta) P_{old}(m' | \theta) \quad \text{and} \quad \frac{\partial \log \zeta}{\partial \beta} = F . \tag{29}$$

This looks like the sequential case, (24), but there is a crucial difference: $\beta \neq \lambda$ and $\zeta \neq Z_2$. In the sequential updating case, the multiplier λ is chosen so that the intermediate P_1 satisfies C_1 while the posterior $P_{new}^{(a)}$ only satisfies C_2. In the simultaneous updating case

---

[5] As mentioned in the previous footnote, whether we update from P_{old} or from P_1 we obtain the same posterior $P_{new}^{(b)}$.

the multiplier β is chosen so that the posterior $P_{\text{new}}^{(b)}$ satisfies both C_1 and C_2 or $C_1 \wedge C_2$. Ultimately, the two distributions $P_{\text{new}}(\theta)$ are different because they refer to different problems: $P_{\text{new}}^{(a)}(\theta)$ refers to a single die, while $P_{\text{new}}^{(b)}(\theta)$ applies to all the dice produced by the factory.[6]

SUMMARY AND FINAL REMARKS

The realization that the ME method incorporates Bayes' rule as a special case has allowed us to go beyond Bayes' rule to process both data and expected value constraints simultaneously. To put it bluntly, anything one can do with Bayes can also be done with ME with the additional ability to include information that was inaccessible to Bayes alone. This raises several questions and we have offered a few answers.

First, it is not uncommon to claim that the non-commutability of constraints represents a *problem* for the ME method. Processing constraints in different orders might lead to different inferences and this is said to be unacceptable. We have argued that, on the contrary, the information conveyed by a particular sequence of constraints is not the same information conveyed by the same constraints in different order. Since different informational states should in general lead to different inferences, the way ME handles non-commuting constraints should not be regarded as a *shortcoming* but rather as a *feature* of the method.

Second, we are capable of processing both data and moments. Is this kind of information of purely academic interest or is it something we might encounter in real life? At this early stage our answer must be tentative: we have given just one example – the die factory – which we think is fairly realistic. However, we feel that other applications (e.g. in econometrics and ecology) can be handled in this way as well.[7, 8]

Finally, is it really true that this type of problem lies beyond the reach of Bayesian methods? After all, we can always interpret an expected value as a sample average in a sufficiently large number of trials. True. We can always construct a large imaginary ensemble of experiments. Entropy methods then become in principle *superfluous*; all we need is probability. The problem with inventing *imaginary* ensembles to do away with entropy in favor of mere probabilities, or to do away with probabilities in favor of more intuitive frequencies, is that the ensembles are just what they are claimed to be, imaginary. They are purely artificial constructions invented for the purpose of handling incomplete information. It seems to us that a safer way to proceed is to handle the available information directly as given (i.e., as expected values) without making additional assumptions about an imagined reality.

Acknowledgements: We would like to acknowledge valuable discussions with C. Cafaro, K. Knuth, and C. Rodríguez.

---

[6] For the sake of completeness, we note that, because of the peculiarities of δ functions, had the constraints been processed sequentially but in the opposite order, first the data C_2, and then the moment C_1, the resulting posterior would be the same as for simultaneous update to $P_{\text{new}}^{(b)}$.

REFERENCES

1. E. T. Jaynes, Phys. Rev. **106**, 620 and **108**, 171 (1957); R. D. Rosenkrantz (ed.), *E. T. Jaynes: Papers on Probability, Statistics and Statistical Physics* (Reidel, Dordrecht, 1983); E. T. Jaynes, *Probability Theory: The Logic of Science* (Cambridge University Press, Cambridge, 2003).
2. J. E. Shore and R. W. Johnson, IEEE Trans. Inf. Theory **IT-26**, 26 (1980); IEEE Trans. Inf. Theory **IT-27**, 26 (1981).
3. J. Skilling, "The Axioms of Maximum Entropy", *Maximum-Entropy and Bayesian Methods in Science and Engineering*, G. J. Erickson and C. R. Smith (eds.) (Kluwer, Dordrecht, 1988).
4. A. Caticha, "Relative Entropy and Inductive Inference", *Bayesian Inference and Maximum Entropy Methods in Science and Engineering*, G. J. Erickson and Y. Zhai (eds.), AIP Conf. Proc. **707**, 75 (2004) (arXiv.org/abs/physics/0311093).
5. A. Caticha and A. Giffin, "Updating Probabilities", *Bayesian Inference and Maximum Entropy Methods in Science and Engineering*, ed. by Ali Mohammad-Djafari (ed.), AIP Conf. Proc. **872**, 31 (2006) (http://arxiv.org/abs/physics/0608185).
6. A. Caticha, "Information and Entropy", presented at the *27th International Workshop on Bayesian Inference and Maximum Entropy Methods in Science and Engineering*, Saratoga Springs, NY, 2007.
7. A. Giffin, "Updating Probabilities with Data and Moments: an Econometric Example", to be presented at the *3rd Econophysics Colloquium*, Ancona, Italy, 2007.
8. A. Giffin, "Updating Probabilities with Data and Moments: an Ecological Example", to be presented at the *7th International Conference on Complex Systems*, Boston, 2007.

APPENDIX: MORE ON THE MULTINOMIAL PROBLEM

Here we pursue the calculation of the posterior (28) in more detail. To be specific we choose a flat prior, $P_{\text{old}}(\theta) = \text{constant}$. Then, dropping the superscript (b),

$$P_{\text{new}}(\theta) = \frac{1}{\zeta_e} \delta(\sum_i^k \theta_i - 1) \prod_{i=1}^k e^{\beta f_i \theta_i} \theta_i^{m'_i}. \tag{30}$$

where ζ_e differs from ζ in (29) only by a combinatorial coefficient,

$$\zeta_e = \int \delta(\sum_i^k \theta_i - 1) \prod_{i=1}^k d\theta_i e^{\beta f_i \theta_i} \theta_i^{m'_i}, \tag{31}$$

and β is determined from (13) which in terms of ζ_e now reads $\partial \log \zeta_e / \partial \beta = F$. A brute force calculation gives ζ_e as a nested hypergeometric series,

$$\zeta_e = e^{\beta f_k} I_1(I_2(\ldots(I_{k-1}))), \tag{32}$$

where each I is written as a sum of Γ functions,

$$I_j = \Gamma(b_j - a_j) \sum_{q_j=0}^{\infty} \frac{\Gamma(a_j + q_j)}{\Gamma(b_j + q_j) q_j!} t_j^{q_j} I_{j+1} \quad \text{with} \quad I_k = 1. \tag{33}$$

The index j takes all values from 1 to $k-1$ and the other symbols are defined as follows: $t_j = \beta(f_{k-j} - f_k)$, $a_j = m'_{k-j} + 1$, and

$$b_j = n + j + 1 + \sum_{i=0}^{j-1} q_i - \sum_{i=0}^{k-j-1} m'_i, \tag{34}$$

with $q_0 = m'_0 = 0$. The terms that have indices ≤ 0 are equal to zero (i.e. $b_0 = q_0 = 0$, etc.). A few technical details are worth mentioning: First, one can have singular points when $t_j = 0$. In these cases the sum must be evaluated in the limit as $t_j \to 0$. Second, since a_j and b_j are positive integers the gamma functions involve no singularities. Lastly, the sums converge because $a_j > b_j$. The normalization for the first die example, (25), can be calculated in a similar way. Currently, for small values of k (less than 10) it is feasible to evaluate the nested sums numerically; for larger values of k it is best to evaluate the integral for ζ_e using sampling methods. A more detailed version of the multinomial example is worked out in [7].

Bayesians *Can* Learn from Old Data

William H. Jefferys

University of Texas at Austin, and University of Vermont

Abstract. In a widely-cited paper, Glymour (*Theory and Evidence*, Princeton, N. J.: Princeton University Press, 1980, pp. 63-93) claims to show that Bayesians cannot learn from old data. His argument contains an elementary error. I explain exactly where Glymour went wrong, and how the problem should be handled correctly. When the problem is fixed, it is seen that Bayesians, just like logicians, *can indeed* learn from old data.

Keywords: Logic, Probability Theory, Bayesian Inference, Problem of Old Data
PACS: 02.10.Ab, 02.50.Cw, 02.50.Tt

GENERAL OVERVIEW

Outline of the Paper. I first review some aspects of standard logic that are relevant to this paper. I then discuss the relationship between standard logic and standard probability theory, and in particular point out the fact that standard probability theory contains standard logic in the particular sense that for any argument that reaches a conclusion using standard logic, there exists a parallel argument (calculation) in standard probability theory that reaches the same conclusion, and furthermore, that any *valid* argument by any method (whether logical or Bayesian) must arrive at the same conclusion.

I then introduce a simple "toy example" that is nonetheless sophisticated enough to reveal the problem with Glymour's claim. The toy example is an extension of the example that Glymour used in his paper. I describe Glymour's argument [1], and use the toy example to show that his reasoning leads to a contradiction with ordinary logic, and therefore must be invalid. I then explain, again in terms of the toy example, exactly where Glymour's argument goes wrong, and how to correct it. I conclude with a summary of what we have learned.

Standard Logic

Standard logic tells us how to combine propositions $A, B, C, \ldots$ with logical operations such as $\wedge, \vee, \neg, \rightarrow, \ldots$ to obtain new and valid propositions. The propositional calculus allows us to calculate, using definite rules, the truth value of any proposition that has been constructed from other propositions using these logical operations, given the truth values of the propositions from which they are constructed.

For example, given propositions A, B, we can calculate the truth value of the proposition $C = A \wedge B$ as follows: C is true if both A and B are true, otherwise it is false.

Likewise, the truth value of the proposition $D = A \to B$ is true if A is false, otherwise it is equal to the truth value of B. That is, if A is true, then B must be true. If A is not true, then it doesn't matter what the truth value of B is, $A \to B$ is true.

An important feature of standard logic is that it is time-independent (Jaynes [2], p. 89). That is, it describes relationships between propositions that are independent of when we may learn the truth or falsity of the propositions themselves. For example, the truth-values of the expressions $\neg A$, $A \wedge B$, $A \vee B$, and $A \to B$ depend only on the truth-values of A and B, and not upon when we happen to learn their truth-values.

Probability and Logic

Probability theory extends the basic notions of standard logic to a regime where the *degree of plausibility* of propositions is no longer just "true" or "false", but may be intermediate between the two. That is, to any proposition we can assign a number in the unit interval $[0, 1]$ that corresponds to our assessment of how likely it is that the proposition is true, where 1 means that we are certain the proposition is true and 0 means that we are certain that it is false. The larger the degree of plausibility, the more likely it is that we would regard the proposition as true.

A theorem of Cox [3, 4] proves that, up to an isomorphism, standard probability theory is the unique extension of ordinary logic to this regime that satisfies certain obvious requirements necessary for the theory to yield consistent results. Jaynes ([2], p. 19) lists a set of three such requirements, which he calls *desiderata*:

1 *If a conclusion can be reasoned out in more than one way, then every possible way must lead to the same result.* An important aspect of this desideratum is that if a conclusion can be obtained using ordinary logic, then a *valid* calculation using probability theory must arrive at the same result. If a purported Bayesian calculation arrives at a result different from one that we can derive using standard logic, it must *ipso facto* be invalid. We will see below that Glymour's calculation fails this test.

2 *The calculation takes into account all of the evidence relevant to the question. It does not arbitrarily ignore some of the information, basing its conclusions only on what remains. It is, as Jaynes says, completely nonideological.* Glymour's calculation fails this test in a subtle way, muddling the issue by failing use standard probability notation to indicate all the information that was taken into account in a calculation. Indeed, this results in a basic confusion of models that turns out to be at the root of the problem with Glymour's calculation.

3 *Equivalent states of knowledge are always represented by equivalent plausibility assignments. That is, if in two situations the state of knowledge is the same, then (except for possible relabeling of the propositions), the calculation must assign the same plausibilities to both.* Glymour's calculation fails this test as well.

It turns out that these three desiderata, together with the assumption that degrees of plausibility are represented by real numbers on the unit interval $[0, 1]$, are sufficient to derive standard probability theory as the unique embodiment of these sensible requirements of plausible reasoning.

In particular it turns out, as a consequence of Jaynes' desideratum #1 and Cox's theorem, that standard probability theory contains standard logic as a subset. This means that for every calculation that can be made using standard logic, there is a corresponding calculation in standard probability theory that will arrive at the *same* result, and no *valid* calculation in standard probability theory can yield a different result.

A Toy Example

We consider a situation where there are precisely two theories under consideration, say T and $\overline{T} = \neg T$, and only two observations of evidence are possible, that is E and $\overline{E} = \neg E$. We furthermore presume that $T \rightarrow E$ and $\overline{T} \rightarrow \overline{E}$. This means that if theory T is true, we must observe evidence E, and if theory $\overline{T}$ is true, then we must observe evidence $\overline{E}$.

For example, let T be the theory of general relativity, and $\overline{T}$ be pure Newtonian mechanics. Let E be the (in this case old) evidence that the motion of Mercury's perihelion is anomalous (cannot be explained under Newtonian mechanics). I assume that we can be certain whether we have observed anomalous perihelion motion or not. Then we see immediately that in this toy example $T \rightarrow E$ and $\overline{T} \rightarrow \overline{E}$.

It is important to recognize that these relationships are *defined by the theory*, independently of any data that may have been observed and independently of when those data may have been observed. The relationships are therefore *time-independent*. Newtonian theory *always* predicts that anomalous perihelion motion will *not* be observed, and general relativity *always* predicts that anomalous perihelion motion *will* be observed. This is a consequence of the theories and mathematics.

If we observe evidence E, then standard logic says $\overline{T} \rightarrow \overline{E}$, so $\neg T \rightarrow \neg E$. It follows that $E \rightarrow T$ and $E \rightarrow \neg \overline{T}$. Hence observing E rules out $\overline{T}$ and confirms T.

Note that this result follows from standard logic. Since standard logic is just a calculus on the truth-values of the propositions, and does not depend on when we observe evidence E, it follows that we can certainly learn from old evidence if we use only logic. But, as pointed out above, Jaynes' desideratum #1, together with Cox's theorem, says that the same result *must* be obtainable by a *valid* application of probability theory. If a calculation using probability theory obtains a different result, it is certainly not a valid calculation.

Translated into the language of probability theory, the result $E \rightarrow \neg \overline{T}$ is equivalent to $P(\neg \overline{T} \mid E) = P(T \mid E) = 1$ and $P(\overline{T} \mid E) = 0$. Any purported Bayesian calculation that does not arrive at this result must be invalid. Note also that when we translate the initial assumptions of this toy example into standard probability notation we can calculate the *likelihood* as $P(E \mid T) = 1$ and $P(E \mid \overline{T}) = 0$ for use when we observe E, and $P(\overline{E} \mid T) = 0$ and $P(\overline{E} \mid \overline{T}) = 1$ for use when we observe $\overline{E}$. Since all of these probability assignments are simply translations of statements of ordinary logic into the language of probability theory, they are time-independent, that is, their values are independent of when we happen to observe the evidence.

GLYMOUR'S ARGUMENT

Glymour argues [1] that the Bayesian cannot learn from old evidence E. This article has generated a lively discussion, e.g., [5, 6, 7, 8, 9, 10, 11, 12]. The argument goes as follows[1]: Since we have observed the old evidence E, Glymour claims that

$$P(E) = 1 \quad ??? \qquad (1)$$

I put question marks here because I believe this equation to be wrong. Nonetheless, if we grant Eq. (1), Glymour's argument goes through easily. Since $P(E) = 1$, it follows from standard probability theory that $P(E \mid X) = 1$ for all propositions X that are not absurd (tautologically false). In particular, $P(E \mid T) = 1$. Therefore, by Bayes' theorem,

$$P(T \mid E) = \frac{P(E \mid T)}{P(E)} P(T) = P(T)$$

and since the posterior probability is equal to the prior probability, we haven't learned anything.

Counterexample to Glymour's Argument

We see immediately that Glymour's calculation fails to satisfy Jaynes' desideratum #1, for we have proved that for our toy problem, knowledge of E together with standard logic leads to the conclusion that T is true and $\overline{T}$ is false, regardless of what we may have thought before we did the calculation. But Glymour's calculation allows for no such conclusion: If for example we had adopted $P(T) = 1/2$, Glymour's calculation tells us that $P(T \mid E) = 1/2$, in blatant contradiction to the calculation from ordinary logic. The equation $P(T \mid E) = 1/2$ says that E does *not* entail T, whereas logic says that E *does* entail T. Since Cox's theorem guarantees that any *valid* calculation using probability theory must arrive at the same conclusion that we got using standard logic, this fact by itself demonstrates that Glymour's argument cannot be valid.

It is not hard to pinpoint the source of the problem, again using the toy example as a guide. If $P(E) = 1$, then it follows that $P(E \mid X) = 1$ for any non-absurd proposition X; in particular, $P(E \mid \overline{T}) = 1$, or translated into the language of logic, $\overline{T} \rightarrow E$. That is, according to Glymour's reasoning, if we have observed E, we must conclude that *Newtonian physics predicts that we will observe anomalous motion of the perihelion of Mercury*. But this is absurd. Newtonian physics predicts unambiguously that we will *not* observe anomalous perihelion motion for Mercury, that is, $\overline{T} \rightarrow \overline{E}$. This is a property of the *theory*, which doesn't depend in any way on what observations may or may not have been made.

The absurdity of this situation is compounded when we realize that Glymour's reasoning transforms *every* theory X into Jaynes' dreaded "Sure Thing®" theory [13], which predicts the observed data E *perfectly*.

---

[1] I have altered Glymour's notation to conform to standard probability theory

We have thus arrived at a contradiction. Glymour's reasoning would require us to conclude that $\overline{T} \to E$, but we know from physics that $\overline{T} \to \neg E$, independent of time or what we may have observed. Therefore, Glymour's reasoning must be erroneous.

The problem arises from Glymour's assertion that $P(E) = 1$. Without that, the rest of his alleged proof fails.

Glymour's Friend

Physicists are familiar with "Wigner's Friend," a thought experiment named for the late physicist Eugene Wigner, that is designed to help us think about when and under what circumstances the "collapse" of states in quantum mechanics takes place. In this thought experiment, Wigner and his "friend" have different states of knowledge, until Wigner's friend informs Wigner of certain facts, so that they end up with the same state of knowledge, and thus should come to the same conclusions. The details of the physics aren't important here, but the idea that people who start out with different states of knowledge will arrive at the same conclusions, once they have the same state of knowledge, is the key idea that I want to carry over to the present problem.

Let me introduce Glymour's friend Tom. Tom is ignorant of E. Therefore, when Glymour explains the toy problem to Tom, Tom can decide on priors and even calculate in advance what he will think when he learns whether E is true or false, using the usual Bayesian machinery. After he has done this, Tom can tell Glymour what his priors are. Suppose the priors are the same as the ones that Glymour has already adopted, and that $P(T) \neq 1$. Then both are starting with the same priors.

Now Glymour informs Tom that E is true. Tom, upon learning this "new" data, recalls his previous calculations, and concludes that $P(T \mid E) = 1$. Glymour performs the calculation that he advocates (since for him the data are "old") and arrives at $P(T \mid E) = P(T) \neq 1$.

This violates Jaynes' desideratum #3, since at this point both parties have the same state of knowledge, yet they have assigned different plausibilities to $T \mid E$. Since the axioms of probability theory, in virtue of Cox's theorem, cannot violate Jaynes' three desiderata when used validly, we have again arrived at a contradiction. Since it is clear that Tom does not view E as "old" data, and therefore is entitled to carry out the standard Bayesian calculation (which gives the same result as the calculation using logic), his conclusions must be correct and Glymour's wrong.

Where Glymour Went Wrong

Jaynes ([2], pp. 473, 484) points out an important fact: *A fruitful source of error and even apparent paradoxes in probability theory is to fail to condition properly and explicitly on all background information used.* All probability is conditional on every relevant piece of background information, and changing the background information changes the probabilities. To make this crystal clear, let $\mathscr{B}$ represent *all* the relevant background information at our disposal, *except* for any knowledge of E. This includes our assumptions about mathematics and physics; for example, $\mathscr{B}$ includes the fact that $\overline{T} \to \overline{E}$.

Viewed from this point of view, the source of Glymour's error becomes embarrassingly obvious. Recall that Eq. (1) was derived in the light of knowledge of the old evidence E and *actually used* that information as background information, even though this dependence was not explicitly noted in the equations. Following Jaynes' advice above, standard notational convention demands that we call out this fact explicitly. If we do this, we obtain the correct Eq. (2):

$$P(E \mid E, \mathscr{B}) = 1 \quad !!! \tag{2}$$

The rest of the proof translates as follows:

$$P(E \mid E, T, \mathscr{B}) = 1 \tag{3}$$

$$P(T \mid E, E, \mathscr{B}) = \frac{P(E \mid E, T, \mathscr{B})}{P(E \mid E, \mathscr{B})} P(T \mid E, \mathscr{B}) \tag{4}$$

But of course, $P(T \mid E, E, \mathscr{B}) = P(T \mid E \wedge E, \mathscr{B}) = P(T \mid E, \mathscr{B})$ by standard logic. Thus we see that when the conditioning that is implicit but unstated in Eq. (1) is explicitly recognized in Eq. (2), what Glymour has actually proved is the (well-known) fact that the Bayesian machinery, quite sensibly, prevents us from using the same evidence twice. He has *not* proved that a Bayesian cannot learn from old evidence, only that he cannot validly manipulate the Bayesian machinery to get additional information out of information that has *already been used*.

We now see that $P(E \mid \mathscr{B})$ and $P(E \mid E, \mathscr{B})$ are entirely different. $P(E \mid E, \mathscr{B})$ has already used evidence E, whereas according to the standard notational convention, $P(E \mid \mathscr{B})$—Glymour's $P(E)$—has *never* used evidence E, not even once. This is because $P(E \mid \mathscr{B})$ is just the *sampling distribution* of E in the mixture model defined by the priors and the likelihood, given that we know $\mathscr{B}$ and *nothing else*. It is a function of the *theory*, which is included in the background knowledge $\mathscr{B}$. It is in fact *entirely ignorant* of our knowledge of E. Thus, there is no reason to suppose that $P(E \mid \mathscr{B}) = 1$, regardless of our state of knowledge of E, and indeed, it usually is not.

Note that the right-hand side of Eq. (4) has its as prior $P(T \mid E, \mathscr{B})$, *not* $P(T \mid \mathscr{B})$. In other words, the prior in Eq. (4) must be constructed from full knowledge of E; it is *not* the same as $P(T \mid \mathscr{B})$, which is (of course) ignorant of E. One cannot substitute $P(T \mid \mathscr{B})$ for $P(T \mid E, \mathscr{B})$ in Eq. (4); the resulting equation is not a valid equation in probability theory.

In order to calculate the value of $P(T \mid E, \mathscr{B})$ for substitution into Eq. (4), we have to start from $P(T \mid \mathscr{B})$ and then apply Bayes' theorem in the usual way, where in this case the right hand side is calculated *unconditioned* on E (which is to say, the right-hand side is ignorant of any knowledge we may have about E). In this case, $P(E \mid \mathscr{B})$ does not know that E has been observed, and is correctly calculated from the priors and the *time-independent* likelihood from the identity:

$$P(E \mid \mathscr{B}) = P(E \mid T, \mathscr{B}) P(T \mid \mathscr{B}) + P(E \mid \overline{T}, \mathscr{B}) P(\overline{T} \mid \mathscr{B}) \tag{5}$$

Thus, in the toy example, where $P(E \mid T, \mathscr{B}) = 1$ and $P(E \mid \overline{T}, \mathscr{B}) = 0$,

$$P(E \mid \mathscr{B}) = P(T \mid \mathscr{B}), \tag{6}$$

which is in general *not* equal to 1.

This tells us the correct way to do the Bayesian calculation, in the case where E has been observed as old data. We still have to assign priors $P(T \mid \mathscr{B})$ and $P(\overline{T} \mid \mathscr{B})$, and this must be done without taking E into account. Although this step might pose some problems of its own (assignation of priors in general requires careful thought), any such problems are unrelated to Glymour's argument, so I will pass over this issue. Suppose, for example, we have assigned $P(T \mid \mathscr{B}) = \alpha$, $P(\overline{T} \mid \mathscr{B}) = 1 - \alpha$, where $\alpha \in (0,1)$. Then the Bayesian calculation goes through in the usual way as follows:

$$P(T \mid E, \mathscr{B}) = \frac{P(E \mid T, \mathscr{B})}{P(E \mid \mathscr{B})} P(T \mid \mathscr{B}) = 1 \tag{7}$$

since in the toy example example $P(E \mid \mathscr{B}) = P(T \mid \mathscr{B})$ and—from the *theory*, not from Glymour's reasoning—$P(E \mid T, \mathscr{B}) = 1$. Thus, independent of α, we obtain the same result as we did using ordinary logic. Thus, Jaynes' desideratum #1 is satisfied: No matter how we do the calculation, whether by ordinary logic or by a *valid* application of probability theory, Cox's theorem guarantees that we *must* arrive at the same result.

Note in particular that Glymour's argument does not use, and in fact denies, the *one key fact* that allows us to calculate the correct result using logic: that $P(E \mid \overline{T}, \mathscr{B}) = 0$. From this fact, we first derive $P(\overline{E} \mid \overline{T}, \mathscr{B}) = 1$, which in turn implies $\overline{T} \wedge \mathscr{B} \to \overline{E}$ and then (since $\mathscr{B}$ is true) $\overline{T} \to \overline{E}$ and $E \to T$. But the correct Bayesian calculation makes full use of that information by using the time-independent likelihood in the calculation of $P(E \mid \mathscr{B})$ to arrive at the *same* result that we got using logic. Glymour's calculation thus violates Jaynes' desideratum #2.

SUMMARY AND CONCLUSIONS

As Jaynes ([2], p. 89) points out, probability theory, like logic, is time-independent. All of the relationships in probability theory are *logical* relationships and have nothing to do with the order in which we happen to learn about the evidence or recognize it in the Bayesian equations. When we calculate $P(T \mid E, \mathscr{B})$ from $P(T \mid \mathscr{B})$, it does not matter when we have actually observed E; the relationship between the two is purely a logical relationship, and the quantities that go into the calculation (likelihoods, priors) are time-independent and will be the same, regardless of when E happens to have been observed. As Tom Loredo observed when I showed him Glymour's argument, "Time plays the same role in probability theory as it does in logic: That is to say, no role whatsoever." [14]

A *valid* Bayesian calculation takes ones knowledge of a particular piece of data into account in just one uniform way, by conditioning on the data. It is essential that this conditioning be called out *explicitly* in the notation, as Jaynes advises. Using data without explicitly calling it out in the notation, as Glymour did, is a reliable route to disaster.

Glymour's error resulted from a failure to follow these basic principles. Using the principles of his argument I was able to derive a contradiction with logic that seems not to have been noticed up to this point, but which is sufficient to demonstrate that Glymour's argument is invalid. The bottom line is that Bayesians can and do learn from old data, when they do the calculation carefully and correctly.

ACKNOWLEDGMENTS

I thank Jim Berger, David van Dyk and especially Rob Pennock and Tom Loredo for their valuable comments and suggestions. I dedicate this paper to the memory of Edwin T. Jaynes. I was not fortunate to know him personally, but I have learned much from his writings.

REFERENCES

1. C. N. Glymour, "Why I Am Not a Bayesian," in *Theory and Evidence*, Princeton, N. J.: Princeton University Press, 1980, pp. 63–93.
2. E. T. Jaynes, *Probability Theory: The Logic of Science*, Cambridge: Cambridge University Press, 2003.
3. R. T. Cox, *American Journal of Physics* **14**, 1–13 (1946).
4. K. S. Van Horn, *International Journal of Approximate Reasoning* **34**, 3–24 (2003).
5. M. Curd, and J. A. Cover, *Philosophy of Science: The Central Issues*, New York: W. W. Norton & Co., 1998, pp. 656–659.
6. J. Earman, *Bayes or Bust?*, Cambridge, MA: MIT Press, 1992, chap. 5.
7. D. Garber, "Old Evidence and Logical Omniscience in Bayesian Confirmation Theory," in *Minnesota Studies in the Philosophy of Science, Volume X: Testing Scientific Theories*, edited by J. Earman, Minneapolis: University of Minnesota Press, 1983, pp. 99–131.
8. C. Howson, *British Journal of the Philosophy of Science* **42**, 547–555 (1991).
9. C. Howson, and P. Urbach, *Scientific Reasoning: The Bayesian Approach*, La Salle, IL: Open Court, 1989, pp. 270–275.
10. R. Jeffrey, "Bayesianism with a Human Face," in *Minnesota Studies in the Philosophy of Science, Volume X: Testing Scientific Theories*, edited by J. Earman, Minneapolis: University of Minnesota Press, 1983, pp. 133–156.
11. R. T. Pennock, *Annals of the Japan Association for the Philosophy of Science* **13**, 1–26 (2004).
12. R. Rosenkrantz, "Why Glymour Is a Bayesian," in *Minnesota Studies in the Philosophy of Science, Volume X: Testing Scientific Theories*, edited by J. Earman, Minneapolis: University of Minnesota Press, 1983, pp. 69–97.
13. E. T. Jaynes, "Bayesian Methods: General Background," in *Maximum Entropy and Bayesian Methods in Applied Statistics*, edited by J. H. Justice, Cambridge: Cambridge University Press, 1985, pp. 1–25.
14. T. Loredo, *Private communication* (2006).

The Marginalization Paradox and the Formal Bayes' Law

Timothy C. Wallstrom

Theoretical Division, MS B213
Los Alamos National Laboratory
Los Alamos, NM 87545
tcw@lanl.gov

Abstract. It has recently been shown that the marginalization paradox (MP) can be resolved by interpreting improper inferences as probability limits. The key to the resolution is that probability limits need not satisfy the formal Bayes' law, which is used in the MP to deduce an inconsistency. In this paper, I explore the differences between probability limits and the more familiar pointwise limits, which do imply the formal Bayes' law, and show how these differences underlie some key differences in the interpretation of the MP.

Keywords: Marginalization paradox, Objective Bayes, Logical Bayes, Improper Priors, Maximum Entropy
PACS: 02.50.-r, 02.50.Tt, 02.70.Rr

INTRODUCTION

The marginalization paradox (MP) is an apparent inconsistency in Bayesian inference that can arise from the use of improper priors. It was discovered in 1972 by Dawid and Stone [2]; together with Zidek, they published a comprehensive analysis in 1973 [3]. We follow Jaynes in referring to these authors as "DSZ."

The MP arises in problems with a particular structure, where there are two different ways of computing the same marginal posterior. When improper priors are used, the two results are usually incompatible; an inconsistency cannot arise with proper priors. The improper inferences are computed as "formal posteriors" using the usual Bayes' law, $\pi(\theta|x) \propto p(x|\theta)\pi(\theta)$, with an improper prior.

The MP is important to objective Bayes because the "noninformative" priors required by the theory are typically improper. By suggesting that improper priors cannot be used consistently, the MP raises questions as to whether noninformative priors exist, and thus as to whether the objective Bayesian approach is tenable. We use the term objective Bayes to describe the approach of Harold Jeffreys and Edwin Jaynes, in which a certain state of information is represented by a unique prior. (Their approach is also known as "logical Bayes"; the term objective Bayes is now often used for a different but related approach in which the prior may depend on the estimand [4]; for a discussion of different approaches, cf. [5].)

Edwin Jaynes strongly contested the view that the MP represented a true inconsistency, and engaged DSZ in a spirited debate; cf. [6] and references therein. The battle lines were drawn in the 1970s, and have changed little since. Neither side managed to convince the other, and the absence of new results has caused the paradox to be largely set aside, even though it has never been completely understood.

Recently, I have shown [1] that the MP can be resolved if probability limits, rather than formal posteriors, are used to define improper inferences. The purpose of this paper is to reconsider the differences between Jaynes and DSZ in the light of this new result. One might assume, in reviewing their debates, that Jaynes and DSZ were separated by an unbridgeable chasm. It is shown, by contrast, that the differences between Jaynes and DSZ hinge on a single assumption, which might at first appear as a mere technicality. This assumption is the type of limit used to define the improper inference. We then make the case that the limit process used to support the use of formal posteriors is unsound. Our analysis relies heavily on the ideas of Mervyn Stone, who first introduced the notion of probability limit, and has long maintained that the paradoxes associated with formal posteriors are due to the inadequacy of the pointwise limit.

We also discuss the impact of these recent findings on the prospects for objective Bayes. One might have hoped that a resolution of the paradox might open the door to a refined theory of improper inference, using probability limits instead of formal posteriors, which would provide a consistent foundation for objective Bayes. Unfortunately, there is strong evidence that probability limits rarely exist. Although we can identify some problems that we can now solve which previously led to inconsistency, it appears that probability limits do not exist for most of the problems previously leading to inconsistencies. Thus, the MP remains a serious challenge to objective Bayes.

THE MARGINALIZATION PARADOX

We briefly describe the key elements of the MP. Let $p(x|\theta)$ be the density function for some statistical model, $\pi(\theta)$ a possibly improper prior, and $\pi(\theta|x)$ the corresponding posterior, as computed formally from Bayes' law. As noted above, if $\pi(\theta)$ is improper, i.e., if its integral is infinite, we call $\pi(\theta|x)$ a *formal* posterior. Now suppose that $x = (y, z)$ and $\theta = (\eta, \zeta)$, that the marginal density $p(z|\theta)$ depends on θ only through ζ, and that the marginal posterior $\pi(\zeta|x)$ depends on x only through z. We denote these functions by $\tilde{p}(z|\zeta)$ and $\tilde{\pi}(\zeta|z)$, respectively.

Intuition would now suggest that

$$\tilde{\pi}(\zeta|z) \propto \tilde{p}(z|\zeta) \pi(\zeta),$$

for some function $\pi(\zeta)$. That is, the marginalized quantities should satisfy a (possibly formal) Bayes' law. In general, however, we find that they do not. The problems in which the inconsistencies arise are ordinary problems of statistical inference, although they need to satisfy certain symmetry properties. However, they are not pathological or contrived in any way.

A schematic of the MP is presented in Figure 1. The paradox is sometimes dramatized by ascribing the different computations to two Bayesians, B_1 and B_2; the routes they take are indicated in the Figure. For numerous examples, we refer the reader to [2, 3].

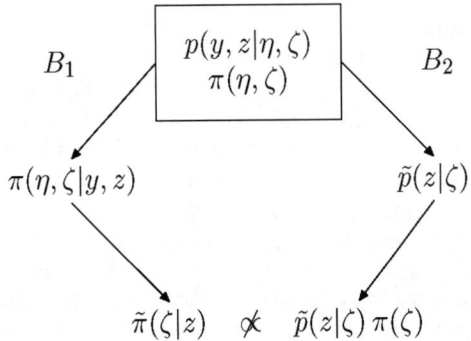

FIGURE 1. Schematic representation of marginalization paradox.

SIGNIFICANCE OF THE MP

Before proceeding to our analysis, we discuss in greater detail the significance of the MP for objective Bayes. One of the perceived weaknesses in the Bayesian position has been the subjectivity inherent in using "personal opinion" to define the prior distribution. Objective Bayes addresses this issue by maintaining that prior distributions can be objectively associated with specific states of information. Inferences are only subjective in the sense that different subjects have different states of information.

Objective Bayes maintains, in particular, that there is a unique numerical representation of "complete ignorance." The operational definition of "complete ignorance" is that our prior beliefs are invariant under certain transformations of the parameter space. For example, if we are ignorant of the scale of the variance, our belief that σ lies in an interval $\Delta\sigma$ is the same as our belief that it lies in $10\Delta\sigma$. These transformations form a mathematical group. It is here that the connection with the MP becomes apparent. If the symmetry group is noncompact, then any group-invariant (*Haar*) measure is improper.

Many of the symmetries of interest in statistics are described by noncompact groups. For example, translational symmetry on R, the real numbers, is expressed by the noncompact translation group, whose invariant measure is Lebesgue measure, with infinite total measure. Similarly, scale invariance is described by the noncompact group of positive reals. The invariant measure is $d\sigma/\sigma$, which diverges. Multivariate analysis typically involves the general linear group, which is also noncompact.

Ignorance priors are often used directly, but they also form the basis for the method of maximum entropy. Here, entropy is really "relative entropy," which is defined relative to some base measure π_0:

$$H(\pi, \pi_0) = -\int \pi(\theta) \log\left[\frac{\pi(\theta)}{\pi_0(\theta)}\right] d\theta.$$

In the absence of constraints representing additional information, maximum entropy is achieved for $\pi(\theta) = \pi_0(\theta)$, which must thus represent a state of complete ignorance. If we cannot define a meaningful ignorance prior, we cannot regard maximum entropy as an "objective" procedure.

Jaynes' clearly recognized the challenge the MP posed for his approach, and opposed it vociferously. A statement of his views can be found in Chapter 15 of [6]. "[The MP]...seemed to threaten the consistency of all probability theory." "It has been able to do far more damage to the cause of scientific inference than any other [paradox]." "Scientific inference thus suffered a setback from which it will take decades to recover."

Jaynes claims to identify several errors in the analysis of DSZ, but his basic criticism is that the analysis uses, at critical junctures, intuitive reasoning about improper quantities, and that the correct result can only be obtained by carefully considering limits of proper quantities, although this analysis does not appear to have been carried out. He claims that a correct analysis will reveal that B_1 and B_2 have used different prior information, so it is only to be expected that their answers will differ.

We will not attempt to address the specific arguments contained in [6] and elsewhere. Instead, we start from a point of agreement between Jaynes and DSZ, and show that the whole interpretation of the paradox hinges on how the specifics of that point are interpreted.

LIMIT CONCEPTS

The point of agreement is this: both Jaynes and DSZ agree that "infinite" quantities must be interpreted as limits of finite quantities. This is the main motif of Jaynes' Chapter 15: "The paradoxes of probability theory." It is expressed, for example, in the statement that "An improper pdf has meaning only as the limit of a well-defined sequence of proper pdfs." [6, p. 487]. DSZ are in complete agreement, and say so explicitly: "'Infinity' finds practical justification only when it can be interpreted as an idealized approximation of the finite." [7, p. 4]

In particular, both Jaynes and DSZ agree that if $\pi(\theta)$ is an improper prior, then in order to define the corresponding posterior, $\pi(\theta|x)$, we need to construct a sequence of proper priors, $\{\pi_n(\theta)\}$, such that $\pi_n(\theta) \to \pi(\theta)$ in some sense, and define $\pi(\theta|x)$ as the limit

$$\pi(\theta|x) \equiv \lim_n \pi_n(\theta|x),$$

where each $\pi_n(\theta|x)$ is the ordinary posterior corresponding to $\pi(\theta|x)$. There is, however, more than one way to define a limit!

Jaynes follows Jeffreys in adopting what may be called pointwise limits. We say that $\pi_{\text{pt}}(\theta|x)$ is a *pointwise limit* for $p(x|\theta)$ and $\{\pi_n(\theta)\}$ if, *for each x*,

$$\int |\pi_n(\theta|x) - \pi_{\text{pt}}(\theta|x)| d\theta \to 0. \tag{1}$$

Jaynes adopts this definition explicitly on p. 471 of [6]. (He does not specify the sense in which the measures $\pi_n(\theta|x) d\theta$ must converge to the measure $\pi_{\text{pt}}(\theta|x) d\theta$, although this is inessential; in (1), we have used convergence in total variation norm.) Jeffreys also adopts this definition, although always implicitly. Thus, for example, Jeffreys writes that "If *in an actual series of observations* the standard deviation is much more than the smallest admissible value of σ, and much less than the largest, the truncation of the distribution makes a negligible change in the results" [8, p. 121] (italics mine). A similar example is worked out explicitly in [9, p. 68], where Jeffreys calculates a posterior as the limit of $\pi_n(\theta|x)$ with x fixed (in our notation).

Pointwise limits lead to the formal posterior, and both Jaynes and Jeffreys justify their use of formal posteriors on this basis. Typically this is done for specific examples, but in fact, the argument can be used to justify the formal posterior for essentially any reasonable prior. Cf. [10] for a general proof under weak assumptions.

An alternative notion of limit is due to M. Stone [11, 12, 13]. We say that $\pi_{\text{prob}}(\theta|x)$ is a *probability limit* for $p(x|\theta)$ and $\{\pi_n(\theta)\}$ if

$$\int \left[\int |\pi_n(\theta|x) - \pi_{\text{prob}}(\theta|x)| d\theta \right] m_n(x) \, dx \to 0, \tag{2}$$

where $m_n(x)$ is the marginal data density, $\int p(x|\theta) \pi_n(\theta) d\theta$. Thus, pointwise limits require that, for any fixed x, the bracketed expression eventually becomes small; probability limits require that the *average* of this expression over $m_n(x)$ eventually becomes small. The intuition behind this definition is that the true prior is close to one of the $\pi_n(\theta)$, and $\pi_{\text{prob}}(\theta|x)$ is an idealization. It is a useful idealization if it would usually be a good approximation in the region where the data is expected.

At first glance, both definitions seem reasonable, and it may seem that the difference could at most be technical. In fact, the difference is profound, as will become apparent when we examine a particular example.

STONE'S EXAMPLE

In this section, I present an example, due to Stone [14], which illustrates the difference between the two limit concepts, and also illustrates some disturbing properties of the pointwise concept. Consider a Gaussian random variable, $X \sim N(\theta, 1)$, and assume that, rather than using the usual uniform prior on θ, we use an exponential: $\pi(\theta) = \exp(a\theta)$. Although no one would use this prior in practice, it illustrates phenomena that arise with realistic priors in more complicated problems [14].

We may approximate the improper exponential prior with the sequence $\pi_n(\theta)$:

$$\pi_n(\theta) \propto \exp\left[-\frac{(\theta - an)^2}{2n}\right] \propto \exp\left[a\theta - \frac{\theta^2}{2n}\right]. \tag{3}$$

Pointwise limits and probability limits both exist for this sequence, but they are different:

$$\pi_{\text{pt}}(\theta|x) \propto \exp[-\tfrac{1}{2}(\theta - x - a)^2], \qquad \pi_{\text{prob}}(\theta|x) \propto \exp[-\tfrac{1}{2}(\theta - x)^2]. \tag{4}$$

The first expression is just the formal posterior; the second result follows from Stone's theorem [13].

To understand the difference between pointwise and probability limits, consider Figure 2, which compares the priors $\pi(\theta)$ and $\pi_n(\theta)$. The likelihood is local, so the posteriors $\pi(\theta|x)$ and $\pi_n(\theta|x)$ will be similar if the priors are nearly proportional in the vicinity of x. Consider the interval $I_n = (-\sqrt[3]{n}, \sqrt[3]{n})$. As $n \to \infty$ the priors, which differ only by the term $\theta^2/n = O(1/\sqrt[3]{n})$ in the exponential, will converge in I_n, and $\pi_n(\theta|x)$

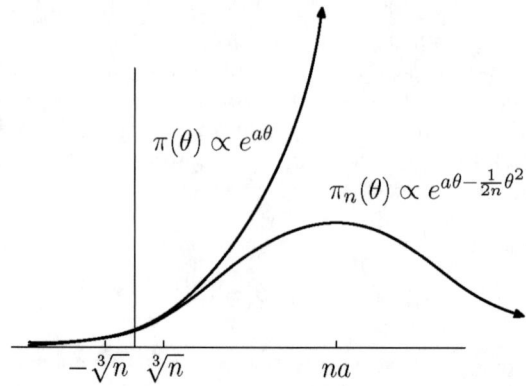

FIGURE 2. Comparison of improper prior and approximation.

will converge to $\pi(\theta|x)$ for each x in I_n. As $n \to \infty$, however, the interval I_n will expand to cover the entire real line. Thus, the pointwise limit of the $\pi_n(\theta|x)$, which is the formal posterior, will exist.

But is this a good reason to regard the formal posterior as the limit of the $\pi_n(\theta|x)$? Consider this from the standpoint of someone whose true prior is $\pi_n(\theta)$, and who seeks an idealized posterior that he can use as a good approximation. As we have just seen, the assertion that $\pi_n(\theta|x) \to \pi_{pt}(\theta|x)$ is based entirely on agreement in the region close to the origin, where the data will almost never be found! On the other hand, in the region where the data *is* expected, $x \sim na$, $\pi_n(\theta|x)$ differs markedly from $\pi_{pt}(\theta|x)$. (Explicit formulas are given in the next section.) In this sense, the formal posterior is poor approximation to $\pi_n(\theta|x)$, for any n.

In the next section, we show that $\pi_{prob}(\theta|x)$, which does not obey the formal Bayes' law, *is* a good approximation in the region where the data expected.

BAYES' LAW: LOCAL BUT NOT GLOBAL

We infer from (2) that where $m_n(x)$ is large,

$$\pi_{prob}(\theta|x) \approx \pi_n(\theta|x) \propto p(\theta|x)\pi_n(\theta). \tag{5}$$

That is, $\pi_{prob}(\theta|x)$ satisfies Bayes' law locally, and approximately. On the other hand, in regions where $m_n(x)$ is small, there is no need for $\pi_{prob}(\theta|x)$ to satisfy Bayes' law, and in general, it will not.

We examine this phenomenon using Stone's example. First, we show that Bayes' law holds locally for $\pi_{prob}(\theta|x)$, as given in Eq. (4); this justifies our assertion that $\pi_{prob}(\theta|x)$ is a probability limit for this problem. It is sufficient to show that $\pi_{prob}(\theta|x) \approx \pi_n(\theta|x)$ when $m_n(x)$ is large. The densities $\pi_n(\theta|x)$ and $m_n(x)$ are easily calculated. Let $\sigma_n^2 =$

$n/(1+n)$. Then

$$\pi_n(\theta|x) \propto \exp\left[-\frac{(\theta - \sigma_n^2(x+a))^2}{2\sigma_n^2}\right]; \quad m_n(x) \propto \exp\left[-\frac{(x-an)^2}{2(1+n)}\right].$$

Let $x = an + \varepsilon$. Then

$$\pi_n(\theta|x) \propto \exp\left[-\frac{(\theta - x + \frac{\varepsilon}{n+1})^2}{2\sigma_n^2}\right].$$

In the region where $m_n(x)$ is large, $\varepsilon = O(\sqrt{n+1})$. As $n \to \infty$, $\varepsilon/(n+1) \to 0$ and $\sigma_n \to 1$, so $\pi_{\text{prob}}(\theta|x) \approx \pi_n(\theta|x)$, as was to be shown.

It is obvious, however, that Bayes' law does not hold globally for $\pi_{\text{prob}}(\theta|x)$:

$$e^{-\frac{1}{2}(x-\theta)^2} \not\propto e^{-\frac{1}{2}(x-\theta)^2} e^{a\theta}$$

DISCUSSION

The presence or absence of a MP is determined by the type of limit we use to define improper inferences. In [1], it is shown that if $\pi(\theta|x)$ is a probability limit, then the marginal, $\pi(\zeta|z)$, is also a probability limit. The MP shows that if $\pi(\theta|x)$ is a formal posterior, then the marginal, $\pi(\zeta|z)$, is not necessarily a formal posterior. Thus, the interpretation of improper inferences as probability limits is internally consistent, at least with respect to the MP, whereas their interpretation as formal posteriors is not.

The use of probability limits, rather than pointwise limits, is already known to resolve other difficulties, such as "strong inconsistency" [14], and "incoherence." [15]. Now that probability limits have been shown to resolve the MP as well, we have a common explanation for most of the key difficulties of improper inference: they arise because the pointwise limit, and the formal posteriors to which they give rise, are fundamentally unsound. Probability limits, by contrast, appear to provide a consistent theory of improper inference.

The local nature of Bayes' law for probability limits is the key to understanding the MP. The requirement that $\pi(\theta|x)$ be a probability limit is essentially equivalent to the requirement that Bayes' law hold locally, where $m_n(x)$ is large, in the sense of (5). The requirement that $\pi(\theta|x)$ be a formal posterior is equivalent to the requirement that Bayes' law hold globally. B_1 and B_2 both make inferences based on the latter requirement. Thus, each of them make inferences that are sometimes erroneous, and so they frequently disagree.

In resolving the MP, we have followed Jaynes in regarding an improper inference as a limit of ordinary inferences, based on proper priors. There is nothing in our procedure that is inconsistent with the rules of probability theory, as developed by Jaynes in Chapter 2 of [6]. These rules, however, do not apply to the case where the prior is improper, nor do they stipulate how the improper case is to be regarded as a limit. In particular, they do not justify the use of formal posteriors. The use of pointwise limits is an additional assumption, the validity of which is open to question. Following Stone, we have argued that pointwise limits are unsound, and that probability limits better capture the intuitive meaning of convergence.

It remains to discuss the implications of this analysis for objective Bayesianism. There are strong indications that the requirement that an improper inference be a probability limit is very restrictive. In group models, Stone has shown that the formal posterior can only be a probability limit if the prior is right Haar measure and the group satisfies a technical condition, known as amenability [13]. Eaton and Sudderth have shown that many of the formal posteriors of multivariate analysis are "incoherent" or strongly inconsistent, and thus cannot be probability limits [16].

Probability limits may be used to construct improper inferences for the translation and scale groups, and these coincide with the formal posteriors. Since the most common applications of improper inference involve these groups, our analysis shows why formal posteriors appear to work in these simple situations. When we get to more complicated problems, however, such as those of multivariate analysis, it appears that probability limits associated with the "ignorance priors" for the relevant symmetries, such as $GL(n)$, do not exist. Thus, the use of probability limits restores the technical viability of objective Bayes to only a very limited domain of improper problems.

ACKNOWLEDGMENTS

I acknowledge support from the Department of Energy under contract DE-AC52-06NA25396.

REFERENCES

1. T. C. Wallstrom, "The Marginalization Paradox and Probability Limits," in *Bayesian Statistics 8*, edited by J. M. Bernardo et al, Oxford University Press, 2007, pp. 669–674.
2. M. Stone, and A. P. Dawid, *Biometrika* **59**, 369–375 (1972).
3. A. P. Dawid, M. Stone, and J. V. Zidek, *J. R. Statist. Soc. B* **35**, 189–233 (1973), with discussion.
4. J. M. Bernardo, *J. R. Statist. Soc. B* **41**, 113–147 (1979).
5. R. E. Kass, and L. Wasserman, *JASA* **91**, 1343–1370 (1996).
6. E. T. Jaynes, *Probability Theory: The Logic of Science*, Cambridge University Press, 2003.
7. A. P. Dawid, M. Stone, and J. V. Zidek, Critique of E. T. Jaynes's 'Paradoxes of Probability Theory', Tech. Rep. 172, University College London (1996).
8. H. Jeffreys, *Theory of Probability*, Oxford Univ. Press, Oxford, 1961, 3rd edn.
9. H. Jeffreys, *Scientific Inference*, Cambridge Univ. Press, Cambridge, 1957.
10. D. L. Wallace, *Ann. Math. Stat.* **30**, 864–876 (1959).
11. M. Stone, *Ann. Math. Stat.* **34**, 568–573 (1963).
12. M. Stone, *Ann. Math. Stat.* **36**, 440–453 (1965).
13. M. Stone, *Ann. Math. Stat.* **41**, 1349–1353 (1970).
14. M. Stone, *JASA* **71**, 114–116 (1976).
15. M. L. Eaton, and D. A. Freedman, *The Bernoulli Journal* **10**, 861–872 (2004).
16. M. L. Eaton, and W. D. Sudderth, *Sankhya A* **55**, 481–493 (1993).

Positive evidence for non-arbitrary assignments of probability

William M Briggs

mattstat@gmail.com; Dept. Mathematics, 214 Pearce Hall, Central Michigan University, Mt. Pleasant, MI 48859

Abstract. How to assign numerical values for probabilities that do not seem artificial or arbitrary is a central question in Bayesian statistics. The case of assigning a probability to the truth of a proposition or event for which there is *no* evidence other than that the event is contingent, is contrasted with the assignment of probability in the case where there is *definte* evidence that the event can happen in a finite set of ways. The truth of a proposition of this kind is frequently assigned a probability via arguments of ignorance, symmetry, randomness, the Principle of Indiffernce, the Principal Principal, non-informativeness, or by other methods. These concepts are all shown to be flawed or to be misleading. The *statistical syllogism* introduced by Williams in 1947 is shown to fix the problems that the other arguments have. An example in the context of model selection is given.

Keywords: Induction; Logical Probability; Principle of Indifference; Probability assignment.

There are (at least) two central foundational problems in statistics: how to objectively justify probability models, and how to objectively assign probabilities to events and to the parameters of probability models. The goal of both of these operations is to insure that they are not arbitrary, or are not guided by the subjective whim of the user, and that they logically follow from the explicit evidence that is given or assumed to be known. First, it is useful to recall, what is often forgotten, that—both deductive and non-deductive—arguments of logic are nothing more than the study *between* statements, and only between the statements explicitly defined.

So, suppose p is a premise and q a conclusion to the argument from p to q: which is an argument that states, "(the proposition) p (is true) therefore (the proposition) q (is true)" (the mathematically succinct way to write this is $p \Rightarrow q$). Logical probability makes statements about the truth of the conclusion q like this:

$$0 \leq \Pr(q|p) \leq 1. \quad (1)$$

Cox [1961], and like those in the logical probability tradition before him [2, 3, 4, 5], states that if the limits 0 or 1 apply to the conclusion q of a given argument with premiss p, then q is, respectively, certainly false or certainly true. When the limits are reached, then the logical connective (between q and p) is said to be *deductive*. If the limits are not reached, then the argument from p tp q is invalid or *non-deductive* and q only probable.

Recall these common definitions: *contingent* means not necessarily true or false, and an observation statement or *event* is some thing that can happen (is not necessarily false or impossible) in the given context (examples will be given below). Inductive arguments—which are arguments from contingent premises which are premises that are, or could have been, observed, to a contingent conclusion about something that has not been, and may not be able to be, observed—are, of course, central in probability. In an earlier paper [6], I examine induction in statistics and probability.

This article surveys the most common arguments used in assigning probabilities to uncertain events where the event can happen in a finite number of (known) ways. These ways are usually assigned equal probability. The usual reasons given for equiprobable assignment are: ignorance, "no reason" or indifference, non-informativeness, symmetry, randomness, and some very well known mathematical arguments. All of these arguments, by no means mutually exclusive, will be shown to be flawed, or to be misleading, or to imply the necessity of subjectivity when it is not needed. Instead, an old argument, called the "statistical syllogism", will be re-introduced. The statistical syllogism avoids the problems inherent in the others, with the added benefit of clearly and completely delineating the information used in a given problem.

IGNORANCE

To stress: logical probability concerns itself with assigning probabilities to the conclusions of arguments with explicitly stated, and fixed, premises. It is easy to assign probability when the argument is deductive: the probability being 0 or 1. For an example of a common, non-inductive (and non-deductive) argument, suppose we have *definite* knowledge, labelled e_c, that M is some non-contradictory contingent statement, proposition, or description of an event, and t any tautology. That is, we *know* that M is not necessarily true or false; we also do not know, we are ignorant, whether M will happen. The argument: $t \wedge e_c \Rightarrow M$ is not valid (and is read "t and e_c, therefore M is true"). The most common tautologies used in cases like this are $t =$"I am ignorant about M, but I know it can be true or false," or $t=$"M will happen or it won't"; both of these ways of writing t implicitly attach the definite knowledge e_c that M is contingent, except that the first mistakenly adds "I am ignorant" since we *know* of M's contingency. Now, it is true that t; or the statement t is always true. A principle of logical probability gives:

$$0 < \Pr(M \text{ (is true)} \,|t, e_c) < 1. \qquad (2)$$

And that is the *best* we can ever do with only the definite knowledge that M is contingent (e.g., Keynes [4]). This point, which has caused much confusion, is well worth reflecting upon, and which is amplified below. It follows from the well known logical fact that it is impossible to argue validly to a contingent conclusion (like M) given a necessarily true or tautologous premiss. This result, known since Aristotle, is not dependent on a particular t; any tautology or necessary truth will do.

Of course, the situation so far is *not* ignorance, since we have already specified that we *know* M is contingent. Suppose instead that somebody asked you, "What is the probability of M?" and refused to tell you anything about M: it may be contingent, it may be necessarily true, or M may even be complete gibberish. Then *no probability at all* can be assigned. If you do assign a probability it is because you are *adding* information that was not given to you, information you suppose that is true, but that may be false. The argument is changed and you cannot say your assignment is based on ignorance.

Some (Bayesian) statisiticians would not like to settle for (2), which is a vague enough statement about M, and would insist that we find some concrete real number r such that

$\Pr(M|t, e_c) = r$. To find this number, there is usually an appeal, to the utterers of (2), to announce some subjective opinion they might have about M, or even about how they would take bets over the truth of M. Not all Bayesians would insist that you must say how you'd bet for or against M. Some try to find r by an argument like the following: "Well, M can be true, or it may be false. So it must be that $\Pr(M) = \frac{1}{2}$." No, it mustn't. The first sentence to this argument is just t, and nothing has been gained. The step from the conclusion (M is true) to the probability statement is therefore arbitrary (as many have felt before; e.g. [7]).

The argument can be modified, by inserting some additional evidence: say, $e_\circ$ = "M is equally like to be true or false", so that we have "$e_\circ$, therefore $\Pr(M|t) = \frac{1}{2}$." This argument is dogmatic; nevertheless, it *is* valid; however, the premiss $e_\circ$ is the same as the conclusion, which isn't wrong, but it is begging the question. This is usually and loosely called a fallacy, but the conclusion *does* follow from assuming the premises are true, therefore the argument *is* valid: it is just of no use.

People will more likely say "Well, M can be true, or it may be false, *and I have no reason to think that it is false or that it is true. I am indifferent.* So it must be that $\Pr(M) = \frac{1}{2}$." This kind of argument is sometimes called the "Principle of Indifference," advanced by Laplace and Keynes [4] and criticized in e.g. [8]. It is the "indifference" or "no reason" clause that is the start of troubles.

NO REASON & INDIFFERENCE

The minor premiss in "Both [possibilities for M] are equally likely" is evidently itself a conclusion from the premiss, "I have no reason to think that M is false or that it is true," or "I am indifferent about M." Now, this argument, in its many forms, has lead a happy life. It, or a version of it, shows up in discussion of priors frequently, and also, of course, in discussions about model selection, e.g. [9]. But it is an argument that should not have had the attention it did. For we can rewrite it like this: "I do not know—I am *ignorant*; I have no reason to know—whether M is true or false, but it can only be true or false (therefore M is true)." The implicit conclusion is usually assigned probability $\Pr(M) = \frac{1}{2}$. The argument, I hope you can see, is not valid and the probability statement is arbitrary. Here's why. This argument *is* valid: "M is true or it is false; therefore, I do not know—I am *ignorant*; I have no reason to know—whether M is true or false, but it can only be true or false." It should now be obvious that this conclusion is nothing more than a restatement of the initial tautology! To be explicit: saying you do not know anything about M, in English, means you *know nothing*, and therefore cannot assign any probability, not even the bounds of (2). But if you are saying you do not know whether it is true or false, this is the same as saying that you *know* that it can be true or false, that is, you *know* $t \wedge e_c$. So, despite our repeated insistence of "ignorance," we are back to the bounds of eq. (2), which is to say, right where we started.

This leaves "indifference", which isn't exactly wrong, but it has unnecessary connotations of subjectivity, and, for some, a certain implication that the probabilities are equal (and so begs the question). The subjectivity is implied in the sense that we are *setting* the probabilities by our will, or that, somehow, our opinions matter as to what the probabilities are (see Franklin [10] for a discussion of how Neyman used a similar trick applied to interpreting classical confidence intervals).

SYMMETRY

Let M represent the fact that I see a head when next I flip this coin. Are you with the majority who insist that the probability of M must be $\frac{1}{2}$? Before you answer, notice that the 'coin flip' M is *entirely* different from any other M' where all you know is that M' is contingent. For example, if instead of a coin flip, suppose M represented the outcome of an experiment where you to open a box and examine some object inside and note whether you can see an 'H'. Now all you know is that M is contingent and can be true or false. Based *solely* on the information you have, you do not know any other possibilities. You do *not* know that an 'H' or some other letter or object might appear. You do not know, even, whether a snake may jump out of the box. If you imply that because the question asked something about an 'H', that the result must be 'H' or some other letter, probably a 'T', then you are *adding* evidence that you were *not* given.

Back to the coin flip. Why is the probability of M $\frac{1}{2}$? Symmetry, perhaps? As in, "It can fall head or tail and there is no reason to prefer—I am indifferent—to head over tail"? But isn't that the same as ignorance, that is, the same as the tautology and knowledge of contingency? It is. Because substitute 'be true' for 'fall head' and 'be false' for 'fall tail' and you are right back at the tautology. Or symmetry as in, "Heads and tails are equally likely because I have no reason to think otherwise"? Again, "no reason to think otherwise" or "Heads and tails are equally likely" or "indifference" are begging the question or can be misleading. The anticlimatic answer for assigning probability to a definite M is the statistical syllogism, as defined by Williams [1947] for the coin flip posed in the familiar form of a syllogistic argument (common example of a syllogism: "All men are mortal, Socrates is a man, etc."):

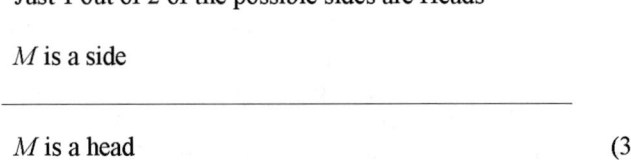

Just 1 out of 2 of the possible sides are Heads

M is a side

M is a head   (3)

This argument, is, of course, invalid in the sense that the conclusion (M is a head) is not entailed by the premises. But we can assign the probability $\Pr(M|e_s) = \frac{1}{2}$ that the conclusion is true, where e_s is the evidence of the two premises (e_s implies e_c). This probability assignment, made explicit in the form of the statistical syllogism, is *derived* from assuming uniform probability across the individual events that make up the "sample space": see a complete discussion in Stove [15] pp. 92-97, who credits Carnap [1950] with the first proof of this. In this case, this sides of a coin, or: $\Pr(H|e_s) = \Pr(T|e_s)$. That is to say, if you are convinced of the probability assigned implied by the statistical syllogism, you must admit the equi-probability of the underlying events.

Symmetry has often been used, and objected to, as a principle to assign probability, e.g. [12, 13, 14]. Arguments based on symmetry tend to be misleading because the examples are always chosen in such a way that they are "physically balanaced" or uniformly symmetric, which gives rise to a certain confusion. For example, Strevens [12] imagines that one side is painted red on a dodecahedral die and asks the probability (in a 'fair' roll) of seeing the red side. He assigns 1/12 because of (physical) symmetry. Hájek [14]—and many, many other authors, invoking something about a "privileged partition"—and argue that Streven's assignment is indeed correct under physical symmetry (one partition of the outcome). But (in another partitioning) they say that you either see the red or you don't, so that under this view, the probability is 1/2. Both probability assignments can't be right, so logical probability itself must be flawed! Well, the "either see red or not" is the tautology, which is very different information than physical symmetry: these two different pieces of information should certainly give different probability assessments, so it is to logical probability's credit—and not its detriment—that it does so. And we have already seen that under the "see red or not" partition (just $t \wedge e_c$), the probability assignment is eq. (2) and not 1/2. Also, all privileged partition arguments have a distinct subjective quality about them: why choose any partition not based on the statistical syllogism unless you are intent on creating difficulties where they do not exist?

Again, Streven's die is physically symmetric, a very strong assumption that is not needed. Consider this example: suppose I have an n-sided object, one side of which is painted red: what's the probability of red? My object may—or may *not*—be physically symmetric. It may be some amorphous blob, no two sides having the same surface area. It may be physically symmetric down to the quark. But you are *not* entitled to say it is physically symmetric without additional evidence. Just as equally, you do not have any evidence that my object is physically *a*symmetric. And so, you can only appeal to the statistical syllogism ("Just 1 side of n etc.").

WHITHER RANDOMNESS?

Suppose there are 10 men in a room and just 9 of these 10 are Schmenges. M is a man in the room. The probability that the conclusion "M is a Schmenge" is true by the statistical syllogism is $\frac{9}{10}$. But if you were to grab a man out of the room randomly: how can you be sure that the probability that he is a Schmenge is $\frac{9}{10}$? Suppose you were to "sample" the men by opening the door and grabbing the nearest man and noting whether or not he is a Schmenge. Or perhaps that doesn't sound "random" enough to you. Instead, you order the men inside to polka madly, to run about and bounce off the walls and to not stop; then you reach in a grab one. This sampling procedure becomes an additional premise, e_r = "Men are arranged in the room randomly."

Here, I take "randomly" to mean, as it can only mean, that "I have no idea—I am ignorant—of how the men are arranged" [17]. To show this, first suppose that *all* we know is that there are men in a room, but *nothing* else. That is, our *only* evidence is e_r, which is just another way of saying, "There are men in the room, and I have no idea who they are or how they are arranged." Tacit in this is the idea that there may be some Schmenges in the room, which, of course, means that there may *not* be any. That is, e_r is equivalent to, "M may be true or it may be false". This is our old friend, the tautology t, which we have already seen adds nothing to the argument that would allow us to assign a definite probability to the conclusion.

It should also not be necessary to say that we do not need to assume anything about infinite "trials" of men in rooms to arrive at the probability of M. Some (objective) Bayesians try this kind of argument in an attempt justify their priors by invoking something called the *Principal Principle*, which states

> that if the objective, physical probability of a random event (in the sense of its limiting relative-frequency in an infinite sequence of trials) were known to be r and if no other relevant information were available, then the appropriate subjective degree of belief that the event will occur on any particular trial would also be r: [8, p. 240].

Ignoring the fact that we can never *know* what happens after an infinite amount of time, and so can *know* r, or that we cannot imagine an infinite number of rooms filled with Schmenges, but pretending that we can, the Principal Principle says "$\Pr(M|\Pr(M) = r) = r$" (it adds the premiss "$\Pr(M) = r$" which is taken to be the 'objective' or physical probability of M), but which we can now see is just begging the question.

MATHEMATICAL ATTEMPTS

The following arguments start with the definite knowledge e that M is contingent and can be decomposed into a finite number of possibilities (like coin flips or rooms of Schmenges) $M_1, M_2, \ldots, M_n, n < \infty$.

First permutation argument (logical probability) [5]: Introduce evidence e which states that either M_1 or M_2 or etc. M_n can be true, but that only one of them can be true. In the case where M is a coin flip, the result can be either M_1="head" or M_2="tail". Thus, $\Pr(M_1 \vee M_2 \vee \ldots \vee M_n|e) = \sum_{i=1}^{n} \Pr(M_i|e)$. We want to assign the probabilities $\Pr(M_i|e)$ for $i = 1\ldots n$. The set of possibilities is $M = \{M_1, M_2, M_3, \ldots M_n\}$. Let π be a permutation on the set $\{1,2\}$. Let $M' = \{M_{\pi(1)}, M_{\pi(2)}, M_3, \ldots M_n\}$. That is, the set M and M' are the same except the first two indexes have been swapped in M'. The evidence e is fixed. Therefore, it must be that $\Pr(M_1|e)_M = \Pr(M_{\pi(2)}|e)_{M'}$ and $\Pr(M_2|e)_M = \Pr(M_{\pi(1)}|e)_{M'}$. Jaynes then makes a crucial step, which is to add evidence to e which states that the evidence is "indifferent" to M_1 and M_2, i.e.

> if it [the evidence] says something about one, it says the same thing about the other, and so it contains nothing that would give [us] *any reason* to prefer one over the other. (p. 39, emphasis mine)

Accepting this for the moment, e then says that our state of knowledge about M or M' is equivalent, including the order of the indexes. Thus, (note the change in indexes) $\Pr(M_1|e)_M = \Pr(M_{\pi(1)}|e)_{M'}$, $\Pr(M_2|e)_M = \Pr(M_{\pi(2)}|e)_{M'}$ and $\Pr(M_j|e)_M = \Pr(M_j|e)_{M'}, j = 3, \ldots, n$. Which implies $\Pr(M_1|e)_M = \Pr(M_2|e)_M$: that is to say, equiprobable prior assignment.

This argument is fine if what Jaynes says in the quotation holds. But we can see in it the presence of two tell-tale phrases, our old friends, "indifferent" and "no reason", which are used, and are needed, to justify the final step. This is just begging the question all over again, for how else could the evidence e be "indifferent"? That is, Jaynes has assumed uniform probability (and thus, the statistical syllogism) as part of the evidence e, which is what he set out to prove.

Second permutation argument (finite exchangeability) [18]: Space does not permit a detailed examination of this argument; details will appear in a future paper.

DISCUSSION

Suppose you are considering M_1 and M_2 as the only competing models for some situation. Then, using the statistical syllogism ("Just M_1 or M_2 etc.") and the logical probability assignments it implies as above, $\Pr(M_1 \vee M_2|e_s) = \Pr(M_1|e_s) + \Pr(M_2|e_s) = 1$ and so $\Pr(M_1|e_s) = \Pr(M_2|e_s) = \frac{1}{2}$. This is the justification for starting with equal probability in model selection. After x is observed, then it is easy (in principle) to calculate $\Pr(M_1|x,e_s)$ and $\Pr(M_2|x,e_s)$.

It is no surprise that this is the same point reached by appealing to the Principle of Indifference, or even the Principle of Maximum Entropy for a finite number of model choices; Jaynes [5]. The statistical syllogism gives the same answers as the Principle of Indifference, but not by the same route and, again, without the hidden assumptions or metaphysical baggage. The built-in question-begging of that principle is gone, and there is no appeal to subjectivity, which many find so distasteful.

In conversation, I have had it pointed out that the same results as the statistical syllogism can be had by appealing the the Principle of Maximum (information) Entropy (MAXENT), or via other complex mathematical arguments. I agree with this. However, the additional complex apparatus of MAXENT, with its own set of axioms and assumptions, is certainly not needed. The uniform probability assumption over events that is used to derive the statistical syllogism is just true; but is it also true that the probability assignment *should* maximize entropy? Maybe, but what do we mean by "should"? If you are trying to convince somebody of the correctness of logical probability, it should be clear that if you introduce MAXENT at an early stage, you are then asking a lot more from your audience.

I attempted to cast light on a few common hidden assumptions in the simplest possible situations. This paper is certainly not a complete answer to the question of how to assign probabilities in an objective way in all situations. The statistical syllogism can clearly be applied to assign priors on probability model parameters when those parameters can take a finite number of values or states. The class of probability models which contain such parameters may or may not be very large, but it is at least not empty, though it of course does not contain the most frequently used probability models, such as those, say, from the exponential family. I make no attempt in this paper to justify, or modify, the use of the statistical syllogism in the case where the number of outcomes is countably or uncountably infinte, as in the case of parameters in models like the normal distribution. [19] is a good starting place for these topics.

But, however simple, the statistical syllogism clearly works and does not suffer from the same flaws as earlier arguments—arguments which may have given the same answers sometimes, but come loaded with hidden assumptions, assumptions which have been barriers to acceptance of Bayesian methods. Too, the statistical syllogism is completely objective and it eliminates any hint of "randomness" and "chance" and the complexity—and mysticism—that these terms imply. To this, much of this paper may seem like quibbling. After all, the results using the statistical syllogism agree with those (at least in these examples) that would be had appealing to "no reason" etc. But this impression of agreement is false. For one, people who would insist, for example, that all probability calculations cannot begin before a properly defined measure space has been carefully laid out, should not quail from a demand for the preciseness of language used in describing such models. More importantly, the terms "no reason" etc. are all improperly *defensive* and are negative. Using them with respect to assigning probabilities naturally creates a certain suspicion in those who hear them that something funny is going on. The terms also over-emphasize, and even use when they should not, subjectivity. With the statistical syllogism, these problems disappear. For one: there is no subjectivity; the probability assignment follows logically from the information given. And the statistical syllogism emphasizes the *definite, positive* knowledge that exists (such as contingency and know number of possible outcomes). People, I believe, would be more inclined to to try to understand Bayesian methods (and be made aware of the multitude of shortcomings of classical probability) if we who promote them are more careful—and justifiably positive—in our language.

ACKNOWLEDGMENTS

I am heavily indebted to an anonymous referee who helped to clarify the arguments used in this paper.

REFERENCES

1. R. T. Cox, *Algebra of Probable Inference*, Johns Hopkins University Press, Baltimore, 1961.
2. M. de Laplace, *A Philosophical Essay on Probabilities*, Dover, Mineola, NY, 1996.
3. H. Jeffreys, *Theory of Probability*, Oxford University Press, Oxford, 1998.
4. J. M. Keynes, *A Treatise on Probability*, Dover Phoenix Editions, Mineola, NY, 2004.
5. E. T. Jaynes, *Probability Theory: The Logic of Science*, Cambridge University Press, Cambridge, 2003.
6. W. M. Briggs, *arxiv.org/pdf/math.GM/0610859* (2006).
7. R. Fisher, *Collected Papers of R.A. Fisher*, University of Adelaide, Adelaide, 1973, vol. 2, chap. The logic of inductive inference, pp. 271–315.
8. C. Howson, and P. Urbach, *Scientific Reasoning: the Bayesian Approach*, Open Court, Chicago, 1993, second edn.
9. J. M. Bernardo, and A. F. M. Smith, *Bayesian Theory*, Wiley, New York, 2000.
10. J. Franklin, *Erkenntnis* **55**, 277–305 (2001).
11. D. Williams, *The Ground of Induction*, Russell & Russell, New York, 1947.
12. M. Strevens, *Noûs* **3:22**, 231–246 (1998).
13. P. Bartha, and R. Johns, *Philosophy of Science* **68**, S109–S122 (2001).
14. A. Hájek, "A philosopher's guide to probability," in *Uncertainty: Multi-disciplinary Perspectives on Risk*, Earthscan, 2007.
15. D. Stove, *The Rationality of Induction*, Clarendon, Oxford, 1986.
16. R. Carnap, *Logical Foundations of Probability*, Chicago University Press, Chicago, 1950.
17. S. Campbell, and J. Franklin, *Synthese* **138**, 79–99 (2004).
18. P. Diaconis, *Synthese* **36**, 271–281 (1977).
19. I. H. Jermyn, *Annals of Statistics* **33**, 583–605 (2005).

Cellular Automata Generalized To An Inferential System

David J. Blower

Cogon Systems
17 S. Palafox Place, Pensacola, FL 32501
djblower@cox.net

Abstract. Stephen Wolfram popularized elementary one-dimensional cellular automata in his book, *A New Kind of Science*. Among many remarkable things, he proved that one of these cellular automata was a Universal Turing Machine. Such cellular automata can be interpreted in a different way by viewing them within the context of the formal manipulation rules from probability theory. Bayes's Theorem is the most famous of such formal rules.

As a prelude, we recapitulate Jaynes's presentation of how probability theory generalizes classical logic using *modus ponens* as the canonical example. We emphasize the important conceptual standing of Boolean Algebra for the formal rules of probability manipulation and give an alternative demonstration augmenting and complementing Jaynes's derivation. We show the complementary roles played in arguments of this kind by Bayes's Theorem and joint probability tables.

A good explanation for all of this is afforded by the expansion of any particular logic function via the *disjunctive normal form* (DNF). The DNF expansion is a useful heuristic emphasized in this exposition because such expansions point out where relevant 0s should be placed in the joint probability tables for logic functions involving any number of variables.

It then becomes a straightforward exercise to rely on Boolean Algebra, Bayes's Theorem, and joint probability tables in extrapolating to Wolfram's cellular automata. Cellular automata are seen as purely deductive systems, just like classical logic, which probability theory is then able to generalize. Thus, any uncertainties which we might like to introduce into the discussion about cellular automata are handled with ease via the familiar inferential path. Most importantly, the difficult problem of *predicting* what cellular automata will do in the far future is treated like any inferential prediction problem.

INTRODUCTION

Jaynes emphasized that probability theory generalizes classical logic. He presented several examples of how inference generalizes deduction in Chapter Two of his book *Probability Theory: The Logic of Science* [4]. His canonical example is the recasting of the *modus ponens* syllogism into Bayes's Theorem. He showed how the "undecidable" logical argument where the consequent is known, but the antecedent is not, can still have an answer within the context of an inference.

Generalizing through inference is a principle that can be applied to cellular automata because they are nothing more than and nothing less than compositions of logic functions on three variables. It is extremely difficult to predict what a system following definite rules will do in the future if we don't generalize the concept of deduction. In this paper, we detail the steps that allow one to see that cellular automata can be generalized just like logic functions. The important conclusion is that prediction *then becomes possible* if CA are inferential systems.

GENERALIZING CLASSICAL LOGIC

Rederiving *modus ponens*

Modus ponens is the well-known classical syllogism that argues: If given the premise that A implies B and, given furthermore that A is in fact TRUE, then B must also be TRUE. In the notation of probability theory, we would write the syllogism just given as $P(B|A, Z)$, the probability that B is TRUE given that we assume A is TRUE and Z is TRUE. Z is a placeholder for whatever logic function we are interested in; here, the implication operator is the chosen function. $P(B|A, Z)$ must then equal 1 to reproduce the result from logic.

We know that the probability of B, assuming that A, as well as the implication indicated by Z, are both true, is given by Bayes's Theorem,

$$P(B \mid A, Z) = \frac{P(BAZ)}{P(AZ)} \qquad (1)$$

$Z \equiv A \to B$ stands for one of the 16 logic functions of two variables. Any Boolean function, such as the implication function, can be expressed in the disjunctive normal form (DNF) [3]. The DNF for implication is,

$$Z = T \equiv A \to B \equiv (A \wedge B) \vee (\overline{A} \wedge B) \vee (\overline{A} \wedge \overline{B}) \qquad (2)$$

This DNF expansion for the $\to$ operator may be expressed in a shorter way. Using this shorter version will make the proofs easier. Making use of the axioms from Boolean Algebra, we arrive at a shorter version of the DNF expansion of the implication operator appropriate for Bayes's Theorem,

$$AB \vee \overline{A}B \vee \overline{A}\,\overline{B} \equiv \overline{A} \vee B \qquad (3)$$

Substitute the DNF expansion for the logic function

Substitute the DNF expansion for the implication operator and then apply any axioms from Boolean Algebra [2] that simplify Bayes's Theorem. After the substitution for Z, the left hand side of Bayes's Theorem now looks like,

$$P(B \mid A, Z) \equiv P(B \mid A, \overline{A} \vee B) \qquad (4)$$

We are now properly positioned to carry out the Boolean operations within Bayes's Theorem as written in Eq. (1) on both the numerator and denominator.

$$P(B \mid A, Z) = \frac{P(B \wedge [\,A \wedge [\,\overline{A} \vee B\,]\,])}{P(A \wedge [\,\overline{A} \vee B\,])} \qquad (5)$$

After working through these Boolean operations within the confines of the probability operator, the goal is reached that shows B must be TRUE given that A and a particular deductive model represented by Z are both TRUE.

$$P(B \mid A, Z) = \frac{P(AB)}{P(AB)} = 1 \tag{6}$$

The formal rules governing probability symbol manipulation have returned the same answer that classical logic provides. This is a necessary, but obviously not sufficient, condition for probability to generalize classical logic.

A joint probability table for *modus ponens*

To augment this purely formal approach, it always helps to have another way of arriving at the same result. To that end, Figure 1 shows a conceivable joint probability table for three variables A, B, and Z. There are some representative numbers inserted into each of the eight cells of the table. Each of these numbers is a legitimate numerical assignment for the probability of the joint occurrence indexed by that cell.[1]

	Z			$\bar{Z}$			
	A	$\bar{A}$		A	$\bar{A}$		
B	1/4 (Cell 1)	1/4 (Cell 2)	1/2	0 (Cell 5)	0 (Cell 6)	0	1/2
$\bar{B}$	0 (Cell 3)	1/4 (Cell 4)	1/4	1/4 (Cell 7)	0 (Cell 8)	1/4	1/2
	1/4	1/2	3/4	1/4	0	1/4	
				1/2	1/2		1

Figure 1: A $2 \times 2 \times 2$ joint probability table illustrating logical implication.

However, given that implication is the particular logic function being used, Z cannot be TRUE if A is TRUE and B is FALSE. That would violate the very definition of what implication means from the classical logic standpoint. Cell 3 indexes this particular setting for the three variables where an F must be assigned. Cell 3, therefore, will have a numerical assignment of $P(ZA\bar{B}) = 0$.

It is easy to numerically verify the results found by the formal symbol manipulation proof. Substitute the numerical assignment appearing in cell 1 of the joint probability table for the numerator and then the sum of the numerical assignments appearing in cells 1 and 3 for the marginal probability represented by the denominator. Thus, Bayes's Theorem works out to,

$$P(B \mid A, A \rightarrow B) = \frac{P(ZAB)}{P(ZAB) + P(ZA\bar{B})} = \frac{1/4}{1/4 + 0} = 1 \tag{7}$$

---

[1] The numerical assignment comes from the Maximum Entropy Principle [1] with the constraint that the implication function inserts 0s in some cells and further constraints on the marginal probabilities.

ELEMENTARY ONE DIMENSIONAL CELLULAR AUTOMATA

A cellular automaton (CA) is a visual way of showing how some system defined by a rule evolves over time. The simplest cellular automata are called *one-dimensional* because they begin with a single line of cells. This starting line of cells is allowed to be arbitrarily long. The cells of a CA are colored either black or white.

The beginning line of cells is transformed at each time step into another line of cells where, again, each cell is black or white. The transformation of every cell takes place at each time step by following some rule which looks at the cell's neighbors at the preceding time step. Figure 2 contains a sketch of a CA as it evolves over time by following Wolfram's so-called Rule 110 [6].

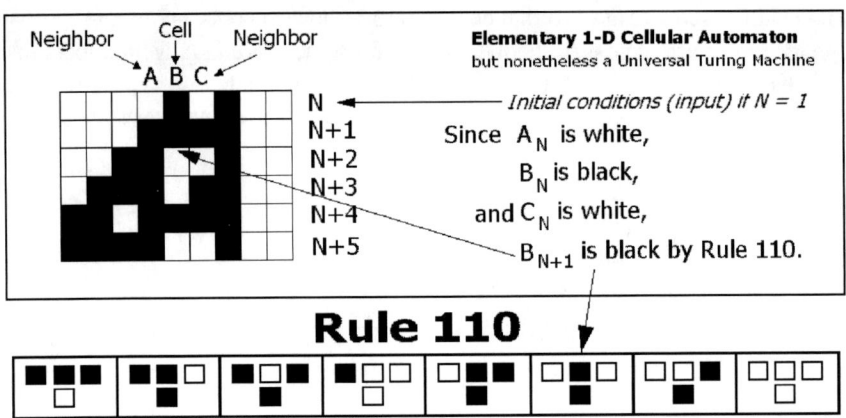

Figure 2: The evolution of a cellular automaton following Rule 110 over five time steps.

In fact, this Rule 110 algorithm for coloring each current time step's cell is none other than a Boolean function involving *three* variables. Rule 110 is presented at the bottom of Figure 2. Let T stand for *black* and F for *white*. The eight possible variable settings are shown in the top row and are the colors of the A_N, B_N and C_N cells at the immediately preceding time step. The functional assignment is shown in the bottom row and is the color for the updated cell B_{N+1}.

So, for example, the top row tells us that if the three cells at the previous time step are all black, then the cell is updated to white. In the initial explanatory example, the cells above B_{N+1} were white, black, and white, and therefore B_{N+1} was colored black.

THREE VARIABLE LOGIC FUNCTIONS

The classical logic functions on two variables can be extended to functions on three variables. The same Boolean definition for a logic function can be used as a template for more than two variables. Let's look at a particularly pertinent example of a three variable function. The general abstract definition for a Boolean function of three variables still applies.

Definition: *An elementary one-dimensional CA is composed from* $f : \mathbf{B}^3 \to \mathbf{B}$

Thus, we will be writing functions on the ordered triples of $\mathbf{B} \times \mathbf{B} \times \mathbf{B}$ that take on values from $\mathbf{B} = \{T, F\}$, such as $f(T,T,T) = F$ and $f(F,F,T) = T$. There are 2^{2^n} possible functions for n variables each of which can take on only one of two values. The implication function, discussed earlier in the *modus ponens* example, is one of the $2^{2^2} = 16$ possible functions for $n = 2$ variables. Now we are going to advance to considering $2^{2^3} = 256$ functions, $f_1(A, B, C)$ through $f_{256}(A, B, C)$, for $n = 3$ variables.

Consider the three variable function explicitly defined in Table 1. This function also defines Rule 110 for cellular automata which is why it is labeled as $f_{110}(A, B, C)$. If

Table 1: A Boolean functional assignment table for three variables. It illustrates one of the 256 possible logic functions; namely the one used as Rule 110 governing the evolution of a cellular automaton.

A	B	C	$f_{110}(A,B,C)$
T	T	T	F
T	T	F	T
T	F	T	T
T	F	F	F
F	T	T	T
F	T	F	T
F	F	T	T
F	F	F	F

we were to perform a DNF expansion on this particular function following the heuristic explanation, we would see that it consists of the disjunction of five terms where the functional assignment takes on the value T. Therefore, following the pattern substitution rule, the DNF for Rule 110 looks like,

$$f_{110}(A,B,C) = AB\overline{C} \vee A\overline{B}C \vee \overline{A}BC \vee \overline{A}B\overline{C} \vee \overline{A}\,\overline{B}C$$

CA VIEWED PROBABILISTICALLY

Now embed our understanding of a CA within the context of the manipulation rules for probability symbols and especially within the context of Bayes's Theorem and a joint probability table. Varying the numerical assignments in these joint probability tables makes it easy to see the difference between a deterministic CA and a probabilistic CA. It is the difference between deduction and inference.

Let $\mathcal{M}_{110}$ stand for the model that implements Rule 110. The DNF expansion tells us where to place the 0s in the table. Then, given this model guiding the evolution of the cellular automaton, together with the colors of three relevant cells at the previous Nth time step, we can ask the question: What is the probability for the color of the cell that is ready to be updated at the next time step?

This general inferential procedure following the formal manipulation rules is exactly what we did before for the *modus ponens* argument. There we conditioned upon the assumed truth of some classical logic function such as $A \to B$. For CA, we condition on the assumed truth of some logic function like Rule 110. And, just as before, the DNF for whatever rule is operative could be inserted so that the relevant Boolean operations could take place.

However, less work is involved, and we arrive at the same answer, if we rely upon a joint probability table. It is important for our argument to demonstrate how easy it is to see what is going on in cellular automata by constructing joint probability tables. Figure 3 shows just such a table consisting of four appropriately labeled variables and sixteen cells. The numbers in the cells are legitimate numerical assignments to joint probabilities as might be made under some model. In fact, they are the ones that can be made under a model reflecting Rule 110. All of the many marginal probabilities are shown as well. It is broken down into four tables of four cells each. The two tables on top are where the updated cell is black and the two tables on the bottom apply when the updated cell is white. In addition to containing the numerical values of the probabilities, each cell is labeled from 1 to 16.

Here is a numerical example using this joint probability table. We want to check whether the probabilistic formulation will return the correct color of a cell in the cellular automaton with certainty, that is, with a probability of 1. Consider the example where the information processor wants to update its state of knowledge that the updated cell will be colored white given its color and the color of its neighbors at the previous times step.

$$P(w \mid b, w, w, \mathcal{M}_{110}) = \frac{P(\overline{B}_{N+1}, A_N, \overline{B}_N, \overline{C}_N \mid \mathcal{M}_{110})}{P(\overline{B}_{N+1}, A_N, \overline{B}_N, \overline{C}_N \mid \mathcal{M}_{110}) + P(B_{N+1}, A_N, \overline{B}_N, \overline{C}_N \mid \mathcal{M}_{110})}$$

Find the cells in the joint probability table that correspond to the two joint probabilities on the right hand side of Bayes's Theorem. This is cell 12 for the numerator and cells 12 and 4 for the denominator. Now substitute their numerical values to find,

$$P(w \mid b, w, w, \mathcal{M}_{110}) = \frac{1/8}{1/8 + 0} = 1$$

We calculate an answer that we had hoped for. It is certain that the updated cell will be colored white under Rule 110.

		A_N				$\bar{A}_N$			
		B_N	$\bar{B}_N$			B_N	$\bar{B}_N$		
B_{N+1}	C_N	0 (1)	1/8 (2)	1/8	C_N	1/8 (5)	1/8 (6)	1/4	3/8
	$\bar{C}_N$	1/8 (3)	0 (4)	1/8	$\bar{C}_N$	1/8 (7)	0 (8)	1/8	1/4
		1/8	1/8	1/4		1/4	1/8	3/8	5/8
$\bar{B}_{N+1}$	C_N	1/8 (9)	0 (10)	1/8	C_N	0 (13)	0 (14)	0	1/8	1/2
	$\bar{C}_N$	0 (11)	1/8 (12)	1/8	$\bar{C}_N$	0 (15)	1/8 (16)	1/8	1/4	1/2
		1/8	1/8	1/4		0	1/8	1/8	3/8	
		1/4	1/4	1/2		1/4	1/4	1/2		
							1/2	1/2		1.00

Figure 3: A joint probability table for the Rule 110 cellular automaton. The numbers placed in the cells make this a deterministic CA.

Now abandon the purely deductive CA and transition to an inferential system. For example, the Bayesian predictive equation is easily derived from the formal rules for probability manipulation.

$$P(F_j \mid \mathcal{D}) = \sum_{k=1}^{\mathcal{M}} P(F_j \mid \mathcal{M}_k) \, P(\mathcal{M}_k \mid \mathcal{D}) \tag{8}$$

We could employ this equation to predict the color of future cells conditioned on what patterns have been observed in some set of data $\mathcal{D}$. The important lesson here is that in order to predict using an inferential system some information must be voluntarily discarded. Here, we discard information that one specific model like Rule 110 is operative and instead average over the predictions made by all the rules. The data tell us how to update our state of knowledge about the relative standing of all the rules being used.

The hard part of developing a good inferential approach to CA is to decide what information available at the microscopic level should be discarded in developing macro-statements. Statistical mechanics has solved this problem for physical systems at equilibrium. One approach I am pursuing is using Information Geometry to project the detailed microscopic probability distributions residing in a high dimensional manifold to a much lower dimensional sub-manifold that somehow captures the essence of what information may be conveniently discarded.

CONCLUSION

Wolfram concluded pessimistically that one could never *predict* the detailed consequences of a CA's evolution far into the future. If we cling irretrievably to the notion of deductive systems, there is, unfortunately, no counter-argument.

But we tried to suggest that when we generalize the deductive process through probability and make it inferential, formerly undecidable statements are transformed into probabilistic statements. If we generalize the elementary CA, essentially what we do is to allow a full range of legitimate numerical assignments into the joint probability table. This implies that we no longer have a deterministic deductive process, but rather an informational inferential process.

The analogy here is to the same reality that had to be faced historically by statistical mechanics and thermodynamics [5]. It quickly became clear that it would be impossible to follow the trajectory of every atom in a gas even though the underlying physics governing the dynamics of each atom and its interaction with other atoms was known. An inferential process dealing with large scale macroscopic averages of the fundamental atomic dynamics was adopted; the deterministic fine scale microscopic information about atomic dynamics simply had to be abandoned.

Similarly, if we abandon the quest of trying to follow the detailed microscopic features of an evolving CA, and concentrate rather on some large grained macroscopic structures that are probabilistic averages, then we might hope to find some computational solution to circumvent the prediction problem. This approach applies to CA and is the same one that any information processor must follow in predicting the consequences of any complicated, detailed, ontological explanation of its world.

REFERENCES

1. Blower, David J. An Easy Derivation of Logistic Regression from the Bayesian and Maximum Entropy Perspective. *Bayesian Inference and Maximum Entropy Methods in Science and Engineering,* ed. by Erickson, G. and Zhai Y., 23rd International Workshop, AIP Conference Proceedings, Volume 707, August 2003.
2. Brown, Frank Markham. *Boolean Reasoning: The Logic of Boolean Equations*, Kluwer Academic Publishers, Norwell, MA, 1990.
3. Garrett, Anthony J. M. Probability Synthesis: How to Express Probabilities in Terms of Each Other. *Maximum Entropy and Bayesian Methods*, Proceedings of the 17th International Workshop on Maximum Entropy and Bayesian Methods of Statistical Analysis, Boise, ID, 1997, ed. by Erickson, G. J., Rychert, J.T. and Smith, C. R., Kluwer Academic, Norwell, MA, 1998.
4. Jaynes, Edwin T. *Probability Theory: The Logic of Science*, Cambridge University Press, New York, NY, 2003.
5. Jaynes, Edwin T. Information Theory and Statistical Mechanics I, *Physical Review*, **106**, pp. 620–630, 1957.
6. Wolfram, Stephen. *A New Kind of Science*, Wolfram Media, Inc., Champaign, IL, 2002.

From Maxent To Machine Learning and Back

Timothy D. Sears

College of Engineering and Computer Science,
The Australian National University, Canberra Australia
tim.sears@anu.edu.au

Abstract. To Jaynes, in his original paper [1], maxent is 'a method of reasoning which ensures that no unconscious arbitrary assumptions have been introduced', while fifty years later, the MAXENT conference home page suggests that the method 'is not yet fully available to the statistics community at large.' In fact, it is possible to see that generalized maxent problems, often in disguise, do play a significant role in machine learning and statistics. Deviations from the classic form of the problem are typically used to incorporate some form of prior knowledge. Sometimes that knowledge would be difficult or impossible to represent with only linear constraints or an initial guess for the density.

To clarify these connections, a good place to start is the classic maxent problem. This can then be generalized until the problem encompasses a large class of problems studied by the machine learning community. Relaxed constraints, generalizations of Shannon-Boltzmann-Gibbs (SBG) entropy and a few tools from convex analysis make the task relatively straightforward. In the examples discussed, the original maxent problem remains embedded as a special case. Providing a trail back to the original maxent problem will highlight the potential for cross-fertilization between the two fields.

Keywords: Maximum Entropy Principle, Machine Learning
PACS: 89.20.-a, 89.20

INTRODUCTION

A major focus of Machine Learning is to extract information from data and use it to automatically perform some task. Only performance on the task really matters. Nevertheless it is helpful to have an understanding of some design principles that will lead to a successful outcome. To a machine learning researcher the following quotation from Jaynes [1] sounds like it might be a useful design principle:

> The principles and mathematical methods of statistical mechanics are seen to be of much more general applicability... In the problem of prediction, the maximization of entropy is not an application of a law of physics, but merely a method of reasoning which ensures that no unconscious arbitrary assumptions have been introduced.

Today the field of machine learning makes use of a variety of principles from various sources. In a few areas, like natural language processing and environmental modeling, a few models have a direct lineage to maxent, *e.g.* [2], while some represent a fully Bayesian approach *e.g.* [3], and others are based on empirical risk minimization and related concepts [4]. A complicating factor is that there is no requirement for task-based machine learning algorithms to explicitly use a probabilistic framework. In effect, producing a decision rule is the main goal. Connections to probabilistic reasoning can therefore be implicit and missing from the presentation of a model.

Informally one can conclude that models based purely on the original maxent principle are in a distinct minority in machine learning. However because maxent is related to maximum likelihood estimation and regularized versions of maximum likelihood, which are still widely used techniques, one could say it lives on. Such an expansive view is adopted here in tracing connections to other modeling techniques. This is not done with the idea of advocating one approach or another, but mainly to sort out just how much the principle is currently embedded into some algorithms. It turns out that by studying a generalized form of maxent, models based on seemingly disparate induction principles, including non-probabilistic ones (*e.g.* support vector machines) start to look more alike than different.

CLASSIC TO GENERAL

To begin with, let us start with the classic maxent problem and generalize it somewhat. In this paper, we will maintain a focus on discrete models throughout, since they are sufficient to highlight the main points. The classic maxent problem is a finite-dimensional optimization problem. Its solution is a vector p that solves

$$\min_{p} \; S(p) := \sum_{i=1}^{N} p_i \log(p_i) \; \text{ subject to } \; Ap = b, \text{ and } p_i \geq 0,$$

where p is a non-negative vector, A is an $M \times N$ matrix, with $M < N$, and the constraint matrix is usually assumed to have full (row) rank and take the form $A = \begin{pmatrix} B \\ \mathbf{1}^{\mathrm{T}} \end{pmatrix}$, *i.e.*, it includes a normalization constraint. The vector b supplies the constraint values, and depending, on the context has been given many names, such as mean, data, average, measurement, and even price vector. Note also that only by including the normalization constraint are we assured that the solution turns out to be a probability vector. Finally, this 'maximum' entropy problem is actually minimizing negative entropy. This is essentially an arbitrary choice that fits better with the development of relative entropy later on.

Using notation from convex analysis it is possible to write this problem down more compactly. More importantly, convex analysis allows us to reason about optimization problems involving non-differentiable functions and hard constraints in a manner similar to the familiar techniques for smooth functions, as long as certain technicalities are addressed along the way. This approach is explained in detail in mathematical references such as [5, 6]. More details specific to maxent can be found in [7]. In the presentation here, an attempt is made to acknowledge the technical issues without overburdening with detail.

The problem can be re-expressed as

$$\min_{p} \; S(p) + \delta_{\{0\}}(Ap - b).$$

Explicit reference to the non-negativity constraint can be dropped by assuming S takes on the value $+\infty$ outside its effective domain. Here also δ_C is the convex indicator function of the set C, which takes on the value 0 on the set and $+\infty$ outside the set. In this case the set in question is simply the singleton set $\{0\}$, since exact matching is required. Our first generalization is to relax the constraints to match within an epsilon-ball around zero, defined by some p-norm. Thus $\{0\}$ is replaced by the set εB_P. Next, note that negative entropy is a special case of KL-divergence, which in turn is a special case of Bregman divergence. Bregman divergence between two distributions is defined as $\Delta F(\boldsymbol{p}, \boldsymbol{p}_0) := F(\boldsymbol{p}) - F(\boldsymbol{p}_0) - \langle \nabla F(\boldsymbol{p}_0), \boldsymbol{p} - \boldsymbol{p}_0 \rangle$, using a convex function F with effective domain equal to the non-negative orthant. (The expression $\langle \boldsymbol{p}, \boldsymbol{q} \rangle$ is notation for $\sum_i p_i q_i$, the inner product of $\boldsymbol{p}$ and $\boldsymbol{q}$). Bregman divergences appear in a number of machine learning algorithms. See *e.g.* [8] for further details. Briefly, the resulting function acts something like a square-distance measure. In this setting the classical objective function can be recovered by specifying $F = S$ and $\boldsymbol{p}_0 = \boldsymbol{1}/N$, the uniform distribution over the state space. The generalized problem is now:

$$\min_{\boldsymbol{p}} \quad \Delta F(\boldsymbol{p}, \boldsymbol{p}_0) + \delta_{\varepsilon B_p}(\boldsymbol{A}\boldsymbol{p} - \boldsymbol{b}) \tag{1}$$

This is only one of many routes for generalization, others will be mentioned later. A key property of the original problem that has been preserved here is that of convexity. Also, the specific form of this *primal problem* makes it well suited to solution via another problem, the *dual*. To set up the dual problem, a special type of duality is used, namely Fenchel duality.

Fenchel duality applies when minimizing the sum of two convex functions as in $\min_x (f(x) + g(x))$. The Fenchel dual problem is given by $\min_{x*} (f^*(x^*) + g^*(-x^*))$, where f^* and g^* are Fenchel conjugates, defined as $f^*(x^*) := \sup_x \langle x, x^* \rangle - f(x)$. Here, the dual problem is constructed by separately conjugating the components of the primal. This modularity is one of the hallmarks of Fenchel duality. A slightly more complicated version of Fenchel duality (*e.g.* , Theorem 3.3.5, Borwein and Lewis [9]) permits the inclusion of a constraint matrix ($\boldsymbol{A}$). The dual problem for (1) turns out to be:

$$\min_{\boldsymbol{\mu}} \quad F^*(A^T \boldsymbol{\mu} + \boldsymbol{p}_0^*) + \langle \boldsymbol{\mu}, \boldsymbol{b} \rangle + \varepsilon \|\boldsymbol{\mu}\|_Q, \tag{2}$$

where the P-norm and Q-norm are dual to each other ($1/P + 1/Q = 1$), and $\boldsymbol{p}_0^* = \nabla F(\boldsymbol{p}_0)$. Because of the shape of $\boldsymbol{A}$, this problem is typically much easier to solve.

Given due regard to the feasibility of the primal problem, strong duality holds under fairly general conditions and the dual problem becomes the preferred problem to solve. The resulting formulation (2) can be seen as a more general form of the maximum *a posteriori* (MAP) problem for exponential families and can be shown to be equivalent to versions involving the log-partition function in the classic version of the problem. The key difference is that the objective does not involve a partition function and instead contains an explicit parameter for the normalization. After solving (almost always numerically) for the optimal dual solution, $\bar{\boldsymbol{\mu}}$, the primal solution, $\bar{\boldsymbol{p}}$, can be recovered via:

$$\bar{\boldsymbol{p}} = \underbrace{\nabla F^*}_{\text{"Family"}} (\overbrace{A^T \bar{\boldsymbol{\mu}} + \boldsymbol{p}_0^*}^{\text{"Score"}})$$

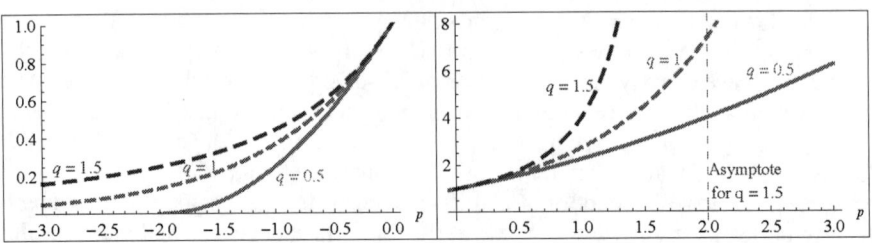

FIGURE 1. Left-hand graph indicates tail behavior. $q > 1 \to$ fat tail. $q < 1 \to$ truncated tail.

The gradient of the conjugate of the entropy function (F) determines the family (∇F^*) of the distribution. In the classic case, the negative entropy function, S, yields the familiar scalar function $\exp(\cdot)$ as the gradient map, hence an exponential family distribution. The score of the model is a linear function of the features (rows) of $\mathbf{A}$. The freedom introduced in (1) can be used to employ a variety of entropy functions, including those based on so-called deformed logarithms [10]. These entropy functions yield the corresponding deformed exponential family as distributions.

An important example of entropies based on deformed logarithms is based on the q-logarithm. This introduces a parameter, q, usually restricted to the interval $(0,2)$. The definition of the q-logarithm, deformed exponential $\exp_q$, and corresponding negative entropy S_q are:

$$\log_q(p) := \frac{x^{1-q}-1}{1-q} \quad \exp_q(p) := (1+(1-q)p)_+^{\frac{1}{1-q}} \quad S_q(\mathbf{p}) = -\sum_{i=1}^{N} p_i \log_q\left(\frac{1}{p_i}\right),$$

where $(\cdot)_+ = \max(\cdot, 0)$. Taking limits in the case $q = 1$, the standard definitions of all three functions are recovered. When used in the generalized maxent problem (1), the solution has the form of a q-exponential family.

The negative entropy function defined above is closely related to Tsallis entropy. A single term of the negative entropy function used here is $-p\log_q(1/p)$ while Tsallis entropy is defined using $p\log_q(p)$ instead. The difference is immaterial with with $0 < q < 2$, since the two formulas are identical under the reparameterization which replaces q with $2-q'$. The version employed here cleanly delivers the $\exp_q$ family distribution as the result.

A graphical comparison of the $\exp_q$ function (Figure 1) suggests how q-exponentials can produce power-law behavior. For values of $q < 1$, exp_q can reach the value zero, in contrast to $q = 1$, while values of $q > 1$ cause the function to approach zero slowly, leading to long-tailed distributions.

Example: The loaded die problem. The loaded die problem [11] is familiar to students of maxent and a useful example to indicate what role q might play in a model. To

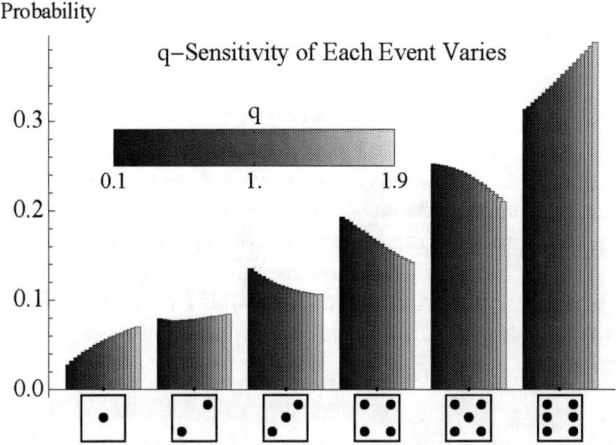

FIGURE 2. The loaded die problem. Mean input value: 4.5 instead of the "fair" value of 3.5

recapitulate, in that problem the matrix and data vector are

$$A = \begin{pmatrix} 1 & 2 & 3 & 4 & 5 & 6 \\ 1 & 1 & 1 & 1 & 1 & 1 \end{pmatrix} \qquad b = \begin{pmatrix} 4.5 \\ 1 \end{pmatrix}.$$

The key feature of the example is that the expectation of the die is set much higher than the fair value of 3.5. Together with normalization, this forms the constraint matrix, clearly ruling out the uniform distribution associated with a fair die. For simplicity we set $\varepsilon = 0$ and vary q. Each value of q produces a different value for the probability of a particular face, while matching the input expectations. The solutions are illustrated in Figure 2. The result confirms the idea suggested by Figure 1: lower values of q depress the probabilities of the extreme events (rolling a 1 or a 6), while higher values of q accentuate them. The classic solution is in the middle of the pack for each face of the die. Figure 2 also illustrates that the sensitivity of the probability to the parameter q varies. From a machine learning perspective it is hard to say much more about this model, since no task has been defined. However, the task of making a two-way market on whether a 2 appears next looks to be somewhat easier than doing so for the other faces. Not surprisingly, the example shows that a different entropy function can yield different results. Apart from potentially introducing 'unconscious arbitrary' assumptions, as quoted at the outset, can something useful be done?

Example: The Dantzig Selector. Recently, Candes and Tao [12] reported that for a regression model $y = X\beta$, the solution to problem $\min_\beta \ ||\beta||_1 + \delta_{\varepsilon B_\infty}(X^T(X\beta - y))$ *exactly* finds the correct set of regressors, with high probability, under special conditions of low noise and a sparse 'true' model $\widehat{\beta}$. The use of the 1-norm as a regularizer is well known to produce sparse models, and this is one case where a sparse model is known to be optimal. Also note, under a change of variables $\beta = (\ I \ \ |-I\ )\ p$, the regularizer can be approached by S_q in the limit as $q \to 0$, and therefore fits the generalized maxent format of (1). The entropy function reflects a very specific kind of prior knowledge, suited to the particular task of discovering the active set of regressors.

Value Regularization

One reason models can be hard to compare is that they are parameterized in widely different ways. This can often disguise the elements they have in common. Since virtually all machine learning algorithms produce outputs, if we can discover a selection criteria for the outputs, the model formulations might be comparable on this basis. This point of view was pursued by Rifkin and Lippert [13], who found that some common techniques in machine learning can be viewed as solving:

$$\min_{y} R(y) + L(y),$$

where y represents an *output* of the model. The two functions involved are a convex regularizer R, which favors smooth outputs, regardless of the data vector, b. Meanwhile the loss L, which typically makes use of the data vector b as a parameter, is used to enforce a close fit to the data. Although maxent is not addressed in [13], it is not hard to work out how the generalized maxent problem (1) can be mapped into this framework. In fact the dual problem is unaffected. For the primal, the first step is a change of variables $y = Ap$. This makes it clear that the "outputs" under discussion are expectations of the rows of A in the generalized maxent context. The loss is then just $L(y) = \delta_{\varepsilon B_p}(y - b)$. Describing the regularizer requires another device from convex analysis, the *image function*, denoted by the compound symbol $A\Delta F$:

$$R(y) = (A\Delta F)(y) := \min_{p} \Delta F(p, p_0) + \delta_{\{0\}}(Ap - y).$$

It should be noted that $A\Delta F$ is also a well-behaved convex function on the set of feasible points for y. Combining the two components yields a new primal problem:

$$\min_{y} \ (A\Delta F)(y) + \delta_{\varepsilon B_p}(y - b). \qquad (3)$$

We are now in a position to compare the maxent model to one of the great workhorses of Machine Learning, the Support Vector Machine (SVM). Two key definitions are required. The first is K, a positive semidefinite kernel matrix, constructed from the training examples. The second is the loss function employed by SVMs, the soft-margin hinge loss: $L(y, b) = \max(0, 1 - yb)$, where y is the output label and b is the training target. Under the value regularization framework (see [13] for details) it can be shown that SVMs solve the primal problem

$$\min_{y} \ \underbrace{\frac{1}{2}\lambda y^T K^{-1} y^T}_{R} + \underbrace{\sum_{i} L(y_i, b_i)}_{L}. \qquad (4)$$

The SVM regularizer is minimized if the output vector, y, is smooth, since K^{-1} is also positive semi-definite. However this is balanced against the SVM loss, which is minimized only when the training output matches the labels. The mathematical techniques used to originate this model and the philosophical orientation of its originators are rather distinct from the maxent approach (see [14]). But as this example makes clear, all those considerations, different as they may be, lead to models that are conceptually not too far apart, differing only in the choice of regularizer and loss function.

DISCUSSION

The use of the maxent principle in Natural Language Processing (NLP), with slight modifications [2] is fairly straightforward to trace. A related set of models could be termed *conditional-empirical models*. They do two things to alter the classic problem. One is to utilize the product rule to factorize the distribution: $p(x,y) = p(y|x)p(x)$. The second is to replace one of the factors, say $p(x)$ with its empirical counterpart. This is a strong modeling assumption, since it places zero weight on a vast portion of the state space. After imposing it, the original problem can be reformulated as a much smaller one, with considerable practical benefits. In machine learning this produces what is referred to as a *discriminative* model, in comparision to a *generative* one.

More complicated factorizations can be expressed using graphical models. Currently, all the examples the author knows of employ exponential families, and are therefore based on SBG entrop. For undirected graphical models the Hammersley-Clifford theorem [16] is fundamental for relating the structure of the graph to the factorizations of the probability density. An initial step toward using other entropy functions is made in [17].

Machine learning algorithms make heavy use of many other *non-smooth* loss functions, besides the SVM-loss. Even the simple norm relaxation employed in (1) is not differentiable at zero. In [18] a table of 20 other common loss functions is presented and only five are smooth. Non-smoothness hinders some analytical statements, but does not represent overwhelming practical problems. Apart from loss functions, the freedom of the choice of regularizer is potentially very wide. Instead of parameterizing the entropy with a single number q, any increasing one-to-one function $\phi(x)$ can be used in constructing a deformed logarithm.The resulting distribution is a ϕ-exponential [10]. Noting that there are other divergence measures (*e.g.* Amari's α-divergence) which can also recover the SBG entropy as a special case, it is clear that the surface has barely been scratched here.

Explicitly including a normalization constraint, rather than embedding it in a partition function, makes it possible to consider models that are near-probabilistic. In a finance setting, it is common to identify prices with probabilities. On the other hand, the task for a model might be to produce *two* sets of prices: bids and offers, with the stipulation that they are based on probabilities that sum to < 1 and > 1, respectively. In machine learning, dropping normalization has led to altogether new algorithms [19].

The foregoing is not meant to suggest that *all* machine learning problems fit into a generalized maxent framework. Many problems lead to non-convex optimization (*e.g.*, missing variables). Nor have fully Bayesian models been brought into the picture. The relationship between maxent and Bayesian statistics remains a topic of ongoing research [20].

Hopefully the reader will agree that convex analysis helps clear away the clutter and exposes what model formulations have in common. Using the language of machine learning the crux of the distinction seems to be in how to choose a regularizer and how to choose a loss. The loss can be viewed as measuring how costly it is to mismatch the data vector, while the regularizer measures how smooth the answer is. Techniques for trading off these two competing goals have been invented independently and repeatedly. As optimization algorithms continue to improve, perhaps we can look forward to machine learning researchers making greater use of entropic regularization variants such as the ones described in this paper, while maxent researchers may benefit from designing loss functions directly as is common in machine learning.

ACKNOWLEDGMENTS

The author wishes to thank Simon Guenter, Kee Siong Ng, Peter Sunehag, Alex Smola and Vishy Vishwanathan for useful comments and discussions.

REFERENCES

1. E. T. Jaynes, *Physical Review* **106**, 620–630 (1957).
2. A. L. Berger, S. A. della Pietra, and V. J. della Pietra, *Computational Linguistics* **22**, 39–71 (1996), ISSN 0891-2017.
3. C. Williams, and C. Rasmussen, *Advances in Neural Information Processing Systems* **8**, 514–520 (1996).
4. V. Vapnik, *Statistical learning theory*, Wiley New York, 1998.
5. J.-B. Hiriart-Urruty, and C. Lemarechal, *Fundamentals of Convex Analysis*, Springer, 2001.
6. R. T. Rockafellar, and R. J.-B. Wets, *Variational Analysis*, Springer, 1998.
7. T. D. Sears, and S. Vishwanathan, Fenchel duality and generalized maxent, in submission (2007).
8. K. Azoury, and M. Warmuth, *Machine Learning* **43**, 211–246 (2001).
9. J. Borwein, and A. Lewis, *Convex Analysis and Nonlinear Optimization: Theory and Examples*, Springer, 2000.
10. J. Naudts, *Journal of Inequalities in Pure and Applied Mathematics* **5 (4) 102** (2004).
11. E. T. Jaynes, *IEEE* **70**, 939–952 (1982).
12. E. Candes, and T. Tao, *Ann. Stat.* **accepted** (2007).
13. R. Rifkin, and R. Lippert, *Journal of Machine Learning Research* **8**, 441–479 (2007).
14. B. Schölkopf, and A. J. Smola, *Learning with Kernels: Support Vector Machines, Regularization, Optimization, and Beyond*, Adaptive computation and machine learning, The MIT Press, 2001.
15. J. Lafferty, A. McCallum, and F. Pereira, "Conditional random fields: Probabilistic models for segmenting and labeling sequence data," in *Proc. 18th International Conf. on Machine Learning*, Morgan Kaufmann, San Francisco, CA, 2001, pp. 282–289.
16. S. Lauritzen, *Graphical models: Clarendon Press*, Clarendon Press, 1996.
17. T. D. Sears, and P. Sunehag, "Induced Semantics for Undirected Graphs," in *26th International Workshop on Bayesian Inference and Maximum Entropy Methods in Science and Engineering*, edited by K. H. Knuth, 2007.
18. C. Teo, Q. Le, A. Smola, and S. Vishwanathan, "A Scalable Modular Convex Solver for Regularized Risk Minimization," in *Conference on Knowledge Discovery and Data Mining*, 2007.
19. G. Lebanon, and J. Lafferty, "Boosting and Maximum Likelihood for Exponential Models," in *NIPS*, 2001, vol. 14.
20. A. Caticha, and A. Giffin, Updating probabilities, arXiv.org:physics/0608185 (2006).

Induced Semantics for Undirected Graphs: Another Look at the Hammersley-Clifford Theorem

Timothy D. Sears* and Peter Sunehag[†]

*College of Engineering and Computer Science,
The Australian National University, Canberra Australia
tim.sears@anu.edu.au
[†]Statistical Machine Learning Program, NICTA
Canberra, Australia
peter.sunehag@nicta.com.au

Abstract. The Hammersley-Clifford (H-C) theorem relates the factorization properties of a probability distribution to the clique structure of an undirected graph. If a density factorizes according to the clique structure of an undirected graph, the theorem guarantees that the distribution satisfies the Markov property and vice versa. We show how to generalize the H-C theorem to different notions of decomposability and the corresponding generalized-Markov property. Finally we discuss how our technique might be used to arrive at other generalizations of the H-C theorem, inducing a graph semantics adapted to the modeling problem.

Keywords: Graphical models, Tsallis statistics, Hammersley-Clifford Theorem
PACS: 89.70, 89.20

INTRODUCTION

Statistical distributions of the q-exponential form can be motivated by a generalization of maximum entropy termed Tsallis entropy. This is one possible generalization of Shannon-Boltzmann-Gibbs (SBG) entropy. An important property of Tsallis entropy is its capability of generating distributions with power-law behaviour and distributions with finite support. One of these is the q-Gaussian distribution. Its density, assuming it is centered around the origin is

$$f(x) = \exp_q(-\beta x^2 - \alpha_q(\beta)),$$

where $\exp_q$ is the q-exponential, defined later in the section on q-analogues. For now, we just note that the usual Gaussian is recovered when $q \to 1$. See [1, 2, 3] for more discussion and references along these lines. Here, our main concern is to judge the compatibility of distributions of the form $\exp_q(\cdot)$ with the tools and techniques of graphical models. We are motivated by the possibility of using joint distributions of q-exponential form to model situations where a collection of random variables correlate more, or less, under dramatic circumstances than they do in an everyday environment. For example, a portfolio of equities is considered to be more correlated in the face of steep market decline.

If we have N independent Gaussian (q=1) random variables, the joint density is

$$\prod_{i=1}^{N} \exp(-\beta_i x_i^2 - \alpha_i) = \exp(-\sum_{i=1}^{N}(\beta_i x_i^2) - (\sum_{i=1}^{N} \alpha_i)).$$

To have a corresponding formula for the q-exponentials, an operation called the q-product($\otimes_q$) which satisfies $\exp_q(x_1) \otimes_q, \ldots, \otimes_q \exp_q(x_N) = \exp_q(\Sigma_{i=1}^{N} x_i)$ must be introduced. The q-product has been studied elsewhere [4, 5]. It leads to a definition of q-independence that says that X_1 and X_2 are q-independent if their joint density, f, q-factorizes, i.e. if for some g and h

$$f(x_1, x_2) = g(x_1) \otimes_q h(x_2).$$

Note when $q \neq 1$, g and h are not the marginals of f. Even so the q-product can be used to create joint distributions from univariate distributions. Recently, a central limit theorem for q-independent variables was proved [5]. Bayesian updating of such distributions is a different matter.

Introducing the idea of q-independence to the setting of graphical models induces a different semantics for the edges of a graph. It is possible to demonstrate this by formulating a q-Markov condition and prove a version of the Hammersley-Clifford theorem that says that a distribution is q-Markov with respect to the graph in question, if and only if it q-factorizes over the maximal cliques of that graph. We will conclude with a brief discussion of how to use this type of graphical model to perform typical tasks involving graphical models, focusing on a corresponding extension of the Viterbi algorithm.

The proof of the Hammersley-Clifford Theorem relies on terminology and results from both number theory and graph theory. We give the necessary elements here as well as on q-analogues of some elementary mathematical operations.

GRAPHICAL MODELS

A graph $\mathcal{G} = (\mathcal{V}, \mathcal{E})$ consists of a set of vertices (or nodes), $\mathcal{V}$, and a set of edges $\mathcal{E}$. An edge is an ordered pair of nodes. A clique, $c \in \mathcal{V}$ is a fully connected subgraph of $\mathcal{G}$. The vertices, $\{a_i\}_{i=1}^{M}$ in a graphical model typically correspond to the variables of a distribution, with density $f(\mathbf{x})$. We will move between these representations assuming the order of the nodes is the same.

We will say a function $\mathcal{T}$-decomposes according to a graph for the transformation $\mathcal{T} : \mathbb{R}^{\mathcal{X}} \to \mathbb{R}^{\mathcal{X}}$ if there are ψ_c such that

$$\mathcal{T}f(\mathbf{x}) = \sum_{c \in \mathcal{C}} \psi_c(\mathbf{x}_c), \tag{1}$$

where $\mathcal{C}$ is the set of cliques of $\mathcal{G}$ and where $\mathbf{x}_c$ is equal to $\mathbf{x}$ with the entries corresponding to $\mathcal{V} \setminus c$ removed.

Let $\mathbf{x}^c$ denote a vector equal to $\mathbf{x}$ with the m-th and n-th entries deleted. We say that a function has the pairwise generalized-Markov property if, whenever the nodes a_m and a_n

do not share an edge, then there exists functions with appropriate domains and ranges, h_1 and h_2, such that

$$\mathcal{T} f(\boldsymbol{x}) = h_1(x_m, \boldsymbol{x}^c) + h_2(x_n, \boldsymbol{x}^c). \tag{2}$$

In the classic formulation of the Hammersley-Clifford Theorem $\mathcal{T} = \log$ is implicitly assumed. Also in this setting decomposition is equivalent to multiplicative factorization. The role of $\mathcal{T}$ is therefore to convert factorization (of some kind) into addition. The generalized-Markov property then asserts the separability of the effects of x_m and x_n when the corresponding nodes are separated. More in-depth background in graph theory, especially as it relates to graphical models is given by [6].

NUMBER THEORY ESSENTIALS

Number theoretic tools play an important role in accounting for all the operations that could be defined on a graph. Although the subject is rather deep, we only require a modest gathering of results for our problem.

The prime numbers will be denoted p_i, which represents the i-th prime number. The number 1 will be considered to be the 0-th prime number.

An arithmetic function is one of the form $g : \mathbb{N} \to \mathbb{C}$. A multiplicative function is an arithmetic function that also satisfies $g(m \cdot n) = g(m)g(n)$ if the greatest common divisor of m and n is 1. If it holds for all m, n the arithmetic function is said to be totally multiplicative. The sum-function of an arithmetic function g is defined as

$$S_g(n) := \sum_{d \mid n} g(d),$$

where the summation notation $\sum_{d \mid n}$ is standard notation for the sum over the divisors of n.

The fundamental theorem of arithmetic states that any $1 < n \in \mathbb{N}$ decomposes into a unique product of powers of primes: $n = p_{i_1}^{\alpha_{i_1}} \cdot, \cdots, \cdot p_{i_M}^{\alpha_{i_M}}$. The function $\lambda(n)$ will be used to denote the number of prime factors of n.

We require a few of the important number theoretic functions. One is the Möbius function: For $n = p_{i_1}^{\alpha_{i_1}} \cdot, \cdots, \cdot p_{i_M}^{\alpha_{i_M}}$,

$$\mu(n) = \begin{cases} 0 & \text{if any } \alpha_i > 1 \\ 1 & \text{if } n = 1 \\ (-1)^{\lambda(n)} & \text{otherwise} \end{cases} \tag{3}$$

The first condition tests whether or not the factorization of n contains a square number.

The constant function will be denoted $1(n)$ or simply 1 when the context makes it clear that a function, not a number, is required. The Dirichlet identity is defined as

$$\varepsilon(n) = \begin{cases} 1 & \text{if } n = 1 \\ 0 & \text{otherwise.} \end{cases}$$

The Dirichlet convolution of two arithmetic functions, f and g, is another arithmetic function. It is denoted $f * g$ and defined as

$$(f*g)(n) = \sum_{d|n} f(d) g(\frac{n}{d}). \tag{4}$$

This operation is commutative and associative. The sum-function, defined earlier, is in fact the Dirichlet convolution $S_f = f * 1$. Two other important identities are: $\mu * 1 = \varepsilon$, and, for all arithmetic functions, f, $f * \varepsilon = f$ holds. Using these, we can easily derive the famous Möbius Inversion theorem:

$$\mu * S_f = \mu * (f * 1) = (\mu * 1) * f = \varepsilon * f = f \tag{5}$$

This theorem plays an important role in the analysis. Our account of the number theoretic tools has been terse. To place them in their proper context one should consult a reference such as [7] or [8].

q-LOGARITHM

The q-logarithm is defined for $q > 0$ as

$$\log_q(p) := \begin{cases} \log(p) & \text{if } q = 1 \\ \frac{p^{1-q}-1}{1-q} & \text{otherwise} \end{cases} \tag{6}$$

We let the notation $(v)_+$ mean v if $v > 0$ and 0 otherwise. The inverse of the q-logarithm is

$$\exp_q(v) = (1 + (1-q)v)_+^{\frac{1}{1-q}}.$$

Using these two functions we can define an analogue to multiplication:

$$x \otimes_q y = \exp_q(\log_q(x) + \log_q(y)) = (x^{1-q} + y^{1-q} - 1)^{\frac{1}{1-q}}$$

if $x^{1-q} + y^{1-q} - 1 > 0$ and otherwise it is 0. It is associative, commmutative and it has 1 as its neutral elements. Under this definition of $\otimes_q$ we have the identities:

$$\exp_q(x+y) = \exp_q(x) \otimes_q \exp_q(y)$$
$$\log_q(x \otimes_q y) = \log_q(x) + \log_q(y) \text{ (whenever the left-hand-side is defined).}$$

Thus a q-exponential does not factorize, but it does "q-factorize".

MAIN RESULT

With the background of the previous section we are ready to state the main result. We let the transformation $\mathcal{T}$ correspond to $\log_q$, leaving q as a parameter. Then it makes sense to refer to the decomposition and generalized Markov properties as q-decomposition and q-Markov properties. Later we will discuss other transformations $\mathscr{T}$ leading to different generalizations of the H-C theorem.

Theorem 1 (q-Hammersley-Clifford) *Let P be a positive measure with non-negative density f. Then f satisfies the pairwise q-Markov property iff f q-factorizes according to the graph $\mathscr{G} = (\mathscr{V}, \mathscr{E})$.*

Proof ($\Leftarrow$-direction) If f q-factorizes according to the graph and $a_m \neq a_n$ do not share an edge, then the expression in (1) can be broken apart into

$$\sum_{\{c|c\in\mathscr{C}, a_m \in c\}} \psi_c(\mathbf{x}_c) + \sum_{\{c|c\in\mathscr{C}, a_n \in c\}} \psi_c(\mathbf{x}_c) + \sum_{\{c|c\in\mathscr{C}, a_m \notin c, a_n \notin c\}} \psi_c(\mathbf{x}_c)$$

The first term can serve as h_1 and the last two terms can serve as h_2 which verifies in the Markov condition (2).

($\Rightarrow$-direction) *Guide:* First we define a map or index from sets of vertices of the graph (equivalently the argument indexes of f) to the natural numbers. This map is simple to define, but special, because its value encodes the membership of the vertices in the set. In other words, knowing only a value of the index, we can determine the identity of all of the vertices which are contained in the set which has that index value. Using this index, we define a generic potential function, ψ, over the index values of each subset of $\mathscr{V}$. Next we employ the results from number theory. The function ψ is seen to be the Dirichlet convolution of the Möbius function μ and another function k. The Möbius inversion theorem provides that k is the sum-function S_ψ, also defined over the index values of all the subsets of $\mathscr{V}$. The inversion theorem provides a representation of k as a sum of potentials taken over *all* all possible subsets of $\mathscr{V}$.

Next we define a special version of f that switches the evaluation of f at a given point $\mathbf{x}$ and an arbitrary alternative point $\hat{\mathbf{x}}$, depending on a supplied set of vertices, b. This function evaluates a vector using the coordinate values of $\mathbf{x}$ corresponding to b and the coordinate values of $\hat{\mathbf{x}}$ corresponding to the complement, b^c. This function must therefore agree with f when $\mathscr{V}$ is the set under consideration. Otherwise it tests the sensitivity of f to arbitrary settings of variables outside the given set of coordinate indexes corresponding to b.

Next we set k to be this modified f evaluated at $\mathscr{V}$, and use the sum function representation of k to express f in terms of a sum of terms involving ψ. In this way, the modified f, which agrees with f at the level of the whole graph, is represented as a sum of local versions of itself. However, this representation contains *all* subsets of $\mathscr{V}$ in the sum. We then use the q-Markov property of f to show that any subset which does not correspond to a clique can be omitted from the sum. This will conclude the proof.

Proof: In the following a and b will be arbitrary subsets of $\mathscr{V}$ unless indicated otherwise, while a_m, a_n represent vertices of the graph and $M = |\mathscr{V}|$.

Define an index map $\iota : \mathscr{P}(\mathscr{V}) \to \mathbb{N}$ as follows. Let $\iota(\emptyset) = 1$. Let us consider 1 as the 0-th prime number. Choose an arbitrary ordering of the nodes $a_i, i = 1 \ldots M$, in $\mathscr{V}$. Define an index map ι as follows. For singleton sets, $b = \{a_i\}$, let $\iota(\{a_i\}) = p_i$, the i-th prime number. For larger subsets, $b = \{a_{i_1} \ldots a_{i_j}\}$, let

$$\iota(b) = p_{i_1} \cdot p_{i_2} \cdots p_{i_{j-1}} \cdot p_{i_j}.$$

This map has certain helpful properties. The range of ι is square-free since no prime appears twice in the definition of $\iota(b)$. If $a \subset b \subset \mathcal{V}$ then $\iota(a)$ divides $\iota(b)$. The Möbius function (3), μ, takes on only the values ± 1 when evaluated on the range of ι, i.e., if $j = \iota(a)$, then $\mu(\iota(a)) = (-1)^{|a|} = (-1)^{\lambda(j)}$, where λ is defined in (3).

Next, define a generic potential function over any subset of $\mathcal{V}$. Let

$$\psi(\iota(a)) = \sum_{j|\iota(a)} \mu\left(\frac{\iota(a)}{j}\right) k(j) = \sum_{j|\iota(a)} (-1)^{\frac{\iota(a)}{j}} k(j),$$

where k is an arithmetic function. This definition is equivalent to $\psi = \mu * k$ where "$*$" denotes Dirichlet convolution (4). The Möbius inversion theorem (5) implies that k is the sum-function of ψ, S_ψ. Thus

$$k(n) = S_\psi(n) = \sum_{j|n} \psi(j). \tag{7}$$

We will fix a particular choice of k in a moment.

Turning to $f(\mathbf{x})$, consider an arbitrary alternative setting of the variables, denoted $\hat{\mathbf{x}}$. We define a *test of settings* as follows: $\mathbf{v} : \mathcal{X} \times \mathcal{X} \times \mathcal{P}(\mathcal{V}) \to \mathcal{X}$

$$\mathbf{v}(\mathbf{x}, \hat{\mathbf{x}}, a) = (v_1 \ldots v_M) \text{ where} \tag{8}$$

$$v_i := \begin{cases} x_i & \text{if } \iota(a_i) | \iota(a) \\ \hat{x}_i & \text{otherwise.} \end{cases} \tag{9}$$

The vector $\mathbf{v}$ matches $\mathbf{x}$ on coordinate indexes that correspond to vertices in a and matches $\hat{\mathbf{x}}$ on the other coordinates. Thus the test $\mathbf{v}$ is not the usual notion of a mixture of values, rather it represents a mixture of settings. In this way $\mathbf{v}$ will permit measurement of the sensitivity of f to settings to groups of different entries in $\mathbf{x}$ outside of a given set. At the top level of the graph, $a = \mathcal{V}$, all such tests are exhausted, since we have $\mathbf{v}(\mathbf{x}, \hat{\mathbf{x}}, \mathcal{V}) = \mathbf{x}$, and therefore $f(\mathbf{v}(\mathbf{x}, \hat{\mathbf{x}}, \mathcal{V})) = f(\mathbf{x})$.

Now we are ready to set k concretely. For $a \subset \mathcal{V}$, let

$$k(\iota(a)) = \log_q f(\mathbf{v}(\mathbf{x}, \hat{\mathbf{x}}, a)).$$

At the top level of the graph, we have (using (7))

$$k(\iota(\mathcal{V})) = \log_q f(\mathbf{x}) = \log_q f(\mathbf{v}(\mathbf{x}, \hat{\mathbf{x}}, \mathcal{V})) = \sum_{j|\iota(\mathcal{V})} \psi(j).$$

Thus we have an expression for f as a sum-function. However, the summand in the last term is evaluated once for *each* subset of $\mathcal{V}$. Next we aim to show if j is not the index of a clique then $\psi(j)$ is zero and the sum can therefore be taken over cliques only.

Let $j = \iota(a)$ for $a \subset \mathcal{V}$, which is not a clique. Then a contains vertices a_m and a_n, which do not have an edge between them. Denote the remainder of the set, $c = a \setminus \{a_m, a_n\}$. Noting that $\iota(a) = p_m \cdot p_n \cdot \iota(c)$, the potential for this subset can be written as

$$\psi(\iota(a)) = \sum_{j|\iota(a)} (-1)^{\lambda\left(\frac{\iota(a)}{j}\right)} k(j)$$
$$= \sum_{j|\iota(c)} (-1)^{\lambda\left(\frac{\iota(a)}{j}\right)} (k(j) - k(j \cdot p_m) - k(j \cdot p_n) + k(j \cdot p_m \cdot p_n)). \quad (10)$$

It is sufficient to show that each term in the sum is zero. Recall the pairwise q-Markov condition (2) as it applies to $f(\boldsymbol{v})$. Let $\bar{\boldsymbol{v}} = \boldsymbol{v}_{\mathcal{V}\setminus\{a_m,a_n\}}$ correspond to $\boldsymbol{v}$ with the m- and n-th entries deleted. Since a_m and a_n are not connected, there exist functions h_1 and h_2 such that

$$\log_q f(\boldsymbol{v}(\boldsymbol{x},\widehat{\boldsymbol{x}},j))$$
$$= \begin{cases} h_1(\widehat{x}_m,\bar{\boldsymbol{v}}) + h_2(\widehat{x}_n,\bar{\boldsymbol{v}}) & \text{if NEITHER } p_m \text{ nor } p_n \text{ is a factor of } j \\ h_1(x_m,\bar{\boldsymbol{v}}) + h_2(\widehat{x}_n,\bar{\boldsymbol{v}}) & \text{if } p_m \text{ IS a factor of } j \text{ and } p_n \text{ is NOT a factor of } j \\ h_1(\widehat{x}_m,\bar{\boldsymbol{v}}) + h_2(x_n,\bar{\boldsymbol{v}}) & \text{if } p_m \text{ is NOT a factor of } j \text{ and } p_n \text{ IS a factor of } j \\ h_1(x_m,\bar{\boldsymbol{v}}) + h_2(x_n,\bar{\boldsymbol{v}}) & \text{if BOTH } p_m \text{ and } p_n \text{ are factors of } i. \end{cases}$$

In (10), we note that each of the conditions above holds exactly once. Therefore the summand is

$$k(j) - k(j \cdot p_m) - k(j \cdot p_n) + k(j \cdot p_m \cdot p_n)$$
$$= h_1(\widehat{x}_m,\bar{\boldsymbol{v}}) + h_2(\widehat{x}_n,\bar{\boldsymbol{v}}) - h_1(x_m,\bar{\boldsymbol{v}}) - h_2(\widehat{x}_n,\bar{\boldsymbol{v}})$$
$$\quad - h_1(\widehat{x}_m,\bar{\boldsymbol{v}}) - h_2(x_n,\bar{\boldsymbol{v}}) + h_1(x_m,\bar{\boldsymbol{v}}) + h_2(x_n,\bar{\boldsymbol{v}}) \quad (11)$$
$$= 0.$$

Since this is true for each term in (10), the proof is complete. ∎

The theorem specializes to the Hammersley Clifford (Theorem 3.9 Lauritzen [6]) exactly when f is assumed to be a probability distribution and $q = 1$.

Remark: A corollary to H-C reduces the decomposition to one over *maximal* cliques, since each clique is a subset of at least one maximal clique.

DISCUSSION

Other Graph Semantics. The proof technique of Theorem 1 can be adjusted to accommodate other graph semantics. One way is to replace $\log_q$ with a member of a larger class of *deformed logarithms*. A variety of non-SBG entropies studied in statistical physics fit into this framework [1]. The families of distributions associated with these entropies are known as phi-exponential families. Let $\phi : [0,\infty) \to [0,\infty)$ be strictly positive and non-decreasing on $(0,\infty)$. Define $\log_\phi$ via

$$\log_\phi(p) := \int_1^p \frac{1}{\phi(y)}\, dy \quad (12)$$

If this integral converges for all finite $p > 0$, then $\log_\phi$ is called a *deformed logarithm*. Note that $\phi(p) = p^q$ recovers $\log_q$.

Dynamic Programming. The method of optimization by combining optimal solutions to simpler subproblems, is often called dynamic programming. The key step to use dynamic programming involves expressing the desired computation as a sum of products in a semi-ring, after which the distributive law can be used to diminish the required number of multiplications [9]. The choice of semi-ring corresponds to the task being performed with the graph. The goal of the Viterbi algorithm is to find the most likely setting of variables, given a joint distribution. For that task it performs dynamic programming using the max-product semi-ring. The q-product discussed here is compatible with a Viterbi-like algorithm since q-factorization is compatible with max i.e., $a \otimes_q \max(b,c) = \max(a \otimes_q b, a \otimes_q c)$. This is true since if $z > 0$, then $x \leq y$ implies $z \otimes_q x \leq z \otimes_q y$.

Conclusion. The theory that has been presented in this article suggests the possibility of using non-exponential-family distributions, such as those derived via generalized maximum entropy, in conjunction with graphical models. To employ these distributions with graphical models requires reinterpretation of the information encoded in the graph adjacency structure and the node distributions, away from conditional independence, and the marginals of the joint distribution, respectively. We believe that in some applications, this may correspond to the type of prior information that is available. In the future we hope to extend the idea to utilize more of the techniques of graphical models and to address the issue of updating.

ACKNOWLEDGMENTS

Thanks to Vishy Vishwanathan, Tiberio Caetano, Alex Smola and Ariel Caticha for discussions and comments.

REFERENCES

1. J. Naudts, *Journal of Inequalities in Pure and Applied Mathematics* **5** (2004).
2. T. D. Sears, and S. Vishwanathan, Fenchel duality and generalized maxent, in submission (2007).
3. M. Gell-Mann, and C. Tsallis, editors, *Nonextensive Entropy*, Sante Fe Institute Studies in the Sciences of Complexity, Oxford University Press, 2004.
4. H. Suyari, M. Tsukada, and Y. Uesaka, "Mathematical Structures derived from the q-product uniquely determined by Tsallis entropy," in *ISIT 2005*, 2005.
5. S. Umarov, C. Tsallis, and S. Steinberg, *Arxiv preprint cond-mat/0603593* (2006).
6. S. Lauritzen, *Graphical models: Clarendon Press*, Clarendon Press, 1996.
7. H. S. Wilf, *generatingfunctionology*, Academic Press, 1994.
8. R. Graham, D. Knuth, and O. Patashnik, *Concrete mathematics*, Addison-Wesley Reading, Mass, 1989.
9. S. M. Aji, and R. J. McEliece, *IEEE Transactions on Inforamtion Theory* **46**, 325–343 (2000).

Origins of the Combinatorial Basis of Entropy

Robert K. Niven

(1) School of Aerospace, Civil and Mechanical Engineering, The University of New South Wales at ADFA, Canberra, ACT, 2600, Australia. Email: r.niven@adfa.edu.au
(2) Niels Bohr Institute, University of Copenhagen, Copenhagen Ø, Denmark.

Abstract. The combinatorial basis of entropy, given by Boltzmann, can be written $H = N^{-1} \ln \mathbb{W}$, where H is the dimensionless entropy, N is the number of entities and $\mathbb{W}$ is number of ways in which a given realization of a system can occur (its statistical weight). This can be broadened to give generalized combinatorial (or probabilistic) definitions of entropy and cross-entropy: $H = \kappa(\phi(\mathbb{W}) + C)$ and $D = -\kappa(\phi(\mathbb{P}) + C)$, where $\mathbb{P}$ is the probability of a given realization, ϕ is a convenient transformation function, κ is a scaling parameter and C an arbitrary constant. If $\mathbb{W}$ or $\mathbb{P}$ satisfy the multinomial weight or distribution, then using $\phi(\cdot) = \ln(\cdot)$ and $\kappa = N^{-1}$, H and D asymptotically converge to the Shannon and Kullback-Leibler functions. In general, however, $\mathbb{W}$ or $\mathbb{P}$ need not be multinomial, nor may they approach an asymptotic limit. In such cases, the entropy or cross-entropy function can be *defined* so that its extremization ("MaxEnt" or "MinXEnt"), subject to the constraints, gives the "most probable" ("MaxProb") realization of the system. This gives a probabilistic basis for MaxEnt and MinXEnt, independent of any information-theoretic justification.

This work examines the origins of the governing distribution $\mathbb{P}$. These include: (a) frequentist-like models; (b) symmetry models; (c) prior MinXEnt models; (d) Kapur-Kesavan inverse models; and (e) game theoretic models. The combinatorial definition and MaxProb are consistent with these different approaches, and the notion of probabilistic inference, yet offer greater utility than traditional MaxEnt / MinXEnt based on the Shannon and Kullback-Leibler functions.

Keywords: MaxEnt; MaxProb; Boltzmann principle; combinatorial; probabilistic inference
PACS: 02.50.Cw, 02.50.Tt, 05.20.-y, 05.70.-a, 05.90.+m, 89.20.-a, 89.70.+c

1. INTRODUCTION

Fifty years ago, Jaynes [1] gave the maximum entropy method (MaxEnt), based on the Shannon entropy [2]:

$$H_{Sh} = -\sum_{i=1}^{s} p_i \ln p_i \qquad (1)$$

where p_i is the (posterior) probability of occurrence of the ith distinguishable state within a system, from s such states. In the MaxEnt method, one maximizes the Shannon entropy of a system, subject to its constraints, to determine the "least informative" or "maximally noncommittal" probability distribution representing the system. From its inception, MaxEnt was advanced as a generic method of inference for the solution of indeterminate problems of all kinds, underpinned by information theory, not merely as an extension of mechanics [1, 3–6]. MaxEnt was later extended into the maximum relative entropy, minimum divergence or minimum cross-entropy method (MinXEnt), involving extremization of the Kullback-Leibler measure [7, 8]:

$$D_{KL} = \sum_{i=1}^{s} p_i \ln \frac{p_i}{q_i} \qquad (2)$$

which allows for unequal prior probabilities q_i. Since that time, MinXEnt and its subsidiary MaxEnt have been successfully applied to the analysis of a vast number of phenomena, throughout most fields of human study [e.g. 6, 9–11], and can rightly be regarded as one of the most important of all human discoveries.

It must be emphasised, however, that the cross-entropy and entropy concepts which underpin MinXEnt and MaxEnt are themselves subject to many different philosophical interpretations. Dominant explanations include the axiomatic basis outlined by Shannon [2], and the information-theoretic ("bits" of information) and coding basis, recognized by Szilard [12] and Shannon [2] [c.f. 13]. These bases led Jaynes, in particular, to consider the Shannon and Kullback-Leibler functions to be the only logically consistent measures of uncertainty, and thus the only ones suitable for analysis. This view has been challenged by many researchers, on the grounds that the above two measures are too narrowly defined and/or inapplicable to many situations. For example, over the past 85 years, many alternative entropy and divergence functions have been introduced [e.g. 14–30]; in most cases, these are incompatible with the Shannon and Kullback-Leibler functions, but have proved *useful* for the analysis of specific classes of systems. Can such measures be explained by some broader philosophical framework? How should we choose the "correct" cross-entropy or entropy function for a given problem? The fact that such questions remain unanswered indicates the need for a unifying philosophical framework, which encompasses (and *explains*) such alternative entropy measures and their connections to information theory.

This study examines one such framework: the combinatorial (or probabilistic) basis of entropy, first given 130 years ago by Boltzmann [31] and subsequently promoted by Planck [32]. This involves the maximization of a governing probability distribution $\mathbb{P}$ or weight $\mathbb{W}$ of a system; this can be viewed as a generalized principle of probabilistic inference, aptly described by Vincze and Grendar & Grendar as the maximum probability ("MaxProb") principle [33, 34]. It also leads to generalized definitions of cross-entropy and entropy, based purely on probability theory [35]. In this study, specific attention is paid to the origins of the governing distribution $\mathbb{P}$, including (a) frequentist-like models (e.g. ball-in-box or urn models); (b) symmetry models; (c) prior MinXEnt models; (d) Kapur-Kesavan inverse models; and (e) game theoretic models. It is shown that the combinatorial basis is consistent with these different approaches, but is more soundly based and offers greater utility than traditional MaxEnt / MinXEnt based on the Shannon and Kullback-Leibler functions.

2. THE COMBINATORIAL BASIS

Owing to a tremendous confusion in terminology - especially amongst physicists - it is first necessary to rigorously define several important terms [c.f. 35]. An *entity* is here taken to be a discrete particle, object or agent within a system, which acts separately but not necessarily independently of the other entities present. A *system* is a collection of entities with a defined boundary, subject to various constraints, which may or may not

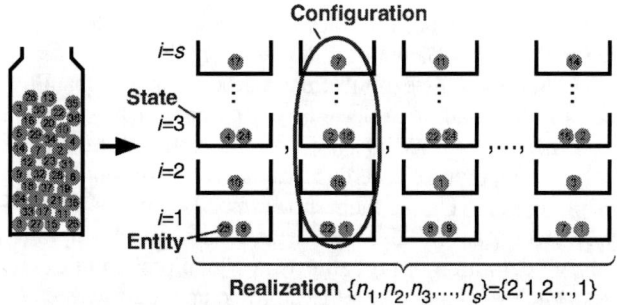

FIGURE 1. Definition of terms used in the combinatorial basis of entropy and cross-entropy.

be open to the exchange of specified entities or substances with an external environment. The entity therefore constitutes the unit of analysis of a system.

Now consider a simple "ball-in-box" model of a system, shown in Figure 1, in which N distinguishable entities (balls) are allocated to s distinguishable non-degenerate states (boxes). As shown:

- A *state* refers to each different category or element of system (e.g. energy levels, sides of a die or alphabetic symbols). The states are therefore properties of, or associated with, each individual entity in the system.
- A *configuration* is a distinguishable permutation or pattern of entities amongst the states of a system (a *complexion*, *microstate* or *sequence*). A configuration is therefore a property of the system as a whole.
- A *realization* is each aggregated arrangement of entities amongst the states of a system, as specified by some rule, for example by the number of entities in each state (a *macrostate*, *outcome* or *type*). In general, a realization will constitute a set of configurations, since several configurations could give the same realization (see Figure 1).

There is such confusion in and sloppy usage of the terms *state*, *microstate* and *macrostate* - severely impairing understanding - that the last two terms should be avoided. In the following, the states are indexed $i = 1,...,s$ (which may be multivariate); n_i denotes the number of entities in the ith state; q_i and $p_i = n_i/N$ respectively denote the prior and posterior probabilities of a entity being in the ith state; and each realization[1] is denoted $\{n_i\}$. Notwithstanding other philosophical differences with Jaynes, the "subjective Bayesian" definition of probabilities, as assignments based on what we know, is adopted here [1, 6].

For the analysis of probabilistic systems, it is possible to delineate a principle which stands out from all others: the *maximum probability* ("MaxProb") principle [31–35]. This can be stated as:

"*A system can be represented by its realization of highest probability.*"

---

[1] A realization can only be denoted $\{p_i\}$ in the asymptotic limits $N \to \infty$ and $n_i \to \infty, \forall i$, since $\{p_i\}$ discards information about the value of N.

This seemingly trivial statement provides a powerful principle for *probabilistic inference*, which is independent of any information-theoretic considerations. This is critical, since in any contradiction between information theory and probability theory - for example, between the distributions inferred by each approach - probability theory must triumph. Like MinXEnt or MaxEnt based on the Kullback-Leibler or Shannon measures, MaxProb is a method of inference (inductive reasoning), which does not give certainty in its predictions. Unlike them, however, MaxProb is founded solely on probability theory. Indeed, MaxProb does not depend upon any asymptotic limits (a feature of the "frequentist" definition of probability, in which probabilities must correspond to measurable frequencies [1, 6]); it can therefore be applied to systems containing finite numbers of entities [29, 30].

Allied to MaxProb is a generalized form of the second law of thermodynamics:

"A system will tend towards its most probable realization."

This provides a purely probabilistic rationale for use of the MaxProb principle, independent of thermodynamics. In effect, if we adopt MaxProb as a principle for probabilistic inference, the above statement is its corresponding ergodic principle, which (on average) explains its success. Of course - as expressed by Jaynes [1] - the concept of ergodicity is not needed for the purpose of inference, since in the absence of other information, we are fully justified in conducting inference without it.

The MaxProb principle also leads to the *combinatorial definition* of entropy, first given by Boltzmann [31] and Planck [32]. This can be written as:

$$H = N^{-1} \ln \mathbb{W}, \tag{3}$$

where $\mathbb{W}$ is the number of ways in which a given realization can occur, referred to as its statistical weight. Maximization of the entropy H of a system, subject to its constraints, therefore selects the realization of highest weight $\mathbb{W}$ (the logarithmic function being a monotonic transformation, which does not alter the position of the extremum). Eq. (3) can be extended to give generalized combinatorial (or probabilistic) definitions of cross-entropy and entropy [35]:

$$D = -\kappa(\phi(\mathbb{P}) + C), \qquad H = \kappa(\phi(\mathbb{W}) + C), \tag{4}$$

where $\mathbb{P} = P(\{n_i\} | \{q_i\}, N, s, I)$ is the probability of a given realization, subject to the prior probabilities $\{q_i\}$, number of entities N, number of states s and background information I; ϕ is a convenient monotonic transformation function; κ is a scaling parameter; and C is an arbitrary constant. This perspective is summarised in Figure 2. If $\mathbb{P}$ or $\mathbb{W}$ satisfy the multinomial distribution or weight:

$$\mathbb{P} = N! \prod_{i=1}^{s} \frac{q_i^{n_i}}{n_i!}, \qquad \mathbb{W} = N! \prod_{i=1}^{s} \frac{1}{n_i!}, \tag{5}$$

then by taking $\phi(\cdot) = \ln(\cdot)$, $\kappa = N^{-1}$ and the asymptotic limits $N \to \infty$ and $n_i \to \infty, \forall i$ (the "Stirling approximation"), D and H converge respectively to the Kullback-Leibler and Shannon functions (1)-(2) [33, 34]. This provides a (well-known) justification for these functions, and their corresponding MinXEnt and MaxEnt principles, as a special case, independently of the arguments used in information theory.

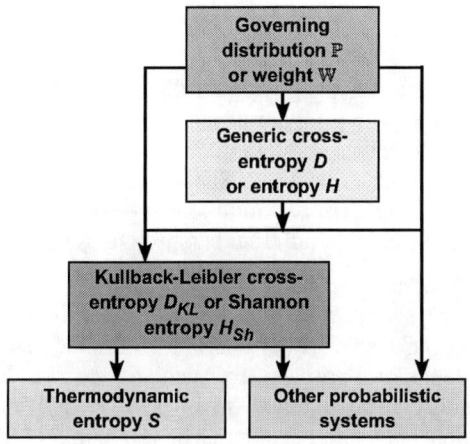

FIGURE 2. Schematic flowchart of the combinatorial basis of entropy and cross-entropy.

In general, however, $\mathbb{P}$ or $\mathbb{W}$ need not be multinomial, nor may they approach an asymptotic limiting form. In such cases, extremization of the cross-entropy or entropy defined by (4), subject to the constraints, gives the most probable (MaxProb) realization of the system (in the non-asymptotic case, due to the effect of quantization, extremization gives an "attractor" distribution which lies close to but not necessarily equal to the MaxProb realization [35]). In consequence, the combinatorial definitions (4) remain consistent with the rules of probability theory, whilst inference using the Kullback-Leibler or Shannon measures may lead to inconsistencies. The combinatorial (or probabilistic) definitions are therefore more broadly applicable than those derived from information-theoretic considerations.

In the foregoing discussion, the astute reader will notice that there may be many different ways to classify the entities and states of a system, and hence to identify its configurations; and many different ways to group the configurations into realizations. We are therefore led to the "subjective" (or "observer-dependent") view of the entropy and cross-entropy concepts, a sentiment vocally defended by Jaynes [1]. This was succinctly expressed by Tseng and Caticha [36]:

"Entropy is not a property of a system ... [it] is a property of our description of a system."

The fact that the thermodynamic entropy S is always defined in the same manner, allowing thermodynamicists to make consistent calculations, should not fool the reader into believing that the entropy concept is "objective"[2].

---

[2] It is important that the symbol S be devoted exclusively to the thermodynamic entropy, since it is a special case of - but is distinct from - the dimensionless Shannon entropy (1).

3. ORIGINS OF THE GOVERNING DISTRIBUTION

We now consider the origins of the governing distribution $\mathbb{P}$ or weight $\mathbb{W}$ used in the combinatorial formulation. The problem of justifying a cross-entropy or entropy function is now replaced by a deeper problem, of how to justify its governing distribution. The ubiquity of the Kullback-Leibler and Shannon measures, in many circumstances, therefore leads to the question: why the multinomial distribution? This question, and the choice of $\mathbb{P}$, is examined from five different perspectives.

(a) Frequentist-like Models

In this approach, one simply asserts a governing distribution $\mathbb{P}$ or weight $\mathbb{W}$ as a probabilistic model of the system under consideration. One may have strong grounds, based on prior knowledge of a problem, for such an assertion; in any case, we should have no "fear of failure" of this method (in Jaynes' words), since if the model gives unsuccessful predictions, we have learnt that it is incorrect. Many such models are available from classical probability theory, for example the *ball-in-box* models of the type represented in Figure 1. Since these arose from frequentist studies, they can be termed "frequentist-like" models, although used here for the purpose of inference.

In the case discussed previously, in which distinguishable balls are allocated to distinguishable boxes in accordance with a set of constant prior probabilities (see Figure 1), one obtains the multinomial distribution (5), and hence the Kullback-Leibler cross-entropy and Shannon entropy functions in the Stirling limits. However, different assumptions lead to different model distributions. If the asymptotic limits are not applied, then from (5), one obtains a non-asymptotic cross-entropy function [c.f. 29]:

$$-D_{KL}^{x} = N^{-1}\ln\mathbb{P} = N^{-1}\left\{\ln N! + \sum_{i=1}^{s} n_i \ln q_i - \sum_{i=1}^{s} \ln n_i!\right\}$$
$$= \sum_{i=1}^{s}\left\{p_i N^{-1}\ln N! + p_i \ln q_i - N^{-1}\ln[(p_i N)!]\right\} \tag{6}$$

This is applicable to systems with finite (small) N. Minimisation of D_{KL}^{x}, subject to the usual constraints $\sum_{i=1}^{s} n_i = N$ and $\sum_{i=1}^{s} n_i f_{ri} = N\langle f_r \rangle$, for $r = 1,...,R$, where f_{ri} is the rth function of each state i and $\langle f_r \rangle$ is its mathematical expectation, gives the "most probable" distribution [c.f. 29]:

$$p_i^{\#} = N^{-1}\left[\psi^{-1}\left(N^{-1}\ln N! + \ln q_i - \lambda_0 - \sum_{r=1}^{R}\lambda_r f_{ri}\right) - 1\right] \tag{7}$$

where $\psi^{-1}(\cdot)$ is the inverse digamma function. Eq. (7) can be viewed as the "attractor" for systems with finite N, which differs from the attractor given by traditional MinXEnt.

If the states are considered to contain g_i distinguishable, degenerate sub-states within each distinguishable state i, then three cases have been examined historically: (i) distinguishable entities; (ii) indistinguishable entities; and (iii) indistinguishable entities, with a maximum of one entity in each state. The resulting distributions were given by Brillouin [37, 38] as, respectively:

$$\mathbb{P}_{MB} = \frac{N!}{G^N} \prod_{i=1}^{s} \frac{g_i^{n_i}}{n_i!}, \tag{8}$$

$$\mathbb{P}_{BE} = \frac{N!(G-1)!}{(G+N-1)!} \prod_{i=1}^{s} \frac{(g_i+n_i-1)!}{n_i!(g_i-1)!}, \tag{9}$$

$$\mathbb{P}_{FD} = \frac{N!(G-N)!}{G!} \prod_{i=1}^{s} \frac{g_i!}{n_i!(g_i-n_i)!}. \tag{10}$$

where $G = \sum_{i=1}^{s} g_i$ is the total degeneracy. The truncated weights and entropy functions corresponding to these distributions, referred to respectively as the Maxwell-Boltzmann, Bose-Einstein and Fermi-Dirac distributions respectively [e.g 16–20, 37–41], played an important role in the development of quantum theory. In the non-asymptotic case, the resulting entropy functions appear to have profound information-theoretic consequences [29, 30].

Recently, a quite different ball-in-box model was considered, in which distinguishable entities are allocated to indistinguishable, equally degenerate states. The statistical weight of each realization $\{n_i\}$ can be expressed as [42]:

$$\mathbb{W}_{D:I(g)} = \frac{N!}{\left(\prod_{i=1}^{k} n_i!\right)\left(\prod_{j=1}^{N} r_j!\right)} \prod_{i=1}^{k} \sum_{\gamma=1}^{\min(g,n_i)} \left\{ {n_i \atop \gamma} \right\} \tag{11}$$

where there are k non-empty states amongst the s states; g is the degeneracy of each state; $\left\{ {n_i \atop \gamma} \right\}$ is a Stirling number of the second kind; and r_j is the number of occurrences of integer j in the set $\{n_i\}$. The combinatorial entropy corresponding to (11), $H_{D:I(g)} = N^{-1} \ln \mathbb{W}_{D:I(g)}$, does not appear to have a straightforward asymptotic form, except in the non-degenerate case $g = 1$ with $k = s$, when it reduces to the Shannon entropy.

Closely related to but distinct from ball-in-box models are *urn models*, in which a container (urn) is set up with a total of M balls, made up of m_i balls of each color i. Balls are then drawn from the urn in accordance with some sampling scheme, recorded and returned to the urn (or the urn modified in some way), and the sampling repeated [c.f. 43, 44]. The asymptotic limits of an infinitely large urn ($M \to \infty$ and $m_i \to \infty, \forall i$), and an infinitely large (smaller) sample ($N \to \infty$ and $n_i \to \infty, \forall i$), are usually applied. Although quite different to the ball-in-box model of Figure 1, an urn model with simple replacement also yields the multinomial distribution [43, 44]. Urn models involving the drawing of balls without replacement, or double replacement, lead respectively to the Fermi-Dirac and Bose-Einstein distributions [43]. Urn models also readily permit the construction of systems in which the prior probabilities are not independently and identical distributed (non-*iid* sampling): e.g. the Pólya distribution, in which after every draw, the ball is returned, and c balls of the same color are also added [45–48]:

$$\mathbb{P}_{Polya} = \frac{N!}{\prod_{i=1}^{s} n_i!} \prod_{i=1}^{s} \frac{m_i(m_i+c)\ldots(m_i+(n_i-1)c)}{M(M+c)\ldots(M+(N-1)c)}, \tag{12}$$

Substituting the *initial* prior probabilities $q_i = m_i/M$ and parameter $\beta = N/M$, this gives analytic cross-entropy measures in the non-asymptotic and asymptotic cases [48]. The resulting "most probable" distribution is intermediate between the Bose-Einstein and Fermi-Dirac distributions, with physical applications.

(b) Symmetry-Based Arguments

One may also choose a governing distribution on the basis of symmetry arguments (related to the "principle of insufficient reason"). For a system made up of tosses of a coin, it is rational to consider the sampling to follow the binomial distribution, with equal prior probabilities of $\frac{1}{2}$ for each face, due to the symmetry of the states (there being no information to suggest that one state should be preferred). Alternatively, as suggested by David Blower at MaxEnt07, one can obtain a binomial distribution by the symmetry of all possible models in the model space (assigning a uniform prior to the models, over the entire spectrum from an all-head to an all-tail model, there being no reason to prefer any model). Applied to systems with more than two states, either argument leads to the multinomial distribution. In this respect, the multinomial distribution plays a role somewhat analogous to a central limit theorem (a "central model theorem"), a point which deserves greater mathematical attention; this may be the reason for the ubiquity of the Kullback-Leibler and Shannon measures. Without symmetry, however, the argument breaks down, and one must adopt some other method to identify the governing distribution.

(c) Prior MinXEnt Models

A third origin of the governing distribution $\mathbb{P}$ is as a result of the application of MinXEnt at a higher level, for example to the set of systems within which the actual system resides. For example, the multinomial distribution can be obtained by MinXEnt based on the Kullback-Leibler cross-entropy, subject to a multinomial prior and mean constraints on each variate [11]. This might then be imposed as a lower-level governing distribution. One can in fact envisage a hierarchy of governing and "most probable" distributions, at different levels of description. In a complex system, in which there is bidirectional feedback, the result will be a mosaic of interconnected probabilistic models (with thanks to the discussion by Tony Bell at MaxEnt07).

(d) Kapur-Kesavan Inverse Models

The governing distribution $\mathbb{P}$ can also be obtained by extension of the arguments of Kapur and Kesavan [11], in which one works backwards from an observed probability distribution $\{p_i^*\}$, prior distribution $\{q_i\}$ (if available) and any constraints, to derive the measure of cross-entropy or entropy applicable to a system. By unravelling of the asymptotic limits, this could (at least in principle) be extended to determine the governing distribution of the system. This avenue of research has not been examined in detail, and deserves greater attention.

(e) Game Theoretic Models

The final method considered here is to derive the governing distribution of a system by analysis of a code-length game between the system ("Nature") and the observer [49, 50]. For a multivariate system of *iid* random variables, which take discrete values, this yields the multinomial distribution at game-theoretic equilibrium [49]. As in case (c), this could then be imposed as the governing distribution at a lower level of description.

4. CONCLUSIONS

This study examines the MaxProb principle, in which a system is represented by its distribution of highest probability. This can be interpreted as a generalized method of probabilistic inference, which does not provide certainty in its predictions, yet is always consistent with the rules of probability theory. In contrast, inference using the Kullback-Leibler cross-entropy or Shannon entropy functions, in cases in which the governing distribution is not multinomial and/or does not satisfy the asymptotic limits, can lead to inconsistencies. The MaxProb principle also gives rise to generalized combinatorial definitions of cross-entropy and entropy, an extension of the idea given by Boltzmann 130 years ago. The cross-entropy or entropy can therefore be *defined* so that its extremization, subject to the constraints, gives the "most probable" ("MaxProb") realization of the system. This provides a purely probabilistic basis for MaxEnt and MinXEnt, which is independent of any information-theoretic justification.

This work examines the origins of the governing distribution $\mathbb{P}$, including by (a) frequentist-like models (e.g. ball-in-box or urn models); (b) symmetry models; (c) prior MinXEnt models; (d) Kapur-Kesavan inverse models; and (e) game theoretic models. It is shown that the combinatorial definition and MaxProb are consistent with these different approaches, and the "subjective Bayesian" definition of probability, yet is more broadly based and offers greater utility than traditional MaxEnt / MinXEnt based on the Shannon and Kullback-Leibler functions.

ACKNOWLEDGMENTS

The author thanks the European Commission for support as a Marie Curie Incoming International Fellow; Marian Grendár, David Blower, Tony Bell and Flemming Topsøe for specific arguments (as detailed above) and stimulating discussions; and the organisers and participants of MaxEnt07.

REFERENCES

1. E.T. Jaynes, Phys. Rev. 106 (1957) 620.
2. C.E. Shannon, Bell Sys. Tech. J. 27 (1948) 379; 623.
3. E.T. Jaynes, in K.W. Ford (ed.), Brandeis University Summer Institute, Lectures in Theoretical Physics, Vol. 3, Benjamin-Cummings Publ. Co. (1963) 181.
4. E.T. Jaynes, IEEE Trans. Systems Science and Cybernetics SSC-4 (1968) 227.
5. E.T. Jaynes, in R.D. Levine, M. Tribus (eds.), The Maximum Entropy Formalism, MIT Press, Cambridge, MA (1978) 15.
6. E.T. Jaynes, (G.L. Bretthorst, ed.) Probability Theory: The Logic of Science, Cambridge U.P., Cambridge, 2003.
7. S. Kullback, R.A. Leibler, Annals Math. Stat. 22 (1951) 79.
8. S. Kullback, Information Theory and Statistics, John Wiley, NY, 1959.
9. R.D. Levine, M. Tribus (eds.), The Maximum Entropy Formalism, MIT Press, Cambridge, MA, 1978.
10. J.N. Kapur, Maximum-Entropy Models in Science and Engineering, John Wiley, NY, 1989.
11. J.N. Kapur, H.K. Kesavan, Entropy Optimization Principles with Applications, Academic Press, Inc., Boston, MA, 1992.

12. L. Szilard, Zeitschrift für Physik 53 (1929) 840.
13. J.E. Shore, R.W. Johnson, IEEE Trans. Information Theory IT-26(1) (1980) 26.
14. R.A. Fisher, Philos. Trans. Royal Soc. London A 222 (1922) 309.
15. R.A. Fisher, Proc. Camb. Philos. Soc. 22 (1925) 700.
16. S.N. Bose, Z. Phys. 26 (1924) 178.
17. A. Einstein, Sitzungsber. Preuss. Akad. Wiss. Phys. Math. Kl (1924) 261.
18. A. Einstein, Sitzungsber. Preuss. Akad. Wiss. Phys. Math. Kl (1925) 3.
19. E. Fermi, Z. Phys. 36 (1926) 902.
20. P.A.M. Dirac, Proc. Roy. Soc. 112 (1926) 661.
21. A. Rényi, Proc. 4th Berkeley Symp. Math. Stat. and Prob. 1 (1961) 547.
22. B.D. Sharma, D.P. Mittal, J. Math. Sci. (Calcutta) 10 (1975) 28.
23. B.D. Sharma, D.P. Mittal, J. Combinat. Inform. Sys. Sci. 2 (1977) 122.
24. C. Tsallis, J. Stat. Phys. 52(1/2) (1988) 479.
25. C. Tsallis, *in* S. Abe, Y. Okamato (eds.), Nonextensive Statistical Mechanics and its Applications, Springer, Berlin, (2001) 3.
26. G. Kaniadakis, Physica A 296(3-4) (2001) 405.
27. G. Kaniadakis, Phys. Rev. E 66(5) (2002) 056125.
28. C. Beck, E.G.D. Cohen, Physica A 322 (2003) 267.
29. R.K. Niven, Phys. Lett. A 342(4) (2005) 286.
30. R.K. Niven, Physica A 365(1) (2006) 142.
31. L. Boltzmann, Wiener Berichte, 76 (1877) 373-435.
32. M. Planck, Annalen der Physik 4 (1901) 553.
33. I. Vincze, Progress in Statistics, 2 (1974) 869-895.
34. M. Grendár, Jr. and M. Grendár, *in* Bayesian Inference and Maximum Entropy Methods in Science and Engineering, A. Mohammad-Djafari (ed.), AIP, Melville (2001) 83-94.
35. R. K. Niven, *cond-mat/0512017*, 2005-2007.
36. C.-Y. Tseng, A. Caticha, Yet another resolution of the Gibbs paradox: an information theory approach, preprint (2002).
37. L. Brillouin, Annales de Physique 7 (1927) 315.
38. L. Brillouin, Les Statistiques Quantiques et Leurs Applications, Les Presses Universitaires de France, Paris, 1930.
39. R.C. Tolman, The Principles of Statistical Mechanics, Oxford Univ. Press, London, 1938.
40. L. Brillouin, J. Appl. Phys. 22(3) (1951) 338.
41. N. Davidson, Statistical Mechanics, McGraw-Hill, NY, 1962.
42. R.K. Niven, CTNEXT07, Catania, Sicily, Italy, July 2007, in submission to AIP.
43. D.R. Jensen, *in* S. Kotz, N.L. Johnson, Encyclopedia of Statistical Sciences, 6: 5200.
44. S. Berg, Urn Models, *in* S. Kotz, N.L. Johnson, *Encyclopedia of Statistical Sciences*, 9: 424.
45. F. Eggenberger, G. Pólya, Über die Statistik verketter Vorgänge Z. Angew. Math. Mech., 1 (1923) 279-289.
46. H. S. Steyn, Proc. Koninklijke Nderlandse Akademie van Wetenschappen, Ser. A, 54 (1951) 23-30.
47. N. L. Johnson, S. Kotz, N. Balakrishnan, Discrete Multivariate Distributions. New York: Wiley, 1997.
48. M. Grendar, R.K. Niven, *cond-mat/0612697*, 2006.
49. F. Topsøe, IEEE Trans. Info. Theory 48(8) (2002) 2368.
50. F. Topsøe, Physica A 340 (2004) 11.

On Shannon-Jaynes Entropy and Fisher Information

Vesselin I. Dimitrov[1]

Idaho Accelerator Center, Idaho State University
1500 Alvin Ricken Dr.
Pocatello, ID 83201, USA

Abstract. The fundamentals of the Maximum Entropy principle as a rule for assigning and updating probabilities are revisited. The Shannon-Jaynes relative entropy is vindicated as the optimal criterion for use with an updating rule. A constructive rule is justified which assigns the probabilities least sensitive to coarse-graining. The implications of these developments for interpreting physics laws as rules of inference upon incomplete information are briefly discussed.

INTRODUCTION

Motivation

Ever since E. Jaynes formulated the Maximum Entropy (MaxEnt) principle and used it to derive Statistical Thermodynamics [1], questions of the following type have developed significance and have been lingering around: To what extend other physics laws can be understood as rules of inference? How cheap can mechanics' first principles be [3]? How much of physics is just computation [2]? Are predictions of physics theories just those containing the least amount of information compatible with the descriptive framework used [4]? Is Quantum Mechanics the only consistent way to manipulate probability amplitudes [5]? And, last but not least, can *all of physics* be derived from information-theoretic principles [6]? In order to be able to even start thinking of answering these and similar questions, one needs a firm conceptual ground allowing unambiguous understanding of the MaxEnt principle itself. Unfortunately, we appear to still lack such an understanding [7]. Therefore, in the present work we set out to revise the fundamentals of the Maximum Entropy paradigm.

The Paradigm and Its Shortcomings

Constructive rule (MaxEnt)

In its original scope [1] the MaxEnt principle is a constructive one - a device for assigning probabilities based on incomplete information. Probability assignments are thought to be adequate representation of one's state of knowledge about the system of interest, and the incomplete information is usually obtained by performing

---

[1] dimivess@isu.edu

measurements on that system. The ultimate goal is to be able to predict the results of all possible measurements on the system from the outcomes of a finite, preferably small, number of measurements. This of course is not a well-posed problem in the usual sense, so the answer is sought in the form of probability distribution compatible with the measurements but otherwise as undedicated as possible. In order to put operational meaning into the vague notion of "as undedicated as possible" it is necessary to agree on a measure of information content (or the lack thereof) in a probability distribution. Once such a measure is selected, "most undedicated" translates into the "one with the least information content" compatible to the available information. For such a measure of the lack of information Jaynes chose to use the Shannon's entropy of a probability distribution $\{p_i, i = 1, \cdots, N\}$, namely $S[p] = -\sum_{i=1}^{N} p_i \ln p_i$, therefore we refer to this form, as well as to its generalizations in what follows, as the *Shannon-Jaynes (relative) entropy*. This choice is by no means unambiguous. Indeed, its usual justification is based on requiring the measure to obey a set of axioms and proving an existence and uniqueness theorem [8,9]. Such an axiomatic approach can be challenged on two counts - the plausibility of one or more of its axioms and the completeness of the proof that the proposed solution is unique. Unfortunately, the Shannon-Jaynes entropy characterization is susceptible to both these challenges. First, neither Shannon nor Khinchin made a very good point demonstrating that their distributivity axiom represents an absolutely necessary property of the measure, as discussed e.g. by Renyi [10]. Second, in proving the uniqueness of the measure Shannon apparently overlooks the circumstance that his argument holds for an arbitrary non-negative base of the logarithm function, including the possibility that different bins of the probability domain have different bases which may even depend on the bin's probability[2]. Thus a more general expression of the form

$$S[p] = -\sum_{i=1}^{N} p_i \log_{\alpha_i} p_i = -\sum_{i=1}^{N} p_i \ln \frac{p_i}{m_i},$$

with $m_i = p_i^{1-\ln \alpha_i}$ playing the role of apriori weights, results from the axioms. Introducing $\{m_i\}$ in place of the logarithm bases $\{\alpha_i\}$ merely facilitates the interpretation of these parameters as bin weights. That such weights are necessary becomes particularly obvious when one tries to pass to the limit of a continuous domain, and their arbitrariness forces a change on the interpretation of the MaxEnt prescription from a rule for assigning probabilities to a rule for updating probabilities.

Updating rule (M.E.)

Being interested mainly in the physics applications of the MaxEnt principle, in what follows we will consider probability distributions with continuous domains. In contrast to Shannon's original expression, the passage to continuum in the above formula is straightforward and results in a change-of-variable invariant measure

$$S[p] = -\int dx p(x) \ln \frac{p(x)}{m(x)}$$

---

[2] Khinchin, on the other hand, explicitly requires that the entropy is maximal for uniform probabilities, which effectively invokes the notorious Principle of Insufficient Reason and is open to all well-known objections to it [Ufink'95].

where $p(x)dx$ is the probability of $x \in [x, x+dx)$ and $m(x) > 0$. This is but the negative of the Kullback-Leibler (KL) distance between the probability distributions $p(x)$ and $m(x)$. In the new interpretation of the MaxEnt principle as an updating rule, which, following [11] I'll call "M.E.", one updates $m(x)$ to $p(x)$ by maximizing $S[p]$ (or, equivalently, minimizing the KL distance between m and p) subject to constraints representing new information. In an attempt to circumvent the shortcomings of the Shannon's argument in the continuous case, Shore and Johnson [12] set out to derive an updating rule by imposing uniqueness and consistency requirements not on the form of the measure but directly on the outcome of the procedure. Although their system of axioms includes a requirement equivalent to simple additivity of the measure for independent subsystems, they surprisingly derived the Shannon-Jaynes form as the unique measure suitable for using in an updating rule. The surprise comes from the fact that, 20 years after Renyi published his paper, it should have been fairly well-known that simple additivity alone was too weak a condition to single out the Shannon-Jaynes relative entropy. Closer examination of Shore and Johnson's proof identifies a statement at the very end of their second Appendix where they argue that two functions, F and H, are equivalent inasmuch as the one is a monotonic function of the other, so one can extremize either one, and they restrict their further considerations to F. What they apparently fail to recognize is that imposing the additivity requirement on F is not the same as imposing it on H and this is exactly what prevents them from obtaining the Renyi's family of relative entropies as further solution of their problem along with the Shannon-Jaynes relative entropy.

Vàn's derivation

Recently P.Vàn proposed a derivation of a constructive rule for the continuous case [13] where he allowed the information measure to depend on the first derivative of the probability density. This kind of dependence is a commonplace in variational calculus, but, apparently due to the discrete domain roots of the problem, has not been considered before in the MaxEnt context. Allowing the derivative to appear in the functional is also the most straightforward way to impose differentiability condition on the probability distribution function - a natural one in view of its physics applications. Regrettably, Vàn's derivation falls into the same trap as Shore and Johnson's in that being based on a simple additivity axiom it derives a measure in the form of a linear combination of the simple Shannon-Jaynes entropy and the Fisher information in p associated with x. Although deficient, Van's derivation is the first to clearly indicate that Fisher information may play an important role in a rigorous MaxEnt approach.

There are further conceptual problems with the MaxEnt approach related to the way new information, obtained from measurements or otherwise, is introduced into the rule. They were first pointed at by Karbelkar [14], addressed by Ufink [7] and are discussed at length by Giffin and Caticha in these Proceedings. We will briefly mention some of those in the following sections.

To summarize, there are number of problems with the current MaxEnt paradigm which seem to justify a fresh, from scratch, approach to the derivation of updating and constructive rules. Such an approach is proposed in the subsequent two sections.

CHARACTERIZATION OF THE UPDATING RULE

The rationale behind the M.E. updating rule is a very simple and sound one: Given that our state of knowledge regarding a system is encoded into a probability density $m(x)$ (designated "a prior") and new information is obtained in the form of an average value of system's observable, update $m(x)$ to $p(x)$ (called "posterior") such that the information distance between $m(x)$ and $p(x)$ is minimal, subject to the average value constraint. If this can be done in more than one equivalent ways the result should be the same. In other words, when updating probability distributions, one must be *conservative* and *consistent*. Clearly, all we need to implement the above desiderata is a proper definition of information distance. As discussed in the Introduction, when attempting to define information content in the usual way one necessarily ends up with information distance instead, so I'll try to axiomatically characterize the information content of a probability distribution in the hope to obtain an information distance measure.

Basic Axioms

1. (Weak) Locality: The information content of a probability distribution $p(x)$ is an increasing function of a linear functional of $p(x)$ and $\nabla p(x)$ [3]:

$$S[p] = g(\int dx \tilde{f}(p,(\nabla p)^2))$$

   where $\tilde{f}(x,z)$ is a function to be determined;

2. Expandability: The information content is invariant upon expansion of the domain with zero probability assignments:

$$\int dx \tilde{f}(p,(\nabla p)^2) = \int dx pf(p,(\nabla p)^2) \text{ with } pf(p,(\nabla p)^2) \xrightarrow[p \to 0]{} 0$$

   where the unspecified $f(x,z)$ replaces $\tilde{f}(x,z)$;

3. Additivity: For independent probability distributions the information content of the joint distribution is the sum of the information contents of the individual distributions:

$$S[p_1(x)p_2(y)] = S[p_1(x)] + S[p_2(y)]$$

   This condition should restrict the particular form of the functions $g(x)$ and $f(x,z)$, hopefully unambiguously identifying them.

Discussion of the Basic Axioms

The justification of the first axiom is given by Shore and Johnson [12]. We simply replace the locality requirement by a weak locality one in that dependence on the first derivatives is allowed. Including first derivatives in the linear functional restricts the original notion of a small variation of the probability density $p(x)$ from $\delta p(x): \max_x |\delta p(x)| = \varepsilon$ to $\delta p(x): \max_x |\delta p(x)| = \varepsilon, \max_x |\delta \nabla p(x)| = \varepsilon'$ thus enforcing differentiability upon $p(x)$. While the relevance of including differential criterion is not

---

[3] Vàn has shown that if one allows higher derivatives the form of the possible terms depends on the dimensionality of the configuration space, which is clearly undesirable. Therefore we restrict ourselves to first derivatives only.

immediately obvious in the discrete case, in the continuous case it cannot be apriori dismissed without further analysis.

The expandability axiom is a straightforward generalization of Khinchin's third axiom to the continuous case. In addition to its original justification we'll see later on that it is also a sufficient condition for the resulting updating rule to be consistent with the product rule of probability theory.

The additivity property is equivalent to what Shore and Johnson call "subset independence". While one can argue whether it is a necessary property of a general information content measure *per se*, it is absolutely essential for an updating rule based on such a measure in order to obtain consistent results when treating independent systems separately or combining them into a joint system.

Consequences of the Basic Axioms

In view of the form of the expression $S[p] = g(\int dx p f(p,(\nabla p)^2))$ there are two possibilities for achieving additivity of $S[p]$: *i)* take $g(x+y) = g(x) + g(y)$ and require that $\int dx p f(p,(\nabla p)^2)$ is additive; or *ii)* take $g(xy) = g(x) + g(y)$ and require that $\int dx p f(p,(\nabla p)^2)$ factorizes. As it turns out, the first possibility is contained as a limiting case in the second one, so we only consider *ii)*. In this case $g(x)$ can be taken as $c\ln(x)$ where c is an arbitrary constant. It is convenient to rewrite the so far unknown function $f(p,(\nabla p)^2)$ in the form $f(q,(\nabla q)^2)$ where $q(x) = \ln p(x)/m(x)$ and $m(x)$ is an arbitrary positive function. The advantage of this form is that for a joint distribution of independent variables both arguments of f decompose to sums: $q(x,y) = q_1(x) + q_2(y)$, $(\nabla q(x,y))^2 = (\nabla_x q_1(x))^2 + (\nabla_y q_2(y))^2$. Thus it becomes easy to see that the factorization requirement

$$\int dx dy p_1 p_2 f(q_1 + q_2, (\nabla_x q_1)^2 + (\nabla_y q_1)^2) = \int dx p_1 f(q_1,(\nabla_x q_1)^2) \times \int dy p_2 f(q_2,(\nabla_y q_1)^2)$$

implies $f(q_1 + q_2, z_1 + z_2) = f(q_1, z_1) \times f(q_2, z_2) \Rightarrow f(q,z) = \exp(\alpha q + \beta z)$ with α and β arbitrary constants, and, finally

$$S[p] = c\ln \int dx p \left[\left(\frac{p}{m}\right)^\alpha \exp[\beta(\nabla \ln \frac{p}{m})^2]\right]$$

This family, depending on three numerical parameters *c*, α, β and one arbitrary positive function *m(x)*, appears to be the most general form consistent with the basic axioms[4]. The presence of *m(x)* however indicates that the original intention of deriving a unique measure of information content, and thus a constructive rule, in this way, is doomed. In view of the fact that, for non-negative α, β and *c*, *S[p]* is non-negative and vanishes only for *p(x)=m(x)*, the alternative interpretation of *S[p,m]* as an information distance measure between the probability densities *p(x)* and *m(x)* suggests itself. Indeed, for particular choices of the parameters *S[p,m]* reduces to known forms of information distance: for β=0 and $c = -\alpha^{-1}$ $S[p,m] = -\alpha^{-1} \ln \int dx p (p/m)^\alpha$ is the Renyi distance, while for $\beta = b\alpha$, $c = a/\alpha$ and $\alpha \to 0$

---

[4] A linear combination over the parameters with arbitrary weights will obviously also satisfy all requirements resulting from the axioms.

$S[p,m] = a\int dx p \ln(p/m) + b\int dx p(\nabla \ln(p/m))^2$ is a Van-type distance; if further $b=0$ it reduces to the Shannon-Jaynes (or KL) distance, and if $a=0$ it turns into a kind of Fisher distance.

To summarize, the axiomatic approach doesn't seem capable of characterizing an information content measure but provides a characterization, albeit ambiguous one, of general purpose information distance measure. With such a characterization at hand, one can proceed to obtain an updating rule imposing goal-specific additional consistency requirements.

Updating Rule: Narrowing Down the Choices

The use of the information distance measure derived above in a M.E. updating rule implies solving the variational problem $\delta\left\{S[p,m] + \lambda \int dx p(x) C(x)\right\} = 0$ where the variation is with respect to $p(x)$ and the Lagrange multiplier λ is chosen such that $p(x)$ is consistent with the new information: $\int dx p(x) C(x) = d$. Instead of minimizing $S[p,m]$ one can equivalently minimize $\exp(c^{-1} S[p,m])$ so the general variational problem becomes

$$\delta\left\{\int dx p f(q, (\nabla q)^2) + \lambda \int dx p(x) C(x)\right\} = 0.$$

Let us now consider the case of two possibly dependent variables x and y, a prior knowledge in the form of a joint probability distribution $m(x,y)$ and new information about y to the effect that $y=y_0$. This is the same as constraining the marginal probability density of y to a delta distribution $p(y) = \delta(y - y_0)$, which can be achieved by imposing infinitely many simple constraints, one for each y value:

$$\int dy \lambda(y) \int dx p(x,y) = \delta(y - y_0).$$

Under the circumstances, any reasonable updating rule must produce a posterior of the form $p(x,y) = m(x|y)\delta(y-y_0) = m(x,y)\delta(y-y_0)/m(y)$ in compliance with probability theory's product formula for conditional probabilities. Working out the variational equation for this case

$$\left[1 + \alpha - 2\beta\Delta q - 2\beta(1+q\alpha)\frac{(\nabla q)^2}{q} - 4\beta^2 \nabla q \cdot \nabla\nabla q \cdot \nabla q - 2\beta \nabla \ln m \cdot \nabla q\right] f = \lambda(y)$$

it is easily seen that a solution with $q = p(x,y)/m(x,y) = q(y)$ is only possible, because of the last term in the brackets, for $\beta = 0$. Thus, requiring consistency with the probability theory's product rule eliminates the derivative terms from the information measure and leaves us with the family of Renyi measures. It is worthwhile noticing that, as mentioned in the Introduction, the fact that it is possible at all to choose the parameters in compliance with probability theory's product rule hinges essentially on the particular form of the functional required in Axiom 2 above.

The Unique Updating Rule

Consider the following situation: There is a common knowledge about two variables x and y which is codified into a joint prior $m(x,y)$. A researcher A acquires a "$g(x)$ meter" and uses it to measure the average value of $g(x)$. He then duly uses M.E. to update his probability distribution from $m(x,y)$ to $p_A(x,y)$. Another researcher, B, is fortunate enough to borrow a "$r(y)$ meter" from a friend and uses it to measure the average value of $r(y)$. Then she similarly updates her prior to $p_B(x,y)$. Both researchers promptly publish the results of their measurements and thus become aware of each other's measured averages. A uses B's result to update his probability distribution $p_A(x,y)$ to $p_{AB}(x,y)$. Analogously, B uses A's result and updates her probability distribution $p_B(x,y)$ to $p_{BA}(x,y)$. Then there is a slacker C who doesn't care to experiment himself but carefully follows the literature and learns about A and B's measurements. Using them simultaneously, he updates his prior $m(x,y)$ to $p_{(AB)}(x,y)$. For the sake of consistency, we require that $p_{AB}(x,y) = p_{BA}(x,y) = p_{(AB)}(x,y)$. Turning to the variational equation resulting from the choice $\beta = 0$ above one observes that the updating amounts to multiplying the prior with a certain function of the constraint:

$$p(x,y) = m(x,y) F_\alpha(\lambda C(x,y))$$

This being the case it is obvious that the first part of our consistency requirement is already satisfied - A and B will end up with identical probability distributions:

$$p(x,y) = m(x,y) F_\alpha(\lambda_1 g(x)) F_\alpha(\lambda_2 r(y))$$

As for the second part, it requires that the function $F_\alpha(z)$ has the property $F_\alpha(z_1 + z_2) = F_\alpha(z_1) F_\alpha(z_2)$ which is only the case for α=0 when it is an exponential. Thus, our three basic axioms plus the two consistency conditions uniquely single out the Shannon-Jaynes relative entropy as the one to use in M.E. updating. One has to be aware though that if in the above example the researchers measured different functions of the same variable, it would be in general impossible to obtain posteriors consistent between the three of them. This is precisely the problem pointed out by Karbelkar and discussed by Giffin and Caticha in these proceedings. M.E. with the Shannon-Jaynes relative entropy is the closest one can get to a consistent inference scheme, but it, too, falls short of being exactly that. Giffin and Caticha advocate for using M.E. with uninformative priors and keeping track of all previously used constraints. When new information becomes available, the corresponding constraint are to be combined with the old ones and the whole set employed to obtain a posterior. This obviously invalidates the fundamental tenet of the updating rule that prior knowledge should be encoded into a prior probability distribution, and thus renders the prior distributions obsolete.

DERIVATION OF THE CONSTRUCTIVE RULE

In the preceding section we obtained a general information distance measure and used arguments derived from its intended use in a M.E. updating procedure to narrow down the choice of the β parameter value to β=0, thus excluding derivative terms. In fact, a much more general justification for this choice can be given by observing that the effective value of β depends on the scale on which the variable x is measured. Indeed, rescaling $x \to sx$ implies $\nabla \to s^{-1} \nabla$ and thus, the ratio p/m being invariant, $\beta \to s^{-2} \beta$. If one is to have a unique information distance measure, the only way to lift the ambiguity is to put β=0. This choice leaves us with a one-parameter family of "reasonable" information distance measures of the Renyi's type. Now we set out to use this family for the purpose of working out a rule for assigning probabilities based on incomplete information - the original Jaynes goal. The departure point is the following pair of axioms.

The Axioms and the Constructive Rule

1. It is possible to meaningfully define a information content $I[p]$ of a probability distribution $p(x)$ so that $I[p] \in [0, \infty)$ and $|I[p_1] - I[p_2]| = L(S_\alpha[p_1, p_2])$ where $L()$ is some convex linear combination over α;
2. The information content decreases upon coarse-graining: $I[p_\sigma] < I[p]$ where $p_\sigma(x)$ is $p(x)$ coarse-grained with a coarse-graining scale σ;

Consider now the amount of the decrease $0 < I[p] - I[p_\sigma] = L(S_\alpha[p, p_\sigma])$. Since the information content cannot go negative, the relation $I[p] - I[p_\sigma] = I[p]F(I[p])$ should hold with some function $F(z)$ subject to the condition $zF(z) \xrightarrow[z \to 0]{} 0$. Inspecting both sides of the relation above it is easily seen that the probability density distribution for which the information content $I[p]$ is minimal can be found among the probability density distributions for which $I[p] - I[p_\sigma]$, or equivalently, the information distance between the original and the coarse-grained probability density $S_\alpha[p, p_\sigma]$, is minimal. Thus, the following rule for assigning probabilities suggests itself: <u>When incomplete information is available, the probability distribution to be assigned is the one least sensitive to coarse-graining subject to all known constraints.</u>

The Variational Equation

In working out the particularities of the constructive rule a great simplification results from utilizing the well-known fact that coarse-graining is statistically equivalent to adding noise. Indeed, the expression for the coarse-grained probability density $p_\sigma(x) = \int dz\, p(x + \sqrt{\sigma}z) f(z)$ where $f(z)$ is a non-negative normalized coarse-graining kernel and σ is the coarse-graining scale, can be interpreted as a marginalization $p_\sigma(x) = \int dz\, p_\sigma(x|z) f(z) = \int dz\, p_\sigma(x, z)$ from a joint probability density $p_\sigma(x, z)$ describing a noisy system. In the limit $\sigma = 0$ there is no coarse-graining and $p_{\sigma=0}(x, z) = p(x) f(z)$. We now consider the information distance between the original and the noisy probability distributions for infinitesimal coarse-graining ($\sigma \ll 1$): $S_\alpha[p_{\sigma=0}, p_\sigma] = \alpha^{-1} \ln \int dx dz\, p_{\sigma=0}(x,z)[p_{\sigma=0}(x,z)/p_\sigma(x,z)]^\alpha$.

Noticing that for isotropic coarse-graining kernel $\int dzz_i f(z) = 0$ and $\int dzz_i z_j f(z) = \chi \delta_{ij}$ where χ is a constant of the order of unity, we obtain by expanding in the powers of $\sigma^{1/2}$

$$S_\alpha[p_{\sigma=0}, p_\sigma] = \alpha^{-1} \int dx dz f(z) p(x)[p(x + \sqrt{\sigma}z)/p(x)]^{-\alpha} =$$
$$= \sigma\chi(\alpha + 1) \int dx\, p^{-1}(x)[\nabla p(x)]^2 + O(\sigma^2)$$

Remarkably, irregardless of the value of α^5, our probability assignment rule calls for a constrained minimization of the Fisher information

$$\int dx\, p^{-1}(x)[\nabla p(x)]^2 + \lambda \int dx\, p(x)C(x) \to \min$$

that would produce the $p(x)$ least sensitive to coarse-graining subject to the known constraints. The corresponding Euler-Lagrange equation is easily obtained by setting $p(x) = \psi^2(x)$ and varying $\psi(x)$ without having to worry about the non-negativity of $p(x)$, to obtain

$$-\Delta\psi(x) + \lambda C(x)\psi(x) = 0$$

DISCUSSION

Our revision of the MaxEnt fundamental principles vindicated the Shannon-Jaynes relative entropy form as the information distance measure which produces the closest thing to a consistent rule for updating probabilities. This does not come as a big surprise, given the documented enormous success of its numerous applications. The analysis, however, also indicates the impossibility of deriving a probability assignment rule from the usual requirements for consistency and uniqueness. New ideas are needed, and the strategy proposed above is not unlike the way the energy is treated in Field Theory: although we cannot define a particular finite form of an information content measure, the mere assumption that such a measure exists allows us to constructively address the question of locating its minimum. The derived constructive rule involves minimizing a particular instance of the Fisher information subject to data constraints. This result is especially interesting in view of our original motivation of establishing a relation between physics laws and information theory. While the role Fisher information plays in many variational principles of physics has been known and extensively discussed for a long time [6,16,17], no plausible explanation of why exactly the Fisher information should be used has been offered. The result of the previous section seems to provide such an explanation. It also can give insight into why physical systems appear to follow this or that particular law. For example, comparing the constructive rule with the Schrodinger equation of Quantum Mechanics $-\Delta\psi(x) + (2m/\hbar^2)[E - U(x)]\psi(x) = 0$, it appears that the ground state probability distribution of quantum systems is obtained by merely constraining the average kinetic energy, or equivalently the temperature, of the system to have a certain value as if quantum systems were in contact with a universal thermostat. The difference from the conventional statistical physics is that instead of the average kinetic energy value, the value of the Lagrange multiplier is prescribed to be proportional to the system's mass, namely $2m/\hbar^2$. Studying the why's and how's of this and similar connections could greatly advance our understanding of the origin of the physics laws and the inner workings of the universe around us.

---

[5] The statement hold even for linear combinations of different alphas, so the resulting rule is independent of whether one uses a Renyi distance, the symmetrized KL distance or even the Bernardo-Rueda divergence [15]

REFERENCES

1. E. T. Jaynes, *Phys.Rev.* **106**, 620 and **108**, 171 (1957)
2. T. Toffoli, in *Complexity, Entropy, and the Physics of Information* (W. H. Zurek ed.), Addison-Wesley p 301, (1990)
3. T. Toffoli, *Superlattices and Microstructures* **23**, p.381 (1998)
4. J. Semitecolos, arXiv.org/abs/quant-ph/0212080)
5. A. Caticha, *Phys.Rev.A,* **A57**, p.1572 (1998)
6. B. R. Frieden, "Physics from Fisher Information" (Cambridge: Cambridge University Press 1998); R. B. Frieden and B. H. Soffer, *Phys. Rev E*, **E52**. p.2274 (1995)
7. J. Uffink, *Studies in History and Philosophy of Modern Physics* **26B**, p.223 (1995) ; **27**, p.47 (1996)
8. C. E. Shannon, *The Bell System Technical Journal*, **27**, p.379, p. 623 (1948)
9. A. I. Khinchin, "Mathematical Foundations of Information Theory" (Dover Publications, 1957)
10. A. Renyi, in *Proc. 4th Berkeley Symp. Math., Stat. and Probability,* p.547 (1960)
11. A. Caticha, "Relative Entropy and Inductive Inference", in Bayesian Inference and Maximum Entropy Methods in Science and Engineering, ed. by G. Erickson and Y. Zhai, *AIP. Conf. Proc.* **707**, 75 (2004); (arXiv.org/abs/physics/0311093).
12. J. E. Shore and R. W. Johnson, *IEEE Trans. Inf. Theory* **IT-26**, 26 (1980); *IEEE Trans. Inf. Theory* **IT-27**, 26 (1981).
13. P. Vàn , *Physica A,* **365**, p.2833 (2006); P. Vàn,, T. Fülöp, *Proc. Roy. Soc.* **A462**, 541 (2006)
14. S. N. Karbelkar, *Pramana J. Phys.*, **26**, p.301 (1986)
15. J. M. Bernardo, R. Rueda, *Int. Stat. Rev.* **70**, p.351 (2002)
16. M. Reginatto, *Phys. Rev. A.* A58, p.1775 (1998)
17. R. R. Parwani, *J. Phys. A: Math. Gen* .**38**, p.6231 (2005)
18. V. I. Dimitrov, "Making Sense of Quantum Mechanics", in preparation.

The Role of Information in the Probabilistic Reconstruction of Quantum Theory

Philip Goyal

Cavendish Laboratory
University of Cambridge

Abstract. In this paper, we explore the possibility that the concept of information may enable a derivation of the quantum formalism from a set of physically comprehensible postulates. Taking the probabilistic nature of measurements as a given, we introduce the concept of information via a novel invariance principle, the Principle of Information Gain. Using this principle, we then show that it is possible to deduce the abstract quantum formalism for finite-dimensional quantum systems from a set of postulates, of which one is a novel physical assumption, and the remainder are based on experimental facts characteristic of quantum phenomena or are drawn from classical physics. The concept of information plays a key role in the derivation, and gives rise to some of the central structural features of the quantum formalism.

Keywords: Foundations of Physics, Quantum Theory, Information Theory

INTRODUCTION

Since its formulation in the mid-1920s, quantum theory has been successfully applied to an ever broadening range of physical phenomena. However, a fundamental question has remained largely unanswered: what does quantum theory tell us about the way that nature operates? One of the main obstacles in formulating a comprehensive answer to this question is that, owing to the fact that the quantum formalism was obtained using a significant amount of mathematical guesswork, the formalism has many features (such as its use of complex numbers) whose physical origin is obscure. Consequently, it is not at all clear what constitutes the actual *physical content* of the theory, a situation that has led to a plethora of often conflicting answers to the above question.

Over the last two decades, a number of authors have expressed the view that our efforts to develop an understanding of quantum theory would be significantly aided by a systematic derivation of the formalism from a set of physically comprehensible assumptions [1, 2, 3]. Furthermore, several authors have proposed that the concept of information may be the key, hitherto missing, ingredient which, if appropriately applied and formalized, might make such a derivation possible [4, 1, 5, 6, 3].

The proposal that information might enable a derivation of the quantum formalism rests, to a significant degree, upon the recognition that the concept of information plays a new and fundamental role in quantum physics. One way to see this is as follows. In classical physics, an experimenter presented with a system in an unknown state can, in principle, perform a ideal measurement upon the system which gives perfect knowledge about the state of the system. Hence, there is no fundamental distinction between the state of the system on the one hand, and an ideal experimenter's *knowledge* of the state on the other. However, if one takes the probabilistic nature of measurement as suggested by quantum theory at face value, so that the state of a system only determines the outcome probabilities of measurements performed upon it, it follows that an ideal measurement (or even a finite number of such measurements performed upon an ensemble of identically-prepared systems) provides only partial knowledge about the unknown state of the state of the system. Hence, in sharp contrast to the situation in classical physics, a fundamental distinction is drawn between the state of the system and the knowledge that the experimenter can possibly have of the state. It is then natural to attempt to *quantitatively* relate the two, and so to ask: 'How much information has been obtained by the experimenter about the state?', whereby the concept of information assumes a fundamental role.

One of the earliest attempts to explore the possible role of information in determining the quantum formalism is due to Wootters [7], who showed that Malus' law can be derived from an intuitively plausible information-theoretical principle using the standard inferential methods of probability theory and the standard Shannon information measure. More recently, other attempts [8, 9, 10, 5] have been made to examine and quantify the gain of information in the measurement process, and which differ in various ways from Wootters' approach, but which lead to Malus' law. However, none of these approaches are able to generalize their results in a physically motivated way to obtain a significant part of the quantum formalism.

In contrast, several other recent approaches [1, 11, 12, 13, 14, 15] which involve the concept of information succeed in deriving a significant fraction of the quantum formalism, but make abstract assumptions of key importance which are given no physical interpretation, and which thereby significantly detract from the understanding of the physical origin of the quantum formalism that can be obtained.

In [16, 17], we have attempted to build upon the insights provided by Wootters' approach, and to derive the abstract quantum formalism, together with the correspondence rules of quantum theory, from assumptions that can be clearly understood as assertions about the physical world. The key information-theoretic postulate is the *Principle of Information Gain*, which expresses the idea that, although different measurements (such as different Stern-Gerlach measurements on a spin) yield different information about the state of a system, they nonetheless provide the same *amount* of information about the state. That is, although different measurements provide different perspectives on a system, none is informationally privileged with respect to any other. In this paper, we outline our approach, and, for reasons of space, focus on the postulates and their physical origin, describing the derivation only briefly. The reader is referred to [16, 17] for a more thorough discussion.

EXPERIMENTAL SET-UP AND POSTULATES

Abstract Experimental Set-up

Consider an experimental set-up where, in each run, a system (from a source of identical systems) undergoes a preparation and an interaction, and is then subject to a measurement. Suppose that the purpose of the set-up is to allow the experimenter

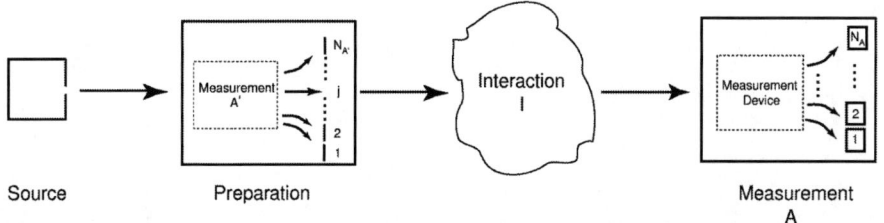

FIGURE 1. An abstract experimental-set up. In each run of the experiment, a physical system (such as a silver atom) is emitted from a source, passes a preparation step, and is then subject to a measurement. The preparation is implemented as a measurement, **A**′, which has $N_{A'}$ possible outcomes, followed by the selection of those systems which yield some outcome j ($j = 1, 2, \ldots, N_{A'}$). The measurement, **A**, has N_A possible outcomes. An interaction, **I**, may occur as indicated between the preparation and measurement.

to investigate how some property (or observable), such as position or spin, of the system is affected by the interaction with the system. Ideally, in such an experiment, the measurement data obtained should be independent of interactions with the system that occur prior to the preparation. For, otherwise, the measurement data could be influenced by conditions that are not under experimental control. In our derivation of the quantum formalism, we shall take such set-ups, which we shall say are *closed* (or have the property of *closure*), as our starting point.

On the experimentally-founded assumptions that the possible outcomes of a measurement performed upon a system are finite in number and the outcome probabilities are determined by the state of the system, it follows that, in a given experimental set-up, the measurement data obtained in repeated trials is theoretically characterized by a finite set of probabilities. Therefore, the closure condition, when applied in the context of these assumptions, requires that these outcome probabilities are independent of the pre-preparation history of the system.

Accordingly, in the abstract experimental set-up illustrated in Fig. 1, we restrict the measurements and interactions that are permitted to those which result in a closed set-up. Further, suppose that measurements **A**, **A**′ are such that a set-up in which **A** is used to implement a preparation, and **A**′ is subsequently performed, without an interaction in the intermediate time, is closed. Then, if closure persists when the roles of **A**, **A**′ are reversed, we shall say that **A**, **A**′ form a *measurement pair*. We shall then define the measurement set $\mathscr{A}$ *generated* by a **A** as the set of all measurements that form a measurement pair with **A**, excluding any measurements that are a composite of other measurements in $\mathscr{A}$. Finally, if an interaction **I** can be introduced into a set-up that uses any two measurements in $\mathscr{A}$ without disturbing the closure of the experiment, we shall say that it is *compatible* with $\mathscr{A}$. We then define the interaction set $\mathscr{I}$ as the set of all such interactions.

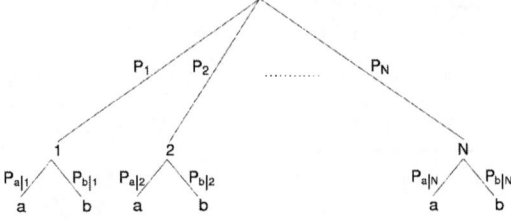

FIGURE 2. (*State Interpretation Postulate*) A probability tree showing the outcomes of measurement **A** when performed on a system in state **S** given by $(\vec{P}, \vec{\chi})$. Outcome i is observed with probability P_i, and outcome a or b is realized with probability $P_{a|i}$ or $P_{b|i}$, respectively. In addition, outcome $+$ or $-$ is realized, determined by the sign of $Q_{a|i}$ if a has been realized, or $Q_{b|i}$ if b has been realized. The outcomes a, b, and $+, -$ cannot be observed by the experimenter.

Statement of the Postulates.

Consider the idealized experiment illustrated in Fig. 1 in which a system passes a preparation step that employs a measurement $\mathbf{A}'$ in measurement set $\mathscr{A}$, undergoes an interaction, $\mathbf{I}$ in the interaction set $\mathscr{I}$, and is then subject to a measurement, $\mathbf{A}$, in $\mathscr{A}$. The abstract theoretical model that describes this set-up satisfies the following postulates.

1. **Measurements**
   1.1 *Outcome Number.* When any measurement $\mathbf{A} \in \mathscr{A}$ is performed, one of N ($N \geq 2$) possible outcomes are observed.
   1.2 *Measurement Representation.* For any given pair of measurements $\mathbf{A}, \mathbf{A}' \in \mathscr{A}$, there exist interactions $\mathbf{I}, \mathbf{I}' \in \mathscr{I}$ such that $\mathbf{A}'$ can, insofar as probabilities of the observed outcomes and insofar as the possible output states of the measurement are concerned, be represented by an arrangement where $\mathbf{I}$ is immediately followed by $\mathbf{A}$ which, in turn, is immediately followed by $\mathbf{I}'$.
2. **States**
   2.1 *State Representation.* With respect to any given measurement $\mathbf{A} \in \mathscr{A}$, the state, $\mathbf{S}(t)$, of a quantum system at time t is given by $(\vec{P}, \vec{\chi})$, where $\vec{P} = (P_1, P_2, \ldots, P_N)$ and $\vec{\chi} = (\chi_1, \chi_2, \ldots, \chi_N)$ are real n-tuples, and where P_i is the probability that the ith outcome of measurement $\mathbf{A}$ is observed.
   2.2 *State Interpretation.* When measurement $\mathbf{A} \in \mathscr{A}$ is performed on a system in state $\mathbf{S}(t)$ and the outcome i is observed, there are additional outcomes that are objectively realized but unobserved (see Fig. 2):
      (i) one of two outcomes, labeled a and b, which are obtained with respective probabilities $P_{a|i} = Q_{a|i}^2$ and $P_{b|i} = Q_{b|i}^2$, where $Q_{a|i} = f(\chi_i)$ and $Q_{b|i} = \tilde{f}(\chi_i)$, where f is not a constant function and $f, \tilde{f}$ have range $[-1, 1]$, and
      (ii) one of two possible outcomes, with values labeled $+$ and $-$, which is given by the sign of $Q_{a|i}$ if a has been realized, or of $Q_{b|i}$ if b has been realized.

2.3 *Information Gain.* When measurement $\mathbf{A} \in \mathscr{A}$ is performed on a system in any given unknown state $\mathbf{S}(t)$, the amount of Shannon-Jaynes information provided by the observed outcomes and the outcomes a and b about $\mathbf{S}(t)$ in n runs of the experiment is independent of $\mathbf{S}(t)$ in the limit as $n \to \infty$.

2.4 *Prior Probabilities.* The prior probability $\Pr(\chi_i|\mathrm{I})$ is uniform for $i = 1,\ldots,N$, where I is the background knowledge of the experimenter prior to performing the experiment.

3. **Transformations**
   3.1 *One-to-one.* If a physical transformation of the system is represented by a map, $\mathscr{M}$, over the state space, $\mathscr{S}$, of the system, then $\mathscr{M}$ is one-to-one.
   3.2 *Continuity.* If a physical transformation is continuously dependent upon the real-valued parameter n-tuple $\boldsymbol{\pi}$, and is represented by the map $\mathscr{M}_{\boldsymbol{\pi}}$, then $\mathscr{M}_{\boldsymbol{\pi}}$ is continuously dependent upon $\boldsymbol{\pi}$.
   3.3 *Continuous Transformations.* If $\mathscr{M}_{\boldsymbol{\pi}}$ represents a continuous transformation, then, for some value of $\boldsymbol{\pi}$, $\mathscr{M}_{\boldsymbol{\pi}}$ reduces to the identity.
   3.4 *Invariance.* The map $\mathscr{M}$ is such that, for any state $\mathbf{S} \in \mathscr{S}$, the observed outcome probabilities, $P'_1, P'_2, \ldots, P'_N$, of measurement $\mathbf{A} \in \mathscr{A}$ performed upon a system in state $\mathbf{S}' = \mathscr{M}(\mathbf{S})$ are unaffected if, in any representation, $(\vec{P}, \vec{\chi}) = (P_i; \chi_i)$, of the state $\mathbf{S}$ written down with respect to $\mathbf{A}$, any arbitrary real constant, χ_0, is added to each of the χ_i.
   3.5 *Consistency* The posterior probability distributions over $\mathscr{S}$ that result from the following two processes coincide in the limit as $n \to \infty$:
      (i) inferring a posterior over $\mathscr{S}$ based upon the objectively realized outcomes when the measurement $\mathbf{A} \in \mathscr{A}$ is performed upon n copies of a system in state $\mathbf{S}$, and then transforming the posterior using $\mathscr{M}$, or
      (ii) inferring a posterior over $\mathscr{S}$ based upon the objectively realized outcomes when the measurement $\mathbf{A} \in \mathscr{A}$ is performed upon n copies of a system in state $\mathscr{M}(\mathbf{S})$,
   3.6 *Temporal Evolution.* The map, $\mathscr{M}_{t,\Delta t}$, which represents temporal evolution of a system in a time-independent background during the interval $[t, t + \Delta t]$, is such that any state, $\mathbf{S}$, represented as $(P_i; \chi_i)$, of definite energy E, whose observable degrees of freedom are time-independent, evolves to $(P'_i; \chi'_i) = (P_i; \chi_i - E\Delta t/\alpha)$, where α is a non-zero constant with the dimensions of action.

The above postulates, together with the Average-Value Correspondence Principle (AVCP), not discussed here, suffice to determine the form of the abstract quantum model for the abstract set-up. From the Outcome Number Postulate, it follows that, when any measurement in $\mathscr{A}$ is performed on the system, one of N possible outcomes is observed. Accordingly, we shall denote the abstract quantum model of such a set-up by $\mathbf{q}(N)$.

Finally, we shall need a further postulate (Composite Systems postulate), which is not discussed here, in order to obtain the tensor product rule that allows one to relate the quantum model of a composite system to the quantum models of its component systems.

ORIGIN OF THE POSTULATES

In our discussion below, we shall divide the postulates according to their origin as follows:

1. *Based on Experimental Observations.* Postulates obtained through direct generalization of experimental facts that are taken to be characteristic of quantum phenomena.
2. *Drawn from Classical Physics.*
    (i) Postulates adopted unchanged from the theoretical framework of classical physics.
    (ii) Postulates obtained from classical physics via a classical–quantum correspondence argument.
3. *Novel Postulates.* Postulates (one informational and one physical postulate) that are based on novel theoretical principles or ideas which cannot obviously be traced to classical physics or experimental observations.

1. Postulates based on Experimental Observations

In experiments where Stern-Gerlach measurements are performed upon microscopic particles such as silver atoms, one observes that:

1. Every Stern-Gerlach measurement yields the same number of possible outcomes (for example, two outcomes in the case of silver atoms), and
2. Any Stern-Gerlach measurement can be implemented using a given Stern-Gerlach measurements flanked either side with suitable magnetic fields.

The Outcome Number and the Measurement Representation postulates can be regarded as direct generalization of these observations.

2. Postulates drawn from Classical Physics

Postulates adopted from Classical Framework

The framework of classical physics assumes that physical transformations of a system are represented by maps over the state space of the system, and that these maps have the properties expressed in the One-to-One, Continuity, and Continuous Transformations Postulates. We adopt these properties unchanged in the quantum framework.

Finally, we note that the classical framework is consistent in sense that, if the framework allows one to make a prediction about a given measurement outcome by following two or more calculational paths, then these paths must yield the same result. This elementary consistency requirement, adopted and applied in a probabilistic framework, leads directly to the Consistency postulate.

Postulates obtained by a Correspondence Argument

A general guiding principle in building up a quantum model of a physical system is that, in an appropriate limit, the predictions of the quantum model of the system stand in some one-to-one correspondence with those of a classical model of the system. In particular, we shall consider a quantum model of a particle and its classical counterpart, and shall require that these models are in one-to-one correspondence in the classical limit as the mass, m, of the particle tends to macroscopic values. Using this correspondence, we shall transpose several elementary properties of the classical model into the quantum model of the particle, and thence, by generalization, into the quantum model of a arbitrary physical system.

Based on experimental data, one finds that, in a quantum experiment, if one prepares the particle using a coarse position measurement, and then performs a coarse position measurement on the particle, the set-up is closed (up to experimental precision) provided that the position measurements used are of sufficiently high resolution. Now, in the classical limit, we expect that the system will behave classically between the preparation and measurement, and one can readily establish that the relevant classical model in this situation is a particle ensemble model, the Hamilton-Jacobi model.

In the Hamilton-Jacobi model, the state of the ensemble is given by $(P(\vec{r},t), S(\vec{r},t))$, where $P(\vec{r},t)$ is the spatial probability density function, and S is the action function. In the case of coarse position measurements with N possible outcomes, we shall use the discretized form of the Hamilton-Jacobi state, $(P_1^{(CM)}, \ldots, P_N^{(CM)}, S_1, \ldots, S_N) = (P_i^{(CM)}; S_i)$, where $P_i^{(CM)}$ is the probability that the position measurement yields a detection at the ith measurement location, and S_i is the classical action at the ith measurement location.

In order that the predictions of the quantum and classical models agree in the classical limit, the quantum state $\mathbf{S}(t)$ ($t > t_0$) must contain degrees of freedom which encode N quantities, which we shall denote $S_1^{(QM)}, \ldots, S_N^{(QM)}$, which, in the classical limit, are equal to the S_i. Equivalently, we shall assume that $\mathbf{S}$ contains N dimensionless real quantities, $\chi_1, \ldots, \chi_N$, such that $S_i^{(QM)} = \alpha \chi_i$, where α is a non-zero constant with the dimensions of action.

From the above discussion, in the quantum model of the particle, which we shall denote $\mathbf{q}^*(N)$, the state, $\mathbf{S}$, is given by $(\vec{P}, \vec{\chi})$, where $\vec{\chi} = (\chi_1, \ldots, \chi_N)$. The State Representation postulate directly generalizes this statement to the abstract model $\mathbf{q}(N)$.

We now observe that the Hamilton-Jacobi model has the following properties, which can be readily verified from the Hamilton-Jacobi equations:

1. *Invariance.* The evolution of the state $(P_i^{(CM)}(t_1); S_i(t_1))$ to the state $(P_i^{(CM)}(t_2); S_i(t_2))$ is such that the $P_i^{(CM)}(t_2)$ are unchanged if an arbitrary real constant, S_0, is added to each of the $S_i(t_1)$.

2. *Temporal Evolution.* In a time-independent background, a state, $(P_i^{(CM)}(t); S_i(t))$ whose observable degrees of freedom are time-independent, evolves in time Δt to the state $(P_i^{(CM)}(t); S_i(t) - E\Delta t)$, where E is the total energy of the system.

Furthermore, from the first property, since the zero-value of the S_i is conventional and therefore has no physical correlate, it follows that, for any $S_1, \ldots, S_N$ and any S_0,

$$\Pr(S_1, \ldots, S_N | \mathrm{I}) = \Pr(S_1 + S_0, \ldots, S_N + S_0 | \mathrm{I}), \tag{1}$$

from which, by marginalization, if follows that, for any S_0, S_1,

$$\Pr(S_1 | \mathrm{I}) = \Pr(S_1 + S_0 | \mathrm{I}), \tag{2}$$

where I represents the state of knowledge of the experimenter prior to performing measurements on the system. Therefore, the prior $\Pr(S_i | \mathrm{I})$ is uniform, which, for convenience, we shall list as a third property:

3. *Prior Probabilities.* The prior $\Pr(S_i | \mathrm{I})$ is uniform ($i = 1, 2, \ldots, N$), where I represents the state of knowledge of the experimenter prior to performing a measurement on the system.

On the assumption of the above correspondence between the Hamilton-Jacobi model and the model $\mathbf{q}^*(N)$, it is now possible to transpose these properties to the model $\mathbf{q}^*(N)$ in the classical limit, and then to generalize the abstract quantum model $\mathbf{q}^N$. This transposition immediately yields the Invariance, Temporal Evolution, and Prior Probabilities postulates.

3. Novel Postulates

Postulate 2.2: State Interpretation.

According to the State Representation postulate, the state $S(t)$, written with respect to some measurement $\mathbf{A} \in \mathscr{A}$, consists of the pair $(\vec{P}, \vec{\chi})$, where $\vec{P}$ contains the probabilities of the observed outcomes, and $\vec{\chi}$ is an ordered set of real-valued degrees of freedom. Hence, the state consists of a mixture of probabilities and degrees of freedom unconnected to probabilities. The State Interpretation postulate is motivated by the aesthetical desideratum that a quantum state consist, as far as possible, of probabilities of events, rather than being such a mixture.

Accordingly, we postulate that χ_i encodes the probabilities of some events, labeled a and b. Hence, when measurement $\mathbf{A}$ is performed on the system, one of $2N$ possible outcomes is obtained, with probabilities determined by the state of the system. Since, by the Outcome Number postulate, an experimenter observes only one of N possible outcomes upon performing measurement $\mathbf{A}$, we are forced to postulate that, for some reason to be investigated later, the outcomes a and b are not observed by the experimenter.

Now, we make the reasonable assumption that the abstract quantum framework being developed is capable of modeling the behavior of a photon when subject to polarization measurements, and that this model will agree with the predictions of electromagnetism under a particle interpretation. Now, an electromagnetic plane wave of constant amplitude moving along the $+z$-direction is described by the vector-valued function $\vec{E} = E_0(\cos\theta \, \vec{i} + \sin\theta \, \vec{j})$, and the information about the polarization of the wave is contained in $(\cos\theta, \sin\theta)$ with respect to polarization measurements in the xy-plane. In the particle interpretation, the probability that a photon will pass through a polarizer whose axis points along the x-axis or y-axis is given by $\cos^2\theta$ or $\sin^2\theta$, respectively. The key feature which we wish to abstract from this example is that, since the map from $(\cos\theta, \sin\theta)$ (the 'state-level') to $(\cos^2\theta, \sin^2\theta)$ (the 'probability-level') is many-to-one, the computed probabilities are *not* the fundamental quantities when describing the state of the photon. Rather, the more fundamental quantities are $\cos\theta$ and $\sin\theta$, which we can regard as square roots of probability in the range $[-1, 1]$, which are squared to obtain probabilities.

To incorporate this two-layered feature into the abstract quantum model, we assume that, following the realization of outcome a or b, one of two outcomes, labeled $+$ and $-$, is obtained. This ensures that one binary-valued degree of freedom is associated with each of the $2N$ possible probabilistically-determined outcomes. Furthermore, we assume that the value of χ_i *determines* whether $+$ or $-$ is obtained via the sign of either $Q_{a|i}$ or $Q_{b|i}$, depending upon whether a or b was obtained, where $P_{a|i} = Q_{a|i}^2$ and $P_{b|i} = Q_{b|i}^2$. In summary, the quantum state consists of the N probabilities $P_1, \ldots, P_N$ and the $2N$ quantities $Q_{a|1}, Q_{b|1}, \ldots, Q_{a|N}, Q_{b|N}$ which encode the probabilities $P_{a|1}, P_{b|1}, \ldots, P_{a|N}, P_{b|N}$ and encode the values of the $2N$ binary-valued degrees of freedom.

In the discussion, we sketch some ideas which help to provide a better physical understanding of this postulate.

Postulate 2.3: Information Gain.

Suppose that, in trial 1 of n runs of an experiment, a measurement **A** is performed on a system in state $\mathbf{S}(t)$, and suppose that trial 2 is identical to trial 1 except that measurement **A'** is performed instead of **A**. The data obtained in trials 1 and 2 provides information (via the Shannon-Jaynes entropy functional) about $S(t)$. If, in one of the two trials 1 and 2, the data obtained yields more information about the state $S(t)$ than in the other trial, this means that one of the two measurements **A** and **A'** is privileged compared to the other insofar as the amount of information that it yields about $\mathbf{S}(t)$. Although this possibility cannot be ruled out *a priori*, we make the intuitively plausible assertion that, although these different measurements provide different perspectives on the system, these perspectives are not informationally privileged.

Now, by the Measurement Representation postulate, trial 2 is equivalent (insofar as the probabilities of the probabilistically-determined outcomes are concerned) to trial $2'$ consisting of n runs of an experiment where a system in state $\mathbf{S}(t)$ is sent through an arrangement consisting of a suitable physical interaction with the system, represented by map $\mathscr{M}$, followed by measurement **A**, followed by another physical interaction. Since the data obtained in trials 2 and $2'$ is statistically identical (as ensured by the Measurement Representation postulate), the amount of information obtained about $S(t)$ in trial 2 is asymptotically equal to the amount of information obtained about $S'(t) = \mathscr{M}(S(t))$ in trial $2'$

But, our assertion that the amount of information obtained about $S(t)$ in trials 1 and 2 is the same therefore implies that the amount of information obtained in trial 1 about $S(t)$

and in trial $2'$ about $S'(t)$ is the same. Hence, we are led to the Information Gain principle. That is, the postulate can be understood as arising from the requirement that no measurement in the measurement set provides an informationally privileged perspective on the system.

DEDUCTION OF THE QUANTUM FORMALISM

In this section, we shall indicate the broad steps which lead to the deduction of the quantum formalism from the above postulates. The reader is referred to [16] for details.

- **Step 1.** Obtain State Space using the States Postulates (2.1–2.4).
    - Postulates 2.1–2.3 imply that:
        * States are represented as unit vectors, $\vec{Q}$, in a $2N$-dimensional Euclidean space.
        * The prior over the unit hypersphere sphere is uniform..
        * After n runs of experiment, the posterior is symmetric Gaussian over one orthant with standard deviation $1/2\sqrt{n}$, and is zero in other orthants.
    - Postulate 2.4 (Prior Probabilities) then implies that the functions $f(\chi_i)$ and $\tilde{f}(\chi_i)$ in the State Interpretation postulate can be taken, without loss of generality, to be $\cos\chi_i$ and $\sin\chi_i$, respectively.
- **Step 2.** Obtain the set of allowable Transformations using the Transformations Postulates (3.1–3.6).
    - The One-to-one and Consistency Postulates imply that transformations are orthogonal transformations of the unit hypersphere.
    - The Invariance postulate implies that the transformations are a particular subset of the orthogonal transformations.
    - This subset of the orthogonal transformations can equivalently be represented by the set of all unitary and antiunitary transformations of a suitably-defined N-dimensional complex vector space.
    - The Parameterized Transformations and Continuous Transformations postulates then imply that continuously-parameterized transformations are either unitary or antiunitary, and that continuous transformation are unitary.
    - The Temporal Evolution postulate, together the the Average-Value Correspondence Principle, gives the explicit form of the temporal evolution operator.
- **Step 3.** Obtain the Hermitian operator representation of measurements using the Measurement Representation postulate (1.2).

DISCUSSION

Owing to the transparency of the assumptions used in the derivation, it is apparent from the derivation that the concept of information plays a substantial role in giving rise to various structural features of the quantum formalism, and can reasonably be said to provide the backbone of the quantum formalism. In particular, the information gain condition directly leads to Q-space, which introduces square-roots of probability, or *real* amplitudes and, via the State Interpretation postulate, leads to a 2N-dimensional Q-space. Furthermore, in conjunction with the Prior Probabilities postulate, the Information Gain postulate leads to the function $f(\chi_i) = \cos(\chi_i)$. Hence, the sinusoidal functions into which the phases in a quantum state enter can be directly traced to the concept of information. Finally, the prior over the unit hypersphere in Q^{2N}-space induced by the imposition of the Information Gain postulate leads, via the Consistency postulate, to the strong constraint that physical transformations can only be represented by orthogonal transformations of the unit hypersphere.

The derivation provides a number of insights into the quantum formalism. First, the derivation given above gives rise to a mathematical structure that is neither more nor less general than the finite-dimensional abstract quantum formalism, and thereby lends support to the view that the quantum formalism is the most general formalism for the description of quantum phenomena in flat space-time.

Second, the use of complex numbers in the quantum formalism is perhaps one of its most mysterious mathematical features. The emergence of complex numbers in the derivation depends on most of the postulates, and so is not easy to unravel, but the role of the Invariance postulate is perhaps the most obvious: its imposition leads to the possibility of representing the set of all possible orthogonal transformations of Q^{2N} by the set of all unitary or antiunitary transformations of a suitably defined complex vector space. This is particularly interesting since it suggests that the importance of complex numbers is directly tied to the set of possible transformations.

Discussion of the State Interpretation postulate

The above derivation rests upon two key postulates, the Information Gain postulate and the State Interpretation postulate, the remainder of the postulates being traceable to elementary experimental observations and to classical physics. Of these two key postulates, the State Interpretation postulate invites the most immediate questions: (i) when a measurement is performed on the system, why is the experimenter able to observe outcomes $1,\ldots,N$, but not the outcomes a,b and $+,-$, and (ii) how one can more intuitively understand the physical meaning of these outcomes. A tentative answer to these questions is as follows.

First, for a system in an eigenstate of energy E, the overall phase of its quantum state (in the complex representation) changes at the rate $E/\hbar$. Consequently, the probabilities $P_{a|i}$ and $P_{b|i}$ are oscillating at frequency $2E/h$, and the signs of $Q_{a|i}$ and $Q_{b|i}$ are switching at frequency E/h. Now, it is reasonable to expect that, if one wishes to observe the actualization of one of the possible outcomes a,b and $+,-$, the measurement performed must have a temporal resolution $\Delta t \ll h/2E$. If the measurement does not have such resolution, it seems plausible to suppose that none of the possible outcomes will be cleanly actualized, leading to the situation where each measurement performed on the system effectively yields *all* of the possible outcomes, forming an unchanging, smeared 'background' to the outcomes $1,\ldots,N$, so that one would remain unaware that other outcomes (other than $1,\ldots,N$) are being actualized.

Now, from the energy-time uncertainty relation $\Delta E \Delta t \geq \hbar/2$, it then follows that ΔE

must be of the order of the rest energy of the system. A measurement of such energy would therefore probably not preserve the identity of the system. Hence, a measurement with the requisite temporal resolution cannot be consistently described within the quantum formalism. Conversely, a measurement which, with high probability, preserves the identity of the system, will have insufficient temporal resolution to resolve the outcomes a, b and $+, -$.

Second, the outcomes a, b and $+, -$ can be graphically represented as follows. Consider a unit circle in Euclidean space, with orthogonal axes labeled a and b, respectively, and consider a unit vector at angle χ_i to the a-axis. Since the prior $\Pr(\chi_i|I)$ is uniform, the vector is equally likely to be pointing in any direction. When the measurement $\mathbf{A}$ is performed, the outcomes a and b register which axis the vector is found to be pointing along, and the additional $+$ and $-$ outcomes determine whether the vector is pointing along the positive or negative direction along the respective axis. From repeated trials, one obtains information about the probabilities $P_{a|i}$ and $P_{b|i}$, from which one can infer that the angle χ_i is one of four possibilities; the signs $+$ or $-$ associated with a and b determine which of the four possibilities is the correct one.

Under temporal evolution, the vector rotates at angular frequency $E/\hbar$ if the system is in an eigenstate of energy E. This picture appears to be closely related to the idea that massive particles can be regarded as energy that is 'trapped' in a region of space and is undergoing rapid, to-and-fro motion (an idea that was shown by Einstein to quantitatively account for the inertia of mass), and related to Hestenes' [18, 19] contention that there is a localized circular motion which accounts for electron spin, but the elucidation of these connections is left to future work.

REFERENCES

1. C. Rovelli, *Int. J. Theor. Phys.* **35**, 1637–1678 (1996), see quant-ph/9609002v2.
2. S. Popescu, and D. Rohrlich (1997), see quant-ph/9709026v2.
3. C. A. Fuchs (2002), see quant-ph/0205039.
4. J. A. Wheeler, "It from bit," in *Proceedings of the 3rd international symposium on the foundations of quantum mechanics, Tokyo*, 1989.
5. J. Summhammer (1999), see quant-ph/9910039.
6. A. Zeilinger, *Found. Phys.* **29**, 631 (1999).
7. W. K. Wootters, *The acquisition of information from quantum measurements*, Ph.D. thesis, University of Texas at Austin (1980).
8. Č. Brukner, and A. Zeilinger, *Phys. Rev. Lett.* **83**, 3354–3357 (1999).
9. Č. Brukner, and A. Zeilinger, *Phys. Rev. A* **63** (2001).
10. Č. Brukner, and A. Zeilinger (2002), see quant-ph/0212084v1.
11. A. Caticha, *Phys. Rev. A* **57**, 1572 (1998).
12. A. Caticha (1999), see quant-ph/9810074v2.
13. R. Clifton, J. Bub, and H. Halvorson (2003), see quant-ph/0211089v2.
14. A. Grinbaum, *Int. J. Quant. Inf.* **1**, 289–300 (2003), see also quant-ph/0306079.
15. A. Grinbaum, *The Significance of Information in Quantum Theory*, Ph.D. thesis, Ecole Polytechnique, Paris (2004), see quant-ph/0410071.
16. P. Goyal (2007), see quant-ph/0702124.
17. P. Goyal (2007), see quant-ph/0702149.
18. D. Hestenes, *Found. Phys.* **15**, 63–87 (1985).
19. D. Hestenes, *Found. Phys.* **20**, 1213–1232 (1990).

From Information Geometry to Newtonian Dynamics

Ariel Caticha and Carlo Cafaro

Department of Physics, University at Albany–SUNY, Albany, NY 12222, USA

Abstract. Newtonian dynamics is derived from prior information codified into an appropriate statistical model. The basic assumption is that there is an irreducible uncertainty in the location of particles so that the state of a particle is defined by a probability distribution. The corresponding configuration space is a statistical manifold the geometry of which is defined by the information metric. The trajectory follows from a principle of inference, the method of Maximum Entropy. No additional "physical" postulates such as an equation of motion, or an action principle, nor the concepts of momentum and of phase space, not even the notion of time, need to be postulated. The resulting entropic dynamics reproduces the Newtonian dynamics of any number of particles interacting among themselves and with external fields. Both the mass of the particles and their interactions are explained as a consequence of the underlying statistical manifold.

Keywords: Information Geometry, Entropic Dynamics, Jacobi Action
PACS: 04.20.Cv,02.50.Tt,02.40.Ky

INTRODUCTION

It is widely assumed that geometry is useful because it describes properties of the real world. Indeed, Euclidean geometry may very well have been the first successful physics theory, the first example of a "law of nature". Later developments such as Riemannian geometry and the theory of fiber bundles have only strengthened this conception: geometry works because it lies at the very core of physics. Thus, it may be surprising, at least at first sight, to find that the same methods of geometry have also turned out to be useful in statistical inference, a separate field that makes no claims to authority on natural phenomena. It could just be a coincidence but perhaps it is not.

Perhaps the laws of physics are deeply geometrical because they are practical rules to process information about the world and geometry is the uniquely natural tool to do just that. This notion, that *the laws of physics are not laws of nature but rules of inference*, seems outrageous but deserves serious attention. The evidence supporting it is already considerable. Indeed, most of the formal structure of statistical mechanics [1] and of quantum theory [2] can already be derived from principles of inference (consistency, probabilities, entropy, etc.).

The objective of this paper is to use well established principles of inference to derive Newtonian dynamics from relevant prior information codified into a statistical model. The challenge, of course, is to accomplish this task without assuming what we want to derive. One must not assume equations of motion or principles of least action, and in particular, one must not assume the concept of momentum and the associated phase space, and not even the notion of an absolute Newtonian time.

The first step is to construct a suitable statistical model of the space of states of a system of particles. A most remarkable fact is that the statistical configuration space is automatically endowed with a geometry and that this "information" geometry turns out to be unique [3][4].

Next we tackle the dynamics: Given the initial and the final states, what trajectory is the system expected to follow? In the usual approach one postulates an equation of motion or an action principle that presumably reflects a "law of nature." For us the dynamics follows from a principle of inference, the method of Maximum Entropy, and we show that with a suitable choice of the statistical manifold the resulting "entropic dynamics" [5][6] reproduces Newtonian dynamics.

The entropic dynamics approach allows us to see familiar notions such as time, mass and interactions from an unfamiliarly fresh perspective. For example, there is no reference to an external time but there is an internal "intrinsic" time that is a measure of the change of the system itself. Thus, the Newtonian universe turns out to be its own clock, and the familiar Newtonian time is not particularly fundamental but merely a convenient definition designed to make motion look as simple as possible. Both the mass of the particles and their interactions are explained in terms of an irreducible uncertainty of their positions; they are features of the underlying statistical manifold.

CONFIGURATION SPACE AS A STATISTICAL MANIFOLD

Let us start with a single particle moving in space: the configuration space is a three dimensional manifold with some unknown metric tensor $g_{ij}(x)$. Our main assumption is that there is a certain fuzziness to space; there is an irreducible uncertainty in the location of the particle. Thus, when we say the particle is at the point x what we mean is that its "true" position y is somewhere in the vicinity of x. This leads us to associate a probability distribution $p(y|x)$ to each point x and the configuration space is thus transformed into a statistical manifold: a point x is no longer a structureless dot but a probability distribution.

Remarkably there is a *unique* measure of the extent to which the distribution at x can be distinguished from the neighboring distribution at $x+dx$. It is the information metric of Fisher and Rao [3]. Thus, physical space, when viewed as a statistical manifold, inherits a metric structure from the distributions $p(y|x)$. We will assume that the originally unspecified metric $g_{ij}(x)$ is precisely the information metric induced by the distributions $p(y|x)$.

In [6] we proposed that a Gaussian model,[1]

$$p(y|x) = \frac{\gamma^{1/2}(x)}{(2\pi)^{3/2}} \exp\left[-\frac{1}{2}\gamma_{ij}(x)(y^i - x^i)(y^j - x^j)\right], \tag{1}$$

where $\gamma = \det \gamma_{ij}$, incorporates the physically relevant information which consists of an estimate of the particle position,

$$\langle y^i \rangle = \int dy\, p(y|x) y^i = x^i, \tag{2}$$

---

[1] We adopt the standard summation convention: repeated indices are summed over.

and of its uncertainty given by the covariance matrix,

$$\langle (y^i - x^i)(y^j - x^j) \rangle = \int dy\, p(y|x)(y^i - x^i)(y^j - x^j) = \tilde{\gamma}^{ij}(x), \qquad (3)$$

where $\tilde{\gamma}^{ij}$ is the inverse of γ_{ij}, $\tilde{\gamma}^{ik}\gamma_{kj} = \delta^i_j$.

Unfortunately the expected values in eqs.(2) and (3) are not covariant under coordinate transformations. Indeed, the transformation $y'^i = f^i(y)$ does not lead to $x'^i = f^i(x)$ because in general $\langle f(y) \rangle \neq f(\langle y \rangle)$ except when uncertainties are small. Our Gaussian model can at best be an approximation valid when $p(y|x)$ is sharply localized in a very small region within which curvature effects are negligible. Fortunately this is all we need for our present purpose.

[As an interesting aside we note that it is possible to devise fully covariant models. Here is an example: Let $\gamma_{ij}(x)$ be a positive definite tensor field and let us use it as if it were a metric tensor, $d\ell^2 = \gamma_{ij} dx^i dx^j$. Let $\ell(x,y)$ be the γ-length along the γ-geodesic from the point x to the point y. The proposed distribution is

$$p(y|x) = \frac{1}{\zeta} \gamma^{1/2}(y) \exp -\frac{\ell^2(x,y)}{2\sigma^2(x)}, \qquad (4)$$

which is a manifestly covariant object: the normalization constant ζ, the γ-length $\ell(x,y)$, the scalar field $\sigma(x)$, and $dy\,\gamma^{1/2}(y)$ are all invariants. From this model we can compute a second metric, the information metric g_{ij}, which need not in general coincide with γ_{ij}. In the limit of small uncertainties (after absorbing σ into γ_{ij}) one recovers eq.(1).]

THE INFORMATION METRIC

The information distance between $p(y|\theta)$ and $p(y|\theta + d\theta)$ where the θ^a are parameters is calculated from (see e.g., [3])

$$d\ell^2 = G_{ab} d\theta^a d\theta^b \quad \text{with} \quad G_{ab} = \int dy\, p(y|\theta) \frac{\partial \log p(y|\theta)}{\partial \theta^a} \frac{\partial \log p(y|\theta)}{\partial \theta^b}. \qquad (5)$$

Consider the 9-dimensional space of Gaussians

$$p(y|x,\gamma) = \frac{\gamma^{1/2}}{(2\pi)^{3/2}} \exp\left[-\frac{1}{2}\gamma_{ij}(y^i - x^i)(y^j - x^j)\right]. \qquad (6)$$

Here the parameters θ^a include the three x^i plus six independent elements of the symmetric matrix γ_{ij}. Eq.(5) gives the information distance between $p(y|x,\gamma)$ and $p(y|x+dx, \gamma+d\gamma)$ as

$$d\ell^2 = G_{ij} dx^i dx^j + G^{ij}_k d\gamma_{ij} dx^k + G^{ijkl} d\gamma_{ij} d\gamma_{kl}, \qquad (7)$$

where

$$G_{ij} = \gamma_{ij}, \quad G^{ij}_k = 0, \quad \text{and} \quad G^{ijkl} = \frac{1}{4}(\tilde{\gamma}^{ik}\tilde{\gamma}^{jl} + \tilde{\gamma}^{il}\tilde{\gamma}^{jk}). \qquad (8)$$

($\tilde{\gamma}^{ik}$ is the inverse of γ_{kj}.) Therefore,

$$d\ell^2 = \gamma_{ij}dx^i dx^j + \frac{1}{2}\tilde{\gamma}^{ik}\tilde{\gamma}^{jl}d\gamma_{ij}d\gamma_{kl}.\tag{9}$$

This is the metric of the full 9-dimensional manifold, but it is not what we need.

What we want is the metric of the embedded 3-dimensional submanifold where $\gamma_{ij} = \gamma_{ij}(x)$ is some function of x. To find the induced metric we cannot just substitute $d\gamma_{ij} = \partial_k \gamma_{ij} dx^k$ into eq.(9) because under a change of coordinates dx^k transforms as a tensor but the ordinary derivative $\partial_k \gamma_{ij}$ does not. In a model of physical space the i indices in x^i cannot be treated independently from the ij indices that appear in γ_{ij} because any transformation that changes the x^i also changes the γ_{ij}. Accordingly, we require that $d\gamma_{ij} = \nabla_k \gamma_{ij} dx^k$ where ∇_k is the covariant derivative and the corresponding induced information metric is

$$g_{ij} = \gamma_{ij} + \frac{1}{2}\tilde{\gamma}^{ac}\tilde{\gamma}^{bd}\nabla_i \gamma_{ab}\nabla_j \gamma_{cd}.\tag{10}$$

Normally one is given a manifold of probability distributions and the problem is to find the corresponding information metric. In order to do physics we are also concerned with the inverse problem: we want to design statistical manifolds with the appropriate geometries. We want to find the covariance field tensor $\gamma_{ij}(x)$ that leads to a given metric tensor $g_{ij}(x)$. Thus, we regard eq.(10) as a set of differential equations for $\gamma_{ij}(x)$. Since $\nabla_k g_{ij} = 0$,[2] a straightforward substitution shows that the solution is

$$\gamma_{ij}(x) = g_{ij}(x).\tag{11}$$

In words: *information distance is measured in units of the local uncertainty*. This beautifully simple but non-trivial result is valid in the low uncertainty regime where eq.(1) holds. The uniqueness of the solution (11), and whether it also holds in high curvature regions, such as near singularities, remains to be ascertained.

ENTROPIC DYNAMICS FOR A SINGLE PARTICLE

The key to the question "Given an initial and a final state, what trajectory is the system expected to follow?" lies in the implicit assumption that there exists a continuous trajectory. A large change is the result of a succession of very many small changes and *therefore* we only need to determine what a short segment of the trajectory looks like. The idea behind entropic dynamics is that as the system moves from a point x to a neighboring point $x + \Delta x$ it must pass through a halfway point [5].

The basic dynamical question can now be rephrased as follows: The system is initially described by the probability distribution $p(y|x)$ and we are given the information that it has moved to one of the neighboring states in the family $p(y|x')$ where the x' lie on the

---

[2] The choice of the Levi-Civita connection is justified in the next section.

plane halfway between the initial x and the final $x + \Delta x$. Which $p(y|x')$ do we select? The answer is given by the method of maximum (relative) entropy, ME. The selected distribution is that which maximizes the entropy of $p(y|x')$ *relative* to the prior $p(y|x)$ subject to the constraint that x' is equidistant from x and $x + \Delta x$. The result is that the selected x' minimizes the distance to x and therefore the three points x, x' and $x + \Delta x$ lie on a straight line.

Since any three neighboring points along the trajectory must line up, the trajectory predicted by entropic dynamics is the geodesic that minimizes the length

$$J = \int_{\lambda_i}^{\lambda_f} d\lambda \left[g_{ij} \dot{x}^i \dot{x}^j \right]^{1/2} \quad \text{with} \quad \dot{x}^i = \frac{dx^i}{d\lambda}, \tag{12}$$

where λ is any parameter that labels points along the curve, $x^i = x^i(\lambda)$.

Incidentally, note that in entropic dynamics there is one family of curves that is singled out as special: these are the minimal-length geodesics. From the purpose of building useful physics models no additional structure is needed and thus none will be introduced. It is therefore natural to use this same family of curves to *define* the notion of parallelism: the minimal-length geodesics are defined to be the straightest curves. This definition leads to the Levi-Civita connection which is equivalent to the condition $\nabla_k g_{ij} = 0$ assumed in the previous section. (See e.g. [7])

The simplest statistical model is a three-dimensional manifold of spherically symmetric Gaussians with constant variance σ_0^2. The corresponding information metric is

$$g_{ij}^{(0)}(x) = \gamma_{ij}^{(0)}(x) = \frac{1}{\sigma_0^2} \delta_{ij}, \tag{13}$$

which we recognize as the familiar metric of flat Euclidean space. It is reassuring that already in such a simple model entropic dynamics reproduces the familiar straight line trajectories that are commonly associated with Galilean inertial motion. But this is too simple; non-trivial dynamics requires some curvature.

We are thus led to consider a slightly more complicated model of spherically symmetric Gaussians where the variance is a non-uniform scalar field $\sigma^2(x)$. It is convenient to write the corresponding information metric as the Euclidean metric eq.(13) modulated by a (positive) conformal factor $\Phi(x)$,

$$g_{ij}(x) = \gamma_{ij}(x) = \frac{\Phi(x)}{\sigma_0^2} \delta_{ij}, \tag{14}$$

with $\sigma^2(x) = \sigma_0^2/\Phi(x)$.[3]

It is convenient to rewrite the length eq.(12) with the metric (14) in the form

$$J = 2^{1/2} \int_{\lambda_i}^{\lambda_f} d\lambda L(x, \dot{x}), \tag{15}$$

---

[3] The effect of $\Phi(x)$ is a local dilation. Since each side of a small triangle at x is dilated by the same factor $\Phi(x)$ its angles remain unchanged. Such angle-preserving transformations are called conformal.

with a "Lagrangian" function

$$L(x,\dot{x}) = [\Phi(x)T_\lambda(\dot{x})]^{1/2} \quad \text{with} \quad T_\lambda(\dot{x}) = \frac{1}{2\sigma_0^2}\delta_{ij}\dot{x}^i\dot{x}^j . \tag{16}$$

The geodesics follow from the Lagrange equations,

$$\frac{d}{d\lambda}\frac{\partial L}{\partial \dot{x}^i} = \frac{\partial L}{\partial x^i} , \tag{17}$$

or

$$\frac{1}{\sigma_0^2}\left(\frac{\Phi}{T_\lambda}\right)^{1/2}\frac{d}{d\lambda}\left[\left(\frac{\Phi}{T_\lambda}\right)^{1/2}\frac{dx^i}{d\lambda}\right] = \frac{\partial \Phi}{\partial x^i} . \tag{18}$$

These rather formidable equations can be simplified considerably once we notice that the parameter λ is quite arbitrary. Let us replace the original λ with a new parameter t given by

$$dt = \left(\frac{T_\lambda}{\Phi}\right)^{1/2} d\lambda \quad \text{or} \quad \frac{d}{dt} = \left(\frac{\Phi}{T_\lambda}\right)^{1/2} \frac{d}{d\lambda} . \tag{19}$$

In terms of the new t the equation of motion simplifies to

$$\frac{1}{\sigma_0^2}\frac{d^2x^i}{dt^2} = \frac{\partial \Phi}{\partial x^i} . \tag{20}$$

From eq.(19) the new t is such that

$$\Phi = T_\lambda\left(\frac{d\lambda}{dt}\right)^2 = T_t \quad \text{where} \quad T_t = \frac{1}{2\sigma_0^2}\delta_{ij}\frac{dx^i}{dt}\frac{dx^j}{dt} . \tag{21}$$

Eqs.(20) and (21) are equivalent to Newtonian dynamics. To make it explicit we introduce a "mass" m and a "potential" $\phi(x)$ through a mere change of notation,

$$\frac{1}{\sigma_0^2} = m \quad \text{and} \quad \Phi(x) = -\phi(x) + E \tag{22}$$

where the constant E reflects the freedom to add a constant to the potential. The result is Newton's equation,

$$m\frac{d^2x^i}{dt^2} = -\frac{\partial \phi}{\partial x^i} , \tag{23}$$

and energy conservation,

$$\frac{1}{2}m\delta_{ij}\frac{dx^i}{dt}\frac{dx^j}{dt} + \phi(x) = E , \tag{24}$$

Thus, the constant E is interpreted as energy.

We have just derived $F = ma$ purely from principles of inference applied to the relevant information codified into a statistical model! From eq.(12) onwards our inference approach is formally identical to the Jacobi action principle of classical mechanics [8] but we did not need to know this. Indeed, by a wild stretch of our historical imagination it is perhaps conceivable that had Newton, Lagrange, and Jacobi known less physics and much more inference they might have invented their subject along these lines. Had history actually followed this unlikely course we might not have used the notions of mass m or potential $\phi(x)$ and instead we would have referred to the particle's "intrinsic" position uncertainty σ_0, and how it is modulated throughout space by the field $\Phi(x)$.

The derivation above serves to illustrate the main idea but suffers from two important limitations. First, it applies to a single particle with a fixed constant energy and this means that we deal with an isolated system. Second, while it is true that we have identified a convenient and very suggestive parameter t, how do we know that it actually represents "true" time? Is t the universal Newtonian time or just a parameter that applies only to one particular isolated particle? The original formulation in terms of the "Jacobi" action, eq.(15), is completely timeless; how and where did time sneak in?

The solution to both these problems emerges as we apply the formalism to the motion of the only system known to be completely isolated: the whole universe. Then the fact that the energy is a fixed constant does not represent a restriction. And further, since the preferred time parameter would be associated to the whole universe, it would not be at all inappropriate to call it the *universal* time.

THE WHOLE UNIVERSE: MANY PARTICLES

To simplify our notation we will consider a universe that consists of $N = 2$ particles. The generalization to arbitrary N is trivial. For the 2-particle system the position $x = (x_1, x_2)$ is denoted by 6 coordinates x^A with $A = 1, 2, \ldots 6$. Let $x^A = (x^{i_1}, x^{i_2})$ with $i_1 = 1, 2, 3$ for particle 1 and $i_2 = 4, 5, 6$ for particle 2. A point in the $N = 2$ configuration space is a Gaussian distribution,

$$p(y|x) = \frac{\gamma^{1/2}(x)}{(2\pi)^{3/2}} \exp\left[-\frac{1}{2}\gamma_{AB}(x)(y^A - x^A)(y^B - x^B)\right]. \tag{25}$$

The simplest model for two (possibly non-identical) particles assigns uniform variances σ_1^2 and σ_2^2 to each particle. The corresponding metric, analogous to eq.(13), is

$$g_{AB}^{(0)} = \gamma_{AB}^{(0)} = m_{AB}, \tag{26}$$

where m_{AB} is a constant 6×6 diagonal matrix,

$$m_{AB} = \begin{bmatrix} \delta_{i_1 j_1}/\sigma_1^2 & 0 \\ 0 & \delta_{i_2 j_2}/\sigma_2^2 \end{bmatrix}, \tag{27}$$

where each entry represents a 3×3 matrix. The metric m_{AB} describes a flat space; the trajectories are familiar "straight" lines and the particles move independently of each other; they do not interact. As before, non-trivial dynamics requires the introduction of curvature and the simplest way to do this is through an overall conformal field $\Phi(x)$ with $x = (x_1, x_2)$. Thus we propose

$$g_{AB}(x) = \gamma_{AB}(x) = \Phi(x) m_{AB} . \tag{28}$$

The equation of motion for the $N = 2$ universe is the geodesic that minimizes

$$J = 2^{1/2} \int_{\lambda_i}^{\lambda_f} d\lambda\, L(x_1, x_2, \dot{x}_1, \dot{x}_2) , \tag{29}$$

where

$$L(x, \dot{x}) = [\Phi(x) T_\lambda(\dot{x})]^{1/2} \quad \text{and} \quad T_\lambda(\dot{x}) = \frac{1}{2} m_{AB} \dot{x}^A \dot{x}^B . \tag{30}$$

The Lagrange equations yield,

$$m_{AB} \left(\frac{\Phi}{T_\lambda}\right)^{1/2} \frac{d}{d\lambda}\left[\left(\frac{\Phi}{T_\lambda}\right)^{1/2} \frac{dx^B}{d\lambda}\right] = \frac{\partial \Phi}{\partial x^A} , \tag{31}$$

which suggests introducing a new parameter t defined by

$$dt = \left(\frac{T_\lambda}{\Phi}\right)^{1/2} d\lambda \quad \text{or} \quad \frac{d}{dt} = \left(\frac{\Phi}{T_\lambda}\right)^{1/2} \frac{d}{d\lambda} . \tag{32}$$

In terms of the new parameter the equations of motion are

$$m_{AB} \frac{d^2 x^A}{dt^2} = \frac{\partial \Phi}{\partial x^A} , \tag{33}$$

which, since m_{AB} is a diagonal matrix, is

$$\frac{1}{\sigma_n^2} \frac{d^2 x^{in}}{dt^2} = \frac{\partial}{\partial x^{in}} \Phi(x_1, x_2) , \tag{34}$$

for each of the particles, $n = 1, 2$. Note that the motion of particle 1 depends on the location of particle 2: *these are interacting particles!*

The new time parameter t, eq.(32), is such that

$$\Phi = T_\lambda \left(\frac{d\lambda}{dt}\right)^2 = T_t \quad \text{where} \quad T_t = \frac{1}{2} m_{AB} \frac{dx^A}{dt} \frac{dx^B}{dt} . \tag{35}$$

As before, the equivalence to Newtonian dynamics is made explicit by a change of notation,

$$\frac{1}{\sigma_n^2} = m_n \quad \text{and} \quad \Phi(x) = -\phi(x) + E . \tag{36}$$

The result is

$$m_n \frac{d^2 x^{in}}{dt^2} = -\frac{\partial}{\partial x^{in}} \phi(x_1, x_2) \quad \text{and} \quad \frac{1}{2} m_{AB} \frac{dx^A}{dt} \frac{dx^B}{dt} + \phi(x_1, x_2) = E . \tag{37}$$

The constant E is the total energy of the universe and there are no restrictions on the energy of individual subsystems.

For the conformal factor $\Phi(x_1, x_2)$ we can choose anything we want. For example,

$$\Phi(x_1, x_2) = -v_1(x_1) - v_2(x_2) - u(x_1, x_2) + E , \qquad (38)$$

so the particles can interact with external potentials v_1 and v_2 and also with each other through $u(x_1, x_2)$.

The definition of time t required taking into account all the particles in the universe. This is in accord with the ephemeris time defined by astronomers. We started with a completely timeless theory, eq.(29), and in fact, no *external* time has been introduced. What we have is a convenient t parameter associated to the change of the total system, which in this case is the whole universe. The universe is its own clock; it measures universal time. Incidentally, note that the reparametrization that allowed us to introduce a Newtonian time was possible only because the same conformal factor $\Phi(x)$ applies equally to all particles.

Entropic dynamics offers a new perspective on the concepts of mass and interactions. To see this note that since γ_{AB} is diagonal the distribution (25) turns out to be a product,

$$p(y|x) = p(y_1|x_1, x_2) p(y_2|x_1, x_2) . \qquad (39)$$

Note that although the model represents interacting particles the distribution is a product: the uncertain variables y_1 and y_2 are statistically independent. The coupling arises through conditioning on $x = (x_1, x_2)$.

Let us focus our attention on particle 1; similar remarks also apply to particle 2. The distribution $p(y_1|x_1, x_2)$ is a spherically symmetric Gaussian,

$$p(y_1|x_1, x_2) \propto \exp\left[-\frac{1}{2\sigma_1^2(x_1, x_2)} \delta_{ij} (y^i - x^i)(y^j - x^j) \right] . \qquad (40)$$

The uncertainty in the position of particle 1 is given by

$$\sigma_1(x_1, x_2) = [\Phi(x_1, x_2) m_1]^{-1/2} . \qquad (41)$$

The mass m_1 is interpreted in terms of a uniform background contribution to the uncertainty. Mass is a manifestation of an uncertainty in location; higher mass reflects a lower uncertainty. On the other hand, interactions arise from the non-uniformity of $\sigma_1(x_1, x_2)$ that depends on the location of other particles through the modulating field $\Phi(x_1, x_2)$. It is worthwhile to note that even though this is a non-relativistic model there already appears a "unification" between mass and (potential) energy: they are different aspects of the same thing, the position uncertainty.

FINAL REMARKS

We emphasize that the model we have proposed does not take into account all the dynamical information that we know is relevant – relativistic and quantum effects have not been included. Our model is very restricted. For example, our model invokes two apparently unrelated metrics. There is the metric δ_{ij} of flat 3-dimensional Euclidean space that appears in the kinetic energies and there is the information metric g_{ij} that accounts for mass and interactions and applies to the curved configuration space. This is a reflection of the fact that a system of N particles is described as a point in a $3N$-dimensional configuration space. A better model would have N points living within the same evolving 3-dimensional space.

Furthermore, we have not provided any rationale for how to choose the modulating field $\Phi(x)$. Just as Newton deliberately refrained from explaining the origin of his inverse square forces – *hypothesis non fingo* – so have we refrained from offering any physical hypothesis about the underlying fuzziness of space. It is reasonable to expect that a derivation of general relativity as an example of entropic dynamics would yield important insights on this matter. Preliminary steps in this direction appeared in [6].

What we have done is to show, by exhibiting an explicit example, that the tools of inference – probability, information geometry and entropy – are sufficiently rich that one can construct entropic dynamics models that reproduce recognizable laws of physics. Perhaps all laws of physics can be derived in this way.

REFERENCES

1. E. T. Jaynes: Phys. Rev. **106**, 620 and **108**, 171 (1957); *E. T. Jaynes: Papers on Probability, Statistics and Statistical Physics*, ed. by R. D. Rosenkrantz (Reidel, Dordrecht, 1983).
2. A. Caticha: Phys. Lett. **A244**, 13 (1998); Phys. Rev. **A57**, 1572 (1998); Found. Phys. **30**, 227 (2000) (arXiv.org/abs/quant-ph/9810074); "From Objective Amplitudes to Bayesian Probabilities" in *Foundations of Probability and Physics-4*, ed. by G. Adenier, C. Fuchs, and A. Khrennikov, AIP Conf. Proc. Vol. 889, 62 (2007) (arXiv.org/abs/quant-ph/0610076).
3. S. Amari and H. Nagaoka, *Methods of Information Geometry* (Am. Math. Soc./Oxford U. Press, Providence, 2000).
4. N. N. Čencov: *Statistical Decision Rules and Optimal Inference*, Transl. Math. Monographs, vol. 53, Am. Math. Soc. (Providence, 1981); L. L. Campbell: Proc. Am. Math. Soc. **98**, 135 (1986).
5. A. Caticha, "Entropic Dynamics" in *Bayesian Inference and Maximum Entropy Methods in Science and Engineering*, ed. by R. L. Fry, AIP Conf. Proc. **617**, 302 (2002). (arXiv.org/abs/gr-qc/0109068).
6. A. Caticha, "Towards a Statistical Geometrodynamics" in *Decoherence and Entropy in Complex Systems* ed. by H.-T. Elze (Springer Verlag, 2004) (arXiv.org/abs/gr-qc/0301061); "The Information geometry of Space and Time" in *Bayesian Inference and Maximum Entropy Methods in Science and Engineering*, ed. by K. Knuth, A. Abbas, R. Morris, and J. Castle, AIP Conf. Proc. **803**, 355 (2006) (arXiv.org/abs/cond-mat/0508108).
7. B. F. Schutz, *Geometrical Methods of Mathematical Physics* (Cambridge U. Press, 1980).
8. C. Lanczos, *The Variational Principles of Mechanics* (Dover, New York, 1986).

Information Geometry and Chaos on Negatively Curved Statistical Manifolds

Carlo Cafaro

Department of Physics, University at Albany–SUNY, Albany, NY 12222, USA

Abstract. A novel information-geometric approach to chaotic dynamics on curved statistical manifolds based on Entropic Dynamics (ED) is suggested. Furthermore, an information-geometric analogue of the Zurek-Paz quantum chaos criterion is proposed. It is shown that the hyperbolicity of a non-maximally symmetric $6N$-dimensional statistical manifold $\mathcal{M}_s$ underlying an ED Gaussian model describing an arbitrary system of $3N$ non-interacting degrees of freedom leads to linear information-geometric entropy growth and to exponential divergence of the Jacobi vector field intensity, quantum and classical features of chaos respectively.

INTRODUCTION

Entropic Dynamics (ED) [1], namely the combination of principles of inductive inference (Maximum relative Entropy Methods, [2]) and Information Geometry (IG) [3], is a theoretical framework constructed on statistical manifolds and it is developed to investigate the possibility that laws of physics, either classical or quantum, might reflect laws of inference rather than laws of nature. This paper is a follow up of a series of the author's works [4, 5]. In this article, we use the ED theoretical framework to explore the possibility of constructing a unifying (classical and quantum) criterion of chaos. We assume the system under investigation has $3N$ degrees of freedom, each one described by two pieces of relevant information, its expectation value and its variance (Gaussian statistical variables). This leads to consider an ED model on a non-maximally symmetric $6N$-dimensional statistical manifold $\mathcal{M}_s$. The manifold $\mathcal{M}_s$ has constant negative Ricci curvature proportional to the number of degrees of freedom of the system, $R_{\mathcal{M}_s} = -3N$. An information-geometric analog of the Zurek-Paz quantum chaos criterion is suggested. It is shown that the system explores statistical volume elements on $\mathcal{M}_s$ at an exponential rate. We define a dynamical information-geometric entropy $S_{\mathcal{M}_s}$ of the system and we show it increases linearly in time (statistical evolution parameter) and is proportional to the number of degrees of freedom of the system. The geodesics on $\mathcal{M}_s$ are hyperbolic trajectories. Using the Jacobi-Levi-Civita (JLC) equation for geodesic spread, it is shown that the Jacobi vector field intensity $J_{\mathcal{M}_s}$ diverges exponentially and is proportional to the number of degrees of freedom of the system. Thus, $R_{\mathcal{M}_s}$, $S_{\mathcal{M}_s}$ and $J_{\mathcal{M}_s}$ are proportional to the number of Gaussian-distributed microstates of the system. This proportionality leads to conclude there exists a substantial link among these information-geometric indicators of chaoticity.

THE ED GAUSSIAN MODEL

Given two probability distributions, how can one define a notion of "distance" between them? The answer to this question is provided by IG. As it is shown in [6] and [7], the notion of distance between dissimilar probability distributions is quantified by the Fisher-Rao information metric tensor. We consider an ED model whose microstates span a $3N$-dimensional space labelled by the variables $\{\vec{X}\} = \{\vec{x}^{(1)}, \vec{x}^{(2)}, ..., \vec{x}^{(N)}\}$ with $\vec{x}^{(\alpha)} \equiv (x_1^{(\alpha)}, x_2^{(\alpha)}, x_3^{(\alpha)})$, $\alpha = 1, ..., N$ and $x_a^{(\alpha)} \in \mathbb{R}$ with $a = 1, 2, 3$. We assume the only testable information pertaining to the quantities $x_a^{(\alpha)}$ consists of the expectation values $\langle x_a^{(\alpha)} \rangle$ and the variance $\Delta x_a^{(\alpha)} = \sqrt{\langle (x_a^{(\alpha)} - \langle x_a^{(\alpha)} \rangle)^2 \rangle}$. The set of these expected values define the $6N$-dimensional space of macrostates of the system. A measure of distinguishability among the states of the ED model is achieved by assigning a probability distribution $P(\vec{X}|\vec{\Theta})$ to each macrostate $\vec{\Theta}$ where $\{\vec{\Theta}\} = \{{}^{(1)}\theta_a^{(\alpha)}, {}^{(2)}\theta_a^{(\alpha)}\}$ with $\alpha = 1, 2, ..., N$ and $a = 1, 2, 3$. The process of assigning a probability distribution to each state provides $\mathcal{M}_S$ with a metric structure. Specifically, the Fisher-Rao information metric defined in (7) is a measure of distinguishability among macrostates. It assigns an IG to the space of states.

THE STATISTICAL MANIFOLD $\mathcal{M}_S$

Consider an arbitrary physical system evolving over a $3N$-dimensional space. The variables $\{\vec{X}\} = \{\vec{x}^{(1)}, \vec{x}^{(2)}, ..., \vec{x}^{(N)}\}$ label the $3N$-dimensional space of microstates of the system. Each macrostate may be thought as a point of a $6N$-dimensional statistical manifold with coordinates given by the numerical values of the expectations ${}^{(1)}\theta_a^{(\alpha)} = \langle x_a^{(\alpha)} \rangle$ and ${}^{(2)}\theta_a^{(\alpha)} = \Delta x_a^{(\alpha)} \equiv \sqrt{\langle (x_a^{(\alpha)} - \langle x_a^{(\alpha)} \rangle)^2 \rangle}$. The available information can be written in the form of the following $6N$ information constraint equations,

$$\langle x_a^{(\alpha)} \rangle = \int_{-\infty}^{+\infty} dx_a^{(\alpha)} x_a^{(\alpha)} P_a^{(\alpha)}\left(x_a^{(\alpha)} \big| {}^{(1)}\theta_a^{(\alpha)}, {}^{(2)}\theta_a^{(\alpha)}\right)$$

$$\Delta x_a^{(\alpha)} = \left[\int_{-\infty}^{+\infty} dx_a^{(\alpha)} \left(x_a^{(\alpha)} - \langle x_a^{(\alpha)} \rangle\right)^2 P_a^{(\alpha)}\left(x_a^{(\alpha)} \big| {}^{(1)}\theta_a^{(\alpha)}, {}^{(2)}\theta_a^{(\alpha)}\right)\right]^{\frac{1}{2}}$$

(1)

where $^{(1)}\theta_a^{(\alpha)} = \langle x_a^{(\alpha)} \rangle$ and $^{(2)}\theta_a^{(\alpha)} = \Delta x_a^{(\alpha)}$ with $\alpha = 1, 2, ..., N$ and $a = 1, 2, 3$. The probability distributions $P_a^{(\alpha)}$ are constrained by the conditions of normalization,

$$\int_{-\infty}^{+\infty} dx_a^{(\alpha)} P_a^{(\alpha)} \left(x_a^{(\alpha)} \Big|^{(1)}\theta_a^{(\alpha)}, ^{(2)}\theta_a^{(\alpha)} \right) = 1. \qquad (2)$$

Information theory identifies the Gaussian distribution as the maximum entropy distribution if only the expectation value and the variance are known. The distribution that best reflects the information contained in the prior distribution $m\left(\vec{X}\right)$ updated by the information $\left(\langle x_a^{(\alpha)} \rangle, \Delta x_a^{(\alpha)}\right)$ is obtained by maximizing the relative entropy

$$S\left(\vec{\Theta}\right) = -\int d^{3N}\vec{X} P\left(\vec{X}\Big|\vec{\Theta}\right) \log\left(\frac{P\left(\vec{X}\Big|\vec{\Theta}\right)}{m\left(\vec{X}\right)}\right), \qquad (3)$$

where $m(\vec{X})$ is the prior probability distribution. As a working hypothesis, the prior $m\left(\vec{X}\right)$ is set to be uniform since we assume the lack of prior available information about the system (postulate of equal *a priori* probabilities). Upon maximizing (3), given the constraints (1) and (2), we obtain

$$P\left(\vec{X}\Big|\vec{\Theta}\right) = \prod_{\alpha=1}^{N} \prod_{a=1}^{3} P_a^{(\alpha)} \left(x_a^{(\alpha)} \Big| \mu_a^{(\alpha)}, \sigma_a^{(\alpha)}\right) \qquad (4)$$

where

$$P_a^{(\alpha)} \left(x_a^{(\alpha)} \Big| \mu_a^{(\alpha)}, \sigma_a^{(\alpha)}\right) = \left(2\pi \left[\sigma_a^{(\alpha)}\right]^2\right)^{-\frac{1}{2}} \exp\left[-\frac{\left(x_a^{(\alpha)} - \mu_a^{(\alpha)}\right)^2}{2\left(\sigma_a^{(\alpha)}\right)^2}\right] \qquad (5)$$

and $^{(1)}\theta_a^{(\alpha)} = \mu_a^{(\alpha)}$, $^{(2)}\theta_a^{(\alpha)} = \sigma_a^{(\alpha)}$. The probability distribution (4) encodes the available information concerning the system. Note that we have assumed uncoupled constraints among microvariables $x_a^{(\alpha)}$. In other words, we assumed that information about correlations between the microvariables need not to be tracked. This assumption leads to the simplified product rule (4). However, coupled constraints would lead to a generalized product rule in (4) and to a metric tensor (7) with non-trivial off-diagonal elements (covariance terms). Such generalizations would require more delicate analysis. Deviations from Gaussian-type information constraints and the presence of a nonvanishing correlation coefficient ($r = \frac{\langle (x_1-\mu_1)(x_2-\mu_2) \rangle}{\sigma_1 \sigma_2}$, r is the correlation coefficient of the two dependent random variables x_1 and x_2) among the random microvariables $x_a^{(\alpha)}$ are some of the new topics appearing in a forthcoming paper [8].

METRIC STRUCTURE OF $\mathcal{M}_s$

The dimensionless line element ds between $P\left(\vec{X}\,|\,\vec{\Theta}\right)$ and $P\left(\vec{X}\,|\,\vec{\Theta}+d\vec{\Theta}\right)$ is given by,
$$ds^2 = g_{\mu\nu}d\Theta^\mu d\Theta^\nu \tag{6}$$
where
$$g_{\mu\nu} = \int d\vec{X}P\left(\vec{X}\,|\,\vec{\Theta}\right)\frac{\partial \log P\left(\vec{X}\,|\,\vec{\Theta}\right)}{\partial \Theta^\mu}\frac{\partial \log P\left(\vec{X}\,|\,\vec{\Theta}\right)}{\partial \Theta^\nu} \tag{7}$$
is the Fisher-Rao metric. Substituting (4) into (7), the metric $g_{\mu\nu}$ on $\mathcal{M}_s$ becomes a $6N \times 6N$ matrix M made up of $3N$ blocks $M_{2\times 2}$ with dimension 2×2 given by,
$$M_{2\times 2} = \begin{pmatrix} \left(\sigma_a^{(\alpha)}\right)^{-2} & 0 \\ 0 & 2\times\left(\sigma_a^{(\alpha)}\right)^{-2} \end{pmatrix} \tag{8}$$
with $\alpha = 1, 2, ..., N$ and $a = 1, 2, 3$. From (7), the "length" element (6) reads,
$$ds^2 = \sum_{\alpha=1}^{N}\sum_{a=1}^{3}\left[\frac{1}{\left(\sigma_a^{(\alpha)}\right)^2}d\mu_a^{(\alpha)2} + \frac{2}{\left(\sigma_a^{(\alpha)}\right)^2}d\sigma_a^{(\alpha)2}\right]. \tag{9}$$

We bring attention to the fact that the metric structure of $\mathcal{M}_s$ is an emergent (not fundamental) structure. It arises only after assigning a probability distribution $P\left(\vec{X}\,|\,\vec{\Theta}\right)$ to each state $\vec{\Theta}$.

CURVATURE OF $\mathcal{M}_s$

The Ricci scalar curvature R is given by,
$$R = g^{\mu\nu}R_{\mu\nu}, \tag{10}$$
where $g^{\mu\nu}g_{\nu\rho} = \delta^\mu_\rho$ so that $g^{\mu\nu} = (g_{\mu\nu})^{-1}$. The Ricci tensor $R_{\mu\nu}$ is given by,
$$R_{\mu\nu} = \partial_\varepsilon \Gamma^\varepsilon_{\mu\nu} - \partial_\nu \Gamma^\varepsilon_{\mu\varepsilon} + \Gamma^\varepsilon_{\mu\nu}\Gamma^\eta_{\varepsilon\eta} - \Gamma^\eta_{\mu\varepsilon}\Gamma^\varepsilon_{\nu\eta}. \tag{11}$$
The Christoffel symbols $\Gamma^\rho_{\mu\nu}$ appearing in the Ricci tensor are defined in the standard way,
$$\Gamma^\rho_{\mu\nu} = \frac{1}{2}g^{\rho\varepsilon}\left(\partial_\mu g_{\varepsilon\nu} + \partial_\nu g_{\mu\varepsilon} - \partial_\varepsilon g_{\mu\nu}\right). \tag{12}$$
Using (9) and the definitions given above, we can show that the Ricci scalar curvature becomes
$$R_{\mathcal{M}_s} = -3N < 0. \tag{13}$$

From (13) we conclude that $\mathcal{M}_s$ is a $6N$-dimensional statistical manifold of constant negative Ricci scalar curvature. A detailed analysis on the calculation of Christoffel connection coefficients using the ED formalism for a four-dimensional manifold of Gaussians can be found in [5]. Furthermore, it can be shown that $\mathcal{M}_s$ is not a pseudosphere (maximally symmetric manifold) since its sectional curvature is not constant. Considerations about the negativity of the Ricci curvature as a *strong criterion* of dynamical instability and the necessity of *compactness* of $\mathcal{M}_s$ in "true" chaotic dynamical systems will appear in [8].

CANONICAL FORMALISM FOR THE ED-GAUSSIAN MODEL

At this point, we study the trajectories of the system on $\mathcal{M}_s$. We emphasize ED can be derived from a standard principle of least action (Maupertuis- Euler-Lagrange-Jacobi-type) [1, 9]. The geodesic equations for the macrovariables of the Gaussian ED model are given by,

$$\frac{d^2\Theta^\mu}{d\tau^2} + \Gamma^\mu_{\nu\rho}\frac{d\Theta^\nu}{d\tau}\frac{d\Theta^\rho}{d\tau} = 0 \qquad (14)$$

with $\mu = 1, 2,...,6N$. Observe that the geodesic equations are *nonlinear*, second order coupled ordinary differential equations.

GEODESICS ON $\mathcal{M}_s$

We seek the explicit form of (14) for the pairs of statistical coordinates $(\mu_a^{(\alpha)}, \sigma_a^{(\alpha)})$. Substituting the explicit expression of the Christoffel connection coefficients into (14), the geodesic equations for the macrovariables $\mu_a^{(\alpha)}$ and $\sigma_a^{(\alpha)}$ associated to the microstate $x_a^{(\alpha)}$ become,

$$\frac{d^2\mu_a^{(\alpha)}}{d\tau^2} - \frac{2}{\sigma_a^{(\alpha)}}\frac{d\mu_a^{(\alpha)}}{d\tau}\frac{d\sigma_a^{(\alpha)}}{d\tau} = 0, \quad \frac{d^2\sigma_a^{(\alpha)}}{d\tau^2} - \frac{1}{\sigma_a^{(\alpha)}}\left(\frac{d\sigma_a^{(\alpha)}}{d\tau}\right)^2 + \frac{1}{2\sigma_a^{(\alpha)}}\left(\frac{d\mu_a^{(\alpha)}}{d\tau}\right)^2 = 0. \qquad (15)$$

with $\alpha = 1, 2,...., N$ and $a = 1,2,3$. This is a set of coupled ordinary differential equations, whose solutions are

$$\mu_a^{(\alpha)}(\tau) = \frac{\left(B_a^{(\alpha)}\right)^2}{2\beta_a^{(\alpha)}} \frac{1}{\cosh\left(2\beta_a^{(\alpha)}\tau\right) - \sinh\left(2\beta_a^{(\alpha)}\tau\right) + \frac{\left(B_a^{(\alpha)}\right)^2}{8\left(\beta_a^{(\alpha)}\right)^2}} + C_a^{(\alpha)},$$

(16)

$$\sigma_a^{(\alpha)}(\tau) = B_a^{(\alpha)}\frac{\cosh\left(\beta_a^{(\alpha)}\tau\right) - \sinh\left(\beta_a^{(\alpha)}\tau\right)}{\cosh\left(2\beta_a^{(\alpha)}\tau\right) - \sinh\left(2\beta_a^{(\alpha)}\tau\right) + \frac{\left(B_a^{(\alpha)}\right)^2}{8\left(\beta_a^{(\alpha)}\right)^2}}.$$

The quantities $B_a^{(\alpha)}$, $C_a^{(\alpha)}$, $\beta_a^{(\alpha)}$ are *real* integration constants and they can be evaluated once the boundary conditions are specified. We are interested in investigating the stability of the trajectories of the ED model considered on $\mathcal{M}_s$. It is known [9] that the Riemannian curvature of a manifold is closely connected with the behavior of the geodesics on it. If the Riemannian curvature of a manifold is negative, geodesics (initially parallel) rapidly diverge from one another. For the sake of simplicity, we assume very special initial conditions: $B_a^{(\alpha)} \equiv \Lambda$, $\beta_a^{(\alpha)} \equiv \lambda \in \mathbb{R}^+$, $C_a^{(\alpha)} = 0$, $\forall \alpha = 1, 2, ..., N$ and $a = 1, 2, 3$. However, the conclusion we reach can be generalized to more arbitrary initial conditions. It is worthwhile noticing that, in our case, $\mathcal{M}_s$ is a geodesically complete manifold since every maximal geodesic is well-defined for all temporal parameters τ. Therefore, $\mathcal{M}_s$ represents a natural setting for *global* questions in this Riemannian geometric framework applied to probability theory and the search for a *weak criterion* of chaos can be carried out [8].

LINEARITY OF THE INFORMATION-GEOMETRIC DYNAMICAL ENTROPY

Recall that $\mathcal{M}_s$ is the space of probability distributions $P\left(\vec{X}\middle|\vec{\Theta}\right)$ labeled by $6N$ statistical parameters $\vec{\Theta}$. These parameters are the coordinates for the point P, and in these coordinates a volume element $dV_{\mathcal{M}_s}$ reads,

$$dV_{\mathcal{M}_S} = \sqrt{g}d^{6N}\vec{\Theta} = \prod_{\alpha=1}^{N}\prod_{a=1}^{3}\frac{\sqrt{2}}{\left(\sigma_a^{(\alpha)}\right)^2}d\mu_a^{(\alpha)}d\sigma_a^{(\alpha)}. \quad (17)$$

The volume of an extended region $\Delta V_{\mathcal{M}_s}(\tau; \lambda)$ of $\mathcal{M}_s$ is defined by,

$$\Delta V_{\mathcal{M}_s}(\tau; \lambda) \stackrel{def}{=} \prod_{\alpha=1}^{N}\prod_{a=1}^{3}\int_{\mu_a^{(\alpha)}(0)}^{\mu_a^{(\alpha)}(\tau)}\int_{\sigma_a^{(\alpha)}(0)}^{\sigma_a^{(\alpha)}(\tau)}\frac{\sqrt{2}}{\left(\sigma_a^{(\alpha)}\right)^2}d\mu_a^{(\alpha)}d\sigma_a^{(\alpha)} \quad (18)$$

where $\mu_a^{(\alpha)}(\tau)$ and $\sigma_a^{(\alpha)}(\tau)$ are given in (16). The quantity that encodes relevant information about the stability of neighboring volume elements is the the average volume $\bar{V}_{\mathcal{M}_s}(\tau; \lambda)$,

$$\bar{V}_{\mathcal{M}_s}(\tau; \lambda) \equiv \langle \Delta V_{\mathcal{M}_s}(\tau; \lambda)\rangle_\tau \stackrel{def}{=} \frac{1}{\tau}\int_0^\tau \Delta V_{\mathcal{M}_s}(\tau'; \lambda)d\tau' \stackrel{\tau \to \infty}{\approx} e^{3N\lambda\tau}. \quad (19)$$

This asymptotic regime of diffusive evolution in (19) describes the exponential increase of average volume elements on $\mathcal{M}_s$. The exponential instability characteristic of chaos forces the system to rapidly explore large areas (volumes) of the statistical manifolds. It is interesting to note that this asymptotic behavior appears also in the conventional description of quantum chaos where the entropy increases linearly at a rate determined by the Lyapunov exponents. The linear entropy increase as a quantum chaos criterion was introduced by Zurek and Paz [10]. In our information-geometric approach a relevant variable that can be useful to study the degree of instability characterizing the ED model is the information-geometric entropy quantity defined as,

$$S_{\mathcal{M}_s} \stackrel{\text{def}}{=} \lim_{\tau \to \infty} \log \bar{V}_{\mathcal{M}_s}(\tau; \lambda). \tag{20}$$

Substituting (18) in (19), equation (20) becomes,

$$S_{\mathcal{M}_s} \stackrel{\tau \to \infty}{\approx} 3N\lambda\tau. \tag{21}$$

The entropy-like quantity $S_{\mathcal{M}_s}$ in (21) is the asymptotic limit of the natural logarithm of the statistical weight $\langle \Delta V_{\mathcal{M}_s} \rangle_\tau$ defined on $\mathcal{M}_s$ and it grows linearly in time, a *quantum feature of chaos*. Indeed, equation (21) may be considered the information-geometric analog of the Zurek-Paz chaos criterion. Zurek and Paz considered a chaotic system, a single unstable harmonic oscillator characterized by a potential $V(x) = -\frac{\lambda x^2}{2}$ (λ is the Lyapunov exponent), coupled to an external environment. In the *reversible classical limit*, the von Neumann entropy of such a system increases linearly at a rate determined by the Lyapunov exponent,

$$S_{\text{quantum}}^{(\text{chaotic})} (\text{Zurek-Paz}) \stackrel{\tau \to \infty}{\sim} \lambda\tau. \tag{22}$$

In general, the von Neumann entropy $S_{\text{quantum}}^{(\text{chaotic})}$ is given by

$$S_{\text{quantum}}^{(\text{chaotic})} = -tr(\hat{\rho} \log_2 \hat{\rho}) = -\sum_j \lambda_j \log_2 \lambda_j \tag{23}$$

where the normalized ($tr(\hat{\rho}) = 1$) density operator $\hat{\rho}$ is defined as

$$\hat{\rho} = \sum_j \lambda_j |\psi_j\rangle \langle\psi_j|, \hat{\rho} |\psi_j\rangle = \lambda_j |\psi_j\rangle. \tag{24}$$

Notice that the consideration of $3N$ uncoupled identical unstable harmonic oscillators characterized by potentials $V_i(x) = -\frac{\lambda_i x^2}{2}$ ($\lambda_i = \lambda_j; i, j = 1, 2, ..., 3N$) would simply lead to

$$S_{\text{quantum}}^{(\text{chaotic})} (\text{Zurek-Paz}) \stackrel{\tau \to \infty}{\sim} 3N\lambda\tau. \tag{25}$$

The resemblance of equations (21) and (25), either in the form or the content is astonishing. A detailed discussion about this result and an additional discussion about a possible connection of (21) to the Kolmogorov-Sinai entropy, one of the most powerful indicators of chaos in classical dynamical systems, will appear in [8].

EXPONENTIAL DIVERGENCE OF THE JACOBI FIELD INTENSITY ON $\mathcal{M}_S$

Finally, we consider the behavior of the one-parameter family of neighboring geodesics $\mathcal{F}_{G_{\mathcal{M}_s}}(\lambda) \equiv \left\{\Theta^{\mu}_{\mathcal{M}_s}(\tau;\lambda)\right\}_{\lambda \in \mathbb{R}^+}^{\mu=1,...,6N}$ where,

$$\mu_a^{(\alpha)}(\tau;\lambda) = \frac{\Lambda^2}{2\lambda} \frac{1}{\cosh(2\lambda\tau) - \sinh(2\lambda\tau) + \frac{\Lambda^2}{8\lambda^2}}, \quad (26)$$

$$\sigma_a^{(\alpha)}(\tau;\lambda) = \Lambda \frac{\cosh(\lambda\tau) - \sinh(\lambda\tau)}{\cosh(2\lambda\tau) - \sinh(2\lambda\tau) + \frac{\Lambda^2}{8\lambda^2}}.$$

with $\alpha = 1, 2, \ldots, N$ and $a = 1, 2, 3$. The relative geodesic spread on a (non-maximally symmetric) curved manifold as $\mathcal{M}_s$ is characterized by the Jacobi-Levi-Civita equation, the natural tool to tackle dynamical chaos [11, 12],

$$\frac{D^2 \delta \Theta^{\mu}}{D\tau^2} + R^{\mu}_{\nu\rho\sigma} \frac{\partial \Theta^{\nu}}{\partial \tau} \delta \Theta^{\rho} \frac{\partial \Theta^{\sigma}}{\partial \tau} = 0 \quad (27)$$

where the Jacobi vector field J^{μ} is defined as,

$$J^{\mu} \equiv \delta \Theta^{\mu} \stackrel{def}{=} \delta_\lambda \Theta^{\mu} = \left(\frac{\partial \Theta^{\mu}(\tau;\lambda)}{\partial \lambda}\right)_{\tau} \delta \lambda. \quad (28)$$

Equation (27) forms a system of $6N$ coupled ordinary differential equations *linear* in the components of the deviation vector field (28) but *nonlinear* in derivatives of the metric (7). When the geodesics are neighboring but their relative velocity is arbitrary, the corresponding geodesic deviation equation is the so-called generalized Jacobi equation [13]. Substituting (26) in (27) and neglecting the exponentially decaying terms in $\delta \Theta^{\mu}$ and its derivatives, integration of (27) leads to the following asymptotic expression of the Jacobi vector field intensity,

$$J_{\mathcal{M}_S} = \|J\| = \left(g_{\mu\nu} J^{\mu} J^{\nu}\right)^{\frac{1}{2}} \stackrel{\tau \to \infty}{\approx} 3N e^{\lambda \tau}. \quad (29)$$

Further details on the derivation of this result for a four-dimensional statistical manifold are in [5]. We conclude that the geodesic spread on $\mathcal{M}_s$ is described by means of an *exponentially divergent* Jacobi vector field intensity $J_{\mathcal{M}_s}$, a *classical* feature of chaos. In our approach the quantity λ_J,

$$\lambda_J \stackrel{def}{=} \lim_{\tau \to \infty} \frac{1}{\tau} \ln \left(\frac{\|J_{\mathcal{M}_S}(\tau)\|}{\|J_{\mathcal{M}_S}(0)\|}\right) \quad (30)$$

would play the role of the conventional Lyapunov exponents. In conclusion, we have shown that,

$$R_{\mathcal{M}_S} = -3N, \; S_{\mathcal{M}_S} \stackrel{\tau \to \infty}{\approx} 3N\lambda\tau, \; J_{\mathcal{M}_S} \stackrel{\tau \to \infty}{\approx} 3N e^{\lambda \tau}. \quad (31)$$

The Ricci scalar curvature $R_{\mathcal{M}_s}$, the information-geometric entropy $S_{\mathcal{M}_s}$ and the Jacobi vector field intensity $J_{\mathcal{M}_S}$ are proportional to the number of Gaussian-distributed microstates of the system. This proportionality leads to the conclusion that there exists a substantial link among these information-geometric measures of chaoticity, namely

$$R_{\mathcal{M}_s} \sim S_{\mathcal{M}_s} \sim J_{\mathcal{M}_S}. \tag{32}$$

Equation (32), together with the information-geometric analog of the Zurek-Paz quantum chaos criterion, equation (21), represent the fundamental results of this work. We believe our theoretical modelling scheme may be used to describe actual systems where transitions from quantum to classical chaos scenario occur, but this will be argued elsewhere [8].

FINAL REMARKS

In conclusion, a Gaussian ED statistical model has been constructed on a $6N$-dimensional statistical manifold $\mathcal{M}_s$. The macro-coordinates on the manifold are represented by the expectation values of microvariables associated with Gaussian distributions. The geometric structure of $\mathcal{M}_s$ was studied. The manifold $\mathcal{M}_s$ is a curved manifold of constant negative Ricci curvature $-3N$. The geodesics of the ED model are hyperbolic curves on $\mathcal{M}_s$. A study of the stability of geodesics on $\mathcal{M}_s$ was presented. The notion of statistical volume elements was introduced to investigate the asymptotic behavior of a one-parameter family of neighboring volumes $\mathcal{F}_{V_{\mathcal{M}_s}}(\lambda) \equiv \{V_{\mathcal{M}_s}(\tau; \lambda)\}_{\lambda \in \mathbb{R}^+}$. An information-geometric analog of the Zurek-Paz chaos criterion was suggested. It was shown that the behavior of geodesics is characterized by exponential instability that leads to chaotic scenarios on the curved statistical manifold. These conclusions are supported by a study based on the geodesic deviation equations and on the asymptotic behavior of the Jacobi vector field intensity $J_{\mathcal{M}_s}$ on $\mathcal{M}_s$. A Lyapunov exponent analog similar to that appearing in the Riemannian geometric approach to chaos [14] was suggested as an indicator of chaoticity. We think this is a relevant result since a rigorous relation among curvature, Lyapunov exponents and Kolmogorov-Sinay entropy is still under investigation [15] and since there does not exist a well defined unifying characterization of chaos in classical and quantum physics due to fundamental differences between the two theories [16].

ACKNOWLEDGEMENT

The author is grateful to Dr. Saleem Ali and Adom Giffin for very useful comments and suggestions. Special thanks go to Prof. Ariel Caticha for clarifying explanations on "Entropic Dynamics" and for his constant support and advice during this work.

REFERENCES

1. A. Caticha, "Entropic Dynamics", AIP Conf. Proc. **617**, 302 (2002).
2. A. Caticha and A. Giffin, "Updating Probabilities", AIP Conf. Proc. **872**, 31-42 (2006).
3. S. Amari and H. Nagaoka, *Methods of Information Geometry*, Oxford University Press, 2000.
4. C. Cafaro, S. A. Ali and A. Giffin, "An Application of Reversible Entropic Dynamics on Curved Statistical Manifolds", AIP Conf. Proc. **872**, 243-251 (2006).
5. C. Cafaro and S. A. Ali, "Jacobi Fields on Statistical Manifolds of Negative Curvature", Physica D (2007), doi: 10.1016/j.physd.2007.07.001.
6. R.A. Fisher, "Theory of statistical estimation" Proc. Cambridge Philos. Soc. **122**, 700 (1925).
7. C.R. Rao, "Information and accuracy attainable in the estimation of statistical parameters", Bull. Calcutta Math. Soc. **37**, 81 (1945).
8. S. A. Ali and C. Cafaro, "Towards an Information Geometrodynamical Approach to Classical and Quantum Chaos", accepted for presentation at the "*Ettore Majorana Centre*", Erice-Italy (November, 2007).
9. V.I. Arnold, *Mathematical Methods of Classical Physics*, Springer-Verlag, 1989.
10. W. H. Zurek and J. P. Paz, "Quantum Chaos: a decoherent definition", Physica D **83**, 300 (1995).
11. M. P. do Carmo, *Riemannian Geometry*, Birkhauser, Boston, 1992.
12. C. W. Misner, K. S. Thorne and J. A. Wheeler, *Gravitation*, Freeman & Co., San Francisco, 1973.
13. C. Chicone and B. Mashhoon, "The generalized Jacobi equation", Class. Quantum Grav. **19**, (2002).
14. L. Casetti et al., "Riemannian theory of Hamiltonian chaos and Lyapunov exponents", Phys. Rev. E **54**, (1996).
15. T. Kawabe, "Indicator of chaos based on the Riemannian geometric approach", Phys. Rev. E **71**, (2005).
16. A. J. Scott et al., "Hypersensitivity and chaos signatures in the quantum baker's map", J. Phys. A **39**, (2006).

Gravity from a Probabilistic Point of View

Pete Martin

Spatial Information Research Centre, Department of Information Science, University of Otago, Dunedin, New Zealand. email: pmartin@infoscience.otago.ac.nz

Abstract. Intended as an introduction of the author's research questions, this paper is a further exploration of "probability as a physical motive", an attempt to entertain an alternative to causal, deterministic explanation in science. According to this approach, explanation need not be an account of what *forces* dynamics; explanation may be found in the correlations of dynamics to *possibilities*.

Uniform distribution of mass (near-zero Weyl tensor of space-time curvature) has been suggested by Penrose as that initial condition which accounts for the second law of thermodynamics, as the physical expression of the "MaxEnt" principle. A distribution of mass with respect to gravity is taken as a certain space-time topography, and inquiry is made into how there might be more ways for space-time topography to be irregular than for it to be flat. The attempt to understand the counter-intuitive circumstance of *uniform* distribution representing *dis*-equilibrium, in the case of gravity, leads to discussion of the Machian question of how a configuration may affect, or even effect, the very space in which it is supposed to reside. This leads to speculation on the idea that even state space might depend on state.

Keywords: Gravity, probability, entropy, phase space.
PACS: 01.70.+w; 04.20.Cv; 02.50.-r

INTRODUCTION

Einstein is said to have remarked, "What really interests me is whether God had any choice in the creation of the world" [1]. I'm inverting the question and proposing, "What really interests me is whether God has anything *but* choice in the creation of the world."

It is to be expected that participants of this MaxEnt conference will be inclined to favor the "observer ignorance" interpretation of probability, and perhaps even to regard thermodynamic entropy as "an anthropomorphic concept" (after Jaynes [2]). Yet I am rather inclined to think that we ought to regard probability (or possibility), not strictly as a tool of inference, but as the fundamental basis of physical dynamics [3].

In the spirit of a workshop submission, this paper is meant to be a contribution to a suggested alternative scientific viewpoint, in which the figure of interest would be probability, interpreted physically, rather than cause, interpreted deterministically. Attention is given to freedom instead of constraint. I see this as consonant with the intent of Jaynes' MaxEnt, where "least-biased inference" depends on the assumption of maximal freedom for an unknown quantity, consistent with known constraints. The radical conjecture underlying my research is that constraints may themselves arise from the realization of freedoms: *choice* may account for *law*.

This is a tentative presentation of some ideas, rather than a traditional presentation of research results. The basic ideas will be presented, leading to a sketch of their attempted application to gravity and their extension to some Machian ideas concerning phase space.[1] The author claims to be neither physicist nor philosopher, but offers these ideas with an invitation for workshop participants' feedback and corrections.

BACKGROUND OF ENTROPY DYNAMICS

This research comes after considerable reflection on the apparent fundamental importance of *difference*, very generally speaking, in dynamics, together with years of observation that life is eminently opportunistic, rather than competitive, as it is traditionally supposed to be. Although the notion of "difference" may seem too vague or general to be of much use, the idea that it is fundamental, both in logic and in physics, has been expressed in various ways by many prominent scholars.

George Spencer-Brown, in a work on the foundations of logic [4], begins simply by drawing an abstract distinction, noting that, "there can be no distinction without motive, and there can be no motive unless contents are seen to differ in value." Although his work is the exposition of a calculus conceived as a foundation of symbolic logic, he states that his theme is that "a universe comes into being when a space is severed or taken apart"—that is, when a distiction is made, or when a difference exists.

Pierre Curie argued (in an 1894 paper) that "it is asymmetry that produces phenomena" (Curie's symmetry principle) [5]. Asymmetry represents *difference*, or disequilibrium, and hence the potential to move toward equilibrium. Symmetry is *in*difference, formally speaking, with respect to some transformation, and, as Rosen has shown, can be understood as a concept parallel to that of entropy, or equilibrium [6].

Sadi Carnot, in an early 19th-century work that is generally regarded as the founding work of thermodynamics, stated that "wherever a difference of temperature exists, motive force can be produced" [7]. *Difference* is of the essence here, I claim, not temperature.

The second law of thermodynamics

My study in "the quantification of spatio-temporal order" (i.e. *difference*) has much to do with the second law of thermodynamics. The second law is commonly understood as the law of increasing entropy (the "running down" of the world) but I am trying to understand it as the law of possibilities being realized, via the equilibration of differences. Since thermodynamic entropy is commonly associated with "disorder", I tentatively borrow the term "order" to refer to disequilibrium, or to the sum of differences which might afford the dynamic opportunity for equilibration processes.

Statistical mechanics was conceived to provide a deterministic explanation for thermodynamics, and in particular for the second law. According to Boltzmann, one of its founders, and others, the essence of the second law of thermodynamics is that systems

---

[1] Ernst Mach questioned absolute frames and suggested that inertia may be the effect of the rest of the mass in the universe on a given mass. By "Machian" I mean more generally the view that configurations might determine laws rather than *vice versa*, that space might depend on what is in it.

progress from less probable states to more probable states [8, 9, 10, 11]. The meaning of "less probable" and "more probable" here refers to the number of specific ways (*microstates*) that an observed *macrostate* may be realized. The definition of a macrostate, or "thermodynamic state" may seem unsatisfactory for its arbitrariness, yet we cannot do science without lumping particular states together; science is essentially perception of symmetries, and we proceed by recognizing *equivalence classes* [6]. A macrostate is thus an equivalence class of particular microstates.

The second law of thermodynamics can be understood in terms of the path of a particular system in state space, or phase space, with respect to equivalent states: however you partition the phase space into equivalence-class regions, defined in terms of any thermodynamic parameters, such as temperature or pressure, which are typically averages, the system is expected to progress from smaller regions to larger regions. For this reason Campbell suggested that entropy could be associated with a geometric *measure* (generalized volume) of phase space [12], actually an extension of Boltzmann's original insight (engraved on his tombstone) that the entropy S of a macrostate should be a simple monotonic function of the number W of consistent microstates: $S = k\ln W$, where k is Boltzmann's constant.

Penrose derives the second law of thermodynamics as a *consequence* of the extraordinarily unlikely state of the early universe (located in an exceedingly small region of the phase space of the universe), exhibiting the spatio-temporal "order" (in my proposed sense of the word) of near-zero *Weyl* tensor of space-time curvature [11]. This describes a uniform distribution of mass, with the greatest sum of *differences* in locations of centers of mass, and hence the greatest gravitational potential. The second law of thermodynamics is thus seen as the physical realization of MaxEnt inference applied to the location of the universe in phase space. The apparent primacy of the "spatial order" of gravity, cosmologically speaking, motivates the inquiry into how it might be understood in terms of configuration (phase) space *possibilities*, without presupposing any *force*.

Causality and explanation in science

The larger intent of my research is to explore the idea of considering possibilities, or opportunities, as the reason that things happen, rather than insisting on explanation based on causal forces (like gravity).

Such an idea is not without precedent. In a 2003 paper Anandan has argued that "there are no fundamental causal laws but only probabilities for physical processes" [13]. Almost sixty years before, Schrödinger had remarked that "physical laws rest on atomic statistics and are therefore only approximate" [14]. It is apparent that the second law of thermodynamics stood as the first sign (in the tradition of modern science) that there is more to scientific explanation than deterministic causation; quantum mechanics stood as the second sign that the very criteria for acceptable explanation should be expanded.

Ancient Greek philosophy recognized four types of cause (Aristotelian cause): *formal* cause, *material* cause, *efficient* cause, and *final* cause. Modern science (of the past several centuries) has focused on *efficient* cause, or "push", to explain what is observed. But the second law of thermodynamics appears more in the role of *final* cause, or "pull", even if a sense of "aim" or "purpose" is not assumed. Like other maximal principles, such as the Le Chatelier-Braun principle, least-action principle, or Maximum Entropy Production (MEP), such explanations seem unsatisfactory to some, insofar as a deterministic mechanism is not described, even though these principles qualify as "laws" insofar as they describe "rhythms or patterns" that are observed in nature [15], and fulfill the Popperian criteria of scientific theory with respect to prediction and refutation.

Recognizing that scientific law is essentially summary of observations, Solomonoff and Chaitin introduced a formal theory of inductive inference based on algorithmic information theory [16, 17]. In this formalism, scientific laws are certain minimal-length "strings"—instructions for generating all the actual or potential observation "strings" about which they speak. Regarded thus as information compression (*describing* rather than explicitly *exhibiting* strings), scientific laws are relieved of the requirement that they contain any mechanistic or temporal reference.

By the latter I refer to the idea that an explanation must provide a temporally prior cause. But as Rosen points out, there need not be any temporal import even in a cause-effect relation; it is rather a relation of logical implication between members of equivalence classes [6]. To the extent that an event at one time logically determines another event at a previous time, we might as well say that the consequent event *caused* the antecedent event. Moreover, if two events bear a necessary-and-sufficient logical relation of mutual implication to each other, one might say that they are both "cause" and "effect" of each other.

While all the other laws of physical dynamics are apparently symmetrical with respect to the direction of time[2], the second law of thermodynamics introduces time asymmetry: movies of individual billiard-ball collisions may be run backward and appear sensible, but reversed movies of so-called "irreversible" processes encompassing the evolution of averages do not appear sensible.

Eddington [18], Popper [19] and Denbigh [20] have argued that irreversibility (time asymmetry) involves more than second-law effects; specifically time asymmetry seems to have to do with (or to coincide with) expansion, or "spreading out", quite generally speaking. This idea is key to my proposed ideas concerning the correlation of dynamics to *possibilities*, thereby providing explanation by reference to rules of expansion in *possibility space* (phase space), instead of by reference to deterministic causal laws that would constrain or motivate processes.

Weizsäcker and others argue that irreversibility (or temporal asymmetry) is "a precondition of experience" [21], more or less fundamental to consciousness and not to be taken up as something to explain.[3] Alternative to such Kantian ideas, that certain notions of time and space are fundamental to thought itself, are such Machian ideas as that of Barbour [22], that time merely emerges from the universal, timeless configuration space that he calls "Platonia".

---

[2] Except perhaps decay of the long-lived kaon particle

[3] In fact these authors suggest that the temporal *symmetry* of dynamics is the figure that ought to be explained, against the ground of temporal *asymmetry* of conscious experience.

MaxEnt and MEP; "spreading out" in phase space

One of the useful ideas that has come out of non-equilibrium thermodynamics is that of Maximum Entropy Production, or MEP (see, for example, [23]). MEP is the tendency for open systems far from equilibrium to respond to imposed differences, or gradients, by evolving to a steady state in which the production of entropy is maximized. Flow induced by an imposed difference acts to organize the system of interest in such a way that it can conduct the flow as rapidly as possible.[4]

Recently the connection between MEP and MaxEnt has been elucidated by Dewar [24]. While I suppose that the intended interpretation of these results was that they prove that MEP is merely a matter of inference in consideration of "observer ignorance", perhaps they rather suggest the information-theoretic basis of physical dynamics, as possibilities being realized—"spreading out" in phase space. The reason I picked gravity as a case in point is that it seems harder, intuitively, to see why things would tend to collect, rather than to disperse, when "left to their own devices".

Often as an introduction to the concept of thermodynamic entropy, textbook authors present the model of a container with a divider in the middle, with some gas on one side and vacuum on the other. When you remove the divider, the gas molecules disperse, and the measure of entropy increases because of this dispersion, this equilibration of a difference and decrease in free energy.[5]

In the case of gravity, however, matter "wants to" collect; there is maximum potential for evolution (maximum free energy) when matter is evenly dispersed, and minimum potential for evolution (minimum free energy) when matter has condensed, say into a black hole. This picture is rather the opposite of the picture of the confined gas molecules that "want to" disperse.

So you can say, well, sure the matter wants to collect, because there is this force of attraction. You presume the force. My idea is, instead, not to presume the force, but to understand how there might be more possible ways for matter to be collected than for it to be evenly distributed. This involves the interpretation of gravity as curvature of space-time. The core idea is that any assumed space may not be absolute; it may be determined by what is in it.

SPACE AS A FUNCTION OF ITS CONTENT

Geographers are acquainted with the idea that the locations of things, for example cities and transportation routes, change the shape of certain abstract geographic spaces which might represent variant notions of "distance" or "area" (such as "travel time" or "resource consumption"), and moreover that the locations of things affect the future locations of things, because of the mutual relation between object configuration and

---

[4] The "organization" of the system may itself be regarded as a difference—potential or gradient—generated by the preceding flow.
[5] This whole tendency toward dispersion depends of course on the presumed "random" (actually, equilibrated) motion of the gas molecules. From the classical point of view nothing would happen in this model if they were frozen, presumed to be somehow motionless!

abstract space. Considered only as a cartographic device for representation of geographic relationships, this idea of relative space may not seem terribly profound, but here I consider the intriguing idea that distribution of matter might affect the most basic space that we habitually presume to know, and hence might affect the future location of matter. This is the first of two radical ideas that I offer for consideration: that the effect of matter on space (or *vice-versa*—geometrodynamics [1]) might be sufficient to "explain" dynamics, from a MaxEnt point of view.

Biologist Stuart Kauffman, considering questions of the evolution of life, advanced the conjecture that "living systems expand the dimensionality of the adjacent possible as rapidly as possible" [25]. The sense of this, I think, is that life, through evolution, maximizes its own opportunities for further evolution. This seems to be akin to the MEP principle insofar as it describes progress in a state space, not just following a trend but following the steepest trend. Although "the adjacent possible" presumably is a notion of proximity in a state space, the Kauffman conjecture, more radically interpreted, might be taken to suggest that life creates its own future, that the entire possibility space expands. This would seem to be nonsense, if the state space were taken, *a priori*, to encompass all possibilities in space-time. But perhaps we can question whether this notion is itself sensible. This is the second of two radical ideas that I offer for consideration: that states might affect their state space, that the field of possibilities might actually be a function of the configuration.

Toward a MaxEnt view of gravity

Following some ideas of Barbour about a timeless configuration space [22], I consider a first (metaphysical) law of dynamics: all configurations exist in some sense, but the more probable (*i.e.* highly represented as members of a class) configurations are the ones that are observed and considered to be reproducible. A general second law of dynamics might then be supposed: whatever configuration is observed now, we expect to observe a more probable configuration at another time. Finally it may turn out (after MEP) that we should expect to observe a natural expedition of this progress from the less probable to the more probable.

Mass shapes space-time, according to Einstein's general relativity, and a configuration of mass may be identified with a certain space-time topography. If there are more ways for space-time to be irregular than for it to be flat, then that is what one should expect to observe. I do not offer here a complete quantitative development of probabilistic geometrodynamics, but simply describe in general terms how this might be imagined.

Apparently we most readily visualize flat three-dimensional space, and hence we visualize curved space as a surface embedded in a space of higher dimension. In this way the gravitationally-curved space in the vicinity of a given mass may be visualized crudely as a depression in an otherwise flat two-dimensional surface. With this model in mind, one can then comfortably imagine that, the depth of the depression being dependent on the quantity and distribution of mass with which it is associated, there is a "stretching" of the gravitationally-curved space (imagined as a surface embedded in a space of higher dimension) as a consequence of the collection of mass, and this provides more location for mass to occupy, from this point of view.

Non-absolute phase space

Certainly one would expect that any imaginable distribution of mass (or equivalent space-time topography) could be considered as a point in a previously-defined phase space. But after the philosophical tradition of Leibniz, Mach, and Einstein, which has questioned the presumption that space and time should be taken as fixed or absolute frameworks, I wish to question the presumption of a fixed or absolute phase space.

Distinguishability of objects and number of states. By associating distributions of mass with space-time topography, implicitly we are not distinguishing between distributions wherein "different" mass of the same quantity is located in the same place. It is not clear whether it makes sense, even in quantitative analysis of probability, to imagine any individual identity attached to theoretically identical "objects".[6]

Moreover, since the number of possible states of a system depends on the number of objects in the system, and the distinguishability of objects may well depend on the state of the system, perhaps "system state" is an inherently self-referential concept. In that case the state space cannot be regarded as fixed or absolute.

Dimensionality and extent of state space. The dimensionality of phase space is taken to be the number of quantities that can vary, while the extent is taken to be the range of possible values for each of the dimensions. As Popper suggested (presumably referring to physical space), spatial extent must be fixed in order to define entropy [27]. Likewise this would be true in phase space. Yet as observers we have nothing but the present universe from which to infer other possible states; perhaps it is not unthinkable that both the dimensionality and extent of state space might depend on the present state.

Types of numbers used to specify state-space components. Clearly the measure of the field of possibilities in possibility space (state space) depends not only on the range, or bounds, for each of the dimensions, but also on the type of number assumed to specify a component. One might assume a discrete model or a continuous model, with state-space components specified by integers, real numbers, or complex numbers for example. Probability density of states between bounds in any dimension of state space would depend on this specification. If this specification could conceivably depend on the evolution of the system in question, then phase space could not be considered to be absolute.

---

[6] In "What is quantum mechanics trying to tell us?"[26], Mermin offers the answer: "Correlations have physical reality; that which they correlate does not."

SUMMARY

The author's research agenda concerning dynamics regarded as equilibration of difference and realization of possibility has been sketched. Some thoughts on how gravity might be regarded in this way were offered, applying the idea that space depends on the configuration of its content. Finally, the suggestion has been introduced that this idea might apply to phase space itself.

ACKNOWLEDGMENTS

The author is grateful to the people of New Zealand for NZIDRS support of this research, and to the University of Otago department of Information Science for financial support toward attendance of MaxEnt2007.

REFERENCES

1. C. Misner, K. Thorne, and J. Wheeler, *Gravitation*, Freeman, San Francisco, 1973.
2. E. Jaynes, *Physical Review* **106**, 620–630 (1957).
3. P. Martin, *Entropy* **9**, 42–57 (2007).
4. G. Spencer-Brown, *Laws of Form*, George Allen and Unwin, London, 1969.
5. P. Curie, *J. Phys (3rd ser.)* **3**, 393–415 (1894).
6. J. Rosen, *Symmetry in Science*, Springer-Verlag, New York, 1995.
7. E. Mendoza, editor, *Reflections on the Motive Force of Fire by Sadi Carnot and other Papers on the Second Law of Thermodynamics by E. Clapeyron and R. Clausius*, Peter Smith, Gloucester, MA, 1977.
8. L. Boltzmann, "The Second Law of Thermodynamics," in *Theoretical Physics and Philosophical Problems*, edited by B. McGuinness, Reidel, Dordrecht, 1886, collection published in 1974.
9. J. W. Gibbs, *Elementary Principles in Statistical Mechanics*, Dover, New York, 1902, dover republication 1960 of Yale University Press publication 1902.
10. P. Ehrenfest, and T. Ehrenfest, *The Conceptual Foundations of the Statistical Approach in Mechanics (English translation 1959)*, Cornell University Press, Ithaca, 1912.
11. R. Penrose, *The Emperor's New Mind*, Oxford University Press, Oxford, 1989.
12. L. Campbell, *IEEE Transactions on Information Theory* **2**, 112–114 (1965).
13. J. Anandan, *International Journal of Theoretical Physics* **42**, 1943–1955 (2003).
14. E. Schrödinger, *What is Life? The Physical Aspect of the Living Cell*, Cambridge University Press, Cambridge, 1944.
15. R. Feynman, *The Character of Physical Law*, M.I.T. Press, Cambridge, MA, 1967.
16. R. Solomonoff, *Information and Control* **7**, 1–22 and 224–254 (1964).
17. G. Chaitin, *IEEE Transactions on Information Theory* **20**, 10–15 (1974).
18. A. Eddington, *The Nature of the Physical World*, University of Michigan Press, Ann Arbor, 1928.
19. K. Popper, *Nature* **177**, 538 (1956).
20. K. Denbigh, *The British Journal for the Philosophy of Science* **40**, 501–518 (1989).
21. T. Görnitz, E. Ruhnau, and C. Weizsäcker, *International Journal of Theoretical Physics* **31**, 37–46 (1992).
22. J. Barbour, *The End of Time*, Oxford University press, Oxford, 2000.
23. A. Kleidon, and R. Lorenz, editors, *Non-equilibrium Thermodynamics and the Production of Entropy*, Springer, Berlin, 2005.
24. R. Dewar, *Journal of Physics A* **36**, 631–641 (2003).
25. S. Kauffman, *Investigations*, Oxford University Press, Oxford, 2000.
26. N. Mermin, *American Journal of Physics* **66**, 753–767 (1998).
27. K. Popper, *Nature* **207**, 233–234 (1965).

METHODS

The concept of Integrated Data Analysis of complementary experiments

R. Fischer* and A. Dinklage[†]

*Max-Planck-Institut für Plasmaphysik, EURATOM Association,
Boltzmannstr. 2, D-85748 Garching, Germany
[†]Greifswald Branch, Wendelsteinstr. 1, D-17493 Greifswald, Germany

Abstract. The Integrated Data Analysis (IDA) concept allows one to combine data from different experiments to obtain improved results. Heterogeneous and complementary experimental data as well as various kinds of physical prior information are easily integrated employing Bayesian probability theory. The concepts of IDA are compared to the traditional approach for data analysis where sequential analysis and iterative schemes are usually found. In contrast to classical inversion techniques IDA needs only forward modeling and a thorough error assessment: The ingredients are given by a model linking the physical quantities of interest to the measured data, a statistical description of the measurements, and a probabilistic description of all nuisance model parameters suffering from uncertainties. In practice, the probabilistic description of systematic measurement and model uncertainties are of major importance to resolve data inconsistencies. Complex error propagation is obtained automatically combining data in a concise probabilistic one-step analysis.

Key Words: Integrated Data Analysis (IDA), Bayesian probability theory, data consistency

INTRODUCTION

A major step in the analysis of experimental data from nuclear fusion is the coherent combination of measurements from different diagnostics. The goal is to replace the usual combination of results from the analysis of individual experimental data by a combination of the measured data sets for a one-step analysis of pooled data. The analysis of the pooled data allows one to obtain a coherent and unique result from exploiting all information/measurements available. Integrating heterogeneous diagnostics by combining measured data instead of combining inferred results automatically considers all correlations involved in the parameters to be inferred. It is the use of these correlations which allows one to extract more information from given data compared to sequential analysis.

Integrated Data Analysis (IDA) in the framework of Bayesian probability theory offers a unified way of combining all available information. The advantages from an integration of the measured data are manyfold: Physical interdependencies of heterogeneous diagnostics are considered from the beginning and no iterative procedure is necessary [1]. The interdependencies also imply the proper treatment of complex error propagation [2]. A quantitative framework for data validation and consistency checks is provided [3]. A measure for signal credibility or if a measurement should be regarded to be faulty can be provided. An one-step analysis allows to build automated procedures for next generation fusion devices which huge amount of data being analyzed automatically [4]. IDA provides off-line to real-time analysis approaches on different time scales for different purposes [5].

The problems arising from the inversion of ill-posed problems from noisy data are mitigated by providing more data and using only forward modeling of the data. Since the probabilistic approach compels one to make quantitative and testable statements about every piece of information entering the analysis, a full documentation of the analysis process is provided. This is the basis for effective maintenance or revisions of data analysis tools. Consequently, the discussion about the validity of arguments or the credibility of uncertainty measures is based on a quantitative formulation [6, 7]. Analyzing measured data to obtain first-interest quantities is conceptually easy to couple with theory codes, e.g. for the evaluation of transport mechanism in plasmas [8]. In addition to the analysis of measured data from a running experiment, IDA provides in the framework of Bayesian experimental design an approach to optimize future experiments and combination of experiments with respect to physical goals already in the construction phase [9, 10].

The effort for the implementation of IDA consists in a thorough assessment and quantification of all sources of data, additional information, and errors and uncertainties in the measured data as well as in the modeling of the data. The probabilistic formulation of the inference problem in the Bayesian framework is straightforward, but one has to be aware that the necessary elaborate description of the different experiments poses a major effort for the physicists in charge. Quantification of the errors in all measured data and quantification of uncertainties in all model nuisance parameters is often a non-trivial task but is of vital importance for a comparable analysis of heterogeneous diagnostics.

IDA is of great value if large amount of data, additional information and interdependencies exist. This is the case in nuclear fusion but can be extended to other experiments or complex systems for which heterogeneous information is available (data from different measurements, model parameters, physical constraints, etc.). The present work compares the concept of IDA with the traditional data analysis scheme exemplified at a typical use case in fusion.

TRADITIONAL DATA ANALYSIS SCHEME

The left panel of figure 1 depicts a typical flow-chart of a traditional approach for data analysis. Different measurement techniques based on different physical effects were applied to the same experiment in order to estimate the same physical parameters (quantities of interest). Usually, the measurements were analyzed separately although an overlap of the quantities of interest exists. In the present case a Thomson scattering measurement providing electron temperature T_e and density n_e profiles of a plasma and an electron cyclotron emission (ECE) measurement providing T_e only were analyzed individually. Both experimental techniques have their advantages and disadvantages such that they complement each other. It should be noted that the independent analysis of heterogeneous measurements generally found in scientific research is traditionally owing to the personalization of hardware and software developments. De-personalization of software, as routinely realized in industry, is of minor importance in science but has

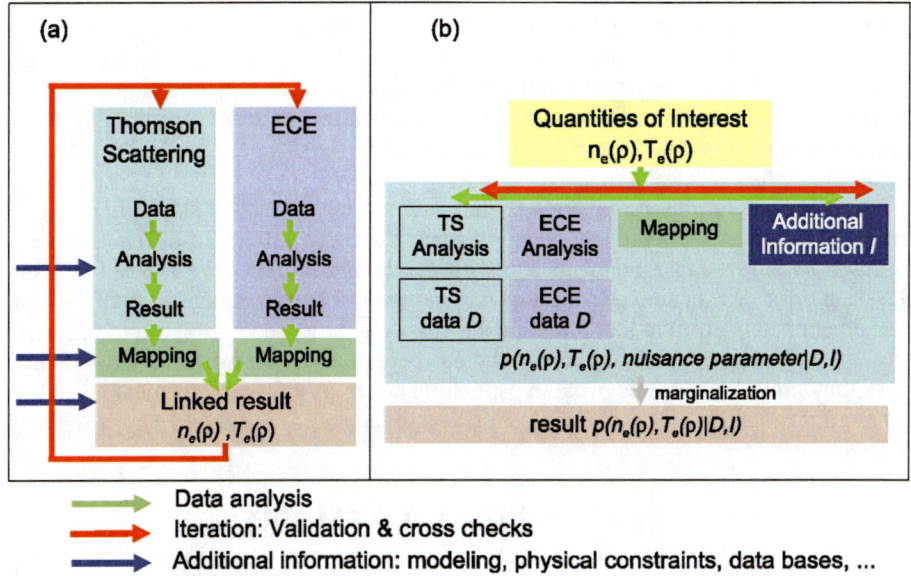

FIGURE 1. Simplified flow-charts for typical data analysis steps inferring electron temperature and density profiles for magnetic confinement fusion experiments from the Thomson scattering and electron cyclotron emission (ECE) diagnostics in (a) the traditional approach and (b) within the IDA concept.

to become more important as the scientific devices increase in complexity.

Since both measurements provide information about the same subset of parameters a combination of the results, e.g. different T_e-profiles, to obtain a unique profile is performed traditionally in a second step. A difficulty of a straightforward combination of the results is given by the fact that the measurements are not performed exactly at the same time and spatial coordinates. Time dependencies or measurements at different spatial positions might corrupt the assumption of having the same values for the quantities of interest. In magnetic confinement fusion devices, measurements at different spatial positions are mapped to a common coordinate system of so called magnetic flux surfaces which are constant pressure surfaces in ideal magneto hydrodynamics [11]. This is depicted in figure 1 as a block which maps the laboratory coordinates $\vec{x}$ of the different measurements on a common magnetic coordinate system $\rho(\vec{x})$. The mapping procedure itself usually suffers from uncertainties since the flux surfaces depend on the plasma pressure and there are usually modeling simplifications [12]. The situation complicates since some of the input quantities of the mapping procedure are the quantities of interest T_e and n_e. T_e and n_e profiles influence the plasma equilibrium and, therefore, the mapping from laboratory to magnetic coordinates. In the traditional approach the interdependencies of the mapping procedure and the different experimental data are solved iteratively. An iterative solution becomes a tedious task if two or more measurements have to be combined with additional information from physical considerations and (uncertain) physical data from other data bases. An automized procedure of the full iteration, e.g. necessary for next generation steady-state fusion devices like W-7X or ITER, appears to be barely feasible.

A severe issue for the combination of measurements is the lack of standardization of error interpretation and treatment which hampers the comparability of different experimental data and results. Statements about estimation uncertainties are at best based on Gaussian error propagation. Additionally, it is often difficult to obtain uncertainties on model parameters given in literature or data bases, e.g. cross sections or atomic data. In subsequent analyses such values are treated as being exactly known although in many cases they provide the leading role in estimation errors. A general agreement about interpretation, quantification and use of errors is still lacking.

INTEGRATED DATA ANALYSIS

The Bayesian approach of IDA provides an alternative scheme for integrating any kind of (uncertain) information. The right panel of figure 1 shows the corresponding flowchart for the two measurements described above. The basic idea is simple: IDA aims to determine the probability of the quantities of interest, given all data and physics assumptions. IDA starts with the quantities of interest, e.g. T_e and n_e, as a function of the relevant coordinates, e.g. the magnetic coordinates ρ. Due to the large number of diagnostics routinely applied to fusion machines the list of physical parameters of interest might become long. Additional parameters of interest are given by, e.g., plasma impurities, particle and energy transport mechanism, the interaction of the hot plasma with the surrounding walls and heating scenarios. Plasma modeling provides additional information which links various of those parameters [8].

Modeling individual diagnostics

With the corresponding subset of the parameter list the measured data of an individual diagnostics is modeled. Modeling of the diagnostics data is usually done independent of IDA and it is often straightforward to implement the already present data descriptive model into the Bayesian framework. First the canonical coordinates ρ have to be mapped on laboratory coordinates of the individual diagnostics $\vec{x}(\rho)$. The mapping procedure which has to be applied here is inverse to the traditional mapping procedure since the point $\vec{x}$ of a diagnostic line of sight needs to be determined for any magnetic coordinate ρ. It is assumed that all parameters necessary for the mapping procedure are provided in the list of quantities of interest. Hence, no iteration is involved in order to obtain a consistent mapping between coordinate systems.

In contrast to classical inversion techniques IDA needs forward modeling only. For a given set of values for the quantities of interest, e.g. T_e and n_e, the calculation of the ideal measured data is usually much simpler to be provided than the solution of the inversion problem. The inverse problem is often ill-posed due to the inevitable noise in the measured data. The forward modeling only has to provide ideal data which means data which would be measured if there would be no statistical measurement noise. The measurement noise enters the probabilistic description with the likelihood probability distribution function (pdf) which quantifies the probability of measuring the actual data given the modeled (ideal) data. The maximum likelihood (ML) principle in orthodox statistics exploits this statistical interpretation and minimizes the misfit between the measured and modeled data with respect to the parameters of interest. Here, the likelihood describes the uncertainty (reliability) of the measured data. It quantifies the plausibility of measuring the data set given the values for the parameters.

An important ingredient of the IDA approach is an elaborate assessment of all uncertainties of the measurement systems. This is necessary to allow for a reliable combination of the heterogeneous diagnostics. The uncertainties arise usually from statistical fluctuations but are often also given by systematic uncertainties. Statistical fluctuations appear in the measured data to be analyzed as well as in data recorded for relative and absolute calibration [2]. Systematic uncertainties may arise due to incomplete/simplified modeling of the physics, due to uncertainties in model (data base) quantities such as cross sections, due to mis-specification of the measurement system or due to non-stationary measurement conditions such as darkening of windows or degeneration of glass fibers. All statistical and systematic uncertainties have to be incorporated in the likelihood pdf or have to be described using prior pdfs with subsequent marginalization of the corresponding nuisance parameters [3]. Another systematic uncertainty can arise from the different measurement techniques of the same physical quantity. An example is given by the different measurement techniques of TS and ECE for determining T_e. TS measures the electron energy distribution in the scattering geometry whereas ECE measures the radiation temperature perpendicular to the magnetic field. So far it is assumed that both temperatures are identical. Furthermore the assumption of thermal equilibrium might be too optimistic resulting in deviations from the Maxwellian velocity distribution.

Additional information

The probabilistic description of the heterogeneous experiments and the mapping procedure have to be complemented by additional information which can easily be provided using prior probability distribution. Examples for additional information arise from simple constraints such as positivity constraints for T_e and n_e up to quite complex constraints with interdependencies of various parameters. In plasma physics monotonicity on the electron pressure $p_e \propto n_e T_e$ is often assumed where the assumption can be relaxed using appropriate prior distributions. Another example is given by the energy balance and particle transport equations which link n_e, T_e and the ion temperature T_i [13].

Please note that such additional informations should enter only once which is not the case if the individual experiments are analyzed separately and the results are combined afterwards. In the present example prior information on the parameters, e.g. T_e, would be used twice if used in the traditional data analysis scheme.

It can be shown that a sequential analysis of data is fully equivalent to a one-step analysis if the full probabilistic approach is applied and the dependencies are considered correctly. Since this is usually not the case for the traditional sequential approach a one-step analysis is to be preferred.

Bayes theorem and marginalization

After a thorough assessment and quantification of all known sources of uncertainties, the likelihood pdfs for the measured data, the prior pdfs for the nuisance parameters and the prior pdfs for additional testable information are multiplied according to Bayes theorem. The advantage of the Bayesian formulation is that any kind of information can be combined since the probabilistic formulation allows any kind of functional form to be multiplied. For example a Gaussian likelihood can be combined with a Poissonian likelihood from another experiment, or a Gamma distribution quantifying prior information about a model parameter, e.g. a cross section.

Subsequent marginalization of all nuisance parameters provide the final result, namely the marginal posterior probability distribution with respect to the quantities of interest which were defined at the beginning. The value of the posterior pdf for the chosen values of the quantities of interest evaluates the probability (plausibility) of the parameter values given the data and information at hand. From this generally multi-dimensional posterior distribution estimates and estimate uncertainties for the quantities of interest can be derived by maximizing the posterior pdf or taking moments of the distribution.

Data consistency and sensitivity study

In a next step the posterior pdf and parameter estimates can be used to check for a consistent description of the measured data. The difference of the measured data point and the modeled data using the estimated parameters weighted with the data uncertainties (residues) indicate if the modeled data and uncertainties are correct. If a consistent model of all data within the quantified uncertainties is not obtained an important ingredient in either the physical model or the uncertainties is missing.

A major step towards a consistent and reliable description of the individual experiments is possible only on the basis of a thorough quantification of all uncertainties known so far. It provides a framework for a quantitative improvement for single measurements. The frequently observed blaming of the *other* experiment for being responsible for systematic deviances is then replaced by a quantitative approach which helps to improve debate culture.

The Bayesian framework does not provide an answer what is missing or wrong in the physical description of the experiments but the probabilistic formulation allows to identify the most crucial issues in the description. Within a sensitivity study of the most important uncertainties one can identify the most promising experimental improvements [3]. Reasons for inconsistent data or wrong assumptions can be identified by appropriate case studies where the description is hypothetically modified. In such a virtual diagnostic the changes in the result can then be checked for being a possible candidate for the inconsistency.

In conclusion, a probabilistic description of systematic measurement and model uncertainties are of major importance to resolve data inconsistencies.

Complex error propagation

Another advantage of the IDA approach is the intrinsic property of complex error propagation. The combination of likelihood and prior probability distributions describing different experiments and additional information contain all interdependencies of parameters both of interest and of nuisance. Marginalization of nuisance parameters results in a propagation of their uncertainties to the quantities of interest. If the interest is in a subset of the original list of quantities, marginalization of the complementary subset yields a propagation of their uncertainties to the subset of quantities of interest. There is no need for applying the familiar Gaussian error propagation laws which assumptions are frequently not fulfilled since the assumption of existing second moments as well as the assumption of an uni-modal probability distribution might fail. Marginalization of uncertain parameters works under any assumption and provides complex error propagation automatically when combining data in a concise probabilistic one-step analysis.

Bayesian probability theory provides thereby a standardization of error interpretation, quantification and use. Statistical and systematic uncertainty are described using probability distributions as a measure for credibility of the information. The χ^2-misfit of data fitting is equivalent to a special case of normally distributed errors.

At first view the complex error propagation produces parameter estimates with larger uncertainties than for the traditional analysis approach. The larger uncertainties result from inclusion of all uncertainties involved. But the uncertainties of the parameter estimates can be even smaller than in the classical approach. This was shown for an example of Thomson scattering data and supplemented electron temperature T_e data from soft X-ray measurements [2]. The accuracy of the electron density estimate n_e increases by 30% although the additional information does not contain information about the density n_e at all. The reason for the increase of the accuracy of n_e is the correlation between T_e and n_e. More information about T_e implies a reduction of uncertainty on n_e. This synergistic effect employing the full correlations due to T_e-n_e interdependencies of the measurements is one of the most convincing arguments in favor of a Bayesian approach for an integrated data analysis although the effect is easily understood *a-posteriori*.

CONCLUSION

The concepts of IDA are compared to the traditional approach for data analysis. IDA provides a framework for combining any kind of measured data and additional information. Heterogeneous and complementary experimental data as well as various kinds of physical prior information can be integrated employing Bayesian probability theory.

An elaborate error analysis of single measurements and modeling is important for a comprehensive analysis of the pooled data. In contrast to classical inversion techniques IDA needs forward modeling only. The probabilistic description of systematic measurement and model uncertainties are of major importance to resolve data inconsistencies. Complex error propagation is obtained automatically combining data in a concise probabilistic one-step analysis.

REFERENCES

1. R. Fischer and A. Dinklage. Integrated data analysis of fusion diagnostics by means of the Bayesian probability theory. *Rev. Sci. Instrum.*, 75:4237–4239, 2004.
2. R. Fischer, A. Dinklage, and E. Pasch. Bayesian modelling of fusion diagnostics. *Plasma Phys. Control. Fusion*, 45:1095–1111, 2003.
3. R. Fischer, C. Wendland, A. Dinklage, S. Gori, V. Dose, and the W7-AS team. Thomson scattering analysis with the bayesian probability theory. *Plasma Phys. Control. Fusion*, 44:1501–1519, 2002.
4. A. Dinklage, R. Fischer, and J. Svensson. Topics and methods for data validation by means of Bayesian probability theory. *Fusion Science and Technology*, 46:355–364, 2004.
5. A. Dinklage, R. Fischer, J. Geiger, G. Kühner, H. Maassberg, J. Svensson, and U. von Toussaint. From off-line to real-time analysis: Accelerating Bayesian analysis codes. In R. Koch and S. Lebedev, editors, *30th EPS Conference on Controlled Fusion and Plasma Physics*, volume ECA 27A, pages P-4.80. Europ. Phys. Soc., Geneva, 2003.
6. A. Dinklage, R. Fischer, M. Hirsch, E. Pasch, A. Weller, and the W7-AS team. Increasing the significance of Thomson scattering data by Bayesian modelling. In R. Koch and S. Lebedev, editors, *30th EPS Conference on Controlled Fusion and Plasma Physics*, volume ECA 27A, pages P-1.52. Europ. Phys. Soc., Geneva, 2003.
7. J. Svensson, A. Dinklage, J. Geiger, and R. Fischer. An integrated data analysis model for the W7-AS stellarator. In R. Koch and S. Lebedev, editors, *30th EPS Conference on Controlled Fusion and Plasma Physics*, volume ECA 27A, pages P-1.65. Europ. Phys. Soc., Geneva, 2003.
8. A. Dinklage, R. Fischer, H. Dreier, J. Svensson, and Yu. Turkin. Integrated approaches in fusion data analysis. In R. Fischer, R. Preuss, and U. von Toussaint, editors, *Bayesian Inference and Maximum Entropy Methods in Science and Engineering*, volume Conf. Proc. 735, pages 43–51, Melville, NY, 2004. AIP.
9. R. Fischer, H. Dreier, A. Dinklage, B. Kurzan, and E. Pasch. Integrated Bayesian experimental design. In K.H. Knuth, A.E. Abbas, R.D. Morris, and J.P. Castle, editors, *Bayesian Inference and Maximum Entropy Methods in Science and Engineering*, volume Conf. Proc. 803, pages 440–447, Melville, NY, 2005. AIP.
10. H. Dreier, A. Dinklage, R. Fischer, M. Hirsch, P. Kornejew, and E. Pasch. Bayesian design of diagnostics: Case studies for Wendelstein 7-X. *Fusion Science and Technology*, 50:262–267, 2006.
11. H. Zohm. Physics of hot plasmas. In A. Dinklage, T. Klinger, G. Marx, and L. Schweikhard, editors, *Plasma Physics - Confinement, Transport and Collective Effects*, volume Lecture Notes in Physics 670, pages 75–93. Springer, Berlin, 2005.
12. J. Svensson, A. Dinklage, J. Geiger, A. Werner, and R. Fischer. Integrating diagnostic data analysis for W7-AS using Bayesian graphical models. *Rev. Sci. Instrum.*, 75:4219–4221, 2004.
13. R. Fischer, A. Dinklage, and Y. Turkin. Non-parametric profile gradient estimation. In *33th EPS Conference on Controlled Fusion and Plasma Physics*. 2006.

Designing Intelligent Instruments

Kevin H. Knuth[a,b], Philip M. Erner[a], Scott Frasso[c]

a. Univ. at Albany, Dept of Physics, Albany NY USA
b. Univ. at Albany, Dept of Informatics, Albany NY USA
c. Northeastern Univ., Dept. of Electrical and Computer Engineering, Boston MA USA
http://knuthlab.rit.albany.edu/

Abstract. Remote science operations require automated systems that can both act and react with minimal human intervention. One such vision is that of an intelligent instrument that collects data in an automated fashion, and based on what it learns, decides which new measurements to take. This innovation implements experimental design and unites it with data analysis in such a way that it completes the cycle of learning. This cycle is the basis of the Scientific Method.
    The three basic steps of this cycle are hypothesis generation, inquiry, and inference. Hypothesis generation is implemented by artificially supplying the instrument with a parameterized set of possible hypotheses that might be used to describe the physical system. The act of inquiry is handled by an inquiry engine that relies on Bayesian adaptive exploration where the optimal experiment is chosen as the one which maximizes the expected information gain. The inference engine is implemented using the nested sampling algorithm, which provides the inquiry engine with a set of posterior samples from which the expected information gain can be estimated. With these computational structures in place, the instrument will refine its hypotheses, and repeat the learning cycle by taking measurements until the system under study is described within a pre-specified tolerance. We will demonstrate our first attempts toward achieving this goal with an intelligent instrument constructed using the LEGO MINDSTORMS NXT robotics platform.

Keywords: intelligent, robotics, experimental design, automation, instrumentation
PACS: 07.05.Bx, 07.05.Dz, 07.05.Fb, 07.05.Hd, 07.05.Kf

INTRODUCTION

Remote science operations are currently being carried out using robotic explorers both on Mars and in deep sea studies here on Earth. These operations, which employ semi-automated systems that can carry out basic tasks such as locomotion and directed data collection, require human intervention when it comes to deciding where to go, which experiment to perform, and precisely where to place the sensors. However, as we expand to explore more remote worlds, we will require that our instruments be increasingly autonomous. The vision we present in this paper is that of an intelligent instrument that collects data in an automated fashion, and based on what it learns, the instrument decides which new measurements to take. The innovation we describe implements automated experimental design and unites the process with automated data analysis in such a way that it completes the cycle of learning.
    Many researchers have worked on the problem of designing intelligent systems. Relevant to our approach are the concepts of cybernetics (Wiener, 1948) and experimental design (Lindley, 1956; Fedorov, 1972), which have been pursued and

FIGURE 1. A photograph of the robotic arm. The end of the arm is equipped with a light sensor that can make point measurements. The robot is built using the LEGO MINDSTORMS NXT system, and is locally controlled with the NXT brick. The NXT brick can communicate with a laptop computer using Bluetooth. The laptop computer (not shown) runs the inference and inquiry code in MATLAB. At the time of the workshop, the MATLAB to NXT communication was not completely operable, and the system was demonstrated via simulations.

unified in various forms by several researchers. Of particular note is the work on cybernetics by Fry (2002), the active data selection approach of MacKay (1992), and maximum entropy sampling and Bayesian experimental design by Sebastiani and Wynn (2000). The maximum entropy sampling approach to experimental design was expounded upon by Loredo (2003) in his work on Bayesian adaptive exploration, which forms the basis of the approach we present here.

The first author of this paper, having been inspired by the work of Cox (1979) and Fry (2002), has been actively developing a calculus for questions (Knuth, 2002, 2003, 2005, 2006) based on bi-valuations on lattices (Knuth, 2007) with an explicit focus on experimental design. However, this framework, which is still in its infancy, is not yet suited for our efforts here. Instead, we employ proven computational technologies.

To create an intelligent instrument, we require three steps: hypothesis generation, experimental design, and data analysis. Hypothesis generation is implemented by programming the instrument with a parameterized model that represents a set of hypotheses that could be used to describe the physical system. Experimental design, which is an act of inquiry, is implemented using Bayesian adaptive exploration (Loredo, 2003), where the optimal experiment maximizes the expected information gain. Finally the data analysis, or inference, is handled using nested sampling

(Skilling, 2005; Sivia & Skilling, 2006), which allows us to test various hypotheses given the newly collected data. At each stage, the instrument will refine its hypotheses and repeat the cycle taking measurements until the system is described within a pre-specified tolerance. In the following sections, we describe our work in the context of a robotic arm solving a characterization problem.

THE EXPERIMENTAL SETUP

Choosing a problem that is at the same time interesting, challenging, and enlightening is extremely difficult. The problem we have chosen is indeed a toy problem, but one that is easily extended to problems encountered in the real world. We consider an instrument that is designed to locate and characterize a white circle on a black field.

The Experimental Problem

We have developed a robotic instrument that is designed to locate and characterize a white circle on a black background. The instrument is equipped with a light sensor which is able to take point measurements. We have purposely designed the system so that the sensor cannot simply scan the visual scene. Such scans result in numerous non-informative measurements that waste time, energy and transmission bandwidth. This limited sensor capability is intentional and will serve to highlight the power of the computational techniques we are developing. In addition, the light sensor has a rather large point spread function, which we will not consider in this initial presentation. Instead, we assume that the light sensor returns a measurement that is normally distributed about the mean light intensity, and ignore "edge-effects".

This is clearly a search problem using an instrument with limited sensor capability. As such, the results here are readily extended to similar problems, such as land mine detection. To characterize the circle, the instrument will continue to take measurements until both the center position of the circle and its radius are known to within a predefined accuracy. Those familiar with information theory will realize that once the white circle has been detected, on average, only a small number of binary questions will be necessary to achieve this. We will show that our results agree with this expectation.

The Robot and its Brains

The instrument is a robotic arm built with the LEGO MINDSTORMS NXT system (Figure 1). The arm has three-degrees of freedom, with the ability to rotate about the vertical axis (z-axis), and at two points about the y-axis (elbow and wrist). This gives the arm access to a large region of the horizontal plane. The light sensor, which is mounted at the end of the arm, is constrained to point vertically downward at all times.

The LEGO MINDSTORMS NXT Brick is the computer that directly controls the motors and sensors of the robot. The Brick is programmed in the NXT-G programming language, which is a variant of LabVIEW. The Brick has been programmed with a simple program that moves the arm from the home position to a position on the plane and records the light intensity. After writing the measurement result to a file, the arm returns to the home position.

The intelligence of the robot lives on a Dell Latitude D610 laptop computer. The software is programmed in MATLAB and operates within the Windows XP operating system. The laptop computer communicates with the robot via a Bluetooth Wireless connection to the LEGO Brick. The MATLAB software interacts with the Brick by reading files, writing files and starting programs on the Brick. To request a measurement at a specified location, the MATLAB software must compute the number of motor rotations for each motor and write these values to a file on the LEGO Brick. MATLAB then starts the motor program on the Brick, which reads this file and implements the instructions. When the robot is finished it creates a file containing the resulting light level value. The MATLAB software then reads this file to obtain the data and begin its analysis and evaluation.

While both the MATLAB and the Brick software are operational, we were unable to implement the MATLAB to NXT communication by the time of the workshop. Instead, our experiments were performed with the files being transferred manually.

INFERENCE AND INQUIRY

To accomplish this task in an intelligent manner, the instrument must be endowed with both an inference engine and an inquiry engine. The inference engine relies on Bayesian methods to infer the circle parameters from the acquired data. The inquiry engine relies on the posterior density over the space of circles to evaluate which measurement is expected to deliver the greatest amount of information. The following subsections describe these two engines.

The Inference Engine

We begin with the problem of using the available data to infer the circle parameters

$$\mathbf{C} = \{(x_o, y_o), r\} \tag{1}$$

where (x_o, y_o) is the circle center coordinates, and r is the circle radius. In this case, the data consist of a set of N light measurements taken at various points on a plane. We will denote these measurements collectively as $\mathbf{D}$, and write them individually as

$$\mathbf{D} = \{d_1, d_2, \ldots, d_N\} \tag{2}$$

recorded at positions

$$\mathbf{X} = \{(x_1, y_1), (x_2, y_2), \ldots, (x_N, y_N)\}. \tag{3}$$

In this initial exploration, the positions are assumed to be known with certainty.

The goal is to explore the posterior probability

$$p(\mathbf{C} | \mathbf{D}, \mathbf{X}, I) = p(\mathbf{C} | I) \frac{p(\mathbf{D} | \mathbf{C}, \mathbf{X}, I)}{p(\mathbf{D} | I)}, \tag{4}$$

where I represents our prior information. From this we can obtain a set of posterior samples, each representing a possible circle. We accomplish this using the nested sampling algorithm, which samples from the prior probability and explores within an ever-contracting hard likelihood constraint (Skilling, 2005; Sivia & Skilling, 2006). There are multiple benefits to this approach. First, the algorithm provides a set of posterior samples, which are later used by the inquiry engine to select measurement locations. Second, nested sampling produces an estimate of the evidence, which can be used in the event that the robot needs to test one model against another. A simple example of this would be if the robot is designed to identify whether the white object is a circle or a square. However, in this initial exploration, we focus only on circles.

Here we keep the probability assignments as simple as possible and assign uniform distributions over reasonable ranges of values

$$p(x_o | I) = (x_{max} - x_{min})^{-1} \tag{5}$$

$$p(y_o | I) = (y_{max} - y_{min})^{-1} \tag{6}$$

$$p(r | I) = (r_{max} - r_{min})^{-1}. \tag{7}$$

The results we present here are based on simulations on a playing field of 20cm x 30cm, so that $r_{min} = 1$cm and $r_{max} = 15$cm. By assigning the prior for the center position to be independent of the prior for the radius, we are stating that the entire circle may not be in the playing field. This poses no problem for this investigation.

The likelihood function is again greatly simplified for these simulations. We do not consider the point-spread function of the light sensor and instead assume that the sensor will record the light intensity directly below the sensor with some Gaussian noise. The likelihood for one measurement d_i can be written as

$$p(d_i | \mathbf{C}, (x_i, y_i), I) \equiv p(d_i | \{(x_o, y_o), r\}, (x_i, y_i), I). \tag{8}$$

$$= \begin{cases} N(d_W, \sigma) & \text{if } (x_i - x_o)^2 + (y_i - y_o)^2 \leq r^2 \\ N(d_B, \sigma) & \text{if } (x_i - x_o)^2 + (y_i - y_o)^2 > r^2 \end{cases}$$

where $N(\cdot, \cdot)$ represents a Normal distribution, d_W is the expected value of a light measurement on the white circle, and d_B is the expected value of a light measurement on the black background. Clearly, this can be made more accurate by working with the point-spread function, however, our aim here is to tie the inference engine to the inquiry engine in real-time.

The nested sampling algorithm samples circles with centers uniformly distributed across the field, and radii uniformly distributed from 1cm to 15cm. The result is a set of weighted samples from which the mean and the variance of the circle parameters can be estimated. From this set of weighted samples, we obtain a set of 150 circles distributed according to the posterior probability.

The Inquiry Engine

This set of 150 circles is then used to examine the space of all possible measurements. This space is the set of locations in the field where the instrument can measure the light intensity. Each one of these possible measurements is a candidate experiment, so that choosing a measurement location is equivalent to designing an experiment. We will show that the fact that these circles are representative of the posterior probability simplifies the necessary computations. However, first we revisit the theory behind Bayesian adaptive estimation (Loredo, 2003).

Consider a proposed experiment E, which corresponds to taking a measurement at position (x_e, y_e). We do not know for certain what we will measure, nor do we know the parameter values of our circle, but we can write the probability of the measurement d_e in terms of the joint probability of d_e and $\mathbf{C}$ as

$$p(d_e | \mathbf{D},(x_e,y_e),I) \equiv \int d\mathbf{C}\, p(d_e, \mathbf{C} | \mathbf{D},(x_e,y_e),I). \tag{9}$$

Using the product rule, we can write

$$p(d_e | \mathbf{D},(x_e,y_e),I) \equiv \int d\mathbf{C}\, p(d_e | \mathbf{C},\mathbf{D},(x_e,y_e),I)\, p(\mathbf{C} | \mathbf{D},(x_e,y_e),I). \tag{10}$$

This can be simplified by observing that, if we knew the circle parameters $\mathbf{C}$, we would not need the data $\mathbf{D}$

$$p(d_e | \mathbf{D},(x_e,y_e),I) \equiv \int d\mathbf{C}\, p(d_e | \mathbf{C},(x_e,y_e),I)\, p(\mathbf{C} | \mathbf{D},(x_e,y_e),I). \tag{11}$$

Probability theory only takes us so far. In this problem, we wish to make a decision, and this requires us to maximize the expected utility according to an assigned utility function: $U(\text{outcome, action})$, so that

$$(\hat{x}_e, \hat{y}_e) \equiv \int dd_e\, p(d_e | \mathbf{D},(x_e,y_e),I)\, U(d_e,(x_e,y_e)), \tag{12}$$

where the location (x_e, y_e) is indicative of the action and the measurement d_e is the outcome. Here we use a utility function based on the information provided by the measurement, so that we will choose the measurement that provides the greatest expected gain in information. Of course, other utility functions could be used that depend on the time it takes for the measurement to be taken, the energy required, etc. Utility functions such as these will surely be important in a fully-functioning automated instrument. Using the Shannon information for our utility function we find

$$U(d_e,(x_e,y_e)) = \int d\mathbf{C}\, p(\mathbf{C} | d_e,\mathbf{D},(x_e,y_e),I)\, \log p(\mathbf{C} | d_e,\mathbf{D},(x_e,y_e),I). \tag{13}$$

By writing the joint entropy for $\mathbf{C}$ and d_e, and writing the integral two ways, one can show (Loredo, 2003) that the optimal experiment can be found by maximizing the entropy of the possible measurements

$$(\hat{x}_e, \hat{y}_e) \equiv \arg\max_{(x_e,y_e)} \left(-\int dd_e\, p(d_e | \mathbf{D},(x_e,y_e),I)\, \log p(d_e | \mathbf{D},(x_e,y_e),I)\right). \tag{14}$$

This entropy can be easily estimated using the ensemble of models sampled from the posterior. For each measurement position (x_e, y_e), we sample from the likelihood function of each sampled model thereby obtaining a set of potential measurements. The entropy of this set is rapidly estimated by constructing a histogram and computing the entropy directly. To enable the robot to consider a variety of positions, at each step we consider a grid on the space of possible measurements and compute the entropy only at the grid points. The alignment of this grid is randomly jittered so that a greater variety of points can be considered during the course of the experiment. With the optimal measurement position identified, the MATLAB software requests this particular measurement from the robotic instrument. Once the measurement is collected, the inference is updated, and the process is repeated until the system has estimated the model parameters with the desired accuracy.

RESULTS

At this point, we are still working on obtaining a fully-functioning Bluetooth connection between the laptop computer running MATLAB and the NXT Brick. While, we have tested the system by manually transmitting the information between the laptop and NXT Brick via a USB connection, in this presentation, we have simulated the process entirely in MATLAB. The result we present here is typical and dramatically demonstrates that the number of measurements required by an intelligent instrument is much smaller than a similar scanning system.

Figure 2A shows the initial stages of the inference-inquiry procedure where the white area of the circle has not yet been located. For this reason, there are large regions of the measurement space that are potentially equally informative. These are indicated by the large regions of essentially equal entropy in Figure 2B.

After several iterations, the robot finds a white area belonging to the circle. The set of sampled models are now close to the true circle (Figure 2C). The entropy map (Figure 2D) shows that the optimal measurement locations are those that are in the region where the models do not agree. This procedure naturally selects a binary question that at each stage rules out half of the models, which results in an extremely rapid convergence dramatically reducing the number of necessary measurements.

CONCLUSION

This work constitutes an initial investigation into designing an intelligent instrument, which not only makes inferences from data, but also decides which measurements to take based on what the instrument has learned. The approach we have employed here relies on Bayesian adaptive exploration, which selects a measurement based on maximizing the entropy of the possible measurements obtained by querying a set of models sampled from the posterior. The results of this initial investigation reduces nicely to viewing the inquiry process as selecting efficient binary questions, which is known to be optimal from an information-theoretic perspective. It should be noted that these binary questions are not hard-wired into the system.

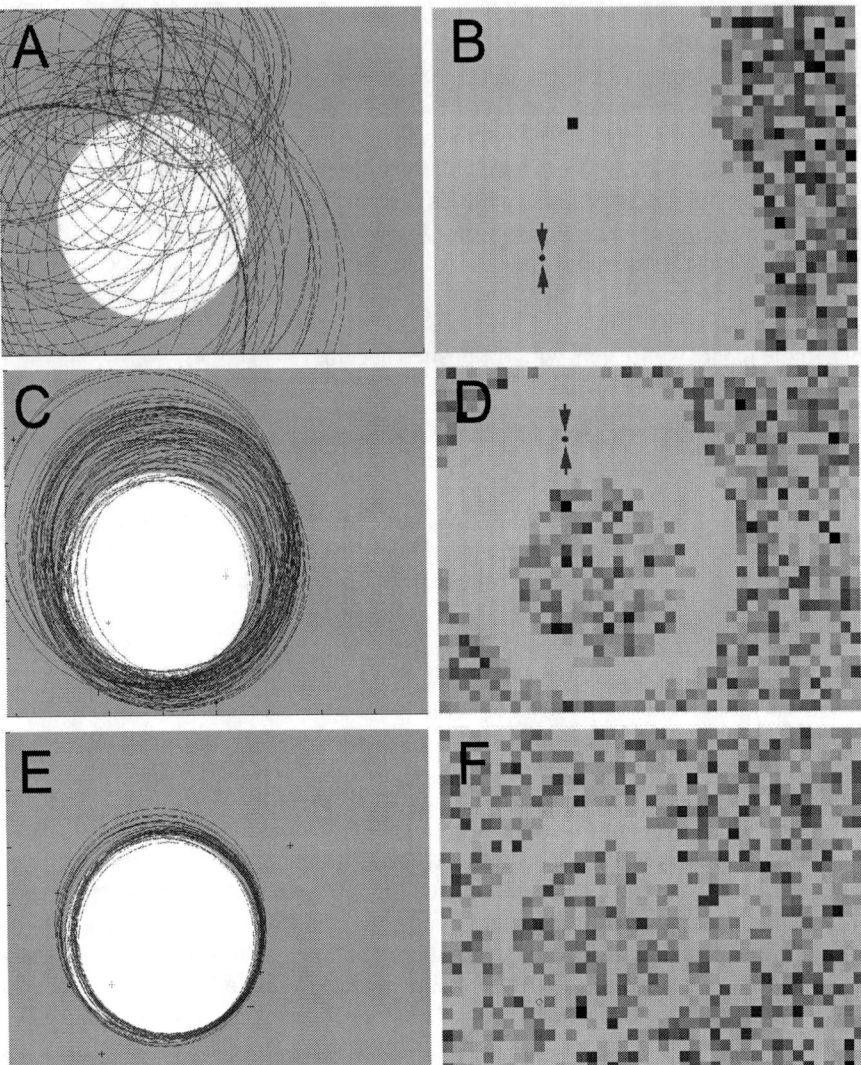

FIGURE 2. The panels on the left show the black playing field with the white circle. Overlaid on this are the set of 150 circles sampled from the posterior. Crosses indicate past measurement positions. The panels on the right show coarse entropy maps where the lighter shades indicate higher entropy. (A) One measurement has been taken (at the edge of the circles in upper right). This measurement has resulted in a set of hypothesized circles sampled from the posterior. (B) Much of the field is still unexplored indicated by the vast region of high entropy. The optimal measurement location is indicated by the dot with two arrows. (C) After 10 measurements, the algorithm is getting close to a solution. (D) Note that the region of high entropy is the region covered by the sampled circles. The chosen location divides the models into two. It will rule out half of the models with an efficient binary question. (E) After 16 measurements, the solution is almost obtained. (F) The corresponding entropy map is now focused on measurements at the edge of the circle.

Instead, they result as a natural application of maximizing the entropy of the potential measurement values given the model, the previous data, and our prior information.

This maximum entropy approximation works as long as the noise level, described by the likelihood function, is independent of the sampling location (Loredo, 2003). This condition will not always hold, and must be considered in future efforts.

Related maximum entropy techniques are finding their way into robotics (Thrun et al., 2005) and promise to enable these automated systems to interact with their environments in an intelligent manner. By creating joint environment-system models, the act of calibration becomes another potential experiment. In such a system, the instrument can decide to interact with the environment via measurement or itself via calibration giving rise to an instrument that actively self-calibrates during an experiment. Such advances are only the beginning.

REFERENCES

Cox R.T. 1979. Of inference and inquiry. In: R. D. Levine & M. Tribus (eds.) *The Maximum Entropy Formalism*, Cambridge:MIT Press, pp. 119–167.

Fedorov V.V. 1972. *Theory of Optimal Experiments*. New York:Academic.

Fry R. L. 2002. The engineering of cybernetic systems. In: R.L. Fry (ed.) *Bayesian Inference and Maximum Entropy Methods in Science and Engineering, Baltimore MD, USA*, AIP Conf. Proc. 617, Melville NY:AIP, pp. 497–528.

Knuth K.H. 2002. What is a question? In: C. Williams (ed.), *Bayesian Inference and Maximum Entropy Methods in Science and Engineering, Moscow ID 2002*, AIP Conf. Proc. 659, Melville NY:AIP, pp. 227–242.

Knuth K.H. 2003. Intelligent machines in the 21st century: Automating the processes of inference and inquiry. *Phil. Trans. Roy. Soc. Lond. A*, Triennial Issue. **361**(1813):2859–73.

Knuth K.H. 2005. Lattice duality: The origin of probability and entropy. *Neurocomputing*. **67**C: 245–274.

Knuth K.H. 2006. Valuations on lattices and their application to information theory. (Invited paper), *Proceedings of the 2006 IEEE World Congress on Computational Intelligence (IEEE WCCI 2006)*, Vancouver, BC, Canada, July 2006.

Knuth K.H. 2007. Lattice Theory, Measures and Probability. In: K.H. Knuth, A. Caticha, J. Center, A. Giffin, C.C. Rodríguez (eds.) *Bayesian Inference and Maximum Entropy Methods in Science and Engineering, Saratoga Springs NY USA*, AIP Conf. Proc., Melville NY:AIP, In Press.

Lindley D.V. 1956. On the measure of information provided by an experiment. *Ann. Math. Statist.* **27**, 986–1005.

Loredo T.J. 2003. Bayesian adaptive exploration. In: G. J. Erickson, Y. Zhai (eds.) *Bayesian Inference and Maximum Entropy Methods in Science and Engineering, Jackson Hole WY, USA*, AIP Conf. Proc. 707, Melville NY:AIP, pp. 330–346.

MacKay D.J.C. 1992. Information-based objective functions for active data selection. *Neural Computation*, **4**(4), 589–603.

Sebastiani P. and Wynn H.P. 2000. Maximum entropy sampling and optimal Bayesian experimental design. *J. Roy. Stat. Soc. B*, **62**, 145–157.

Sivia D.S. and Skilling J. 2006. *Data Analysis A Bayesian Tutorial*, (2nd Edition). Oxford Univ. Press:Oxford.

Skilling J. 2005. Turning ON and OFF, In: K.H. Knuth, A.E. Abbas, R.D. Morris, J.P. Castle (eds.) *Bayesian inference and Maximum Entropy Methods in Science and Engineering, San Jose, California, USA*, AIP Conf. Proc 803, Melville NY:AIP, pp. 3–24.

Thrun S., Burgard W., Fox D. 2005. *Probabilistic Robotics*, MIT Press:Cambridge.

Wiener N. 1948. *Cybernetics or Control and Communication in the Animal and the Machine*, Cambridge:MIT Press.

Deconvolution using thin-plate splines

Udo v. Toussaint and Silvio Gori

Max-Planck-Institut für Plasmaphysik, 85748 Garching, Germany

Abstract. The ubiquitous problem of estimating 2-dimensional profile information from a set of line integrated measurements is tackled with Bayesian probability theory by exploiting prior information about local smoothness. For this purpose thin-plate-splines (the 2-D minimal curvature analogue of cubic-splines in 1-D) are employed. The optimal number of support points required for inversion of 2-D tomographic problems is determined using model comparison. Properties of this approach are discussed and the question of suitable priors is addressed. Finally, we illustrated the properties of this approach with 2-D inversion results using data from line-integrated measurements from fusion experiments.

Keywords: Tomography, reconstruction, inverse problem, thin-plate splines
PACS: 4230W,0250F,0250P

INTRODUCTION

The problem of the reconstruction of a two-dimensional emission profile from line-integrated measurements is an ubiquitous one. In the original paper by Radon [1] it was shown that the reconstruction can be done by means of the Radon transform. However the presence of measurement noise and limited amounts of data prevent a straightforward deconvolution of the line-integrated measurements, leading to meaningless results. Therefore a large number of regularized methods for the inversion procedure have been proposed [2, 3, 4, 5]. In most cases the profile to be reconstructed is expanded in orthogonal function systems up to a given order and adapted to predefined boundary conditions. These approaches give reasonable results - provided the profile to be reconstructed is sufficiently symmetric. In irregular geometries or for localized intensity peaks the required expansion order is too large to provide regularization. However, examples from the field of fusion research which display these properties are soft X ray imaging systems and the bolometry diagnostic. Here we propose to exploit the favorable property of local smoothness of minimal curvature surfaces together with an adaptive distribution of support points to tackle the underdetermined and ill-posed inversion problem.

TOMOGRAPHIC DIAGNOSTICS IN FUSION EXPERIMENTS

Bolometry. The precise measurement of power balance in fusion experiments has always been of utmost importance for a understanding of particle and energy transport. Presently, only one technique for measuring the total radiated power is widely used in fusion experiments with magnetically confined plasma: Bolometry. Usually the temperature change induced by the plasma radiation is detected using the temperature dependent conductivity of small wires attached to an absorbing foil exposed to the plasma. Recently

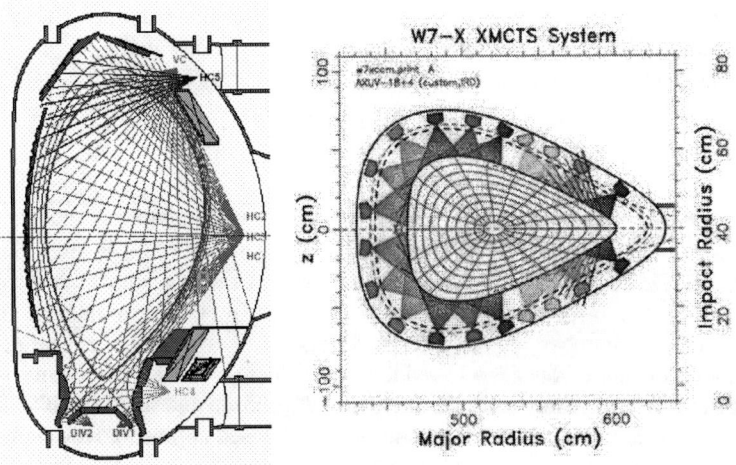

FIGURE 1. a) Sensor configuration of the bolometry line integrating sensor arrays at ASDEX Ugrade(left). b) On the right hand side the intended design of the sensor configuration of the soft x-ray line integrating sensor arrays at W7-X is shown.

also silicon photodiodes have been used which have the advantage of a much better time resolution of about $10^{-6}s$ compared to $2*10^{-5}s$ of foil based systems but on the expense of sensitivity. The sensor arrays are located a few mm behind precision pinholes, so that each detector is detecting a defined conic section of the plasma. The signal level of the detectors is around 10^{-5}A for 1MW total radiated power [6] and approximately 100 lines of sight are available for deconvolution. A typical setup is shown in Fig. 1a.

Soft-X-Ray. The X ray emission from a plasma yields informations on a variety of parameters such as temperature, density and impurities. Since typical plasmas of interest are optically thin in the soft X ray region of the emission spectrum, reabsorption of the emitted radiation can be neglected. A typical set-up (the one shown in Fig. 1b is a setup developed for W7-X) consists of several sets of imaging diode arrays. Every detector diode, usually ion-implanted silicon diodes, samples radiation from a plasma region located in the viewing cone formed by the diode and the collimator aperture. One advantage of the Soft X-ray diagnostic is the combination of a good spatial resolution (large number of viewing chords) with a very good time-resolution (sampling rate>$10^6 s^{-1}$). Therefore the soft X-ray diagnostic provides important informations about the magnetohydrodynamic phenomena in fusion experiments.

THE TOMOGRAPHY PROBLEM

To calculate spatial profiles from the line integrated measurements the inversion of the Radon transform is required. However the direct inversion will not provide useful profiles because the measured data contains noise and the amount of data is often very limited. Consider the following discretized tomography model where $\underline{s}$ is a K-dimensional vector of observed signals recorded by detector k:

$$s_i = O_i[E] = \sum_j Q_{ij} E_j + \varepsilon_i, \qquad (1)$$

where $\underline{\underline{Q}}$ is a (IxJ) known matrix of the proportion of the emission E_j accumulated in detector i, $\underline{E}$ is a vector of J unknown emission intensities to be recovered with the property that $E_j > 0$, and the signal recorded by sensor i is distorted by Gaussian noise ε_i. The objective is to recover $\underline{E}$ from the noisy data $\underline{s}$. Since the number of unknowns J outnumbers the number of data points K this problem is underdetermined and a regularization criterion is needed to reduce the problem to a well-posed one. In Ertl et al[7] the classical maximum-entropy formalism was chosen and subsequently Golan and Dose[8] developed a generalized maximum-entropy based approach. However, in many tomographic problems there is additional prior information available: The profile to be reconstructed is locally smooth - it exhibits local spatial correlations. These properties are explicitly not captured by entropic priors which are invariant under permutations. Therefore we propose a different approach based on the property of local smoothness.

THIN-PLATE SPLINES

A thin plate spline is the function $f(\underline{x}), \underline{x} \in R^d$ that minimises the curvature on a domain Ω([9]):

$$\int_\Omega d\underline{x} \sum_{|\underline{v}|=2} \binom{2}{\underline{v}} (D^{\underline{v}} f(\underline{x}))^2, \qquad (2)$$

where $\underline{v} = (v_1, \ldots, v_d)$ is a d dimensional multi-index, and $|\underline{v}| = \sum_{i=1}^d v_i$. In two dimensions ($d = 2$) the thin plate spline (TPS) can also be considered as generalization of the interpolating cubic spline in one dimension [10]. It is a commonly used basis function for modeling smooth coordinate transformations in computer vision and morphing applications. We indicate the TPS with $t(\underline{r})$, where $\underline{r} = (x, y)$ corresponds to the position on the grid in the detector field. The exposition of the basic thin-plate spline theory is along the lines of Guglielmetti[11]. The shape of the interpolating TPS surface will be given by the minimum curvature condition of eq.2. Often *radial* basis functions are used to represent f as they allow an analytical solution. More specifically, the TPS is a weighted sum of translations of radially symmetric basis functions augmented by a linear term (see ref. [9], [12]), of the form

$$t(\underline{r}) = K(\underline{r}) + \sum_{l=1}^N \lambda_l f(\underline{r} - \underline{r}_l), \qquad \underline{r} \in R^2 \qquad (3)$$

where $K(\underline{r}) = c_0 + c_1 x + c_2 y$ is the added plane. N is the number of support points. The real-valued weight is characterized by λ_l. $f(\underline{r} - \underline{r}_l)$ is a radial basis function, a real function of positive real values depending on the distance between the grid points $\underline{r}$ and the

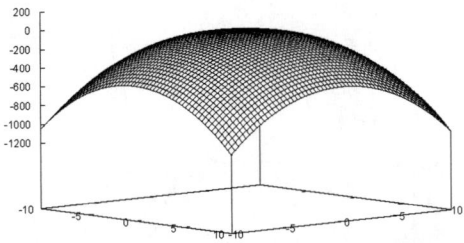

FIGURE 2. The thin plate spline radial basis function $t(r) = r^2 \ln(r^2)$

support points r_l.
Given the scattered data points r_i and the data $z_i = z(r_i)$, the TPS satisfies the interpolation conditions

$$t(r_i) = K(r_i) + \sum_{l=1}^{N} \lambda_l f(r_i - r_l) = z_i, \qquad i = 1,\ldots,N \qquad (4)$$

with

$$f(r) = r^2 \ln(r^2). \qquad (5)$$

and it minimises

$$\|t\|^2 = I[t] = \iint_{\mathbb{R}^2} (t_{xx}^2 + 2t_{xy}^2 + t_{yy}^2) dx dy, \qquad (6)$$

the 2-d case of Eq.2. $\|t\|^2$ is a measure of the bending energy of t. In other words, given a set of data points, a weighted combination of TPSs centered about each data point gives the interpolation function that passes through the points exactly while minimizing the so-called "bending energy". In order to fit the TPS to the data, it is necessary to solve for the weights and the planes' coefficients so that it is possible to compute the local TPS amplitude:

$$t(r_i) = t(r_i, N, \{r_l, z_l, l = 1,\ldots,N\}) \qquad (7)$$

which is a function of z_l, the given emission intensity at the knot positions r_l. The TPS interpolant is defined by the coefficients, c_i of the plane $K(r)$ and the weights λ_l of the basis functions. Given the interpolation values $z = (z_1,\ldots,z_N)$, we search for the weights λ_l and c_i so that the TPS satisfies:

$$t(r_l) = z_l, \qquad l = 1,\ldots,N \qquad (8)$$

and in order to have a converging integral, the following conditions need to be satisfied:

$$\sum_{l=1}^{N} \lambda_l = \sum_{l=1}^{N} \lambda_l x_l = \sum_{l=1}^{N} \lambda_l y_l = 0 \qquad (9)$$

The coefficients of the TPS, λ_l, and the plane, c_i, can be found by solving the linear system, that may be written in matrix form as:

$$\begin{pmatrix} \underline{\underline{F}} & \underline{\underline{Q}} \\ \underline{\underline{Q}}^T & 0 \end{pmatrix} \begin{pmatrix} \underline{\lambda} \\ \underline{c} \end{pmatrix} = \begin{pmatrix} \underline{z} \\ 0 \end{pmatrix} \quad (10)$$

where the matrix components are:

$$F_{ij} = f(\underline{r}_i - \underline{r}_j)$$

$$\underline{z} = (z_1, \ldots, z_N)^T$$

$$0 = (0,0,0)^T$$

$$\underline{\lambda} = (\lambda_1, \ldots, \lambda_N)^T$$

$$\underline{c} = (c_0, c_1, c_2)^T$$

$$\underline{\underline{Q}} = \begin{pmatrix} 1 & x_1 & y_1 \\ 1 & x_2 & y_2 \\ \vdots & \vdots & \vdots \\ 1 & x_N & y_N \end{pmatrix}$$

After having solved for $(\underline{\lambda}, \underline{c})^T$, the TPS can be evaluated at any point and total curvature $I[t]$ is easily accessible using the relation

$$I[t] = \underline{\lambda}^T \underline{\underline{F}} \underline{\lambda}. \quad (11)$$

The Bayesian Framework

Once the number N of support points, the location of the support points $\underline{R} = (\underline{r}_1^T, \ldots, \underline{r}_N^T)$ and the respective intensity $\underline{z} = (z_1, \ldots, z_N)$ is fixed the thin-plate spline model M is completely determined and fully characterized by the set $N, \underline{R}, \underline{z}$:

$$t(\underline{r}) = t(\underline{r}|N, \underline{R}, \underline{z}). \quad (12)$$

To compute the probability for the different models we start with Bayes theorem:

$$p(M|D,I) = \frac{p(M|I)p(D|M,I)}{p(D|I)} = \frac{p(M|I) \int d\underline{z}\, p(D|M,\underline{z},I) p(\underline{z}|M,I)}{p(D|I)}. \quad (13)$$

The likelihood term $p(D|M,\underline{z},I) = p(D|t,I)$ for N_d measured data points is given by

$$p(D|t,I) = \prod_{i=1}^{N_d} \frac{1}{\sqrt{2\pi}\sigma_i} \exp\left(-\frac{1}{2}\left(\frac{d_i - O_i[t]}{\sigma_i}\right)^2\right), \quad (14)$$

for Gaussian distributed noise with zero mean and variance σ^2. The function O_i describes the weighted summation of the intensities of the thin-plate spline within the viewing cone of sensor i (cf. Eq. 1). The first term in the nominator of Eq. 13 is the prior

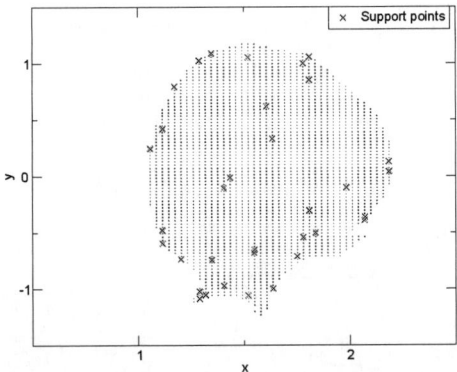

FIGURE 3. The grid indicating the possible support point locations is adapted to the shape of the plasma vessel and the best distribution of the support points is indicated by stars.

probability for the model M. It factors into the prior for the number of support points N and the prior for the support points $p\left(\underline{R}|N,I\right)$. The discretized grid for the support points provides a maximum number N_G of possible locations (see Fig.3), which cannot be exceeded. The uninformative prior is therefore (a minimum number of 4 support points is needed to define a TPS)

$$p(M|I) = p\left(\underline{R}|N,I\right) p(N|I) = \frac{1}{(N_G - 3)\binom{N_G}{N}}. \quad (15)$$

Finally we need to assign the prior for the z-values of the support points. As testable information we use the global curvature $I[t]$ of the TPS. The maximum entropy principle yields as prior probability

$$p(t|\mu,I) = \frac{1}{Z(\mu)} \exp(-\mu I[t]), \quad (16)$$

where we had to introduce μ as scale parameter with a Jeffreys' prior $p(\mu|I) \propto 1/\mu$. However, since t is completely determined by the set $N,\underline{R},\underline{z}$ we can rewite the previous equation as prior for $\underline{z}$:

$$p(\underline{z}|\mu,M,I) = p\left(\underline{z}|\mu,N,\underline{R},I\right) = \frac{1}{Z\left(\mu,N,\underline{R}\right)} \exp\left(-\mu I\left[t(\underline{z})\right]\right). \quad (17)$$

The normalization $Z\left(\mu,N,\underline{R}\right)$ along with the prior probability of the model M tends to keep the number of support points as small as possible. Since we use optimization algorithms to compute the most likely vector $\underline{z}$ we avoid the cumbersome marginalization of the hyperparameter (which is more easily accessible using MCMC methods) and instead look for the most likely value μ^* of the hyperparameter using

$$p(\mu|D,M,I) = p(\mu|I) p(D|\mu,M,I) = p(\mu|I) \int d\underline{z}\, p(D|\mu,\underline{z},M,I) p(\underline{z}|\mu,M,I), \quad (18)$$

followed by an approximation with a multi-variate Gaussian distribution around the maximum :

$$p(D|\underline{z},\mu,M,I)p(\underline{z}|\mu,M,I) \approx p(D|\underline{z}^*,\mu,M,I)p(\underline{z}^*|\mu,M,I)exp\{-\frac{1}{2}\Delta\underline{z}^T H \Delta\underline{z}\} \quad (19)$$

where $\Delta\underline{z} = \underline{z} - \underline{z}^*$, $\underline{z}^*$ being the position of the maximum of the integrand in eq. 18, and the Hessian matrix elements $H_{ij} := -\frac{\partial^2 ln[p(D|z,\mu,M,I)p(z|\mu,M,I)]}{\partial z_i \partial z_j}$. Using this approximation the integration in Eq.18 can be carried out, yielding

$$p(\mu|D,M,I) = p(\mu|I)p(D|\mu,\underline{z}^*,M,I)p(\underline{z}^*|\mu,M,I)\sqrt{\frac{(2\pi)^N}{\det(H)}} \quad (20)$$

followed by an 1-D line scan in μ to determine μ^*. Using μ^* in Eq.17 we have all the necessary quantities to evaluate Eq. 13 and to obtain the model probability $p(M|D,I)$. As final step simple sampling over different model orders (the number of support points is varied between 4 and N_G) and the locations of the support points is performed resulting in a distribution of model probabilities depending on N and $\underline{R}$.

APPLICATIONS

The TPS approach has successfully been applied to a number of different soft X-ray and bolometry tomographic measurements. Here we display mock data which closely resemble the experimental observations and the measurement uncertainties. As can be seen in Fig. 4a two high emission spots are located in the lower parts of the plasma vessel (the divertor) where the plasma touches the wall. The intensity in the main chamber varies only slightly but at the top of the plasma vessel there is an extended area, emitting slightly more than its environment. Additionally there is a hot spot visible slightly below the midplane on the right hand side. The reconstruction, displayed in Fig. 4b is displaying all the key features and does not introduce any spurious ringing despite the peaked intensity profile and the edge location of most of the emission maxima. The difference between the mock profile and the reconstruction is mostly due to the slightly smoother reconstruction of the profiles. In Fig.5 the measured line-integrated signal is compared to the intensity derived from the reconstructed profile. The agreement is within the statistical uncertainties without overfitting tendency. The underlying spatial distribution of the support points is displayed in Fig.3. Most support points are in or close to the divertor therefore adaptively enhancing the flexibility of the reconstruction where the profile has the largest variations.

SUMMARY

We presented a Bayesian analysis for the 2-d profile reconstruction from line-integrated chord measurements which augments the usual approaches by the concept of data dependent adaptive resolution and local smoothness. All structures in the reconstruction

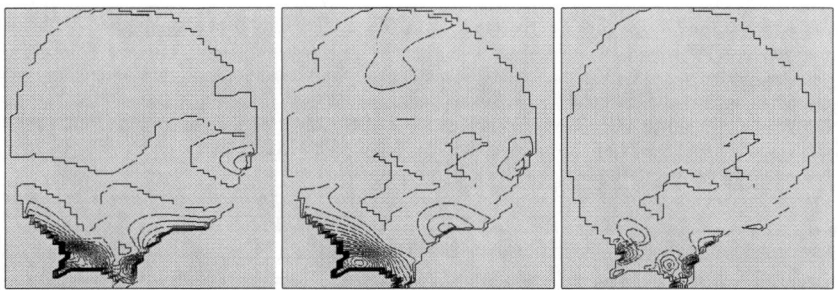

FIGURE 4. a) On the left: Mock 2-D intensity profile in agreement with experimental observations b) Mid dle: reconstructed intensity profile using thin-plate splines as basis functions. c) Right hand side: Difference between true profile and reconstructed intensity (same absolute scale).

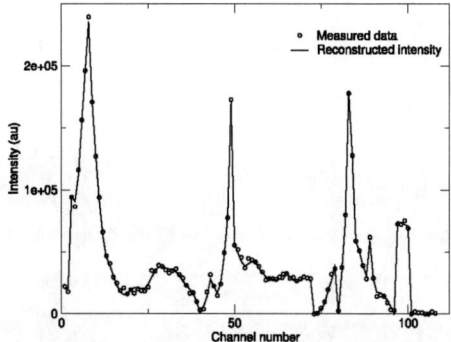

FIGURE 5. Sensor data and intensities obtained from the reconstruction

are supported by data constraints. This has been demonstrated by a mock bolometry example with realistic uncertainties. The straightforward next step is the addition of a suitable MCMC-method (see [13]) to improve on the presently used evidence approximations and the integration into the framework of Bayesian Neural Networks [14] to achieve real-time deconvolution capabilities with $\approx 10^3$ deconvolutions/s.

REFERENCES

1. Radon J., Berichte Sächsische Akademie der Wissenschaften **69**,262 (1917).
2. Deans S., The Radon Transform and Some of its applications, Wiley, New York (1983).
3. Granetz R. S., Smeulders P., Nucl. Fusion **28**, 457 (1988).
4. Cormack M. A., J. Appl. Phys. **35**, 2908 (1964).
5. Fuchs G., Miura Y., Mori M., Plasma Phys. Control. Fusion **36**, pp. 307-316 (1994).
6. Boivin R. L., Goetz J. A., Marmar E. S., Rice J. E., Terry J. L., Rev. Sci. Instr. **70**(1), pp. 260-264 (1999).
7. Ertl K., von der Linden W., Dose V., Weller A., Nucl. Fusion **36**(11), pp.1477-1488 (1996).
8. Golan A., Dose V., J.Phys.A.,**34**, pp.1271-1283 (2001).

9. Wahba G., *Spline Models for Observational Data*, Society for Industrial and Ap plied Mathematics, Pennsylvania (1990).
10. von der Linden W., Dose V., Fischer R., Proceedings of the Maximum Entropy Conference 1996, M. Sears, V. Nedeljkovic, N.E. Pendock, S. Sibisi, eds., 154 (1996).
11. Guglielmetti F., Fischer R., Dose V., Proceedings of the Maximum Entropy Conference 2004, R. Fischer, R. Preuss and U. von Toussaint, eds., AIP 735, p. 111-118 (2004).
12. Meinguet J., J.Appl.Math.Phys. (ZAMP), 30, pp.292-304 (1979).
13. Preuss R., von Toussaint U., Proceedings of the Maximum Entropy Conference 2007 (these proceedings).
14. von Toussaint U., Gori S., Dose V., Neural Networks **19**(10), pp.1550-1557 (2006).

Comparison of Numerical Methods for Evidence Calculation

R. Preuss and U. von Toussaint

Max-Planck-Institut für Plasmaphysik
EURATOM Association, D-85748 Garching, Germany
preuss@ipp.mpg.de

Abstract. Model comparison requires the determination of integrals over the posterior probability function. We present a variety of numerical methods for this calculation. As working examples serve Gaussian structures in various dimensions.

INTRODUCTION

Bayesian model comparison is inevitably associated with the calculation of the prior predictive value or evidence which involves integration over the posterior probability density function. Most of the time this integral has no analytical solution and one is referred to either approximative or numerical approaches or a mixture of both. The integration methods employed have to cope with all the cumbersome features of a function which could sparsely populate a large parameter space, consisting of broad and narrow peaks, involving large and small scales and finally be spread such that the integral weight between the structures is zero (at least according to numerical means). We present a comparison of a variety of numerical methods featuring Laplace approximation, trapezoidal rule, importance sampling, VEGAS (from Numerical Recipes [1]), thermodynamic integration scheme (thin-MCMC), and nested sampling [2]. This choice is far from being complete and simply arises from the fact that we have longtime experience with most of the methods (apart from [1] and [2]). For a further, much more sophisticated method for the integration of ill-conditioned problems, i.e. *Perfect Tempering* we refer to the paper of M. Daghofer, published in the proceedings of the 2004 MaxEnt conference [3].

STATEMENT OF THE PROBLEM

The posterior probability function is composed of likelihood and prior probability density function. If the data is normally distributed the likelihood will be of Gaussian shape. Moreover, if the information gain from an experiment is large the likelihood will be much more structured than the prior and be of dominating role in the posterior. So the posterior will be of Gaussian-like character as well. The exercises of this paper shall therefore consist of the integration of Gaussian peaks in K-dimensions. We look for the integral

$$I = \int p(\bm{x}|\sigma)\mathrm{d}\bm{x} \quad , \tag{1}$$

with the following choice of integrands: The simplest case shall consist of a single Gaussian peak. This problem should be feasible for any method.

$$p_1(\bm{x}|\sigma) = \frac{1}{\left(\sqrt{2\pi}\sigma\right)^K} \exp\left\{-\frac{\bm{x}^T\bm{x}}{\sigma^2}\right\} \quad . \tag{2}$$

Without loss of generality we set $\sigma=0.3$. To simulate multi-modal posteriors, the next case consists of two Gaussians of equal height and width.

$$p_2(\bm{x}) = \frac{1}{2}\frac{1}{\left(\sqrt{2\pi}\sigma\right)^K}\left[\exp\left\{-\frac{(\bm{x}-\bm{d}_1)^T(\bm{x}-\bm{d}_1)}{\sigma^2}\right\} + \exp\left\{-\frac{(\bm{x}-\bm{d}_2)^T(\bm{x}-\bm{d}_2)}{\sigma^2}\right\}\right] \quad . \tag{3}$$

The σ is set to 0.3 and 0.03, respectively. With the first setting, the integrand is still of small weight between the two peaks and therefore would allow "path following" methods to pass from one mode to the other. For $\sigma=0.03$ this is not possible since the peaks are fully separated according to numerical means.

The Monte Carlo method cannot decide between weight originating in one peak or the other and therefore gives a correct result even if the samples are coming erroneously from the same peak. To disclose such failure we investigate another setup with two peaks of different height (10:1) and width (1:2).

$$p_3(\bm{x}) = \frac{1}{10+2^K}\frac{1}{\cdot(\sqrt{2\pi}\sigma)^K}\left[10\exp\left\{-\frac{(\bm{x}-\bm{d}_1)^T(\bm{x}-\bm{d}_1)}{\sigma^2}\right\} + \exp\left\{-\frac{(\bm{x}-\bm{d}_2)^T(\bm{x}-\bm{d}_2)}{(2\sigma)^2}\right\}\right] \quad . \tag{4}$$

σ is again set to 0.3 and 0.03, respectively. For the rest of the paper $\bm{d}_1$ consist of K numbers 2, $\bm{d}_2$ of numbers -2.

DESCRIPTION OF THE METHODS AND RESULTS

In the following we give only a brief description of the employed numerical methods. Please refer to the literature for deeper insights. Some of these methods require a preceding MCMC run in order to determine the covariance of the parameters. The covariances are naturally provided if the expectation values of the parameters are needed anyway and determined beneficially with the Metropolis algorithm which does not require the norm. The MCMC samples are separated into so-called *bins* from which the expectation values and the variances are calculated. Each bin is preceded by burn-in sampling with randomly chosen starting values. The computer code was run on a 2.1GHz processor. The indicated running time is given for comparison reasons only.

TABLE 1. Laplace approximation. The results for the other test cases containing two peaks were clearly senseless.

Dimension	1	2	4	8	16	32
1 peak, $\sigma=0.3$	0.99	0.99	1.02	1.00	0.99	0.94
time[s]	0.26	0.41	0.70	1.41	3.45	10.1

Laplace approximation

This is also called steepest descent method or saddle-point approximation, where the term "Laplace approximation" is reserved for the real space. It constitutes a simple and powerful approximation to the integral of Eq. (1) if the integrand has a single mode only (regardless of dimension).

As mentioned above most posterior probability distributions resemble a Gaussian shape. It is therefore a good approximation to employ a Taylor series at the maximum of the integrand (first order is zero)

$$p(\boldsymbol{x}|\sigma) \sim \exp\{\Phi(\boldsymbol{x})\} \quad . \tag{5}$$

$$\Phi(\boldsymbol{x}) = \Phi(\boldsymbol{x}_{max}) - \frac{1}{2}(\boldsymbol{x}-\boldsymbol{x}_{max})^T \mathbf{H}(\boldsymbol{x}-\boldsymbol{x}_{max}) \quad , \tag{6}$$

with the Hessian matrix

$$H_{ij} = \frac{\partial^2 \ln p(\boldsymbol{x}|\sigma)}{\partial x_i \partial x_j} \quad . \tag{7}$$

The second order is just a Gaussian integral solvable analytically:

$$p(x|\sigma) = \frac{\text{const}}{\sqrt{\det \mathbf{H}}} \exp\{\Phi_0\} \quad . \tag{8}$$

One merely has to find the maximum of Φ in x-space and take advantage of a previous MCMC run in determining the Hessian matrix during parameter estimation from the covariances of the parameters:

$$\mathbf{H} = \mathbf{C}^{-1} \quad , \tag{9}$$

with covariance matrix

$$C_{ij} = \langle (x_i - \langle x_i \rangle)(x_j - \langle x_j \rangle) \rangle \quad . \tag{10}$$

The results are given in table 1. The desired value of 1 is reproduced as good as the previous Monte Carlo run was (10 times 2000 samples plus burn-in of altogether 2000 samples). The time given is therefore the running time of exactly this MCMC run. The time for the calculation of the Laplace approximation itself is negligible. So fast and satisfying the result for p_1 is, so utterly devastating it is for integrands with more than one mode (not shown). Nevertheless, for its simple use it should be part of every evidence calculation program, but be regarded as for diagnostic reasons only.

TABLE 2. Mesh integration. The number of points to calculate were $N_{mesh}=15$ for $\sigma=0.3$ and $N_{mesh}=150$ for $\sigma=0.03$. The question mark means "not calculated due to lack of sufficient time".

Dimension	1	2	4	8
1 peak, $\sigma=0.3$	1.00	0.99	0.98	0.97
time[s]	0.000	0.000	0.04	901
2 peaks, $\sigma=0.3$	1.00	1.00	1.00	1.00
time[s]	0.000	0.004	0.032	2041
2 peaks, $\sigma=0.03$	1.00	0.99	0.98	?
time[s]	0.000	0.024	432	≈ 6935 years
2 different peaks, $\sigma=0.3$	1.00	1.00	1.00	0.99
time[s]	0.000	0.004	0.088	1995
2 different peaks, $\sigma=0.03$	1.00	0.99	0.99	?
time[s]	0.004	0.052	446	≈ 7160 years

Trapezoidal rule (integration on a mesh)

We employ simple trapezoidal integration over the parameter space.

$$p(\boldsymbol{x}|\sigma) = \sum_{i=1}^{N} ... \sum_{j=1}^{N} p(x_{1i},...,x_{Kj}|\sigma)\Delta x_1 \cdot ... \cdot \Delta x_K \qquad (11)$$

One can think of more sophisticated algorithms, refined in adjusting the integration grid automatically according to the integral weight. The accuracy of the result may be controlled by increasing the grid density and comparing the outcome with the step before. However, for larger numbers of parameters all these mesh integration techniques fail due to the curse of dimension (see table 2). In many cases it is possible to run a new programmed code for smaller dimensions where mesh integration still works. In order to detect errors in coding, the recommended procedure is to check the results with other evidence calculation methods and then proceed to the actual problem with its larger number of dimensions, i.e. parameters. Note: numerical problems may occur if the numbers in the exponent become too large (small), so we actually sum over $p(\boldsymbol{x}|\sigma) - p(\boldsymbol{x}_{max}|\sigma)$.

Importance sampling

The idea is to generate samples from a simpler function easy to sample from.

$$I = \int p(\boldsymbol{x}|\sigma)d\boldsymbol{x} = \int \frac{p(\boldsymbol{x}|\sigma)}{g(\boldsymbol{x})} g(\boldsymbol{x})d\boldsymbol{x} \qquad (12)$$

For the function $g(\boldsymbol{x})$ we employ a Gaussian with widths from the covariances generated by the preceding MCMC run already utilized for the Laplace approximation. 100000 sampling steps are performed to create each entry to table 3. While being excellent for the one peak problem, importance sampling fails rapidly for the other problems. It is

TABLE 3. Importance sampling. The entry "##" means an obviously erroneous result.

Dimension	1	2	4	8	16	32
1 peak, $\sigma=0.3$	1.00	1.00	1.00	1.00	1.00	1.00
time[s]	0.084	0.124	0.192	0.344	0.552	0.808
2 peaks, $\sigma=0.3$	0.99	0.98	0.94	##	##	##
time[s]	0.156	0.208	0.276	##	##	##
2 peaks, $\sigma=0.03$	1.04	0.50	##	##	##	##
time[s]	0.444	0.676	##	##	##	##
2 different peaks, $\sigma=0.3$	1.00	0.29	0.62	0.96	0.99	##
time[s]	0.156	0.208	0.284	0.436	0.548	##
2 different peaks, $\sigma=0.03$	1.00	0.28	##	##	##	##
time[s]	0.224	0.848	##	##	##	##

simply harmed from the fact that already the MCMC run for the determination of the covariances produces wrong results getting stuck in a particular peak.

VEGAS algorithm

The algorithm invented by Peter Lepage is freely available from e.g. Numerical Recipes [1] and reportedly "widely used for multidimensional integrals that occur in elementary particle physics". It works accordingly to the importance sampling scheme, however separates the target function into a multidimensional weight function g

$$p(\boldsymbol{x}) = g_1(x_1)g_2(x_2)...g_K(x_K) \quad . \tag{13}$$

The implementation as an additional integration method is simple and straight forward. Table 4 shows the problems of Monte Carlo methods with peaky structures as in the importance sampling case, however it scores somewhat better. Expecially for the one peak problem it would be possible to get reasonable results for dimensions larger than 4 if the integration integral would be confined to a smaller range around the maximum.

TABLE 4. Results obtained with the VEGAS algorithm.

Dimension	1	2	4	8	16	32
1 peak, $\sigma=0.3$	1.00	1.00	1.00	##	##	##
time[s]	0.33	0.55	0.66	##	##	##
2 peaks, $\sigma=0.3$	1.00	1.00	0.99	##	##	##
time[s]	0.50	0.77	1.21	##	##	##
2 peaks, $\sigma=0.03$	1.00	1.00	##	##	##	##
time[s]	0.50	0.90	##	##	##	##
2 different peaks, $\sigma=0.3$	1.00	1.00	0.99	0.95	##	##
time[s]	0.50	0.68	1.08	2.68	##	##
2 different peaks, $\sigma=0.03$	1.00	1.00	##	##	##	##
time[s]	0.64	0.70	##	##	##	##

TABLE 5. Thermodynamic integration.

Dimension	1	2	4	8	16	32
1 peak, σ=0.3	1.00	0.97	0.96	0.95	0.90	0.83
time[s]	5	11	25	64	185	598
2 peaks, σ=0.3	0.99	0.97	0.96	0.95	1.27	449
time[s]	9	20	45	122	365	1133
2 peaks, σ=0.03	1.70	##	##	##	##	##
time[s]	12	##	##	##	##	##
2 different peaks, σ=0.3	1.04	0.64	0.69	0.93	0.94	##
time[s]	9	20	44	118	337	##
2 different peaks, σ=0.03	2.92	##	##	##	##	##
time[s]	12	##	##	##	##	##

However, since we pretend to have no knowledge about the structure of the integrand, we stay with the range of { -4,4 } as for all other case in this paper.

Thermodynamic integration scheme

At the Maxent workshop 1997 in Boise John Skilling suggested to employ a formalism, borrowed from statistical physics, to compute the prior-predictive value, the so-called 'thermodynamic integration' scheme[4]: Define the function

$$Z(\lambda) = \int \Lambda^\lambda(\boldsymbol{x})\Pi(\boldsymbol{x})\, d\boldsymbol{x} \quad , \tag{14}$$

with $Z(\lambda=0) = 1$ and $Z(\lambda=1)$ as the desired quantity. Commonly the function Λ comprises terms from the likelihood. Π is the normalized prior. Here we chose $\Lambda = p(\boldsymbol{x}|\sigma)$ and $\Pi=1/8$ within $[-4,4]$ and 0 otherwise. The derivative with respect to λ gives

$$\begin{aligned}\frac{\partial \ln Z(\lambda)}{\partial \lambda} &= \int \ln \Lambda(\boldsymbol{x})\rho_\lambda(\boldsymbol{x})\, d\boldsymbol{x} \\ &= \langle \ln \Lambda(\boldsymbol{x})\rangle_\lambda \quad ,\end{aligned} \tag{15}$$

with

$$\rho_\lambda(\boldsymbol{x}) = \frac{\Lambda^\lambda(\boldsymbol{x})\Pi(\boldsymbol{x})}{\int \Lambda^\lambda(\boldsymbol{x}')\Pi(\boldsymbol{x}')\, d\boldsymbol{x}'} \tag{16}$$

as the new sampling density. Both sides of Eq. (15) are integrated over λ:

$$\begin{aligned}\int_0^1 \langle \ln \Lambda(\boldsymbol{x})\rangle_\lambda\, d\lambda &= \int_0^1 \frac{\partial \ln Z(\lambda)}{\partial \lambda}\, d\lambda \tag{17}\\ &= \ln Z(\lambda=1) - \ln Z(\lambda=0) \tag{18}\\ &= \ln \boldsymbol{I} \quad . \tag{19}\end{aligned}$$

To obtain the prior predictive value one therefore has to calculate the integral on the l.h.s. in (17) where the expectation value $\langle \ln \Lambda(\boldsymbol{x})\rangle_\lambda$ is accessible by Markov chain Monte

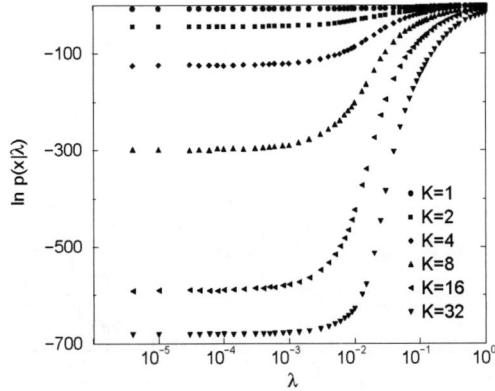

FIGURE 1. $\langle \ln \Lambda(x) \rangle_\lambda$ for the case of two equal peaks with $\sigma=0.3$ in various dimensions.

Carlo techniques[5]. A plot of the latter for the integrand of a two peak case in various dimensions is given in Fig. 1. The λ-axis is on the logarithmic scale. As can be seen, one has to be careful to approach steadily $\lambda \to 0$.

The results are shown in table 5. In comparison with the other methods, the thermodynamic integration scheme works best. However, for the most difficult exercise with the peaky structures completely separated, it fails as well.

Nested sampling

Nested sampling, a recently proposed method [2, 6, 7] for evidence computation tabulates the likelihood function in a probabilistic 'sorted' manner. It samples a collection of n objects x from the prior distribution subject to the constraint that only objects with a likelihood value above an evolving threshold $L(x) > L^*$ are accepted. For the threshold L^* the lowest likelihood value of the collection is used, then the respective object is discarded and a new object is sampled from the prior within the restriction of the constrained likelihood. The worst object of this new sample gives the next threshold L^* and the process is iterated until convergence. The key idea is that the sequence of iterated objects contains (probabilistic) information about the enclosed prior volume $\delta\chi$, which in turn allows the estimate of the corresponding contribution to the total evidence $Z \approx \sum_k z_k$, with $z_k = \delta\chi_k \times L_k$. At the same time the random samples allow the computation of all interesting posterior distributions. Nested sampling is a new sampling method different from the standard techniques. However it does not solve the curse of dimensions. If there is a small hidden likelihood peak in some corner of the prior space the probability of detecting it is low, like in all other MCMC approaches. Nevertheless the different characteristics of this approach makes nested sampling an important tool complementing the conventional suite of MCMC techniques. Taking into account the simple structure (no implementation issues) it is highly recommended to be used as a standard tool for evidence computations.

TABLE 6. Nested sampling.

Dimension	1	2	4	8	16	32
1 peak, $\sigma=0.3$	1.01	0.94	0.95	1.02	1.03	0.7
time[s]	90	160	210	270	490	920
2 peaks, $\sigma=0.3$	0.96	0.85	0.53	0.32	0.13	0.012
time[s]	144	360	410	566	680	1200
2 peaks, $\sigma=0.03$	0.96	0.64	0.32	0.13	0.04	##
time[s]	200	256	270	400	700	1200
2 different peaks, $\sigma=0.3$	0.99	0.86	0.65	0.55	0.25	0.06
time[s]	205	246	290	560	740	1150
2 different peaks, $\sigma=0.03$	0.85	0.73	0.52	0.11	0.05	##
time[s]	265	305	350	560	700	1200

CONCLUSION

In conclusion we have to admit that for integrands in dimensions larger than 10 with sparsely distributed multi-modal structures showing no integral weight in-between, one is left with *Perfect Tempering* [8] as seemingly the only method capable of performing such integrations. However, as it is with any sophisticated method in the numerical analysis business, *Perfect Tempering* needs a lot of experience to sail around the pitfalls of Monte Carlo methods (among them omission of important contributions to the integral from undiscovered parameter space, correlated samples or erroneous bookkeeping). Apart from that, on 'second place', the thermodynamic integration scheme and nested sampling work best, though showing their problems with peaky structures as well. Still, the bunch of the presented methods can come close and should be used as standard diagnostic tools monitoring the outcome all the time.

REFERENCES

1. Press, W., et al., *Numerical Recipes in Fortran 77: The Art of Scientific Computing (Vol. 1 of Fortran Numerical Recipes), 2nd edition*, Cambridge University Press, Cambridge, 2001.
2. Skilling, J., http://www.inference.phy.cam.ac.uk/bayesys/.
3. Daghofer, M., and von der Linden, W., "Perfect Tempering," in *Bayesian Inference and Maximum Entropy Methods in Science and Engineering*, edited by R. Fischer, R. Preuss, and U. von Toussaint, AIP, Melville, N.Y., 2004, vol. 735 of *AIP Conference proceedings*, p. 355.
4. Neal, R. M., *Probabilistic inference using markov chain monte carlo methods*, Dept. of Computer Science, University Toronto, 1993.
5. von der Linden, W., Preuss, R., and Dose, V., "The prior predictive value," in *Maximum Entropy and Bayesian Methods*, edited by W. von der Linden et al., Kluwer Academic, Dordrecht, 1999, p. 319.
6. Skilling, J., "Nested Sampling," in *Bayesian Inference and Maximum Entropy Methods in Science and Engineering*, edited by R. Fischer, R. Preuss, and U. von Toussaint, AIP, Melville, N.Y., 2004, vol. 735 of *AIP Conference proceedings*, p. 395.
7. Skilling, J., *Bayesian Analysis*, **1**, 833 (2006).
8. Daghofer, M., and von der Linden, W., Perfect tempering (2004), these proceedings, and references therein.

Regression for Proportion Data

Julian L. Center, Jr.

Creative Research Corp., 385 High Plain Rd., Andover, MA 01810, jcenter@ieee.org

Abstract. Many applications deal with the relative proportion of items that are assigned to a number of categories. In some of these applications, it is natural to model the measurements as having a multinomial distribution.

Here we explore a one-to-one mapping that transforms proportion data to a vector of real numbers and show how to accurately approximate a multinomial distribution with a Gaussian distribution. Finally, we explore how Nadaraya-Watson and Gaussian process regression techniques can be adapted to deal with proportion data.

Keywords: proportion data, compositional data, categorical data, log-normal

INTRODUCTION

Proportion data, also called compositional data [1][6], is a type of categorical data. Each observation consists of a vector $\mathbf{r}$ whose components are nonnegative and sum to one. A component of $\mathbf{r}$ represents the proportion of something that matches one of a chosen number of categories. For example, in geology, the observation may be the proportion of each of five types of minerals that are present in a rock sample. In medicine, the response to a specific treatment may be characterized by the proportions of patients whose conditions either improve, stay the same, or worsen.

Proportion data is often derived by categorizing items in a sample of size S and counting the number in each category. For example, in climate research, fossil pollen in lake sediment samples is examined to infer climate history [3]. Present day pollen samples from known climates are used as training data to calibrate a model relating climate to the proportion of plant species. For each sample, proportion data is typically derived by sorting 400 pollen grains into 14 categories and counting the number in each category. For problems such as this, a multinomial distribution is often used to model the distribution of counts.

In this type of problem, we are trying to determine the probability relationship between a vector of conditions $\mathbf{x}$ (e.g. climate variables) and an underlying proportion vector $\mathbf{y}$ (e.g. plant species proportions) which relates to a response vector $\mathbf{r}$ (e.g. pollen count proportions) in a sample of size S. To do this we have a set of training data $\mathcal{T} = \{(\mathbf{r}_n, S_n, \mathbf{x}_n) : n = 1, \cdots, N\}$ and a measurement model $p(\mathbf{r}|S, \mathbf{y})$. Our objective is to accurately approximate the probability distribution $p(\mathbf{y}|\mathbf{x}, \mathcal{T})$ for all allowable values of $\mathbf{y}$ and $\mathbf{x}$. Then, for any new measurement $\mathbf{r}$ derived from a sample of size S taken under conditions $\mathbf{x}$, we will be able to compute $p(\mathbf{r}|S, \mathbf{x}, \mathcal{T}) = \int p(\mathbf{r}|S, \mathbf{y}) p(\mathbf{y}|\mathbf{x}, \mathcal{T}) d\mathbf{y}$.

In this paper, we show how to use a multidimensional log-normal distribution to model proportion data and, if desired, to approximate closely the likelihood function corresponding to a multinomial distribution. Using this approach, we can adapt a variety of Bayesian regression techniques to approximate $p(\mathbf{y}|\mathbf{x}, \mathcal{T})$. In particular, we show how to implement Gaussian process regression [2][7] and a form of kernel regression known as a Nadaraya-Watson model [2]. We also show how to use Expectation Propagation to more accurately match a multinomial distribution for the training data. Similar methods have been applied to Gaussian process classification [4][8].

COORDINATE TRANSFORMATION

We can take advantage of a large body of previous research in regression if we transform from a proportion vector to a vector of real numbers. Because its elements are constrained to be non-negative and sum to one, a $(d+1)$-dimensional proportion vector is confined to the d-dimensional simplex $\mathbb{S}^d$ embedded in $\mathbb{R}^{d+1}$. We can establish a one-to-one (bijective) mapping between $\mathbb{S}^d$ and $\mathbb{R}^d$ by using the two mappings

$$\mathbf{sm} : \mathbb{R}^d \to \mathbb{S}^d \; defined\; by\; \mathbf{sm}(\mathbf{f}) = \left[\mathbf{1}_{(d+1)}^T \exp\left(\mathbf{T}^T\mathbf{f}\right)\right]^{-1} \exp\left(\mathbf{T}^T\mathbf{f}\right)$$

$$\mathbf{clr} : \mathbb{S}^d \to \mathbb{R}^d \; defined\; by\; \mathbf{clr}(\mathbf{y}) = \mathbf{T}\ln(\mathbf{y})$$

Here $\mathbf{1}_{(d+1)}$ is the $(d+1)$-dimensional vector of all ones, and $\mathbf{T}$ is a $d \times (d+1)$-dimensional matrix that satisfies

$$\mathbf{TT}^T = \mathbf{I}_d, \quad \mathbf{T1}_{(d+1)} = \mathbf{0}, \quad \mathbf{T}^T\mathbf{T} + \frac{1}{d+1}\mathbf{1}_{(d+1)}\mathbf{1}_{(d+1)}^T = \mathbf{I}_{(d+1)}$$

Those readers in the machine learning community may recognize **sm** as the *softmax* activation function, while those in the statistical community may recognize **clr** as the *centered log ratio* linkage function. Actually, we have modified both of these functions from their usual forms by the inclusion of the matrix **T**. It was included because the usual definition of the softmax function is a many-to-one mapping from $\mathbb{R}^{d+1}$ to $\mathbb{S}^d$ that is totally insensitive to adding a multiple of $\mathbf{1}_{(d+1)}$ to its argument. Including **T** in these definitions makes the pair of functions a one-to-one mapping and simplifies the regression computations. It is always relatively easy to find **T** by the Gram-Schmidt process.

MEASUREMENT MODELS

We need to model the relationship between a proportion measurement **r** from a sample of size S and its underlying proportion vector **y**. That is, we need to specify $p(\mathbf{r}|S,\mathbf{y})$, and in particular, we need the associated likelihood function for training. We consider two different but related models: the multinomial and the log-normal.

The Multinomial Model. If we assume that $\mathbf{y}_n$ represents a discrete probability distribution and that we determine $\mathbf{r}_n$ by drawing S_n independent samples from this distribution, then we are led to the the multinomial distribution

$$p(\mathbf{r}_n|S_n,\mathbf{y}_n) = \mathcal{M}(S_n\mathbf{r}_n|\mathbf{y}_n) \triangleq \frac{S_n!}{\prod_i (S_n[\mathbf{r}_n]_i)!} \prod_i ([\mathbf{y}_n]_i)^{S_n[\mathbf{r}_n]_i}$$

Here $[\mathbf{r}_n]_i$ represents the ith component of the vector $\mathbf{r}_n$ and $[\mathbf{y}_n]_i$ indicates the ith component of $\mathbf{y}_n$. Of course, for training purposes, we view this as the likelihood function for $\mathbf{y}_n$ and treat it as an unnormalized probability distribution. Since we are going to apply regression methods to the transformed variables $\mathbf{f}_n = \mathbf{clr}(\mathbf{y}_n)$, we really need the distribution for $\mathbf{f}_n$, which we can get by taking into account the Jacobian of the transformation [5]. This Jacobian is $\prod_i [\mathbf{y}_n]_i$. Therefore, the portion of the log-likelihood function corresponding to $\mathbf{f}_n$ is (within an additive constant)

$$\ell_n(\mathbf{f}_n) = (S_n + d + 1)\bar{\mathbf{r}}_n^T \ln(\mathbf{y}_n)$$

where

$$\bar{\mathbf{r}}_n = \frac{S_n \mathbf{r}_n + \mathbf{1}_{(d+1)}}{(S_n + d + 1)}$$

We can approximate the likelihood function for a single measurement with a Gaussian distribution by using Laplace's method [5]. The value of $\mathbf{f}_n$ that maximizes the log-likelihood is $\mathbf{m}_n \triangleq \mathbf{T} \ln(\bar{\mathbf{r}}_n)$, and the Laplace approximation is

$$q_n(\mathbf{f}_n) = z_n \mathcal{N}(\mathbf{f}_n | \mathbf{m}_n, \mathbf{V}_n) = z_n |2\pi \mathbf{V}_n|^{-\frac{1}{2}} \exp\left[-\frac{1}{2}(\mathbf{f}_n - \mathbf{m}_n)^T \mathbf{V}_n^{-1} (\mathbf{f}_n - \mathbf{m}_n)\right]$$

where

$$z_n = |2\pi \mathbf{V}_n|^{\frac{1}{2}} \frac{S_n!}{\prod_i (S_n[\mathbf{r}_n])!} \exp[\ell_n(\mathbf{m}_n)]$$
$$\mathbf{V}_n^{-1} = (S_n + d + 1)\mathbf{T}\left[\mathbf{Diag}(\bar{\mathbf{r}}_n) - \bar{\mathbf{r}}_n \bar{\mathbf{r}}_n^T\right]\mathbf{T}^T$$

The Log-Normal Model. This leads us to the more general log-normal model

$$q_n(\mathbf{f}_n) = z_n \mathcal{N}(\mathbf{f}_n | \mathbf{m}_n, \mathbf{V}_n)$$

Clearly, by appropriately choosing the mean $\mathbf{m}_n$, the covariance matrix $\mathbf{V}_n$, and the weighting factor z_n, we can match the Laplace approximation to the likelihood function of a single measurement. But we can do much more with with the log-normal model by manipulating the parameters $z_n, \mathbf{m}_n$, and $\mathbf{V}_n$.

REGRESSION MODELS

Now we use Bayesian regression techniques to accurately approximate $p(\mathbf{f}|\mathbf{x},\mathcal{T})$. These techniques start by prescribing a family of models. A particular model in the family is determined by a particular set of parameters represented by θ. Each model corresponds to a conditional probability distribution $p(\mathbf{f}|\mathbf{x},\theta)$. The prior probabilities of the models are represented by a probability distribution on the model parameters $p(\theta)$. We use the rules of probability theory to determine the evidence for a model $z(\theta) = p(\mathcal{T}|\theta)$. Then we can compute $p(\mathbf{f}|\mathbf{x},\mathcal{T}) = \int p(\mathbf{f}|\mathbf{x},\theta,\mathcal{T}) p(\theta|\mathcal{T}) d\theta$ where $p(\theta|\mathcal{T}) = Z(\theta)p(\theta)/Z$ and $Z = \int Z(\theta) p(\theta) d\theta$. Of course, in most cases, the integrals above must be approximated by a finite weighted sum over the model parameters or by Markov Chain Monte Carlo (MCMC) methods.

Many regression models are possible. We will consider only two: the Nadaraya-Watson model and the Gaussian process model.

The Nadaraya-Watson Model. The Nadaraya-Watson model [2], also known as kernel regression, can be derived by starting with a Parzen density estimator for the joint distribution of **f** and **x**.

$$p(\mathbf{f},\mathbf{x}|\theta) = \sum_{n=1}^{N} \beta_n p(\mathbf{f},\mathbf{x}|n,\theta_n)$$

Here $p(\mathbf{f},\mathbf{x}|n,\theta_n)$ is a probability distribution, called a shape function, that is associated with training data n. To keep things simple, we use a shape function that is the product of independent normal distributions in **f** and **x**. Furthermore, we center the **x** distributions on the locations of the training data and use the same variances for all the shape functions. That is,

$$p(\mathbf{f},\mathbf{x}|n,\theta_n) = \mathcal{N}\left(\mathbf{f}|\widehat{\mathbf{f}}_n,\mathbf{B}\right)\mathcal{N}(\mathbf{x}|\mathbf{x}_n,\mathbf{D})$$

This implies that

$$p(\mathbf{x}|\theta) = \sum_{n=1}^{N}\beta_n\mathcal{N}(\mathbf{x}|\mathbf{x}_n,\mathbf{D})$$

$$p(\mathbf{f}|\mathbf{x},\theta) = \frac{p(\mathbf{f},\mathbf{x}|\theta)}{p(\mathbf{x}|\theta)} = \frac{1}{p(\mathbf{x}|\theta)}\sum_{n=1}^{N}\beta_n\mathcal{N}(\mathbf{x}|\mathbf{x}_n,\mathbf{D})\mathcal{N}\left(\mathbf{f}|\widehat{\mathbf{f}}_n,\mathbf{B}\right)$$

For a particular value of the parameters, we can determine the distribution for a measurement by

$$p(\mathbf{r}|S,\mathbf{x},\theta) = \frac{1}{p(\mathbf{x}|\theta)}\sum_{n=1}^{N}\beta_n\mathcal{N}(\mathbf{x}|\mathbf{x}_n,\mathbf{D})\int p(\mathbf{r}|S,\mathbf{f})\mathcal{N}\left(\mathbf{f}|\widehat{\mathbf{f}}_n,\mathbf{B}\right)d\mathbf{f}$$

If we use a log-normal measurement model, we can solve the integrals analytically to get

$$p(\mathbf{r}|S,\mathbf{x},\theta) = \frac{1}{p(\mathbf{x}|\theta)}z\sum_{n=1}^{N}\beta_n\mathcal{N}(\mathbf{x}|\mathbf{x}_n,\mathbf{D})\mathcal{N}\left(\mathbf{m}|\widehat{\mathbf{f}}_n,\mathbf{B}+\mathbf{V}\right)$$

where z, **m**, and **V** come from the log-normal model. If we use a multinomial measurement model and do not wish to use its Laplace approximation, we can use Monte Carlo sampling to approximate the integrals.

When the number of training points N is relatively small, we can simplify the model family by choosing $\beta_n = \frac{1}{N}$ and $\widehat{\mathbf{f}}_n = \mathbf{m}_n$. Then the only parameters to be determined are **B** and **D**. However, when N is large, the number of computations needed to compute $p(\mathbf{r}|S,\mathbf{x},\theta)$ becomes overwhelming. One approach to solving this problem is to choose a subset of K training points to act as *knots*. To simplify notation, we assume that the training set has been reordered so that the knots are the first K points. Now we set $\beta_n = 0$ for $n > K$ and treat all of the summations in the formulas above as running only to K. The selection of knots is then included in the model parameters.

The evidence for a particular model is given by

$$Z(\theta) = \prod_{n=1}^{N} p(\mathbf{r}_n | S_n, \mathbf{x}_n, \theta)$$

Since the each model is based on a mixture of Gaussians, choices for the prior distribution for the model parameters have been explored in previous research. Space limitations preclude going into details here.

The Gaussian Process Model. A Gaussian process is a probability distribution on real-valued functions characterized by a mean function $\mu(\mathbf{x})$ and a covariance function $k(\mathbf{x}_1, \mathbf{x}_2)$. For any collection of points $\mathbf{x}_1, \mathbf{x}_2, \cdots, \mathbf{x}_m$, the corresponding function values have a joint Gaussian distribution with covariance matrix $\mathbf{C}$ where $[\mathbf{C}]_{ij} = k(\mathbf{x}_i, \mathbf{x}_j)$.

To apply Gaussian process regression to proportion data, we model the prior distribution for each component of the latent function $\mathbf{f}(\mathbf{x})$ as a Gaussian process. For simplicity, we assume that these processes have zero mean and are independent of each other, but share a common covariance function $k(\mathbf{x}_1, \mathbf{x}_2)$.

We are going to use the training data to estimate the values of the latent function $\mathbf{f}(\mathbf{x})$ at a collection of K knots chosen from the training set. That is, we are going to use the rules of probability to determine the posterior probability distribution for the values of $\mathbf{f}$ at the knots. Thus, we choose a family of models parameterized by the covariance function $k(\mathbf{x}_1, \mathbf{x}_2)$ and the selected knots. In the following, we show how to determine $p(\mathbf{f}|\mathbf{x}, \theta, \mathcal{T})$ for a particular choice of model parameters θ. This is called *kriging* in the geostatistics community [6]. To simplify notation, we will often suppress the dependence on θ, but it is always implied.

As before, we simplify notation by assuming that the knots are shuffled to the front of the training set. We build a vector $\mathbf{g}$ by reordering the components of the latent function values at the knots so that we index through the knot locations first. That is,

$$[\mathbf{g}]_{(i-1)K+k} \triangleq [\mathbf{f}(\mathbf{x}_k)]_i \text{ where } i \in \{1, 2, \cdots, d\} \text{ and } k \in \{1, 2, \cdots, K\}$$

Under our assumptions, the prior probability distribution for $\mathbf{g}$ is $\mathcal{N}(\mathbf{0}, \mathbf{G})$ where

$$\mathbf{G} \triangleq \mathbf{I}_d \otimes \mathbf{C}$$
$$[\mathbf{C}]_{jk} \triangleq k(\mathbf{x}_j, \mathbf{x}_k) \text{ for } j, k \in \{1, 2, \cdots, K\}$$

Here $\otimes$ indicates the Kronecker product of two matrices. Therefore, the notation above indicates that $\mathbf{G}$ is a block diagonal matrix with $\mathbf{C}$ in every diagonal block.

If we know $\mathbf{g}$, then the Gaussian process allows us to estimate the value of the latent function at any point $\mathbf{x}$ by

$$p(\mathbf{f}(\mathbf{x})|\mathbf{g}) = \mathcal{N}[\mathbf{f}(\mathbf{x})|\mathbf{H}(\mathbf{x})\mathbf{g}, U(\mathbf{x})\mathbf{I}_d]$$

where

$$\mathbf{H}(\mathbf{x}) \triangleq \mathbf{I}_d \otimes \left[\mathbf{k}(\mathbf{x})^T \mathbf{C}^{-1}\right]$$
$$[\mathbf{k}(\mathbf{x})]_k \triangleq k(\mathbf{x}, \mathbf{x}_k) \text{ for } k \in \{1, 2, \cdots, K\}$$
$$U(\mathbf{x}) \triangleq k(\mathbf{x}, \mathbf{x}) - \mathbf{k}(\mathbf{x})^T \mathbf{C}^{-1} \mathbf{k}(\mathbf{x})$$

We can express this by the equation

$$\mathbf{f}(\mathbf{x}) = \mathbf{H}(\mathbf{x})\mathbf{g} + \mathbf{u}(\mathbf{x})$$

where $\mathbf{u}(\mathbf{x}) \sim \mathcal{N}[\mathbf{0}, U(\mathbf{x})\mathbf{I}_d]$ and $\mathbf{u}(\mathbf{x})$ is independent of $\mathbf{g}$. In particular, the values of the latent function at the training points can be expressed as $\mathbf{f}_n = \mathbf{H}_n\mathbf{g} + \mathbf{u}_n$ where $\mathbf{H}_n \triangleq \mathbf{H}(\mathbf{x}_n)$, $\mathbf{u}_n \sim \mathcal{N}(\mathbf{0}, U_n\mathbf{I}_d)$, and $U_n \triangleq U(\mathbf{x}_n)$. Note that if $\mathbf{x}_n$ is one of the knots, i.e., $n \leq K$, then $\mathbf{u}_n = \mathbf{0}$ and $\mathbf{H}_n$ is a $d \times dK$ sparse matrix that simply selects the appropriate elements of $\mathbf{g}$. To simplify computations, we deviate slightly from the pure Gaussian process model and assume that $\mathbf{u}_i$ is independent of $\mathbf{u}_j$ for $i \neq j$.

Using the log-normal measurement model, we have

$$p(\mathbf{r}_n|S_n, \mathbf{g}) = z_n \int \mathcal{N}(\mathbf{f}_n|\mathbf{m}_n, \mathbf{V}_n) \mathcal{N}(\mathbf{f}_n|\mathbf{H}_n\mathbf{g}, U_n\mathbf{I}_d) d\mathbf{f}_n = z_n \mathcal{N}(\mathbf{H}_n\mathbf{g}|\mathbf{m}_n, \mathbf{R}_n)$$

where $\mathbf{R}_n \triangleq \mathbf{V}_n + U_n\mathbf{I}_d$. Since we have chosen $p(\mathbf{g}) = \mathcal{N}(\mathbf{g}|\mathbf{0}, \mathbf{G})$, we have

$$q(\mathbf{g}) \triangleq p(\mathcal{T}, \mathbf{g}|\theta) = \left[\prod_n z_n \mathcal{N}(\mathbf{H}_n\mathbf{g}|\mathbf{m}_n, \mathbf{R}_n)\right] \mathcal{N}(\mathbf{g}|\mathbf{0}, \mathbf{G})$$

Thus everything is Gaussian, and we can show, by completing squares in the exponent, that $p(\mathbf{g}|\mathcal{T}, \theta)$ takes the form $\mathcal{N}(\mathbf{g}|\widehat{\mathbf{g}}, \mathbf{P})$ and that the model evidence can be computed as

$$Z(\theta) = p(\mathcal{T}|\theta) = \left[\prod_n z_n \mathcal{N}(\mathbf{0}|\mathbf{m}_n, \mathbf{R}_n)\right] \mathcal{N}(\mathbf{g}|\mathbf{0}, \mathbf{G}) [\mathcal{N}(\mathbf{0}|\widehat{\mathbf{g}}, \mathbf{P})]^{-1}$$

To avoid inverting a large matrix, we can determine $\widehat{\mathbf{g}}$ and $\mathbf{P}$ incrementally by the following algorithm, which is essentially the Kalman filter algorithm. (1) Start with $\widehat{\mathbf{g}} \Leftarrow \mathbf{0}$ and $\mathbf{P} \Leftarrow \mathbf{G}$. (2) For $n = 1$ to N, iterate

$$\mathbf{K} \Leftarrow \mathbf{P}\mathbf{H}_n^T (\mathbf{H}_n\mathbf{P}\mathbf{H}_n + \mathbf{R}_n)^{-1}$$
$$\widehat{\mathbf{g}} \Leftarrow \widehat{\mathbf{g}} + \mathbf{K}(\mathbf{m}_n - \mathbf{H}_n\widehat{\mathbf{g}})$$
$$\mathbf{P} \Leftarrow \mathbf{P} - \mathbf{K}\mathbf{H}_n\mathbf{P}$$

If we believe that the log-normal measurement model is correct, then we are finished after one pass through all the training data. We can determine the probability distribution of a measurement $\mathbf{r}$ at $\mathbf{x}$ by

$$p(\mathbf{r}|S, \mathbf{x}, \mathcal{T}) = z\mathcal{N}\left[\mathbf{m}|\mathbf{H}(\mathbf{x})\widehat{\mathbf{g}}, \mathbf{V} + U(\mathbf{x}) + \mathbf{H}(\mathbf{x})\mathbf{P}\mathbf{H}^T(\mathbf{x})\right]$$

where $\mathbf{V}$, $\mathbf{m}$, and z come from the log-normal measurement model.

If we believe that the measurement model is really multinomial, we can get a more accurate approximation using the Expectation Propagation (EP) algorithm [2][4][7][8]. As before we approximate the joint distribution $p(\mathcal{T}, \mathbf{g}|\theta)$ by the form

$$q(\mathbf{g}) \triangleq \prod_n z_n \mathcal{N}(\mathbf{H}_n\mathbf{g}|\mathbf{m}_n, \mathbf{R}_n) \mathcal{N}(\mathbf{g}|\mathbf{0}, \mathbf{G})$$

but our aim is to adjust the z's, $\mathbf{m}$'s, and $\mathbf{R}$'s to minimize the Kullback-Leibler divergence $KL(p||q)$. We do this by iteratively choosing a measurement n and minimizing $KL(p^*||q^*)$ where

$$p^*(\mathbf{g}) = \frac{p(\mathbf{r}_n|S_n,\mathbf{g})}{z_n \mathcal{N}(\mathbf{H}_n\mathbf{g}|\mathbf{m}_n,\mathbf{R}_n)} q(\mathbf{g})$$

$$q^*(\mathbf{g}) = \frac{z_n^* \mathcal{N}(\mathbf{H}_n\mathbf{g}|\mathbf{m}_n^*,\mathbf{R}_n^*)}{z_n \mathcal{N}(\mathbf{H}_n\mathbf{g}|\mathbf{m}_n,\mathbf{R}_n)} q(\mathbf{g})$$

Note that the Gaussian process model establishes that

$$p(\mathbf{r}_n|S_n,\mathbf{g}) = \int \mathcal{M}[S_n\mathbf{r}_n|\mathrm{sm}(\mathbf{f}_n)] \mathcal{N}(\mathbf{f}_n|\mathbf{H}_n\mathbf{g},U_n\mathbf{I}_d)\,d\mathbf{f}_n$$

It is fairly easy to show that we can minimize $KL(p^*||q^*)$ by choosing z_n^*, $\mathbf{m}_n^*$, and $\mathbf{R}_n^*$ so that the moments of $q^*(\mathbf{g})$ match those of $p^*(\mathbf{g})$ up to second order. To approximately match the zeroth order moment, we draw M samples $\mathbf{h}^{(j)} \sim \mathcal{N}\left(\mathbf{H}_n\widehat{\mathbf{g}}, \mathbf{H}_n\mathbf{P}\mathbf{H}_n^T\right)$ and compute

$$z_n^* = \frac{1}{M}\sum_{j=1}^{M} \frac{p\left(\mathbf{r}_n|S_n,\mathbf{h}^{(j)}\right)}{\mathcal{N}\left(\mathbf{h}^{(j)}|\mathbf{m}_n,\mathbf{R}_n\right)} \text{ where}$$

$$p\left(\mathbf{r}_n|S_n,\mathbf{h}^{(j)}\right) = \int \mathcal{M}[S_n\mathbf{r}_n|\mathrm{sm}(\mathbf{f})]\mathcal{N}\left(\mathbf{f}|\mathbf{h}^{(j)},U_n\mathbf{I}_d\right)d\mathbf{f}$$

If n is one of the knots, then $p\left(\mathbf{r}_n|S_n,\mathbf{h}^{(j)}\right) = \mathcal{M}\left(S_n\mathbf{r}_n|\mathbf{h}^{(j)}\right)$; otherwise, we approximate it by Monte Carlo sampling

$$p\left(\mathbf{r}_n|S_n,\mathbf{h}^{(j)}\right) \approx \frac{1}{T}\sum_{k=1}^{T} \mathcal{M}\left[S_n\mathbf{r}_n|\mathrm{sm}\left(\mathbf{h}^{(j)}+\mathbf{u}^{(k)}\right)\right] \text{ with } \mathbf{u}^{(k)} \sim \mathcal{N}(0,U_n\mathbf{I}_d)$$

Now to determine $\mathbf{m}_n^*$ and $\mathbf{R}_n^*$, we can use the same $\mathbf{h}^{(j)}$ samples to compute

$$\widehat{\mathbf{h}} = \frac{1}{z_n^*}\frac{1}{M}\sum_{j=1}^{M}\mathbf{h}^{(j)}\frac{p\left(\mathbf{r}_n|S_n,\mathbf{h}^{(j)}\right)}{\mathcal{N}\left(\mathbf{h}^{(j)}|\mathbf{m}_n,\mathbf{R}_n\right)}$$

and

$$\mathbf{W} = \frac{1}{z_n^*}\frac{1}{M}\sum_{j=1}^{M}\mathbf{h}^{(j)}\mathbf{h}^{(j)T}\frac{p\left(\mathbf{r}_n|S_n,\mathbf{h}^{(j)}\right)}{\mathcal{N}\left(\mathbf{h}^{(j)}|\mathbf{m}_n,\mathbf{R}_n\right)} - \widehat{\mathbf{h}}\widehat{\mathbf{h}}^T$$

To get q^* to have the same moments as p^*, we choose

$$\mathbf{R}_n^{*-1} = \mathbf{R}_n^{-1} + \mathbf{W}^{-1} - \left(\mathbf{H}_n\mathbf{P}\mathbf{H}_n^T\right)^{-1}$$

$$\mathbf{m}_n^* = \mathbf{R}_n^*\left[\mathbf{R}_n^{-1}\mathbf{m}_n + \mathbf{W}^{-1}\widehat{\mathbf{h}} - \left(\mathbf{H}_n\mathbf{P}\mathbf{H}_n^T\right)^{-1}\mathbf{H}_n\widehat{\mathbf{g}}\right]$$

Now we can update $\widehat{\mathbf{g}}$ and $\mathbf{P}$. If $\mathbf{R}_n^{*-1} = \mathbf{R}_n^{-1}$ then the error covariance $\mathbf{P}$ does not change and we update the estimate of $\mathbf{g}$ by

$$\widehat{\mathbf{g}} \Leftarrow \widehat{\mathbf{g}} + \mathbf{P}\mathbf{H}_n^T \mathbf{R}_n^{-1} (\mathbf{m}_n^* - \mathbf{m}_n).$$

Otherwise, we use

$$\begin{aligned} \mathbf{R}_\Delta &\Leftarrow \left(\mathbf{R}_n^{*-1} - \mathbf{R}_n^{-1}\right)^{-1} \\ \mathbf{K} &\Leftarrow \mathbf{P}\mathbf{H}_n^T \left(\mathbf{H}_n \mathbf{P}\mathbf{H}_n^T + \mathbf{R}_\Delta\right)^{-1} \\ \mathbf{P} &\Leftarrow \mathbf{P} - \mathbf{K}\mathbf{H}_n \mathbf{P} \\ \widehat{\mathbf{g}} &\Leftarrow \widehat{\mathbf{g}} + \mathbf{K}\left[\mathbf{R}_\Delta \left(\mathbf{R}_n^{*-1}\mathbf{m}_n^* - \mathbf{R}_n^{-1}\mathbf{m}_n\right) - \mathbf{H}_n \widehat{\mathbf{g}}\right] \end{aligned}$$

Finally, we replace the parameters for measurement n by $z_n \Leftarrow z_n^*, \mathbf{m}_n \Leftarrow \mathbf{m}_n^*$, and $\mathbf{R}_n \Leftarrow \mathbf{R}_n^*$ and go to the next iteration.

The EP algorithm is not guaranteed to converge, but if it does, it will improve the fit to the data. We can also approximate the model evidence using the final values of the z_n's, $\mathbf{m}_n$'s, and $\mathbf{R}_n$'s.

CONCLUSION

We have introduced a one-to-one transformation, based on the softmax function, which allows known Bayesian regression techniques to be applied to the regression of proportion data. We have briefly explored the Nadaraya-Watson kernel regression method and examined in somewhat more detail Gaussian process regression. We also considered the application of the Expectation Propagation technique to find a more accurate approximation when the measurements are multinomial.

REFERENCES

1. J. Aitchison, *The Statistical Analysis of Compositional Data*, Chapman & Hall, Ltd. (1986).
2. C. Bishop, *Pattern Recognition and Machine Learning*, Springer (2006).
3. M. Haslett, et.al., "Bayesian Palaeoclimate Reconstruction," *Journal of the Royal Statistical Society*, Series A, Vol 169, No. 3, pp. 1-36 (2005).
4. H.-C. Kim and Z. Ghahramani, "Bayesian Gaussian Process Classification with the EM-EP Algorithm," *IEEE Trans. on Pattern Analysis and Machine Intelligence*, Vol. 38, No. 12, pp. 1948-1959 (2006).
5. D. MacKay, "Choice of Basis for Laplace Approximation," *Machine Learning*, Vol. 33, No. 1 (1998).
6. V. Pawlowsky-Glahn and R. Olea, *Geostatistical Analysis of Compositional Data*, Oxford U. Press (2004).
7. C. Rasmussen and C. Williams, *Gaussian Processes for Machine Learning*, MIT Press (2006).
8. M. Seeger and M. Jordan, "Sparse Gaussian Process Classification with Multiple Classes," Technical Report TR 661, Dept. of Statistics, Univ. of California at Berkeley (2004).

Inverse covariance simplification for efficient uncertainty management

A. Jalobeanu* and J. A. Gutiérrez[†]

*LSIIT (CNRS - Univ. Strasbourg), Illkirch, France – email: jalobeanu@lsiit.u-strasbg.fr
[†] INNN, Mexico City, Mexico – email: jgutierrez@innn.edu.mx

Abstract. When it comes to manipulating uncertain knowledge such as noisy observations of physical quantities, one may ask how to do it in a simple way. Processing corrupted signals or images always propagates the uncertainties from the data to the final results, whether these errors are explicitly computed or not. When such error estimates are provided, it is crucial to handle them in such a way that their interpretation, or their use in subsequent processing steps, remain user-friendly and computationally tractable. A few authors follow a Bayesian approach and provide uncertainties as an inverse covariance matrix. Despite its apparent sparsity, this matrix contains many small terms that carry little information. Methods have been developed to select the most significant entries, through the use of information-theoretic tools for instance. One has to find a Gaussian pdf that is close enough to the posterior pdf, and with a small number of non-zero coefficients in the inverse covariance matrix. We propose to restrict the search space to Markovian models (where only neighbors can interact), well-suited to signals or images. The originality of our approach is in conserving the covariances between neighbors while setting to zero the entries of the inverse covariance matrix for all other variables. This fully constrains the solution, and the computation is performed via a fast, alternate minimization scheme involving quadratic forms. The Markovian structure advantageously reduces the complexity of Bayesian updating (where the simplified pdf is used as a prior). Moreover, uncertainties exhibit the same temporal or spatial structure as the data.

Keywords: Uncertainties, Inverse Covariance Matrix, Covariance Selection, Gaussian Markov Models, Bayesian Updating, Signal Processing, Image Processing

1. INTRODUCTION

In order to describe a spatially structured object such as an image or signal, while preserving information related to its underlying statistical variability, a covariance matrix can be associated to the object parameters. Even if it only represents second order statistics, it already provides an estimate of the uncertainty associated to each variable, as well as a measure of the interaction between variables. In this work, we focus on multivariate Gaussian distributions as a practical tool to handle uncertain knowledge; in some cases they might provide an exact parametrization of probabilistic objects, however in general they are used as approximations to probability density functions (pdfs) around their mode or their mean. In the case of Bayesian inference [1], they are commonly used to provide an approximate parametric description of the posterior pdf of the parameters of interest, given the observed data. The inverse covariance matrix (or precision matrix) is then given by taking the second derivatives of the log-pdf at the optimum.

By analogy, a diagonal precision matrix can be interpreted as a confidence measure, allowing us to appreciate the quality of the object as an image or signal estimate. Meanwhile, non-diagonal elements relate to dependencies and warn us that the actual number of independent elements might be lower than the number of variables. Somehow it is also a measure of quality, less intuitive than the diagonal-based confidence measure, nonetheless valuable when it comes to physical interpretation. Indeed, highly correlated objects may suggest an inadequate parametrization (e.g. insufficient data with respect to the number of model parameters).

When uncertain objects are combined into a single one (as it is the case in data fusion [2], or integrated data analysis [3]), correctly managing uncertainties is paramount. If the covariance matrix is diagonal, its elements act as weights and a simple weighted average is being computed. If there are correlations between variables, they will also act as weights, and the result can be very different from a variance-weighted average.

In most image or signal processing problems, the inverse covariance matrices produced by Bayesian inference algorithms are particularly large. Even if they are sparse, there are at least two good reasons to simplify them further: a) they are not user-friendly and the storage requirements can be extremely high (e.g. hyperspectral image processing), b) when they are propagated to be combined with other objects of the same type, there is an uncontrolled growth in the computational complexity if all non-diagonal terms are conserved, especially if the combination is performed recursively.

How can we simplify an inverse covariance matrix efficiently while losing a minimum amount of information, so that simplified uncertainties can be propagated and combined without affecting the results significantly?

2. THE COVARIANCE SIMPLIFICATION PROBLEM

2.1. State of the art

A method for covariance selection was first proposed by Dempster [4], consisting of setting to zero some of the elements of the inverse covariance matrix Σ^{-1} in order to get a sparser version of this matrix. Covariance selection means choosing which elements of the precision matrix shall be removed, which corresponds to cutting connections in the graphical model, assuming that the related variables do not interact. For a vector random variable denoted by X of size n, X_i and X_j are conditionally independent if and only if $\Sigma^{-1}_{ij} = 0$. Let us denote the approximate, simplified matrix by $\tilde{\Sigma}^{-1}$.

To quantify the information loss, a natural choice is the conditional entropy or Kullback-Leibler divergence [5] between the approximate and reference distributions:

$$D_{KL}(p|\tilde{p}) = \int_{\mathbb{R}^n} p(X) \log \frac{p(X)}{\tilde{p}(X)} dX \tag{1}$$

where we consider two Gaussian distributions p and $\tilde{p}$ of same mean μ, respectively having inverse covariance matrices Σ^{-1} and $\tilde{\Sigma}^{-1}$; for instance p is expressed as:

$$p(X) = \frac{(\det \Sigma^{-1})^{1/2}}{(2\pi)^{n/2}} e^{-\frac{1}{2}(X-\mu)^t \Sigma^{-1}(X-\mu)} \tag{2}$$

where Σ^{-1} is symmetric positive definite. In this case we have:

$$D_{KL}(p|\tilde{p}) = \frac{1}{2}\left(-\log\left(\det \tilde{\Sigma}^{-1}\right) + \operatorname{tr}\left(\tilde{\Sigma}^{-1}\Sigma\right) - n\right) \tag{3}$$

The idea is to minimize the distance between the new pdf $\tilde{p}$ and the given pdf p subject to constraints, and stop when a desired level of sparsity (e.g. prescribed number of nonzero elements) has been reached. Of course other distance measures could be used, however their soundness is questionable. Anyhow, calculating and optimizing such expressions appears as a computationally intensive task when n is large. An additional difficulty arises when the positive-definite constraint is introduced. Several strategies have been explored: a first order algorithm and block coordinate descent [6], algebra-based decomposition [7], Markov Chain Monte Carlo methods based on Metropolis-Hastings sampling [8], or even Hadamard product-based factor decomposition [9].

A particular class of methods seeks to minimize the information loss under constraints. Markov properties have been used as constraints in [10] under a block-circulant assumption, making computations easier since matrices are diagonalized by the Fourier transform. We also enforce Markov properties in the approach proposed in this paper.

2.2. Naive approaches

When a precision matrix is transformed, one has to make sure it is still positive definite. Setting arbitrary non-diagonal elements to zero is the simplest and fastest way to perform the simplification, but the matrix might not be positive definite anymore. Let us consider the commonly used diagonal dominance property as defined in [10] (which is a sufficient condition for positive definiteness). If the precision matrix already satisfies this property, then a truncated matrix (i.e. some non-diagonal elements set to zero) also satisfies it; one can decide to use this basic simplification method and still have a multivariate Gaussian distribution, though not optimal with respect to information loss. However, in general, this property might not be satisfied (it is a sufficient, but not necessary condition) neither by the original matrix nor by the truncated one, so there is no guarantee of obtaining a consistent result with such a simple method.

Before getting into the details of the new method, let us clarify a point that is often misunderstood. A popular way of analyzing covariance matrices consists of computing their eigenvalues; one can choose to select only the m largest eigenvalues and set the others to zero to simplify the matrix. However, one still needs to store the eigenvectors somehow. Unfortunately, each vector is the same size n as the object and we can easily find examples where Σ^{-1} is very sparse but has many non-negligible eigenvalues (consider for instance a circulant matrix, with only 3 nonzero diagonals; depending on their values, the Fourier transform can have many large coefficients, as a consequence of the Fourier uncertainty principle). Thus, eigenanalysis is not helpful for compressing the precision matrix.

3. THE PROPOSED ALGORITHM

3.1. Covariance-constrained optimization scheme

Some of the entries of the approximate precision matrix $\tilde{\Sigma}^{-1}$ are required to be zero. The other entries are to be determined – this is the purpose of the algorithm. If we constrain the corresponding entries of the approximate covariance matrix $\tilde{\Sigma}$ to be equal to a prescribed set of variances and covariances, then $\tilde{\Sigma}^{-1}$ is fully defined, since $\tilde{\Sigma}^{-1}\tilde{\Sigma} = I$ (we have n^2 constraints and n^2 unknowns, including the entries of interest). The naive solution to this problem consists of alternate projections of $\tilde{\Sigma}^{-1}$ and $\tilde{\Sigma}$ on their respective constraint subspaces. Unfortunately this is not guaranteed to converge, and one can easily find counterexamples, therefore another method has to be devised.

Formally, if Ω is the set of indices related to nonzero entries of $\tilde{\Sigma}^{-1}$, we set:

$$\tilde{\Sigma}^{-1}_{ij} = x_{ij} \text{ if } (i,j) \in \Omega \quad \tilde{\Sigma}^{-1}_{ij} = 0 \text{ if } (i,j) \notin \Omega \quad (4)$$

$$\tilde{\Sigma}_{ij} = \Sigma_{ij} \text{ if } (i,j) \in \Omega \quad \tilde{\Sigma}_{ij} = z_{ij} \text{ if } (i,j) \notin \Omega \quad (5)$$

where x and z are unknown; Σ and 0 are the constraints. The solution can be found by minimizing the following function of x and z:

$$(\hat{x},\hat{z}) = \arg\min_{x,z} \left\| \tilde{\Sigma}^{-1}(x)\tilde{\Sigma}(z) - I \right\|^2 \quad (6)$$

where the squared norm refers to the sum of squares of the matrix elements.

This can be solved by an iterative, alternate optimization scheme (with respect to x given z, then z given x). Each optimization step consists of solving a linear system, which can be done exactly by matrix inversion, or approximately (but faster) using a conjugate gradient [11] stopped after a few iterations, as we did in our implementation. At each iteration, z and x are updated as follows:

$$z^{k+1} = \arg\min_{z} \left(\frac{1}{2} z^t (A_{x^k}) z + (B_{x^k})^t z \right) \quad (7)$$

$$x^{k+1} = \arg\min_{x} \left(\frac{1}{2} x^t (A'_{z^{k+1}}) x + (B'_{z^{k+1}})^t x \right) \quad (8)$$

The matrices A, A' and the vectors B, B' are obtained from the coefficients a and a':

$$A_x = \sum_{ij} (a_x)_{ij}(a_x)^t_{ij} \quad \text{and} \quad B_x = -\sum_i (a_x)_{ii} \quad (9)$$

$$A'_z = \sum_{ij} (a'_z)_{ij}(a'_z)^t_{ij} \quad \text{and} \quad B'_z = -\sum_i (a'_z)_{ii} \quad (10)$$

Here, a_{ij} and a'_{ij} are vectors; the dimension of a is the number of unknown entries x_l, and the dimension of a' is the number of variables z_l. Most elements of a and a' are equal to zero; the only non-zero elements of a_x are equal to x_u and the non-zero elements of a'_z are equal to either Σ_v or z_w, where the indices u, v and w depend on the topology and dimensionality of the problem setting (the detailed expressions of a and a' are not given in this paper but they can be provided upon request).

The initial problem to solve was a system of bilinear equations; we transformed it into a least-squares problem which is more convenient to handle, despite the non-convexity of the function to be minimized, given in Eqn. (6). Indeed, the proposed alternate scheme takes advantage of the quadratic behavior of this function when one of the variables is fixed. As a consequence, a fast convergence is achieved in practice (we observed that it is often achieved in less than 5 iterations, as illustrated in Fig. 4, but we do not have any formal proof yet).

3.2. Fast precision matrix inversion

Instead of inverting the full precision matrix Σ^{-1} to get the required near-diagonal entries of the covariance matrix, we rather invert small blocks B composed of the elements directly interacting with the entries to be computed. Indeed, if we consider the fluctuations of one particular variable, we assume that the fluctuations of all variables not directly interacting with it have a negligible influence, which reminds of the mean field approximation used in statistical physics. Interactions are quantified by the non-diagonal entries of Σ^{-1}. A special case is a diagonal matrix for which elements are inverted separately since there are no interactions.

To determine the variance for each spatial location p, the first step of the inversion consists of selecting the elements $\Sigma^{-1}_{pr} \neq 0$ to construct a submatrix B^p. If we want the covariance Σ_{pq} where q is a nearest neighbor of p, the minimal submatrix required for the inversion is given by the union of elements from both B^p and B^q. The total number of elements depends on the topology, the dimensionality and the range of interaction; e.g. in 1D for a target defined as a first order Markov chain and given interactions limited to the second order (range 2) the size of the matrix is 6×6, see Fig. 1 for an illustration. In 2D, for a target Markov Random Field (MRF) with only 4 nearest neighbors and a given MRF with 8 nearest neighbor interactions, the required size is 12×12, see Fig. 2. These are just special cases having an illustrative purpose; in practice the graphical model encoding the structure of Σ^{-1} has to be used to determine the minimum size.

FIGURE 1. Graphical models capturing the dependency structure of Gaussian pdfs, with a 1D spatial structure (chain). Left: initial pdf with Σ^{-1}; right: simplified pdf with $\tilde{\Sigma}^{-1}$ (first order Markov chain).

Since we only need a few elements of the inverse B^{-1} (covariance Σ_{pq} and variance Σ_{pp}, it is less time consuming to solve a linear system $Bx = e$ than invert the whole matrix. If e is the j-th unit vector (i.e. $e_i = 1$ if $i = j$, 0 otherwise) then the solution is the n-th row of B. We choose j such that this row contains the elements we need to compute. Then we only have to solve for n variables, which can be done efficiently though

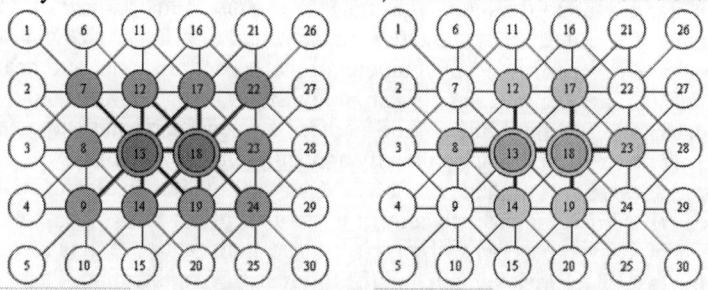

FIGURE 2. Graphical models capturing the dependency structure of Gaussian pdfs, with a 2D spatial structure (lattice). Left: initial pdf with Σ^{-1} and interactions up to 8 neighbors; right: simplified pdf with $\tilde{\Sigma}^{-1}$ and interactions limited to 4 neighbors (first order MRF: thick edges). Shaded nodes form the minimal subset required for matrix inversion to compute the covariance between the 2 central nodes.

a conjugate gradient algorithm [11] in less than n iterations. Stopping this algorithm after only a few iterations allows to reduce the complexity from $O(n^3)$ (regular matrix inversion) to almost $O(n^2)$ (a few matrix multiplications).

Once the inversion has been performed for each block B, the covariance-constrained simplification algorithm is applied in order to obtain a matrix A such that $AB = I$. We use a block sweeping and averaging method to build the large size matrix $\tilde{\Sigma}^{-1}$ from each A (there is a single estimated interaction term pq for each block, and two corresponding diagonal terms pp and qq which are averaged out); see Fig. 3 for an illustration.

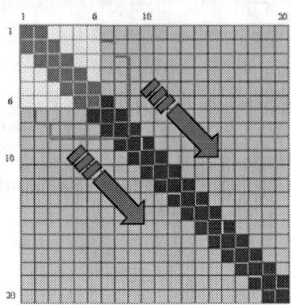

FIGURE 3. A large matrix is swept from top left to bottom right to be simplified block by block.

4. RESULTS AND DISCUSSION

To test the proposed method, we first consider the 1D case: a chain of variables with an interaction range of 2, the goal being to reduce it to a first order Markov chain (nearest neighbor interactions only). The input matrices correspond to realistic Gaussian distributions (arising for instance from signal processing techniques such as filtering, denoising, resampling...). The convergence was achieved in 3 to 5 iterations with respect to the energy to be minimized or the entries of the precision matrix. Monitoring the evolution of the Kullback-Leibler divergence leads us to the same conclusion. Examples of plots showing the evolution of these quantities with the number of iterations are shown on Fig. 4. Due to the interaction range, we chose a matrix size of 6×6 which is a minimum. Other tests (not shown) were carried out on larger size matrices, and a block sweeping was performed to compute all the needed diagonal and near-diagonal elements of $\tilde{\Sigma}^{-1}$, the iterative simplification algorithm being applied blockwise. For a 20×20 matrix, we saw no significant difference between the result computed this way, and the result on the simplification directly applied to the 20×20 matrix.

Experiments were also carried out for a 2D Gaussian random field with an 8-neighbor connectivity (see Fig. 2) in order to convert it to a 4-neighbor Markov random field. The same speed of convergence was observed, and no noticeable difference was detected between the fast, blockwise version and the full matrix simplification. These were only preliminary results and more extensive tests will need to be performed from real data (precision matrices coming from image processing algorithms).

There is no guarantee that the precision matrix obtained by the proposed iterative algorithm be positive definite. This point is still under investigation. However, in all our

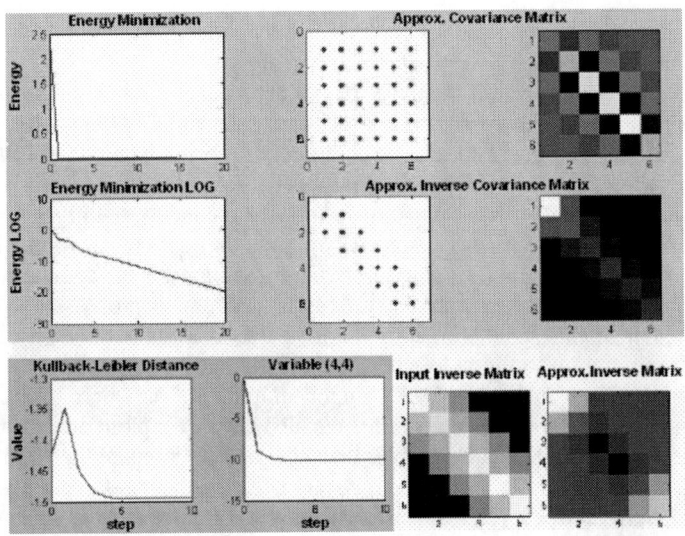

FIGURE 4. Top: Example of simplified covariance and precision matrices, and energy plots. Bottom: example showing the input and simplified precision matrices (from 5 to 3 diagonals as in Fig. 1), and convergence monitoring using the Kullback-Leibler distance and one inverse covariance element.

experiments, the resulting matrices were always checked positive definite.

A few authors suggested to factorize the precision matrix through a Cholesky decomposition; this could be used to parametrize this matrix by setting $\tilde{\Sigma}^{-1} = VV^t$ where V is a lower triangular matrix. However this would only be efficient in 1D, since the sparsity of V does not correspond to the sparisty of $\tilde{\Sigma}^{-1}$ in general. Markovian approximations of Gaussian fields can be found in the literature but only for stationary fields [10], the non-stationary case still being an area of active research.

5. CONCLUSION

We designed a simple yet powerful algorithm to simplify an inverse covariance matrix related to a multivariate Gaussian distribution, by using two constraints: the low order Markov structure of the resulting graph, and the values prescribed for variance and nearest neighbor covariances. This allows us to conserve information from the input pdf explicitly, by forcing the simplified model to have the same near-diagonal covariances. Compared to other (usually more complex) methods, the new technique yields predictable results regarding these uncertainty measures, which is valuable when the quantities of interest include, for instance, the variances (diagonal elements) as error measures on estimated variables, and the correlation coefficient in order to assess the dependencies between neighboring model variables.

Moreover, enforcing a stationary Markov structure enables us to store the simplified precision matrix entries using the same spatial structure as the original model – as extra images for a processed image, preserving the spatial location. Indeed, all uncertainties can be stored as a vector-valued image, which is no more difficult to handle than multiband images that are nowadays commonly encountered in astronomy or planetary imaging. The interpretation and post-processing of spatial data that has already been processed is greatly improved if the uncertainties are used as well, which is made easier by the simplification and the user-friendly storage achieved by the proposed algorithm. Even if uncertainties on the result can be computed for virtually any kind of processing pipeline, they are seldom used in subsequent tasks because of their complexity (topology and dimensionality), and error propagation is at best done independently for each model parameter or pixel using only the variances. We hope that our contribution will encourage the use of uncertainty and interaction terms by reducing their complexity without compromising the information they carry. This way, complex processing tasks can be split into smaller modules and efficiently implemented as a chain, the error propagation ensuring a minimum information loss between the modules.

In the future, we need to overcome the major limitation of our approach, which is the absence of positive definiteness constraint. We also have to explore other ways such as the minimization of the Kullback-Leibler distance, and to compare the results; indeed, our method is designed to conserve local covariances but does not explicitly minimize the information loss. Moreover, extensive tests have to be carried out in order to confirm the quick convergence of the iterative procedure, and to find rigorous ways of assessing the convergence, before applying it to massive data sets in astronomy or remote sensing.

ACKNOWLEDGEMENTS

This work was partially funded by the French Research Agency (ANR) as part of the SpaceFusion project (*"Jeunes Chercheurs 2005" program, grant # JC05_41500*).

REFERENCES

1. A. Gelman, J.B Carlin, H.S Stern, and D.B Rubin. *Bayesian Data Analysis*. Chapman & Hall, 1995.
2. A. Jalobeanu, J.A. Gutiérrez, and E. Slezak. Multi-source data fusion and super-resolution from astronomical images. *Statistical Methodology (Special Issue on Astrostatistics), submitted*, 2007.
3. R. Fischer and A. Dinklage. Integrated data analysis of fusion diagnostics by means of the Bayesian probability theory. *Rev. Sci. Instrum.*, 75, 2004.
4. A. P. Dempster. Covariance selection. *Biometrics*, 28(1), 1972.
5. S. Kullback and R. A. Leibler. On information and sufficiency. *Annals of Mathematical Statistics*, 22, 1951.
6. O. Banerjee, A. d'Aspremont, L. El Ghaoui, and G. Natsoulis. Convex optimization techniques for fitting sparse gaussian graphical models. In *Proc. of 23rd Intl. Conf. on Machine Learning*, 2006.
7. B. Jones and M. West. Covariance decomposition in undirected Gaussian graphical models. *Biometrika*, 92, 2005.
8. F. Wong, C.K. Carter, and R. Kohn. Efficient estimation of covariance selection models. Technical Report 2003-12, Statistical and Applied Mathematical Sciences Institute, 2003.
9. A. Kavcic and J.M.F. Moura. Matrices with banded inverses: Inversion algorithms and factorization of Gauss-Markov processes. *IEEE Trans. on Information Theory*, 46(4), 2000.
10. H. Rue and L. Held. *Gaussian Markov Random Fields: Theory and Applications*. Chapman & Hall / CRC, 2005.
11. W.H. Press, S.A. Teukolsky, W.T. Vetterling, and B.P. Flannery. *Numerical Recipes in C: The Art of Scientific Computing*. Cambridge University Press, 2nd edition, 1993.

Propagation of Statistical Information Through Non-Linear Feature Extractions for Robust Speech Recognition

R. F. Astudillo*, D. Kolossa† and R. Orglmeister**

*ramon@astudillo.com
†d.kolossa@ee.tu-berlin.de
**reinhold.orglmeister@tu-berlin.de

Abstract. Automatic speech recognition systems often rely on statistical noise suppression methods to increase their recognition performance in non-stationary noisy environments. However, even with a good approximation of the noise power spectrum, the estimated clean signal contains residual noise along with artifacts introduced by speech estimation inaccuracies. In this paper, we show that this can be compensated by propagating a measure of the uncertainty of estimation through the feature extraction process and combining it with missing feature techniques directly in the feature domain.

Keywords: Unscented Transform, Complex Gaussian Distribution, Rice Distribution, MMSE-LSA Bayesian Estimation, Missing Feature Techniques

BRIEF INTRODUCTION TO AUTOMATIC SPEECH RECOGNITION

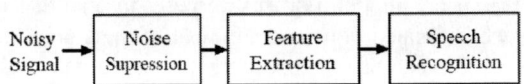

FIGURE 1. *Basic elements in automatic speech recognition.*

Figure 1 shows the standard process followed to recognize speech using an automatic speech recognizer (ASR), that is, the process of outputting a written transcription of the human message contained in a given audio signal. For this purpose, most real world speech recognition applications require a noise suppression preprocessing step to eliminate or reduce the effect of interfering signals in the recognition process. The noise suppression step is followed by the feature extraction step, in which the signal is transformed into a domain more suitable for recognition (class differentiation). In this feature domain recognition itself takes place. Previously trained statistical models that represent abstract language components, typically phonemes or triphones, are used for this purpose. As the noise suppression method, minimum mean square error Bayesian estimation of the log spectral amplitude (MMSE-LSA), better known as Ephraim-Malah filter[Ephraim and Malah(1985)], is widely used[1] due to its ability to avoid musical noise. This and other methods are usually combined with feature extraction methods which imitate the way humans process speech (e.g. mel-cepstra or RASTA-PLP) and are remarkably effective in providing robustness against inter-speaker variability. As a statistical model for the speech features, the well known combination of hidden Markov models (HMM) with Gaussian mixture models (GMM) as output distributions at each emitting state is the most widely used alternative[2].

UNCERTAINTY IN ASR

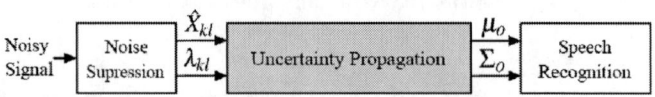

FIGURE 2. *Use of uncertainty in automatic speech recognition.*

Like many other statistical noise suppression methods, the Ephraim-Malah filter provides an estimation of the clean short-time spectrum of the signal given the noisy signal spectrum and a reasonably good estimation of the noise power spectrum. However, even with a good estimation of the noise spectrum, complete elimination of noise and artifacts in speech is not guaranteed. MMSE-LSA Bayesian estimation relies on a set of assumptions about noise and original signal that are not fully met in reality (Gaussianity of noise, for example). Furthermore, some of the estimated statistical parameters, like the a priori SNR[Ephraim and Malah(1984)], are approximated via recursive formulas that introduce additional artifacts in the speech which may result in poor recognition results even for high SNRs. The method presented here ameliorates this problem by using a measure of how uncertain the estimated signal is, propagating this measure into the feature domain and combining the resulting statistical information with the statistical parameters of the speech model. For this purpose, instead of using a deterministic representation of each estimated Fourier coefficient of the signal, a complex Gaussian distribution is used. The mean of this distribution is set to be equal to the deterministic value of the Fourier coefficient obtained from the MMSE-LSA estimator $\hat{X}_{kl}$ and its variance per real dimension λ_{kl} is set proportional to the uncertainty derived from the estimation process. By propagating the mean and covariance through the feature extraction (as shown in Fig. 2) we obtain a statistical description of each feature in the domain where the speech model was trained. We can then combine statistical information of both features and model parameters to ignore or reestimate noisy features in an optimal way by using missing feature techniques[B. Raj(2004)][Kolossa et al.(2005)] or uncertainty decoding[Droppo et al.(2002)].

---

[1] Although Ephraim-Malah filter assumes the signal to be corrupted by additive noise, it has shown positive results with other types of noises.
[2] Other alternatives to HMMs like linear dynamic models, or more general Bayesian networks are showing promising results.

PROPAGATION OF UNCERTAINTY THROUGH THE FEATURE EXTRACTION STAGE

Piecewise Uncertainty Propagation

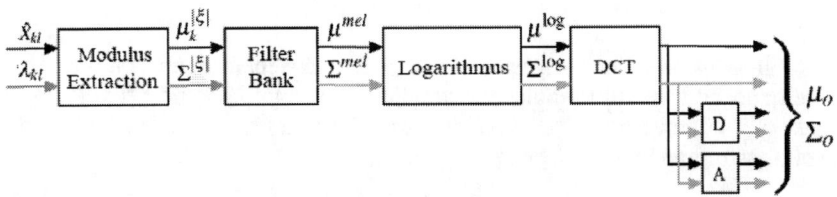

FIGURE 3. *Uncertainty propagation for mel-cepstral, delta and acceleration features.*

Feature extraction in speech recognition generally involves various non linear transformations and therefore finding an analytical solution to propagate first and second order moments through these transformations may be a very complicated task. These calculations would be also specific for each feature extraction type[1]. The alternative use of optimized Pseudo-Montecarlo methods, like the unscented transform (UT)[Julier and Uhlmann(1996)], provide a general solution to the propagation problem, valid for any type of feature extraction, but are slower and inaccurate. An additional problem of these methods is the high number of dimensions of the initial random variable to be considered, which makes this approach computationally intensive. The alternative that showed better results is a combination of both analytical and UT methods in a piecewise approach. In this approach, the initial transformations involving high dimension feature vectors are calculated analytically. Luckily, for many feature extraction methods those transformations are similar and often involve the extraction of the modulus from the short-time Fourier transform (STFT), optional computation of the power spectrum, and the compression of the signal by means of a filter bank that emulates the frequency response of the cochlea (e.g Mel, Bark scales). After these two transformations, the dimensionality of the features has been reduced in an order of magnitude and therefore applying Pseudo-Montecarlo methods efficiently is now possible. In the next sections the piecewise approach to uncertainty propagation will be exemplified with the mel-cepstral feature extraction, depicted in Figure 3. A similar procedure may be nevertheless applied for other feature extractions like for example RASTA-PLP.

Uncertainty Model for Estimated Speech

The output of the Ephraim-Malah filter is a matrix of complex values $\hat{X}$ that correspond to the estimated STFT transform of the clean signal X. Each element $\hat{X}_{kl}$ of the

---

[1] For example, the case of the propagation from linear spectral to cepstral domain was studied in[Gales(1995)].

matrix corresponds to a given frequency bin k and time unit l. To model each of these elements as a random variable ξ_{kl} rather than as a deterministic value, we use a complex Gaussian distribution (1) of mean equal to $\hat{X}_{kl}$ and variance λ_{kl} proportional to the uncertainty of estimation for that element

$$f_\xi(x \mid \hat{X}_{kl}, \lambda_{kl}) = \frac{1}{\pi \lambda_{kl}} \exp\left(-\frac{|x - \hat{X}_{kl}|^2}{\lambda_{kl}}\right). \quad (1)$$

Since all transformations are repeated for each time-step, from now on frame index l will be removed from all formulas for simplicity. Therefore μ^* and Σ^* will be used to denote expectation vector and covariance matrix for each frame, where $*$ will denote the type of transformation.

Extraction of the Modulus

Usually only the modulus of the STFT is relevant for speech recognition and is therefore calculated at the beginning of the feature extraction. In the case of considering each frequency bin k as a complex Gaussian distribution instead of as a deterministic value, the probability distribution can be integrated to obtain the marginal distribution of the modulus $|\xi_k|$. The resulting distribution is a Rice distribution with mean and covariance[1] corresponding to

$$\mu_k^{|\xi|} = \sqrt{\frac{\pi \lambda_k}{4}} \cdot L_2^1\left(-\frac{|\hat{X}_k|^2}{\lambda_k}\right) \quad (2)$$

$$\Sigma_{kk}^{|\xi|} = \lambda_k + |\hat{X}_k|^2 - \left(\mu_k^{|\xi|}\right)^2. \quad (3)$$

Where $L_2^1(x)$ is the Laguerre polynomial that can be expressed in terms of modified Bessel functions.

Filter Bank Compression and other linear transformations

The filter bank transformation is a linear transformation that can be expressed as a multiplication of a fixed matrix A of elements a_{mk} by each frame of $NFFT$ bins [Kolossa et al.(2005)]

$$\hat{X}_m^{mel} = \sum_{k=1}^{NFFT} a_{mk} \cdot |\hat{X}_k|. \quad (4)$$

---

[1] Since the correlation between bins in the frequency domain can be ignored, it is sufficient to consider each bin of each frame as a random variable defined by its mean μ_{kl} and variance per real dimension λ_{kl}. This is not the case for the rest of the transformations where each frame has to be considered as a vector valued random variable defined by its mean vector μ_l and covariance matrix Σ_l.

Due to linearity of the expectation operator, we can compute the filterbank transformed mean and covariance by using

$$\mu^{mel} = A \cdot \mu^{|\xi|} \tag{5}$$

$$\Sigma^{mel} = A \cdot \Sigma^{|\xi|} \cdot A^T. \tag{6}$$

It must be taken into account that after the filterbank transformation, the covariance matrix is no longer diagonal. The same applies to other linear transformations like the discrete cosine transform (DCT), performed after the logarithm transformation, and the delta (D) and acceleration (A) coefficients.

Logarithm Transformation by using Unscented Transform

The computation of the exact distribution after the filterbank transformation is complicated and therefore approximations are used to propagate information through the logarithm transformation[Gales(1995)][Kolossa et al.(2005)]. An alternative, also valid for any other type of non-linear transformation, is the use of the unscented transform (UT). This Pseudo-Montecarlo method propagates a N-dimensional random variable by transforming a set of $2N+1$ so called points and estimating the mean and covariance by using weighted averages according to

$$\mu^{\log} = \frac{\kappa}{(N+\kappa)} \cdot \log(\mu^{mel}) + \frac{1}{2(N+\kappa)} \cdot \sum_{i=1}^{2N} \log(S_i) \tag{7}$$

$$\Sigma^{\log} = \frac{1}{2(N+\kappa)} \cdot \sum_{i=1}^{2N} (\log(S_i) - \mu_{\log}) \cdot (\log(S_i) - \mu_{\log})^T. \tag{8}$$

The sigma points S_i are deterministically chosen to lie on the $\sqrt{N}$th covariance contour of the initial distribution[1] according to

$$S_i = \mu^{mel} + \left(\sqrt{(N+\kappa) \cdot \Sigma^{mel}}\right)_i \tag{9}$$

$$S_{i+N} = \mu^{mel} - \left(\sqrt{(N+\kappa) \cdot \Sigma^{mel}}\right)_i. \tag{10}$$

The additional parameter κ allows higher moments of the distribution to be considered, therefore the UT is not limited to Gaussian random variables[2]. Setting $\kappa = 3 - N$ will minimize the mean squared error up to the fourth order. An additional advantage of the UT is the easy consideration of the non diagonal covariance matrix for any type

---

[1] $(\ )_i$ corresponds here to the ith row of the matrix square root.
[2] It is, in fact, applicable to any symmetric, unimodal distribution[Julier and Uhlmann(1996)].

of non-linear transformation. The relatively low dimensionality of the features after the filter bank transformation (around 20 features per frame) ensures that the UT can be implemented efficiently.

GENERATION OF UNCERTAINTY

Ideally, uncertainty λ should be modeled as a function of the error between the clean signal X and the signal $\hat{X}$ estimated in the noise suppression step, the equation

$$\lambda_{kl} = |\hat{X}_{kl}||X_{kl} - \hat{X}_{kl}|, \tag{11}$$

showed excellent performance for a wide range of SNRs. Since X is not available we need to make an estimation of λ in terms of Ephraim-Malah parameters. An initial approach to this was to consider the estimated noise power as a measure of uncertainty[Kolossa et al.(2005)], but since this is calculated by recursive averaging it is not able to fully capture strong changes in the uncertainty in case of highly non-stationary noise. In order to include this we can consider the amount of change inflicted by the Ephraim-Malah filter on the noisy signal Y with

$$\lambda_{kl} = |\hat{X}_{kl}||Y_{kl} - \hat{X}_{kl}|. \tag{12}$$

USE OF UNCERTAINTY

Once the statistical information for each frame μ_o, Σ_o has been propagated to the feature domain we can use it to yield a probabilistic model $P(o|\mu_o,\Sigma_o)$ for the frame observation o. By combining this with the statistical information of the speech model, it is possible to re-estimate or eliminate noisy features in an optimal way. Different techniques like uncertainty decoding[Droppo et al.(2002)], marginalization[B. Raj(2004)], imputation[B. Raj(2004)] or modified imputation [Kolossa et al.(2005)] can be used for this purpose. Uncertain speech recognition tests carried out in the spectral domain showed superior performance of the marginalization technique whereas in the mel-cepstral domain, modified imputation showed the better results. The marginalization technique integrates all features with an uncertainty above a given threshold out of the state probability distribution[B. Raj(2004)], this is equivalent to removing those features from state mean, covariance and feature vector and performing the usual recognition. Modified imputation estimates the optimal frame observation for each state q of the speech model by maximizing the probability of the observed frame o given its uncertainty information μ_o, Σ_o and the HMM probability distribution of state q defined by μ_q, Σ_q. Applying Bayes theorem this can be reduced to

$$\hat{o}_q = \arg\max_o (P(o|\mu_o,\Sigma_o,\mu_q,\Sigma_q)) = \arg\max_o (P(o|\mu_o,\Sigma_o)P(o|\mu_q,\Sigma_q)). \tag{13}$$

Assuming multivariate Gaussian distributions[1] for both state probability distribution and observation, this maximization results in the estimator

$$\hat{o}_q = (\Sigma_o^{-1} + \Sigma_q^{-1})^{-1}(\mu_q \Sigma_q^{-1} + \mu_o \Sigma_o^{-1}). \qquad (14)$$

Rather than having to establish a certain threshold to eliminate or replace features, like it is necessary for marginalization and imputation, modified imputation provides a smooth output depending on the Σ_o/Σ_q ratio[2].

EXPERIMENTS

The use of uncertainty in mel-cepstral domain with delta and acceleration coefficients was compared to the use of uncertainty in spectral domain to address the benefits of using missing feature in non-linear feature domains. Tests were carried out in highly non-stationary environments. To ensure a proper approximation of the noise spectrum, minima controlled recursive averaging (IMCRA) noise power estimation[Cohen(2003)] was used. The test data comprised 200 files from 10 different speakers of the TI-DIGITS database (half female, half male). The two non-stationary noise samples, wind against windshield and crowded street noise were added at segmental SNRs between -30 and 30dB, with a resolution of 5dB. The standard MMSE-LSA algorithm[Ephraim and Malah(1985)] was used in combination with IMCRA to perform the noise suppression. Both, approximations of the error after MMSE-Noise suppression acording to eq.(12) along with ideal estimations given by Equation (11) were used to test the efficiency of the proposed method. The ASR architecture consisted of 6-mixture phoneme-level HMMs, trained on the whole TI-DIGITS training set.

TABLE 1. Test Results for Reference, Noisy, Processed and Missing Feature Recognition

	Features	WC[%]	WA[%]	WC[%]	WA[%]
Referece		98.76	98.76	98.76	98.76
Wind Noise		-15dB SNR		5dB SNR	
Noisy	MFCC	59.66	28.44	94.74	87.94
MMSE-LSA	MFCC	69.86	34.78	96.45	75.27
MI (Approx.)	MFCC	72.64	46.68	91.65	88.72
MI (Ideal)	MFCC	79.60	51.93	97.53	94.28
Optimal Spect.	SPEC	59.81	44.98	84.08	44.05
Street Noise		-15dB SNR		5dB SNR	
Noisy	MFCC	52.40	22.87	95.83	92.43
MMSE-LSA	MFCC	61.51	36.63	97.06	92.43
MI (Approx.)	MFCC	69.71	22.72	96.14	94.90
MI (Ideal)	MFCC	80.99	48.53	97.06	96.45
Optimal Spect.	SPEC	52.55	19.23	88.24	58.89

---

[1] Tests were carried out with Gaussian mixture models (GMM) where $\hat{o}_q$ was estimated independently for each mixture.
[2] Diagonal covariance matrices for uncertainty and state probability distribution assumed.

Table 1 shows representative results for the SNR range and the different methods tested. Two coefficients were used to measure speech recognition quality. These are based on the number of deleted (d), substituted (s), inserted (i) and the total number of words (n) found when comparing transcriptions output by the recognizer and the original transcriptions by means of dynamic programming.

$$WC = \frac{n-d-s}{n} \qquad WA = \frac{n-d-s-i}{n} \qquad (15)$$

Word correctness (WC) represents the percentage of words correctly identified. Word accuracy (WA) also takes into account the percentage of words found that were not present in original transcriptions (typically silence frames misidentified as speech). The modified imputation method showed the best results for both feature extractions in mel-cepstral domain and a clear superiority compared to the best results using any of the tested methods in spectral domain.

CONCLUSIONS

A method for robust speech recognition has been presented that allows the combination of missing features and non-linear feature extraction methods to compensate the errors produced by MMSE-LSA estimation in non-stationary noisy environments. The introduced piecewise propagation of uncertainty combines analytic solutions and Pseudo-Montecarlo methods to propagate efficiently first and second order statistical information through feature extractions that simulate human perception. The benefits of employing uncertainty in mel-cepstral domain against spectral domain, where this techniques are usually applied and no uncertainty propagation is needed, were also addressed.

REFERENCES

[Ephraim and Malah(1985)] Y. Ephraim, and D. Malah, *Acoustics, Speech, and Signal Processing [see also IEEE Transactions on Signal Processing], IEEE Transactions on* **33**, 443–445 (1985).

[Ephraim and Malah(1984)] Y. Ephraim, and D. Malah, *Acoustics, Speech, and Signal Processing [see also IEEE Transactions on Signal Processing], IEEE Transactions on* **32**, 1109–1121 (1984).

[B. Raj(2004)] R. S. B. Raj, M.L. Seltzer, Reconstruction of missing features for robust speech recognition, Tech. rep., Mitsubishi Electric Research Laboratories (2004).

[Kolossa et al.(2005)] D. Kolossa, A. Klimas, and R. Orglmeister, "Separation and robust recognition of noisy, convolutive speech mixtures using time-frequency masking and missing data techniques," in *Applications of Signal Processing to Audio and Acoustics, 2005. IEEE Workshop on*, 2005, pp. 82–85.

[Droppo et al.(2002)] J. Droppo, A. Acero, and L. Deng, "Uncertainty decoding with SPLICE for noise robust speech recognition," in *Acoustics, Speech, and Signal Processing, 2002. Proceedings. (ICASSP '02). IEEE International Conference on*, 2002, vol. 1, pp. I–57–I–60 vol.1.

[Julier and Uhlmann(1996)] S. Julier, and J. Uhlmann, A general method for approximating nonlinear transformations of probability distributions, Tech. rep., University of Oxford, UK (1996).

[Gales(1995)] M. J. F. Gales, *Model-Based technique for noise robust speech recognition*, Ph.D. thesis, Gonville and Caius College (1995).

[Cohen(2003)] I. Cohen, *Acoustics, Speech, and Signal Processing [see also IEEE Transactions on Signal Processing], IEEE Transactions on* **11** (2003).

Maximum a Posteriori Maximum Entropy Signal Denoising

Abd-Krim Seghouane* and Luc Knockaert[†]

*National ICT Australia
Canberra Research Laboratory
Locked Bag 8001, Canberra ACT 2601, Australia
emai:Abd-krim.seghouane@nicta.com.au
†Department Intec-Imec
Ghent University
St. Pietersnieuwstraat 41, B-9000 Gent, Belgium
email:luc.knokaert@intec.ugent.be

Abstract. When fitting wavelet based models, shrinkage of the empirical wavelet coefficients is an effective tool for signal denoising. Based on different approaches, different shrinkage functions have been proposed in the literature. The shrinkage functions derived using Bayesian estimation theory depend on the prior used on the wavelet coefficients. However, no simple and direct method exists for the choice of the prior. In this paper a new method based on maximum entropy considerations is proposed for the construction of the prior on the wavelet coefficients. The new shrinkage function is obtained by coupling this prior to maximum a posteriori arguments. A comparison with classical shrinkage functions is given in a simulation example of image denoising in order to illustrate the effectiveness of the proposed thresholding method.

Keywords: Signal denoising, wavelet thresholding, Bayesian estimation
PACS: 07.50.Qx

INTRODUCTION

Signals are often corrupted by noise in their acquisition or transmission. The goal of a denoising operation is to separate an observed signal into a noiseless signal that retains the biggest part of important signal features and a remaining noise. One important denoising approach is based on data projection on an orthogonal family of bases functions. Due to their energy compaction property, wavelet bases functions are widely used in signal and image denoising. Crudely, the projection of a signal on a wavelet bases yields a large number of small basis coefficients and a small number of large basis coefficients. The representation of the noiseless signal is then chosen based on a proper choice of the nonzero bases coefficients.

A denoising algorithm that use the wavelet transform consist of three steps: calculate the wavelet transform of the noisy signal, modify the noisy wavelet coefficients according to some rule and compute the inverse wavelet transform using the modified coefficients. One of the most popular method of wavelet denoising is soft thresholding [1]. Due to its effectiveness and simplicity, it is widely used in signal and image processing. Its main idea is to eliminate the wavelet basis coefficients with an absolute value smaller than a threshold T and subtract the value T from all other coefficients. The method of hard thresholding known also as VisuShrink, introduced in [2] removes the additive noise only by eliminating the wavelet basis coefficients with an absolute value smaller than the threshold T and not affecting the remaining ones. Hard and soft thresholds are obtained by solving a minmax problem in the estimation of the expected value of the reconstruction error. The suggested optimal threshold for the wavelet basis coefficients is the well known universal threshold $\sigma_n\sqrt{2\log(N)}$, where σ_n^2 is the variance of the additive noise and N is the data length. A number of other techniques have been introduced in the literature, for example, SureShrink [3] is a data-driven subband adaptive technique with superior performance. Recently, BayesShrink [4] which is also a data-driven subband adaptive technique has been proposed and outperforms VisuShrink and SureShrink.

The basic idea of the method is to model wavelet coefficients with prior probability densities. Then the problem can be expressed as the estimation of free noise coefficients using Bayes techniques. Therefore, two problems arise: 1) what kind of prior probability densities represent the free noise wavelet coefficients ? In this paper, we do not assume any prior density, but we estimate it by maximizing the entropy [5]. The second problem is: 2) what is the corresponding estimator or shrinkage function ? to derive it here, we use the MAP philosophy.

The rest of the paper is organized as follows. In Section II, the idea of Bayesian wavelet denoising is described and the link between Laplace prior and soft thresholding is illustrated. The prior that maximizes the entropy and the associated shrinkage function are derived in Section III. The performance of the proposed estimator are analyzed in a simulation example in Section VI. A conclusion is given in Section V.

THE BAYESIAN APPROACH TO WAVELET DENOISING

In the denoising problem, the data are assumed to be of the form

$$g_i = x_i + \varepsilon_i \quad i = 1,...,N \tag{1}$$

where the ε_i are i.i.d (independent and identically distributed) Gaussian noise samples. Typically, N is an integer power of 2. We observe the g_i's (a noisy version of a signal) and want to estimate the original data signal x_i as accurately as possible according to some criteria. In the wavelet domain, using orthogonal bases, (1) is equivalent to

$$y_i = w_i + n_i \quad i = 1,...,N \tag{2}$$

where the y_i's are the noisy wavelet coefficients, w_i's the noiseless wavelet coefficients and n_i's are i.i.d Gaussian noise samples. Our objective is to estimate the w_i's from the noisy wavelet coefficients y_i's. Shrinkage functions or estimators can be derived using the Bayesian approach by imposing a particular prior structure on the w_i's. In the prior the coefficients are mutually independent.

The MAP estimator for (2) is

$$\hat{w}_i(y) = \arg\max_w p(w_i/y_i). \tag{3}$$

Using Bayes rule, one gets

$$\hat{w}_i(y) = \arg\max_w [p(y_i/w_i)p(w_i)]$$

$$= \arg\max_w [p_n(y_i - w_i) p(w_i)]. \tag{4}$$

From the assumption on the noise, p_n is zero mean Gaussian with variance σ_n^2, i.e.,

$$p_n(n_i) = \frac{1}{\sigma_n \sqrt{2\pi}} \exp\left(-\frac{n_i^2}{2\sigma_n^2}\right). \tag{5}$$

Note that (4) is also equivalent to

$$\hat{w}_i(y) = \arg\max_w [\log(p_n(y_i - w_i)) + \log(p(w_i))] \tag{6}$$

Let us define $f(w_i) = \log(p(w_i))$, by using (5), (6) becomes

$$\hat{w}_i(y) = \arg\max_w \left(-\frac{(y_i - w_i)^2}{2\sigma_n^2} + f(w_i)\right). \tag{7}$$

This is equivalent to solving the following equation

$$\frac{(y_i - \hat{w}_i)}{\sigma_n^2} + f'(\hat{w}_i) = 0. \tag{8}$$

If $p(w)$ is $N(0, \sigma^2)$, then $f(w) = -\log(\sqrt{2\pi}\sigma) - w^2/2\sigma^2$ and the estimator can be written as

$$\hat{w}_i(y) = \frac{\sigma^2}{\sigma^2 + \sigma_n^2} y_i \tag{9}$$

If $p(w)$ is Laplacian

$$p(w) = \frac{1}{\sqrt{2}\sigma} \exp\left(-\frac{\sqrt{2}|w|}{\sigma}\right) \tag{10}$$

then $f(w) = -\log(\sigma\sqrt{2}) - \sqrt{2}|w|/\sigma$, and the estimator will be

$$\hat{w}_i(y) = \text{sign}(y_i)\left(|y_i| - \frac{\sqrt{2}\sigma_n^2}{\sigma}\right)_+ \tag{11}$$

where, $(g)_+$ is defined as

$$(g)_+ = \begin{cases} 0 & \text{if } g < 0 \\ g, & \text{otherwise.} \end{cases}$$

Equation (11) corresponds to the soft thresholding estimator

$$\hat{w}_i(y) = \text{soft}\left(y_i, \frac{\sqrt{2}\sigma_n^2}{\sigma}\right).$$

The shrinkage functions associated to the Gaussian and Laplace prior are displayed in figure 1.

In practice, the prior $p_w(w)$ is rarely known in advance and as it can be seen from the previous developments different priors will lead to different shrinkage functions or estimators. Therefore, having a method that allow us to choose this prior is attractive.

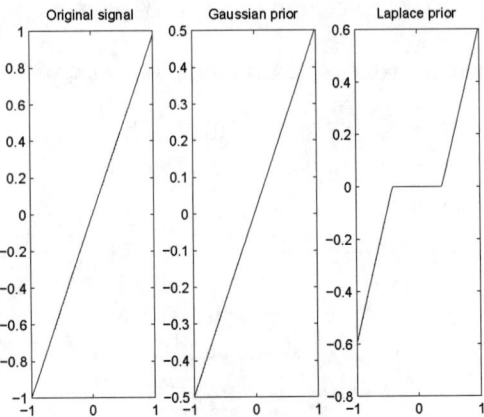

FIGURE 1. Gaussian and Laplace corresponding shrinkage functions.

THE PROPOSED APPROACH

Since $p_w(w)$ is a priori unknown, and if we do not want to assume any prior probability density, we have to estimate it. The proposed approach here, is to take $p_w(w)$ that maximizes the entropy S which is given from (2) as

$$\begin{aligned} S &= -\int\int p(y,w)\ln p(y,w)dydw \\ &= -\int\int p(y/w)p(w)\ln p(y/w)dydw \\ &\quad -\int\int p(y/w)p(w)\ln p(w)dydw \\ &= \int p(w)[H(w)-\ln p(w)]dw, \end{aligned} \qquad (12)$$

where

$$H(w) = -\int p(y/w)\ln p(y/w)dy.$$

The entropy S is always maximized subject to a constraint set defined by statistical knowledge of some averages over $p(w)$

$$E[g_l(w)] = -\int g_l(w)p(w)dw = \varepsilon_l \quad l=1,...,L$$

This is an isoperimetric problem from calculus of variations. We solve it by introducing Lagrange multipliers λ_l, $l=1,...,L$ and setting up variations with respect to $p(w)$ to zero. This yields

$$\ln p(w) = H(w) - \sum_{l=1}^{L} \lambda_l g_l(w)$$

and
$$p(w) = \exp\{H(w)\}\exp\left\{-\sum_{l=1}^{L}\lambda_l g_l(w)\right\}. \tag{13}$$

This probability function with particular constraints and the MAP estimator, can now be used to derive a new shrinkage function for wavelet denoising.
From (2), we have
$$H(w) = \frac{1}{2} + \ln(\sqrt{2\pi}\sigma_n).$$

The constraint used here is
$$E_{p(w)}\{|w|\} = |y|,$$

with this constraint the prior is
$$p(w) = \frac{\exp(-\lambda g(w))}{\int \exp(-\lambda g(w))dw}.$$

Following Buckigham's π theorem the argument of the exponential function $-\lambda g(w)$ should not have any dimension. Therefore, if $g(w) = |w|$ then $\lambda = \alpha/|y|$, since w and y have the same dimension. To estimate α, we use the constraint on the prior
$$\frac{\int |w|\exp(-\alpha|w|/|y|)dw}{\int \exp(-\alpha|w|/|y|)dw} = \frac{|y|}{\alpha} = |y|$$

then $\alpha = 1$.
By inserting this new prior in (8) we obtain the following estimator
$$\hat{w}_i = \begin{cases} y_i - \frac{\sigma_n^2}{y_i} & \text{if } |y_i| > \sigma_n^2 \\ 0, & \text{if } |y_i| \leq \sigma_n^2. \end{cases} \tag{14}$$

Under the constraint
$$E_{p(w)}\{w^2\} = y^2,$$

and following the same steps of derivation of (14) we obtain
$$\hat{w}_i = \frac{y_i}{1 + \sigma_n^2/y_i^2}. \tag{15}$$

The shrinkage functions associated to these constraints are displayed in figure 2.
It can be seen from the figure that the shrinkage function associated to the constraint of first order (14) is similar to the soft thresholding shrinkage function.
This estimator proposed in (14) is used in the following simulation example on a problem of image denoising.

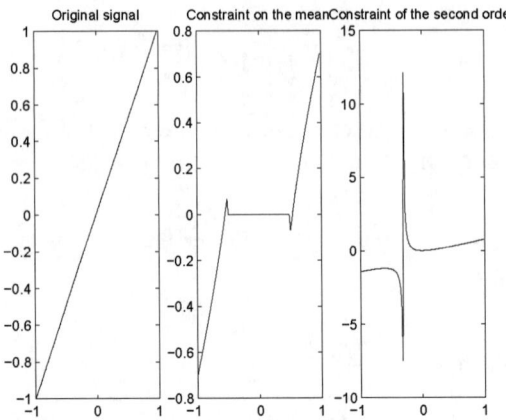

FIGURE 2. First and second order constraints corresponding shrinkage functions.

SIMULATION EXAMPLE

In order to experimentally demonstrate the effectiveness of the proposed shrinkage, the proposed estimator has been applied to the denoising problem of the standard 256×256 image *Barbara*. The image was corrupted by a white gaussian noise $N(0,30)$ in the first experiment and by a uniform noise in the second experiment. The resulting images from these experiments were named image 1 and image 2 in table 1. The wavelet decomposition was a eight level decomposition using Daubechie's two-tap filters [6]. Denoising using wavelet consists in finding the wavelet coefficients that best represents the image, or the set of wavelet coefficients from which an inverse wavelet transform will produce the best approximation to original image in the L_2 sens.

The method proposed in this paper has been compared to the standard hard and soft thresholding approaches. and a Gaussian noisy version are displayed on figure 1. On Table 1 we have presented the results of the proposed denoising method applied on the two noisy images compared to the hard and soft thresholding methods. For each image, we computed the normalized mean square error (mse) between the reconstructed image and the original image. The error presented in table. 1 has been obtained by averaging the mse over 100 noisy images in each of the two cases. The simulations show that the

TABLE 1. Performance of the method by the normalized mean square error (nmse).

Image number	hard threshol.	soft threshol.	proposed
Image 1	0.0719	0.0567	0.0534
Image 2	0.0675	0.0526	0.0497

proposed method has smaller MSE then the other methods, particularly in the Gaussian noise case.

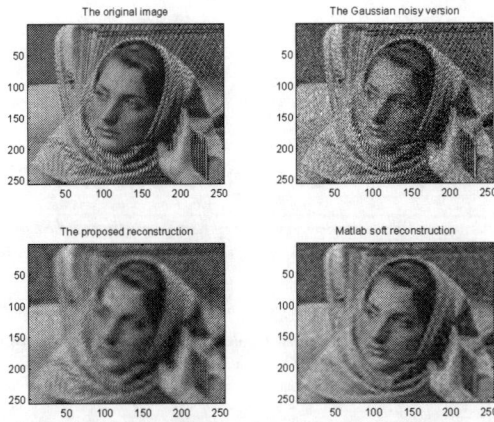

FIGURE 3. Illustration of the different images.

CONCLUSION

In this paper, a new shrinkage function to signal denoising is proposed. The method used to derive this proposed shrinkage function is based on the Bayesian approach and the maximum a posteriori estimation. The difference with existing methods, is that the construction of the used prior on the unknown noiseless wavelet coefficients is based on maximum entropy considerations. The new shrinkage function is therefore obtained by coupling the maximum entropy principle to the maximum a posteriori estimation. The simulation results illustrate that the proposed method is effective in comparison to soft thresholding.

ACKNOWLEDGMENT

National ICT Australia is funded by the Australian Department of Communications, Information Technology and the Arts and the Australian Research Council through Backing Australia's Ability and the ICT Center of Excellence Program.

REFERENCES

1. D. L. Donoho, *IEEE Transactions on Information Theory* **41**, 613–627 (1995).
2. D. L. Donoho, and I. M. Johnstone, *Biometrika* **81**, 425–455 (1994).
3. D. L. Donoho, and I. M. Johnstone, *Journal of the American Statistical Association* **90**, 1200–1224 (1995).
4. B. Y. S. Chang, and M. Vetterli, *IEEE Transactions on Image Processing* **9**, 1532–1546 (2000).
5. L. Knockaert, *IEEE Transactions on Signal Processing* **45**, 1553–1559 (1997).
6. I. Daubechies, *Ten lectures on wavelets*, vol. 61, SIAM, CMBS-NSF Regional Conference Series in Applied Mathematics, Philadelphia, 1992.

FBST for Unit Root Problems

M.Diniz*,†, C.A.B.Pereira* and J.M.Stern*,†

University of Sao Paulo, Institute of Mathematics and Statistics
†*diniz_marcio@yahoo.com.br   jstern@ime.usp.br*

Abstract. This paper presents the Full Bayesian Significance Test for unit roots in auto-regressive time series, and compares it to other approaches on a benchmark of 14 econometric series.
Keywords. ARMA models, e-values, FBST, Unit roots.

INTRODUCTION

Testing for unit roots in ARMA time series models is a problem that presents well known and documented difficulties for standard Bayes Factor methodologies, see [1], [2], [4], [6], [10], and [13] to [16]. In [2, p.159], the authors state:

"Testing for unit root is a Bayesian framework in one of the most controversial topics in the economic literature. There are several reasons for this:

- First ... [the use of] information that is not contained in the likelihood function and this violates the likelihood principle to which Bayesians stick.

- Secondly, the unit root hypothesis is a point hypothesis and Bayesians do not like testing point hypothesis because it is not natural to compare an interval which receives a positive probability with a point null hypothesis of zero mass.

- Finally, classical and Bayesian unit root tests do not *give the same answer. This is a striking example where it is* not *possible to recover the classical results using a non-informative prior."*

We will show that the FBST, or Full Bayesian Significance Test, presented in [9], easily overcomes all these difficulties, see also [7], and [17]. Moreover, the FBST e-values are computed following the absolutely standard form of FBST formalism, using non-informative priors, and in strict observance of the likelihood principle. Finally, the FBST analysis agrees with the classic analysis on a benchmark of 14 time series commonly used in the econometric literature.

The first section describes the problem and the general model used to test for unit roots and derives the posterior distribution used in the present work to perform the FBST. Afterwards, we describe the numerical procedures to calculate the e-values. Concluding, we compare the classical and bayesian procedures, pointing the FBST advantages.

FBST REVIEW

The FBST was specially designed to give an epistemic value, or value of evidence, supporting a sharp hypothesis H. This support function is the *e*-value, ev(H). Furthermore, the e-value has many necessary or desirable properties for a statistical support function, such as:

(I) Give an intuitive and simple measure of significance for the hypothesis in test, ideally, a *probability* defined directly in the original or *natural parameter space*.

(II) Have an intrinsically geometric definition, independent of any non-geometric aspect, like the particular parameterization of the (manifold representing the) null hypothesis being tested, or the particular coordinate system chosen for the parameter space, i.e., be an *invariant* procedure.

(III) Give a measure of significance that is smooth, i.e. *continuous and differentiable*, on the hypothesis parameters and sample statistics, under appropriate regularity conditions of the model.

(IV) Obey the *likelihood principle*, i.e., the information gathered from observations should be represented by, and only by, the likelihood function.

(V) Require *no ad hoc artifice* like assigning a positive prior probability to zero measure sets, or setting an arbitrary initial belief ratio between hypotheses.

(VI) Be a *possibilistic* support function, where the support of a logical disjunction is the maximum support among the support of the disjuncts.

(VII) Be able to provide a *consistent* test for a given sharp hypothesis.

(VIII) Be able to provide *compositionality* operations in complex models.

(IX) Be an *exact* procedure, i.e., make no use of "large sample" asymptotic approximations when computing the *e*-value.

(X) Allow the incorporation of previous experience or expert's opinion via (subjective) *prior distributions*.

The objective of this section is to provide a very short review of the FBST theoretical framework, summarizing the most important statistical properties of its support function, the *e*-value. It also summarizes the logical (algebraic) properties of the *e*-value, and its relations to other classical support calculi, including possibilistic calculus and logic, paraconsistent and classical. Further details, demonstrations of theoretical properties, comparison with other statistical tests for sharp hypotheses, and an extensive list of references can be found in the author's previous papers.

Let $\theta \in \Theta \subseteq R^p$ be a vector parameter of interest, and $L(\theta|x)$ be the likelihood associated to the observed data x, a standard statistical model. Under the Bayesian paradigm the posterior density, $p_n(\theta)$, is proportional to the product of the likelihood and a prior density, The (null) hypothesis H states that the parameter lies in the null set, defined by inequality and equality constraints given by vector functions g and h in the parameter space.

$$p_n(\theta) \propto L(\theta|X) p_0(\theta) , \quad \Theta_H = \{\theta \in \Theta \,|\, g(\theta) \leq \mathbf{0} \wedge h(\theta) = \mathbf{0}\} .$$

From now on, we use a relaxed notation, writing H instead of Θ_H. We are particularly interested in sharp (precise) hypotheses, i.e., those in which $\dim(H) < \dim(\Theta)$, i.e. there is at least one equality constraint.

The FBST defines $\operatorname{ev}(H)$, the *e*-value, the epistemic value or value of (presented or observed) evidence supporting (in favor of) the hypothesis H, and $\overline{\operatorname{ev}}(H)$, the *e*-value against H, as

$$s(\theta) = \frac{p_n(\theta)}{r(\theta)}, \quad s^* = s(\theta^*) = \sup_{\theta \in H} s(\theta), \quad \widehat{s} = s(\widehat{\theta}) = \sup_{\theta \in \Theta} s(\theta),$$

$$T(v) = \{\theta \in \Theta \mid s(\theta) \leq v\}, \quad W(v) = \int_{T(v)} p_n(\theta) d\theta, \quad \text{ev}(H) = W(s^*),$$

$$\overline{T}(v) = \Theta - T(v), \quad \overline{W}(v) = 1 - W(v), \quad \overline{\text{ev}}(H) = \overline{W}(s^*) = 1 - \text{ev}(H).$$

The function $s(\theta)$ is known as the posterior surprise relative to a given reference density, $r(\theta)$. $W(v)$ is the cumulative surprise distribution. The surprise function was used, among other statisticians, by Good, Evans and Royall. Its role in the FBST is to make ev(H) explicitly invariant under suitable transformations on the coordinate system of the parameter space, see next section.

The tangential (to the hypothesis) set $\overline{T} = \overline{T}(s^*)$, is a Highest Relative Surprise Set (HRSS). It contains the points of the parameter space with higher surprise, relative to the reference density, than any point in the null set H. When $r(\theta) \propto 1$, the possibly improper uniform density, $\overline{T}$ is the Posterior's Highest Density Probability Set (HDPS) tangential to the null set H. Small values of $\overline{\text{ev}}(H)$ indicate that the hypothesis traverses high density regions, favoring the hypothesis.

In the FBST the role of the reference density, $r(\theta)$ is to make $\overline{\text{ev}}(H)$ explicitly invariant under suitable transformations of the coordinate system. Invariance, as used in statistics, is a metric concept. The reference density can be interpreted as a compact and interpretable representation for the reference metric in the original parameter space. This metric is given by the geodesic distance on the density surface. The natural choice of reference density is an uninformative prior, interpreted as a representation of no information in the parameter space, or the limit prior for no observations, or the neutral ground state for the Bayesian operation. Standard (possibly improper) uninformative priors include the uniform and maximum entropy densities.

Let us consider the cumulative distribution of the e-value against the hypothesis, $\overline{V}(c) = \Pr(\overline{\text{ev}} \leq c)$, given θ^0, the true value of the parameter. Under appropriate regularity conditions, for increasing sample size, $n \to \infty$, we can say the following:
- If H is false, $\theta^0 \notin H$, then $\overline{\text{ev}}$ converges (in probability) to 1, that is, $\overline{V}(0 \leq c < 1) \to 0$.
- If H is true, $\theta^0 \in H$, then $\overline{V}(c)$, the confidence level, is approximated by the function

$$QQ(t,h,c) = Q\left(t - h, Q^{-1}(t,c)\right), \quad \text{where}$$

$$Q(k,x) = \Gamma(k/2, x/2)/\Gamma(k/2, \infty), \quad \Gamma(k,x) = \int_0^x y^{k-1} e^{-y} dy,$$

$t = \dim(\Theta)$, $h = \dim(H)$ and Q(k,x) is the cumulative chi-square distribution with k degrees of freedom. Figure 2.2 portrays $QQ(t,h,c)$ Q$(t-h,Q^{-1}(t,c))$ for $t = 2\ldots 4$ and $h = 0\ldots t-1$.

Under the same regularity conditions, an appropriate choice of threshold or critical level, $c(n)$, provides a consistent test, τ_c, that rejects the hypothesis if $\overline{\text{ev}}(H) > c$.

The empirical power analysis developed in [7] and [18], provides critical levels that are consistent and also effective for small samples.

THE AUTO REGRESSIVE TIME SERIES MODEL

The $AR(1)$ process
$$y_t = \phi y_{t-1} + \varepsilon_t$$
where $\varepsilon_t \sim i.i.d.(0, \sigma^2)$, has a unit root if $\phi = 1$. In this case its mean and its variance do not exist. If $|\phi| < 1$, then the mean of y_t is zero and its variance $\sigma^2/(1-\phi^2)$ and the process has a strong tendency to return to its mean value after a shock. However, if the process has a unit root, a shock has an everlasting effect. This can be seen if y_t is expressed as the cumulated sum of past errors, each with the same weight. Therefore, test for a unit root consists in testing the precise hypothesis $H_0 : \phi = 1$.

The economic and econometric literature has given great importance to the development of unit root tests in the past two and a half decades. It is very important to know if, for instance, economic recessions have permanent consequences for the level of future GNP, or instead represent just a temporary downturn with the output lost eventually made up during recovery. Nelson and Plosser, [10], argued that many economic series are better characterized by unit roots than by deterministic trends.

However, in the development of the tests difficulties arised because the asymptotic distribution of the ordinary least squares estimators presents a discontinuity at $\phi = 1$. The ADF test is the most used in unit root tests and assumes, in its more general form, that the data generating process has a constant, a deterministic trend and follows an $AR(p)$ structure with $i.i.d.$ errors. Below we introduce this model assuming gaussian disturbances to develop the bayesian inference.

The $AR(p)$, or order p auto-regressive time series model with white Gaussian noise and deterministic intercept and trend, is written as:
$$y_t = \mu + \delta\, t + \phi_1 y_{t-1} + \ldots + \phi_p y_{t-p} + \varepsilon_t$$
where $\varepsilon_t \sim N(0, \sigma^2) \; \forall t = 1, \ldots, T$. This series can also be written in the differenced or correction form:
$$\Delta y_t = \mu + \delta\, t + \Gamma_0 y_{t-1} + \Gamma_1 \Delta y_{t-1} + \ldots + \Gamma_{p-1} \Delta y_{t-p+1} + \varepsilon_t$$
where $\Delta y_t = y_t - y_{t-1}$, $\Gamma_0 = \phi_1 + \ldots + \phi_p - 1$ and $\Gamma_i = -\sum_{j=i+1}^{p} \phi_j$, for $i = 1, \ldots, p-1$.

If using this parametrization, the series has a unit root if $\Gamma_0 = 0$. The ADF tests this hypothesis against $\Gamma_0 \leq 0$, but if $\Gamma_0 \geq 0$ the process is non-stationary.

This model can also be written in standard regression form, using the parameter vector $\theta = [\beta, \sigma]$, where β is a vector with all the linear parameters, $Y_p = [y_1 \ldots y_p]$ is the vector of the first p observations, and Y is the vector of all remaining observations:
$$Y = X\beta + e \text{ , where}$$

$$\theta = \begin{bmatrix} \mu \\ \delta \\ \Gamma_0 \\ \Gamma_1 \\ \ldots \\ \Gamma_{p-1} \end{bmatrix}, \; Y = \begin{bmatrix} \Delta y_{p+1} \\ \Delta y_{p+2} \\ \ldots \\ \Delta y_T \end{bmatrix}, \; X = \begin{bmatrix} 1 & 1 & y_p & \Delta y_p & \ldots & \Delta y_2 \\ 1 & 2 & y_{p+1} & \Delta y_{p+1} & \ldots & \Delta y_3 \\ \ldots & \ldots & \ldots & \ldots & \ldots & \ldots \\ 1 & T & y_{T-1} & \Delta y_{T-1} & \ldots & \Delta y_{T-p+1} \end{bmatrix}$$

The dimensions of these matrices are, respectively, $p+2 \times 1$ for β, $T-p \times 1$ for Y, and $T-p \times p+2$ for X.

Using the matrix regression form, it is easy to see that the ML estimator of β, the predicted ML observations, and the sum of squared errors is given by

$$\widehat{\beta} = (X'X)^{-1}X'Y, \quad \widehat{Y} = X\widehat{\beta}, \text{ and}$$

$$e'e = (Y-X\beta)'(Y-X\beta) = (Y-\widehat{Y})'(Y-\widehat{Y}) + (\beta-\widehat{\beta})'X'X(\beta-\widehat{\beta}).$$

Using the standard non-informative prior $f(\beta,\sigma) \propto 1/\sigma$, the model posterior can be written as:

$$f(\beta,\sigma|Y,Y_p) \propto \sigma^{-(T-p+1)} \exp\left(-\frac{1}{2\sigma^2}\left((Y-\widehat{Y})'(Y-\widehat{Y}) + (\theta-\widehat{\theta})'X'X(\theta-\widehat{\theta})\right)\right)$$

NUMERICAL EXPERIMENTS AND RESULTS

After the model derived above we tested for unit roots 14 U.S. macroeconomic time series first mentioned in Nelson and Plosser, [10]. Here we use the extended series, used in Schotman and van Dijk, [14].

The following table shows the e-values and ADF test for the aforementioned econometric time series. The ADF, Augmented Dickey and Fuller test, based on the Frisch-Waugh-Lovell theorem, is arguably the most used unit root test in econometrics. We have used the computer procedure described in James MacKinnon, at Queen's University, [8]. All numerical time series follow the specification in Bauwens et al. [2], so that the results are comparable.

As can be seen from the posterior expression, the conditional posteriors are $\pi(\theta|\sigma,Y,Y_p) \propto N(\widehat{\theta}, \sigma^2 V)$ and $\pi(1/\sigma^2|\theta,Y,Y_p) \propto \Gamma(T-p/2, B)$, where $B = 0.5(Y-\widehat{Y})'(Y-\widehat{Y}) + (\theta-\widehat{\theta})'X'X(\theta-\widehat{\theta})$ and $V = (X'X)^{-1}$. For the FBST computations, several solvers can be used in the optimization step, as [3] or [5], and standard Monte Carlo sampling is used in the integration step, see [7].

In table 1 we can see that the non-stationary posterior probability is quite distant from the ADF p-value. These results were highlighted by Sims, [15] and Sims and Uhlig, [16]. Considering the simplest $AR(1)$ model, they argued that, once classical inference is based on the distribution of $\widehat{\phi}|\phi=1$, it reaches counterintuitive conclusions because the referred distribution is skewed. Bayesian inference, they conclude, uses the distribution of $\phi|\widehat{\phi},y_1\ldots,y_T$ which is not skewed.

Phillips, [13] claims that the difference in results between classical and bayesian approaches is due to the flat prior that puts much weight on the stationary region. He proposed the use of Jeffreys priors, which restored the conclusions drawn by the classical test. Phillips argued that the flat prior was, actually, informative when used in time series models like those for unit root tests. He made simulations that show the

" [the use of a] flat prior has a tendency to bias the posterior towards stationarity. ... even when [the estimator] is close to unity, there may still be a non negligible downward bias in the [flat] posterior probabilities".

TABLE 1. Unit root tests for Nelson and Plosser data

Series	start	p	trend	ADF	p-value	$P(\Gamma_0 \geq 0|Y)$	e-value
Real GNP	1909	2	yes	-3.52	0.044	0.0005	0.045
Nominal GNP	1909	2	yes	-2.06	0.559	0.0238	0.542
Real GNP per capita	1909	2	yes	-3.59	0.037	0.0004	0.039
Industrial prod.	1860	2	yes	-3.62	0.032	0.0003	0.031
Employment	1890	2	yes	-3.47	0.048	0.0004	0.049
Unemployment rate	1890	4	no	-4.04	0.019	0.0001	0.023
GNP deflator	1889	2	yes	-1.62	0.778	0.0584	0.771
Consumer prices	1860	4	yes	-1.22	0.902	0.1154	0.984
Nominal wages	1900	2	yes	-2.40	0.377	0.0106	0.361
Real wages	1900	2	yes	-1.71	0.739	0.0475	0.724
Money stock	1889	2	yes	-2.91	0.164	0.0029	0.157
Velocity	1869	2	yes	-1.62	0.779	0.0620	0.784
Bond yield	1900	4	no	-1.35	0.602	0.0962	0.941
Stock prices	1871	2	yes	-2.44	0.357	0.0103	0.363

TABLE 2. MLE under $H_0 : \Gamma_0 = 0$

Parameters	Real GNP	Ind. Prod.	GNP def.	Wage
μ	0.01543	0.049427	0.00187	0.01494
δ	0.00011	-0.00014	0.00027	0.00020
Γ_1	0.33146	0.03636	0.44992	0.46687
σ	0.05558	0.09682	0.04364	0.05545

TABLE 3. Standard error of MLE under $H_0 : \Gamma_0 = 0$

Parameters	Real GNP	Ind. Prod.	GNP def.	Wage
μ	0.01320	0.01806	0.00902	0.01247
δ	0.00028	0.00024	0.00016	0.00024
Γ_1	0.10895	0.08966	0.09163	0.09661

TABLE 4. MLE - unrestricted model

Parameters	Real GNP	Ind. Prod.	GNP def.	Wage
μ	0.81849	0.05221	0.09086	0.39792
δ	0.00567	0.00718	0.00112	0.00309
Γ_0	-0.17631	-0.17658	-0.03164	-0.06494
Γ_1	0.41106	0.12432	0.46979	0.50130
σ	0.05193	0.09252	0.04329	0.05392

Tables 2 and 3 display some ML estimators and the respective standard errors assuming unit roots. Table 4 and 5 show the ML estimators for the same series for the unrestricted model. Table 6 and 7 give the number of series which rejected the unit root hypothesis in 100 generated samples assuming that there was (table 6) or not (table 7) a unit root. We used three criteria to reject the hypothesis: the ADF asymptotic p-value for 5% significance, the exact ADF p-value for 5% significance and the e-value set in 0.05.

It is important to remember that finite sample critical values for unit root tests depend

TABLE 5. Standard error of MLE - unrestricted model

Parameters	Real GNP	Ind. Prod.	GNP def.	Wage
μ	0.23279	0.01727	0.05667	0.16301
δ	0.00163	0.00206	0.00056	0.00125
Γ_0	0.05104	0.04941	0.01990	0.02756
Γ_1	0.10436	0.08915	0.09175	0.09522

TABLE 6. Simulated series rejecting H_0 in hundred generated assuming H_0

Series	$<ADF_{5\%}(\infty)$	$<ADF_{5\%}(ex.)$	$ev<0.05$
Real GNP	4	3	3
Ind. Prod.	4	4	4
GNP def.	7	6	6
Wage	4	4	4

TABLE 7. Simulated series rejecting H_0 in hundred generated assuming the unrestricted model

Series	$<ADF_{5\%}(\infty)$	$<ADF_{5\%}(ex.)$	$ev<0.05$
Real GNP	73	67	64
Ind. Prod.	85	82	84
GNP def.	20	18	18
Wage	29	27	27

on the assumption that the error terms are $NID(0,\sigma^2)$ once these values were generated by simulations that use this assumption. The asymptotic critical values are valid much more generally, since they do not require normality or homoskedasticity. Therefore, for small samples, it is safer to rely on asymptotic critical values.

Table 6 shows that the FBST, even using the flat prior, has a power similar to the ADF test. Hence, the argument used by Phillips to criticize conclusions based on posterior probabilities when flat priors were used is not valid for the FBST.

As mentioned in the first section, Bayes Factor tests for unit roots have had many difficulties to deal with time series presented in the field of econometrics. Several alternative Bayes Factor tests have been proposed in order to overcome these difficulties. However, their performance is still in question. For example, [1] concludes:

"In two Monte Carlo simulations, however, we find that the 'objective' Bayesian test have relatively low power in distinguishing between plausible alternatives, making it difficult to draw any conclusions concerning long-run [performance]. We conclude that, at least for the 'objective' Bayesian test, the Bayesian approach is not necessary better than the classical ADF approach."

Based on simulation studies, [6] suggests that practitioners must assign a high probability to the value to be tested in order to get high power when using Bayes Factor tests, although this means to increase the non-stationary weight when testing for unit root.

There have also been other tests based on or using specially designed priors, that show a better performance. However, the use of such priors departs from some basic paradigms of Bayesian statistics, like the Likelihood Principle. Moreover, these techniques have to be fine tuned to each particular problem type or application.

In contrast, the FBST e-value derivation and implementation is straightforward from its general definition, using absolutely no ad hoc artifice, like a special prior, or a measure on the hypothesis set induced by some special parameterization, or an arbitrary initial likelihood ratio. It respects the Likelihood Principle and does not need to eliminate nuisance parameters.

REFERENCES

1. F.W.Ahking (2004). The Power of the "Objective" Bayesian Unit-Root Test. University of Connecticut Department of Economics Working Paper 2004-14.
2. L.Bauwens, M.Lubrano, J.F.Richard (1999). *Bayesian Inference in Dynamic Econometric Models*. Oxford University Press.
3. E.G.Birgin, R.Castillo, J.M.Martinez (2004). Numerical Comparison of Augmented Lagrangian Algorithms for Nonconvex Problems. *Computational Optimization and Applications*, 31, 1-56. Software and documentation available at *www.ime.usp.br/~egbirgin/tango/* .
4. C.Conigliani, F.Spezzaferri (2002). A Robust Bayesian Approach for Unit Root Testing. To appear in *Econometric Theory*, 23, 3, 2007.
5. W.L.Goffe, G.D.Ferrier, J.Rogers (1994). Global Optimization of Statistical Functions with Simulated Annealing. *Journal of Econometrics*, 60, 65–99.
6. Z. I. Kalaylioglu (2002). *Frequentist and Bayesian Unit Root Tests in Stochastic Volatility Models*, PhD Thesis, NC State University.
7. M.Lauretto, C.A.B.Pereira, J.M.Stern, S.Zacks (2003). Full Bayesian Significance Test Applied to Multivariate Normal Structure Models. Comparing Parameters of Two Bivariate Normal Distributions Using the Invariant Full Bayesian Significance Test. *Brazilian J. Probability and Statistics*, 17, 147-168.
8. J.G.MacKinnon (1994). Approximate Asymptotic Distribution Functions for Unit-Root and Cointegration Tests. *Journal of Business and Economic Statistics*, 12, 167-176.
9. M.R.Madruga, C.A.B.Pereira, J.M.Stern (2003). Bayesian Evidence Test for Precise Hypotheses. *Journal of Statistical Planning and Inference*, 117,185–198.
10. C.R.Nelson, C.I.Plosser (1982). Trends and Random Walks in Macroeconomics Time Series: Some Evidence and Implications. *Journal of Monetary Economics*, 10, 139–162.
11. C.A.B.Pereira, J.M.Stern, (1999). Evidence and Credibility: Full Bayesian Significance Test for Precise Hypotheses. *Entropy Journal*, 1, 69–80.
12. C.A.B.Pereira, J.M.Stern (2001). Model Selection: Full Bayesian Approach. *Environmetrics* 12, 6, 559-568.
13. P.C.B.Phillips (1991). To Criticize the Critics: An Objective Bayesian Analysis of Stochastic Trends. *Journal of Applied Econometrics*, 6, 333–364.
14. P.C.Schotman, H.K. van Dijk (1993). Posterior Analysis of Possibly Integrated Time Series with an Application to Real GNP. in P.Caines, J.Geweke, M.Taqqu (eds.), *New Directions in Time Series Analysis, Part II*. Springer-Verlag, New York.
15. C.A.Sims (1988). Bayesian Skepticism on Unit Root Econometrics. *Journal of Economics Dynamic and Control*, 12, 463–474.
16. C.A.Sims, H.Uhlig (1991). Understanding Unit Rooters: A Helicopter Tour. *Econometrica*, 59, 1591-1600.
17. J.M.Stern (2007). Cognitive Constructivism, Eigen-Solutions, and Sharp Statistical Hypotheses. *Cybernetics and Human Knowing*, 14,1, 9-36.
18. J.M.Stern, S.Zacks (2002). Testing Independence of Poisson Variates under the Holgate Bivariate Distribution. The Power of a New Evidence Test. *Statistical and Probability Letters*, 60, 313–320.

The Problem of Separate Hypotheses via Mixture Models

Marcelo de Souza Lauretto*,[†], Silvio R. de Faria Jr.*, Basilio B. Pereira**, Carlos A. B. Pereira* and Julio M. Stern*,[†]

*University of Sao Paulo, Institute of Mathematics and Statistics
[†]lauretto@ime.usp.br   jstern@ime.usp.br
**University of Rio de Janeiro, Medical School and COPPE

Abstract. This article describes the Full Bayesian Significance Test for the problem of separate hypotheses. Numerical experiments are performed for the Gompertz vs. Weibull life span test.

Keywords: FBST, Life span, Model selection, Separate hypotheses, Significance tests, Reliability.

INTRODUCTION

An important problem in statistics inference consists of deciding which of m alternative models, $f_k(x, \psi_k)$, more adequately fits a given dataset. When the candidate models f_k have distinct (unrelated) functional forms, it is usual to call them "separate" models (or hypotheses). Many discriminate models have been developed, which counterpoise a (null) model $f_1(x, \psi_1)$ against one alternative model $f_2(x, \psi_2)$, providing a measure of evidence in data favoring model 1 over model 2 [1,18]. However, these methods are not capable of give a straight answer when neither candidate model individually describes well the data. Non-parametric tests (e.g. Goodness-of-fit and Kolmogorov-Smirnov), on the other hand, have a comparatively slow convergence rate.

In this article we analyze this problem in the context of mixture models, see [14]. The basic distribution of this statistical model is a weighted sum of two or more candidate pdf's. Deciding if the data comes from a specific distribution is to test if the other distributions weights equal 0. Under this formulation, if neither model describes adequately the data, the test is capable of give a direct answer – a high evidence against all candidate models. As a numerical example we use a classical problem in reliability analysis, the Gompertz vs. Weibull life span, see [11,12].

The Fully Bayesian Significance Test (FBST) is presented by Pereira and Stern [19] as a coherent Bayesian significance test. The FBST is intuitive and has a geometric characterization. In this article the parameter space, Θ, is a subset of R^n, and the hypothesis is defined as a further restricted subset defined by vector valued inequality and equality constraints: $H : \theta \in \Theta_H$ where $\Theta_H = \{\theta \in \Theta | g(\theta) \leq 0 \wedge h(\theta) = 0\}$. For simplicity, we often use H for Θ_H. We are interested in precise hypotheses, with $\dim(H) < \dim(\Theta)$. $f(\theta)$ is the posterior probability density function.

The computation of the evidence measure used on the FBST is performed in two steps: The optimization step consists of finding f^* and $\widehat{f}$, the constrained (over H) and unconstrained maxima of the posterior. The integration step consists of integrating the

CP954, *27th International Workshop on Bayesian Inference and Maximum Entropy Methods in Science and Engineering*,
edited by K. H. Knuth, A. Caticha, J. L. Center, Jr., A. Giffin, and C. C. Rodriguez
© 2007 American Institute of Physics 978-0-7354-0468-7/07/$23.00

posterior density over the Tangential Set, $\overline{T}$ where the posterior is higher than anywhere in H, i.e., $\overline{T} = \{\theta \in \Theta : f(\theta) > f^*\}$, $f^* = \max_H f(\theta) = f(\theta^*)$, $\widehat{f} = \max_\Theta f(\theta) = f(\widehat{\theta})$, $\overline{\mathrm{Ev}}(H) = \Pr(\theta \in \overline{T}\,|\,x) = \int_{\overline{T}} f(\theta) d\theta$.

$\overline{\mathrm{Ev}}(H)$ is the evidence against H, and $\mathrm{Ev}(H) = 1 - \overline{\mathrm{Ev}}(H)$ is the evidence supporting (or in favor of) H. Intuitively, if $\overline{\mathrm{Ev}}(H)$ is "large", $\overline{T}$ is "heavy", and the hypothesis set is in a region of "low" posterior density, meaning a "strong" evidence against H.

Let us consider the cumulative distribution of the evidence value against the hypothesis, $\overline{V}(\tau) = \Pr(\overline{\mathrm{Ev}} \leq \tau)$, given θ^0, the true value of the parameter. Under appropriate regularity conditions, for increasing sample size, $n \to \infty$, we can state the following:
- If H is false, $\theta^0 \notin H$, then $\overline{\mathrm{Ev}}$ converges (in probability) to one, that is, $\overline{V}(\tau) \to \delta(1)$.
- If H is true, $\theta^0 \in H$, then $\overline{V}(\tau)$, the confidence level, is approximated by the function $\overline{W}(t,h,\tau) = \mathrm{Chi2}\left(t - h, \mathrm{Chi2}^{-1}(t,c)\right)$, where $t = \dim(\Theta)$, $h = \dim(H)$ and $\mathrm{Chi2}(k,x)$ is the cumulative chi-square distribution with k degrees of freedom.

Hence, for large n, to reject H with a level of significance δ, we set $\tau = \overline{W}^{-1}(t,h,1-\delta)$, i.e. set τ such that $\overline{W}(t,h,\tau) = 1 - \delta$.

Several FBST applications and examples, efficient computational implementation, interpretations, and comparisons with other techniques for testing sharp hypotheses, can be found in the authors' papers in the reference list. For a FBST review see the on line document [21].

WEIBULL AND GOMPERTZ DISTRIBUTIONS

In this paper we analyze the Gompertz vs. Weibull life span model selection problem. For the importance and interpretation of this problem see [11].

The Weibull hazard and probability density functions, for a failure time $x \geq 0$, given the shape and characteristic life (or scale) parameters, $\beta > 0, \gamma > 0$, are:

$$h_W(x|\beta,\gamma) = \beta x^{\beta-1}/\gamma^\beta , \quad f_W(x|\beta,\gamma) = (\beta x^{\beta-1}/\gamma^\beta)\exp(-(x/\gamma)^\beta) .$$

The Gompertz hazard and probability density functions, for a failure time $x \geq 0$, given the parameters, $\alpha > 1, \lambda > 0$, are:

$$h_G(x|\alpha,\lambda) = \lambda \alpha^x , \quad f_G(x|\alpha,\lambda) = \lambda \alpha^x \exp(-(\alpha^x - 1)\lambda/\log\alpha) .$$

The Gompertz distribution exhibits a strong nonlinear correlation between the parameters α and λ, see Figure 1A. This correlation explains the *compensation law of mortality*, which states that higher values for the parameter α are compensated by lower values of parameter λ in different populations of a given species: $ln(\lambda) = ln(M) - B\alpha$, where B and M are universal species-specific invariants, see [11]. As a result, the Gompertz density in its original form is not log-concave. As we shall discuss later, we use adaptive samplers for the parameters, which depend on the shape of the density function – preferably log-concave distributions. In order to separate the parameters α and λ, diminishing this nonlinear dependence and enhancing the shape of density function for sampling, we adopt the reparameterization $u = 1/\log\alpha$ and $v = \log(\log\alpha)/\lambda$, suggested by Meeker and Escobar [17], see Figure 1B.

The log-likelihoods of Weibull and (reparameterized) Gompertz models and their respective gradients (used for maximum likelihood estimation) are:

$$L_W(\beta,\gamma|X) = n\log\beta - n\beta\log\gamma + (\beta+1)\sum_j \log x - \sum_j (x_j/\gamma)^\beta ,$$
$$dL_W/d\beta = n/\beta - n\log\gamma + \sum_j \log x_j - \sum_j (x_j/\gamma)^\beta \log(x_j/\gamma) ,$$
$$dL_W/d\gamma = -n\beta/\gamma + \beta/\gamma \sum_j (x_j/\gamma)^\beta ,$$
$$L_G(u,v|X) = -n\log u - nv + \sum_j x_j/u + n/\exp(v) - \sum_j \exp(x_j/u - v) ,$$
$$dL_G/du = -n/u - \sum_j x_j/u^2 + \sum_j x_j/u^2 \exp(x_j/u - v) ,$$
$$dL_G/dv = -n - n/\exp(v) + \sum_j \exp(x_j/u - v) .$$

MIXTURES OF SEPARATE MODELS

Given a dataset $X = \{x_1, x_2 \ldots x_n\}$ and distinct alternative probability densities, $f_1(X|\psi_1), f_2(X|\psi_1), \ldots, f_m(X|\psi_m)$, where ψ_k are (vector) parameters, the problem of interest is to measure the evidence in favour of each model for fitting the dataset. In this paper, the consider a general model including all candidate distributions, where the choice of a specific distribution is a special case. The origin of this model comes in the work of Cox [7], who suggested that, in the presence of two alternative models, the p.d.f of data could be taken proportional to

$$f(x|w, \psi_1, \psi_2) \propto f_1(x|\psi_1)^{w_1} f_2(x|\psi_2)^{w_2} , \quad w > 0 | w\mathbf{1} = w_1 + w_2 = 1 ,$$

Then, deciding if the model 1 is adequate to describe the data is to test the hypothesis $H_1 : w_1 = 1$ against the hypothesis $w \neq 1$. Atkinson [2] developed this idea for some distributions of the exponential class, writing the density as

$$f(x|w, \psi_1, \psi_2) = \frac{f_1(x|\psi_1)^{w_1} f_2(x|\psi_2)^{w_2}}{\int f_1(y|\psi_1)^{w_1} f_2(y|\psi_2)^{w_2} dy} .$$

In this paper, we consider that the p.d.f. of data is a convex linear combination of the fixed candidate densities: denoting $\theta = [w, \psi_1, \ldots \psi_m]$,

$$f(x|\theta) = w_1 f_1(x|\psi_1) + \ldots + w_m f_m(x|\psi_m) , \quad w \geq 0 | w\mathbf{1} = 1 .$$

The likelihood then is

$$f(X|\theta) = \prod_{j=1}^n \sum_{k=1}^m w_k f_k(x_j|\psi_k) .$$

Here it is important to remember some key concepts of mixture models. In mixture analysis for unsupervised classification, we assume that the data come from one or more subpopulations (classes), distributed under distinct densities. The evidence in favor of the existence of more than one subpopulation will be higher if some subsets of data are more adequately fitted by a particular component of the mixture, where other subsets are better fitted by another components. In order to detect this situation, the mixture model must be able to infer the (probability of) data classifications. The real classifications are considered non observable and, for this reason, called *hidden* or *latent* variables. The problem of deciding if one single candidate distribution fits adequately the data is analogous to decide the number of components in a traditional mixture model, and the behavior of the system will be also similar: if the candidate model does not fit well the data, some observed points may be better described by a particular component of mixture, where the remaining will be better fitted by other components.

A sample j of class $k = c(j)$ is distributed with density $f_k(x_j | \psi_k)$. The boolean classification matrix Z indicates whether or not x_j is of class k, i.e. $z_j^k = 1$ iff $c(j) = k$. Conditioning on the latent variables we can rewrite:

$$f(x_j|\theta) = \sum_{k=1}^m f_k(x_j|\theta, z_k^j) f(z_k^j|\theta) = \sum_{k=1}^m w_k f_k(x_j|\psi_k),$$
$$f(X|\theta) = \prod_{j=1}^n f(x_j|\theta) = \prod_{j=1}^n \sum_{k=1}^m w_k f_k(x_j|\psi_k).$$

Given the mixture parameters, θ, and the observed data, X, the conditional classification probability matrix, $P = f(Z|X, \theta)$, is given by:

$$p_k^j = f(z_j^k|x_j, \theta) = \frac{f_k(z_j^k, x_j|\theta)}{f(x_j|\theta)} = \frac{w_k f_k(x_j|\psi_k)}{\sum_{k=1}^m w_k f(x_j|\psi_k)}.$$

We use y_k for the number of samples of class k, i.e. $y_k = \sum_j z_j^k$, or $y = Z\mathbf{1}$.

The density for the "completed" data, X, Z, is:

$$f(X,Z|\theta) = \prod_{j=1}^n f_{\psi_{c(j)}}(x_j|\psi_{c(j)}) f(z_j^k|\theta) = \prod_{k=1}^m \left(w_k^{y_k} \prod_{j|c(j)=k} f_k(x_j|\psi_k)\right).$$

In the remaining of this section we discuss the FBST formulation for the Weibull vs. Gompertz mixture model. The conjugate prior for a multinomial distribution is a Dirichlet distribution:

$$M(y|n,w) = n!/(y_1!\ldots y_m!)\, w_1^{y_1}\ldots w_m^{y_m},$$
$$D(w|y) = \Gamma(y_1+\ldots+y_k)/(\Gamma(y_1)\ldots\Gamma(y_k)) \prod_{k=1}^m w_k^{y_k-1},$$

with $w > 0$ and $w\mathbf{1} = 1$. Prior information given by $\dot{y}$, and observation y, result in the posterior parameter $\ddot{y} = \dot{y} + y$. Here we take the non-informative prior given by $\dot{y} = \mathbf{1}$. We also consider a improper uniform prior for (β, γ, u, v). Therefore, the posteriori is

$$f(\theta|X) \propto f(X|\theta) = \prod_{j=1}^n \left(p_j^1 w_1 f_W(x_j|\beta,\gamma) + p_j^2 w_2 f_G(x_j|\alpha,\lambda)\right),$$
$$p_j^1 = \frac{w_1 f_W(x_j|\beta,\gamma)}{w_1 f_W(x_j|\beta,\gamma) + w_2 f_G(x_j|\alpha,\lambda)}, \quad p_j^2 = 1 - p_j^1.$$

The hypotheses of interest are $H_1 : w_1 = 1 \wedge w_2 = 0$ and $H_2 : w_1 = 0 \wedge w_2 = 1$. The FBST procedure for testing $H_k, k = 1, 2$ consists of two steps:
- Estimate the maximum of the log-likelihood L_k^* under H_k, which corresponds to the maximum log-likelihood under the corresponding single component distribution.

- Estimate the e-value supporting the hypothesis H_k, that is, the ratio

$$\text{Ev}(H_k) = \frac{\int_{T_k} f(\theta|X)d\theta}{\int_\Theta f(\theta|X)d\theta} \;,\; T_k = \{\theta \in \Theta | L(\theta) \le L_k^*\} \;.$$

Notice that since the likelihood normalization constant is the same for both numerator and denominator, so it is cancelled and can therefore be ignored in the computational procedure. For the optimization step, we used the Algencan-Tango solver, which source code and detailed description are freely distributed (see internet link at the reference), see [4,5].

In order to perform the integration over the posterior measure, we used a Gibbs sampling Markov Chain Monte Carlo algorithm, MCMC. Given the current vector parameter θ^i, we compute P. Given P, we draw Z from $f(z_j|p_j)$, a simple multinomial distribution. Given the latent variables, Z, we separate the samples of classes 1 and 2. In the Weibull component, we draw a parameter value $[\beta^{i+1}, \gamma^{i+1}]$ with density proportional to the partial likelihood $\prod_{j|c(j)=1} f_W(x_j|\beta,\gamma)$. The same idea is applied to draw the Gompertz parameters $[\alpha^{i+1}, \lambda^{i+1}]$. Given $\ddot{y} = Z\mathbf{1} + \dot{y}$, we can draw a new weight vector $[w_1^{i+1}, w_2^{i+1}]$ using a Dirichlet distribution $D(w|\ddot{y}_1, \ddot{y}_2)$. At the end of iteration (i), we have a new vector parameter $\theta^{i+1} = [w_1^{i+1}, w_2^{i+1}, \beta^{i+1}, \gamma^{i+1}, \alpha^{i+1}, \lambda^{i+1}]$, and can begin iteration $(i+1)$.

We do not know a direct method to draw the parameters from the Weibull or Gompertz likelihood. For this purpose we used the adaptive sampler HITRO, see [13,20,22]. HITRO combines the multivariate Ratio-of-Uniforms method with the Hit-and-Run sampler. The Ratio-of-Uniforms transformation maps the region below the p.d.f f, i.e. $G(f) = \{(x,y) : 0 < y < f(x)\}$ into the region

$$A(f) = A_{r,m}(f) = \left\{ (u,v) : 0 < v < f\left(\frac{u}{v^r} + m\right)^{1/(rn+1)} \right\}$$

by means of the transformation

$$(u,v) \mapsto (x,y) = \left(\frac{u}{v^r} + m,\; v^{rn+1}\right) \;.$$

The vector m must be a point near the mode (in our implementation, we set m as the mode). The method relies on the theorem that, if (u,v) is uniformly distributed over $A(f)$, then $x = u/v^r + m$ has probability density function $f(x)/\int f(z)dz$. The Hit-and-run sampler is used for generating points (u,v) uniformly over $A(f)$.

NUMERICAL EXPERIMENTS AND FINAL REMARKS

We run some numerical experiments in order to evaluate the FBST performance on our problem of separate models. The experiments were based on the IBGE data bank for the mortality of Brazilian male population in the year of 2005, available on line at http://www.ibge.gov.br /home/estatistica/populacao/tabuadevida/2005/default.shtm. We used the mortality rate table from ages 5 to 80, hence avoiding the early infancy or burn-in period, see [3,17].

The experiments were based on simulated data, drawn from four distributions, the parameters of which have always been chosen to provide the best fit to the IBGE data bank. The distributions fitted were: (1)-Weibull, (2)-Gompertz, (3)-Gamma, and (4)-Beta (rescaled), see Figure 2. Our main interest was to measure the convergence rate of correct decisions, concerning the acceptance / rejection of the Weibull vs. Gompertz hypotheses, when using the FBST on the mixture model. Of course, in cases (1) and (2) we want to accept the correct hypothesis and reject the false one, whereas in cases (3) and (4) we want to reject them both.

As acceptance / rejection threshold, we adopted the critical level τ according to criterion presented in section 1, with a significance level of 5%. Since the mixture model and the restricted model have 5 and 2 degrees of freedom, respectively, we have $\tau = \overline{W}^{-1}(5,2,0.95) = 0.83$. Therefore, we reject H if $\overline{Ev}(H) > 0.83$, or equivalently if $Ev(H) < 0.17$. Using each of the four fitted distributions we generated 500 samples of size $n = 30, 50, 75, 100, 150, 200, 300, 400$ and 500.

We have compared the performance of the FBST with the Kolmogorov–Smirnov (KS) test, [9]. In this test, the goodness of fit measure is taken to be the Kolmogorov distance $D_n^* = D(F_n, F^*) = sup_x |F_n(x) - F^*(x | \theta)|$, where F_n denotes the sample (empirical) distribution and F^* denotes the theoretical distribution to be tested. Due to difficulty in estimate θ which minimizes $D(F_n, F^*)$, it is usually adopted the maximum likelihood estimator for θ. Kolmogorov and Smirnov demonstrated in 1930's that, if the null hypothesis $F(X) = F^*(X | \theta)$, then $lim_{n \to \infty} Pr(\sqrt{n} D_n^* \leq t) = 1 - 2\sum_{i=1}^{\infty} (-1)^{i-1} \exp(-2i^2 t^2)$. The distribution at the right side of this equation allows one to compute the significance (p-value) of D_n^*. For a meaningful comparison, we also used a 5% significance level.

The hole batch of 500 simulations for each of the 4 cases and 9 sample sizes, took about 2 day of computation on a Intel Pentium server, or about 10 seconds per test. Computing time was dominated by *Hitro*, a flexible and robust but generic subroutine. Hence, its substitution by a tailor made and more efficient sampler could enhance the program computational performance.

Figure 3 summarized the correct decision rates in the numerical simulations. The Weibull distribution can approximate very well a Gamma distribution. This explains the relatively slow convergence in the decision to reject the Weibull hypothesis in the simulations from the Gamma.

As expected, the FBST had a good performance. Moreover its implementation is straightforward, following the guidelines presented in [19,21]. It would be interesting to replace the Kolmogorov-Smirnov benchmark with a parametric alternative, like some form of jump MCMC. However, as far as the authors know, none is available at this time. The authors intend to collaborate with other research groups in order to develop and implement such algorithms.

Acknowledgments: The authors are grateful for the support of CAPES - Coordenação de Aperfeiçoamento de Pessoal de Nível Superior, CNPq - Conselho Nacional de Desenvolvimento Científico e Tecnológico, and FAPESP - Fundação de Amparo à Pesquisa do Estado de São Paulo.

REFERENCES

1. M.A.Araujo, B.B.Pereira (2007). A Comparison of Bayes Factors for Separated Models: Some Simulation Results. *Communications in Statistics-Simulation and Computation* 36(2), 297–309.
2. A.C.Atkinson (1970). A Method for Discriminating Between Models. *J. R. Statist Soc. B* 32, 323–354.
3. R.E.Barlow, F.Prochan (1981). *Statistical Theory of Reliability and Life Testing Probability Models.* Silver Spring: To Begin With.
4. E.G.Birgin, J.M.Martnez and M.Raydan (2000). Nonmonotone Spectral Projected Gradient Methods on Convex Sets. *SIAM Journal on Optimization*, 10, 1196-1211. Software and documentation available at *www.ime.usp.br/~egbirgin/tango/* .
5. E.G.Birgin , J.M.Martnez (2002). Large-scale Active-Set Box-Constrained Optimization Method with Spectral Projected Gradients. *Computational Optimization and Applications*, 23, 101-125.
6. W.Borges, J.M.Stern (2006). Evidence and Compositionality. p.307-315 in J.Lawry et al., *Soft Methods for Integrated Uncertainty Modelling.* NY: Springer.
7. D.R.Cox (1962). Further Results on Tests of Separated Families of Hypotheses. *J. R. Statist Soc. B* 24, 406-423.
8. D.R.Cox and D.Oakes (1984). *Analysis of Survival Data.* Monographs on Statistics and Applied Probability, Chapman & Hall.
9. M.H.DeGroot (1986). *Probability and Statistics.* Addison–Wesley.
10. B.Dodson (1994). *Weibull Analysis.* Milwaukee: ASQC Quality Press.
11. L.A.Gavrilov and N.S.Gavrilova (1991). *The Biology of Life Span: A Quantitative Approach.* New York: Harwood Academic Publisher.
12. L.A.Gavrilov and N.S.Gavrilova (2001). The Reliability Theory of Aging and Longevity. *J. Theor. Biol.* 213, 527–545.
13. R.Karawatzki, J.Leydold, K.Pötzelberger (2005). Automatic Markov Chain Monte Carlo Procedures for Sampling from Multivariate Distributions. Department of Statistics and Mathematics Wirtschaftsuniversität Wien Research Report Series. Report 27, December 2005. Software available at *http://statistik.wu-wien.ac.at/arvag/software.html*.
14. M.S.Lauretto, J.M.Stern (2005). FBST for Mixture Model Selection. Maxent'2005, *AIP Conf. Proc.* 803, 121–128.
15. M.R.Madruga, L.G.Esteves, S.Wechsler (2001). On the Bayesianity of Pereira-Stern Tests. *Test*, 10, 291–299.
16. M.R.Madruga, C.A.B.Pereira, J.M.Stern (2003). Bayesian Evidence Test for Precise Hypotheses. *Journal of Statistical Planning and Inference*, 117, 185–198.
17. W.Q.Meeker and L.A.Escobar (1998). Statistical Methods for Reliability Data. Wiley Series in Probability and Statistics.
18. B.B.Pereira (2005). Separate Families of Hypotheses. in P.Armitage, T.Colton (eds.) *Encyclopedia of Biostatistics* (2nd.ed.). 7, 4881-4886. NY: Wiley.
19. C.A.B.Pereira, J.M.Stern, (1999). Evidence and Credibility: Full Bayesian Significance Test for Precise Hypotheses. *Entropy Journal*, 1, 69–80.
20. R.L.Smith (1984). Efficient Monte Carlo Procedures for Generating Points Uniformly Distributed over Bounded Regions. *Operations Research 32*, 1296–1308.
21. J.M.Stern (2007). Cognitive Constructivism, Eigen-Solutions, and Sharp Statistical Hypotheses. *Cybernetics and Human Knowing*, 14,1, 9-36.
24. I.Vaduva (1984). Computer Generation of Random Vectors Based on Transformation of Uniformly Distributed Vectors. In *Probability Theory, Proc. 7th Conf.*, Brasov/Rom, 589–598.
25. J.C.Wakefield, A.E.Gelfand, A.F.Smith (1991). Efficient Generation of Random Variates via Ratio-of-Uniforms Method. *Statist. Comput.* 1,2, 129-133.

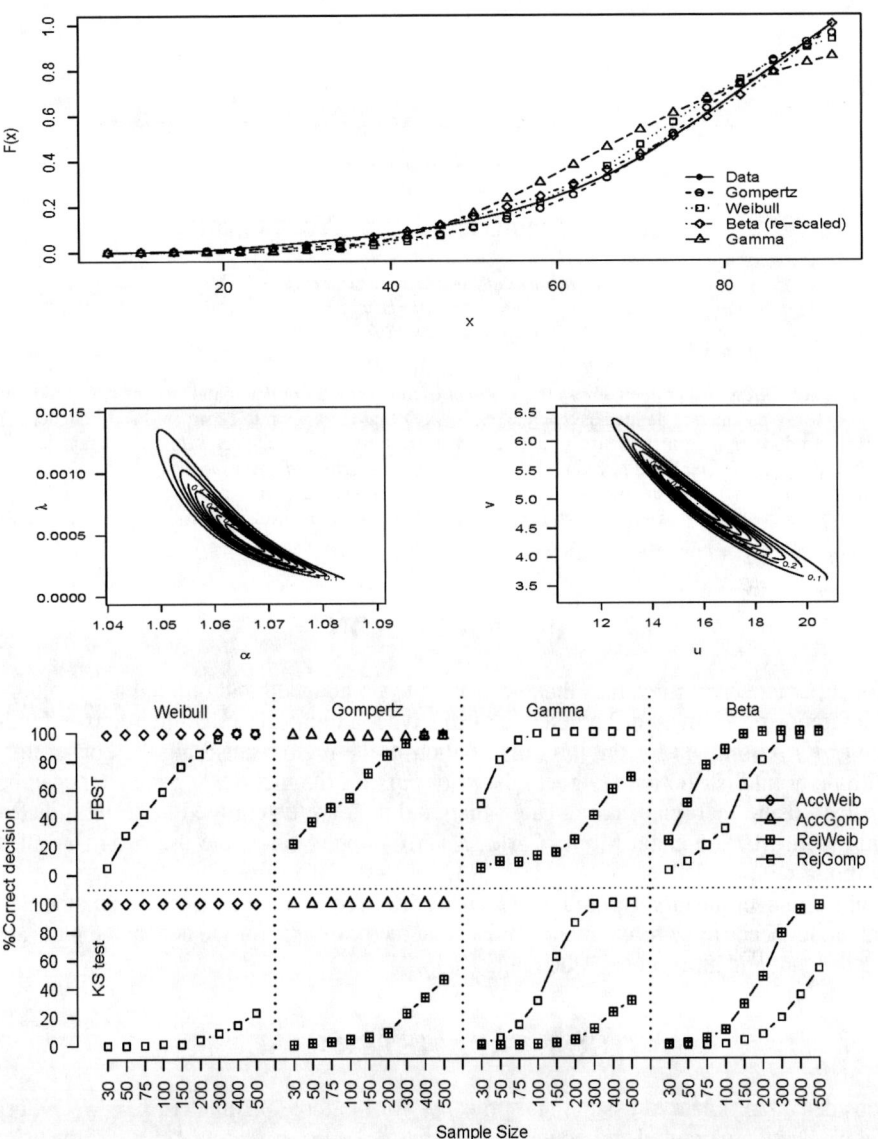

Figure 1: IBGE Brazilian mortality rates and fitted distributions.
Figure 2A,B: Contour plots for Gompertz density and reparameterization.
Figure 3: Correct decision rates on numerical simulations.

Assigning Priors for Ordered and Bounded Parameters

Paul M. Goggans and Chung-Yong Chan

Department of Electrical Engineering
University of Mississippi
University, MS 38677

Abstract. In this paper, we address the problem of assigning a prior probability density function (*pdf*) for N parameters defined as $\mathbf{x} = \{x_1, x_2, \cdots, x_N\}$ that are known to be ordered and bounded. We introduce transformation functions that relate the original parameters $\mathbf{x}$ to new parameters $\mathbf{u} = \{u_1, u_2, u_3, \ldots, u_N\}$ that have independent and uniform prior *pdfs* over the range $(0,1)$. Results are initially derived for the three-parameter case and then extended to the N-parameter case. The transformation presented here is compared with an alternative transformation developed by Skilling.

Keywords: Prior Probability Density Function, Re-Parametrization

INTRODUCTION

Model parameters with values that are known to be bounded and ordered are common in inference problems. For parameters of this type, a uniform probability density function (*pdf*) is assigned for the joint distribution in the region satisfying the constraints. With the joint distribution assigned, the prior *pdfs* for the individual parameters can be obtained using marginalization. These marginal *pdfs* are determined to be Beta distributions. In Markov chain Monte Carlo (MCMC) applications, the use of binary slice sampling with the Hilbert Curve [1, 2, 3] requires that the prior *pdf* for the parameters be uniform over the unit hypercube. Because of this requirement, the original mathematical model needs to be re-parameterized so that the prior *pdfs* for the new parameters are independent and uniform over the range $(0,1)$.

THREE-PARAMETER CASE

Consider the problem of assigning a prior *pdf* for the three parameters $\mathbf{x} = \{x_1, x_2, x_3\}$ that we know are ordered and bounded such that $x_{min} < x_3 < x_2 < x_1 < x_{max}$. If the prior information I tells us nothing further about $\mathbf{x}$ then the principle of indifference suggests assigning the joint prior *pdf* to be constant in the region $x_{min} < x_3 < x_2 < x_1 < x_{max}$ and 0 outside this region. This assignment yields the following expression for the joint prior *pdf* for the parameters:

$$p(\mathbf{x}|I) = \begin{cases} 6\Delta x^{-3} & \text{for } x_{min} < x_3 < x_2 < x_1 < x_{max} \\ 0 & \text{otherwise} \end{cases} \tag{1}$$

where $\Delta x = x_{max} - x_{min}$. The prior probabilities assigned to the individual parameters can be determined using marginalization as follows:

$$p(x_1|I) = \begin{cases} \int_{x_{min}}^{x_1} \int_{x_{min}}^{x_2} \frac{6}{\Delta x^3} dx_3 dx_2 = \frac{3}{\Delta x^3}(x_1 - x_{min})^2 & \text{for } x_{min} < x_1 < x_{max} \\ 0 & \text{otherwise,} \end{cases} \quad (2)$$

$$p(x_2|I) = \begin{cases} \int_{x_2}^{x_{max}} \int_{x_{min}}^{x_2} \frac{6}{\Delta x^3} dx_3 dx_1 = \frac{6}{\Delta x^3}(x_2 - x_{min})(x_{max} - x_2) & \text{for } x_{min} < x_2 < x_{max} \\ 0 & \text{otherwise,} \end{cases} \quad (3)$$

and

$$p(x_3|I) = \begin{cases} \int_{x_3}^{x_{max}} \int_{x_3}^{x_1} \frac{6}{\Delta x^3} dx_2 dx_1 = \frac{3}{\Delta x^3}(x_{max} - x_3)^2 & \text{for } x_{min} < x_3 < x_{max} \\ 0 & \text{otherwise.} \end{cases} \quad (4)$$

In numerical Bayesian analysis, it is often desirable to re-parameterize models so that the prior *pdfs* for the new parameters $\mathbf{u} = \{u_1, u_2, u_3, \ldots, u_N\}$ are independent and uniformly distributed on $(0,1)$. That is, so that $p(\mathbf{u}|I) = U(u_1)U(u_2)U(u_3)\cdots U(u_N)$ where

$$U(u_n) = \begin{cases} 1 & \text{for } 0 < u_n < 1 \\ 0 & \text{otherwise.} \end{cases} \quad (5)$$

In the current problem, this means finding the functions $x_1 = f_1(\mathbf{u})$, $x_2 = f_2(\mathbf{u})$, and $x_3 = f_3(\mathbf{u})$ such that (1) results when $p(\mathbf{u}|I) = U(u_1)U(u_2)U(u_3)$.

The process of finding the transformation functions can be broken into two steps. In the first step, we define functions that map from independent parameters $\mathbf{y} = \{y_1, y_2, y_3\}$ to the region $x_{min} < x_3 < x_2 < x_1 < x_{max}$. For $0 < y_1 < 1$, $0 < y_2 < 1$, and $0 < y_3 < 1$, suitable functions that accomplish this mapping are $x_1 = x_{min} + \Delta x y_1$, $x_2 = x_{min} + \Delta x y_1 y_2$, and $x_3 = x_{min} + \Delta x y_1 y_2 y_3$. Because the parameters $\mathbf{y}$ are independent, $p(\mathbf{y}|I) = p_{y_1}(y_1) p_{y_2}(y_2) p_{y_3}(y_3)$. The prior *pdf* for the original parameters can be written in terms of $p(\mathbf{y}|I)$ using the expression

$$p(\mathbf{x}|I) = p(\mathbf{y}|I) \begin{vmatrix} \frac{dy_1}{dx_1} & \frac{dy_1}{dx_2} & \frac{dy_1}{dx_3} \\ \frac{dy_2}{dx_1} & \frac{dy_2}{dx_2} & \frac{dy_2}{dx_3} \\ \frac{dy_3}{dx_1} & \frac{dy_3}{dx_2} & \frac{dy_3}{dx_3} \end{vmatrix} \quad (6)$$

so that here

$$p(\mathbf{x}|I) = p_{y_1}\left(\frac{x_1 - x_{min}}{\Delta x}\right) p_{y_2}\left(\frac{x_2 - x_{min}}{x_1 - x_{min}}\right) p_{y_3}\left(\frac{x_3 - x_{min}}{x_2 - x_{min}}\right) \frac{1}{\Delta x(x_1 - x_{min})(x_2 - x_{min})}. \quad (7)$$

Choosing

$$p_{y_1}(y_1) = \begin{cases} 3y_1^2 & \text{for } 0 < y_1 < 1 \\ 0 & \text{otherwise,} \end{cases} \quad (8)$$

$$p_{y_2}(y_2) = \begin{cases} 2y_2 & \text{for } 0 < y_2 < 1 \\ 0 & \text{otherwise,} \end{cases} \quad (9)$$

and
$$p_{y_3}(y_3) = U(y_3) \tag{10}$$

yields $p(\mathbf{x}|I)$ as given by (1). The second step in finding the transformation functions is to find functions that transform uniformly distributed parameters into parameters with the *pdfs* given above. These functions are the inverse cumulative distribution functions so that $y_1 = u_1^{1/3}$, $y_2 = u_2^{1/2}$, and $y_3 = u_3$. Combining these transformations with the transformations from $\mathbf{y}$ to $\mathbf{x}$ yields the desired transformation functions. These functions are

$$x_1 = f_1(\mathbf{u}) = x_{min} + \Delta x u_1^{1/3}, \tag{11}$$

$$x_2 = f_2(\mathbf{u}) = x_{min} + \Delta x u_1^{1/3} u_2^{1/2}, \tag{12}$$

and

$$x_3 = f_3(\mathbf{u}) = x_{min} + \Delta x u_1^{1/3} u_2^{1/2} u_3. \tag{13}$$

N-PARAMETER CASE

The results for the three-parameter case can be extended to the N-parameter case where $\mathbf{x} = \{x_1, x_2, \cdots, x_N\}$, $x_{min} < x_N < \cdots < x_2 < x_1 < x_{max}$ and

$$p(\mathbf{x}|I) = \begin{cases} N! \Delta x^{-N} & \text{for } x_{min} < x_N < \cdots < x_2 < x_1 < x_{max} \\ 0 & \text{otherwise.} \end{cases} \tag{14}$$

In the N-parameter case, the marginal prior *pdf* assigned to the parameter x_n is given by the expression

$$p(x_n|I) = \begin{cases} \frac{N! \Delta x^{-N}}{(N-n)!(n-1)!}(x_n - x_{min})^{N-n}(x_{max} - x_n)^{n-1} & \text{for } x_{min} < x_n < x_{max} \\ 0 & \text{otherwise.} \end{cases} \tag{15}$$

The Beta distribution given above is a polynomial of order $N-1$. This distribution is unimodal with mode, mean, and standard deviation given by the following expressions:

$$\hat{x}_n = x_{min} + \Delta x \frac{N-n}{N-1}, \tag{16}$$

$$\mu_n = x_{min} + \Delta x \frac{N-n+1}{N+1}, \tag{17}$$

and

$$\sigma_n = \frac{\Delta x}{N+1} \sqrt{\frac{n(N-n+1)}{N+2}}. \tag{18}$$

The marginal prior *pdfs* for the five-parameter case with $x_{min} = 0$ and $x_{max} = 1$ are illustrated in Figure 1.

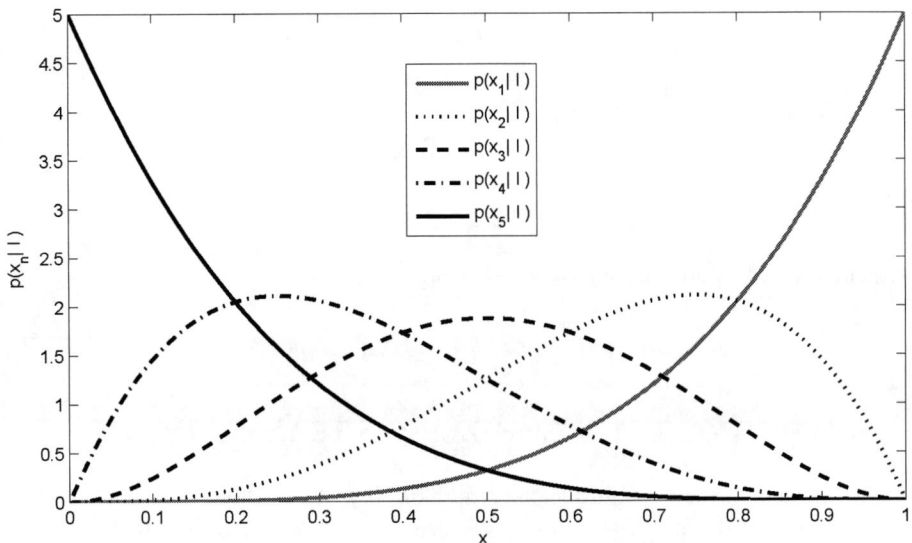

FIGURE 1. Marginal prior *pdfs* for the five parameter case.

For the N-parameter case the functions that transform $\mathbf{u}$ to $\mathbf{x}$ are

$$x_1 = f_1(\mathbf{u}) = x_{min} + \Delta x u_1^{1/N},$$
$$x_2 = f_2(\mathbf{u}) = x_{min} + \Delta x u_1^{1/N} u_2^{1/(N-1)},$$
$$\vdots$$
$$x_N = f_N(\mathbf{u}) = x_{min} + \Delta x u_1^{1/N} u_2^{1/(N-1)} \cdots u_{N-1}^{1/2} u_N.$$

The equations above can be written compactly as

$$x_n = f_n(\mathbf{u}) = x_{min} + \Delta x \prod_{i=1}^{n} u_i^{1/(N-i+1)} \quad \text{for } 1 \leq n \leq N. \tag{19}$$

INCREASING VALUES WITH INCREASING INDEX

The results presented so far have the parameter values decreasing with an increasing parameter index. In some situations it is convenient to define the parameter index so that the parameter values increase with an increasing parameter index. In this case, (14) through (17) are modified as follows:

$$p(\mathbf{x}|I) = \begin{cases} N! \Delta x^{-N} & \text{for } x_{min} < x_1 < x_2 < \cdots < x_N < x_{max} \\ 0 & \text{otherwise,} \end{cases} \tag{20}$$

$$p(x_n|I) = \begin{cases} \frac{N!\Delta x^{-N}}{(N-n)!(n-1)!}(x_n - x_{min})^{n-1}(x_{max} - x_n)^{N-n} & \text{for } x_{min} < x_n < x_{max} \\ 0 & \text{otherwise,} \end{cases} \quad (21)$$

$$\hat{x}_n = x_{min} + \Delta x \frac{n-1}{N-1}, \quad (22)$$

and

$$\mu_n = x_{min} + \Delta x \frac{n}{N+1}. \quad (23)$$

Equation (18) remains unchanged and (19) becomes

$$x_n = f_n(\mathbf{u}) = x_{min} + \Delta x \prod_{i=1}^{N-n+1} u_{N-i+1}^{1/(N-i+1)} \quad \text{for } 1 \leq n \leq N. \quad (24)$$

COMPARISON WITH SKILLING'S TRANSFORMATION

In the manual for the program BayeSys [5], Skilling gives an alternative transformation from $\mathbf{u}$ to $\mathbf{x}$ for the case where $x_{min} < x_1 < x_2 < \cdots < x_N < x_{max}$. The transformation functions are

$$x_n = f_n(\mathbf{u}) = x_{min} - \frac{a\Delta x}{r}\sum_{i=1}^{n} \log u_i \quad \text{for } 1 \leq n \leq N \quad (25)$$

where

$$r = -\sum_{i=1}^{N} \log u_i$$

and

$$a = [\mathrm{P}(r,N)]^{1/N}.$$

In the expression above, P is the incomplete gamma function defined as follows:

$$\mathrm{P}(r,N) = \int_0^r dx \frac{e^{-x}x^{N-1}}{(N-1)!}.$$

To compare the mapping performed by the present transformation with the mapping in Skilling's transformation, a two-parameter case $\mathbf{u} = \{u_1, u_2\}$ is considered. In the comparison, nine points that are located on the edges of or inside the unit square are considered. The points and their corresponding mappings are given in Table 1.

The present transformation provides a one-to-one mapping for all points in or on the unit square with the exception of points with $u_2 = 0$ which are all mapped to the origin. The mapping in Skilling's transformation is one-to-one for all points in or on the unit square. For the conditions, $u_1 = 0$, $u_2 = 0$, or $u_1 = u_2 = 1$, Skilling's transformation will encounter numerical problems because of the terms $\log u_i$ or $1/r$ in (25). For points inside the unit square, the mappings in both transformations are one-to-one and neither has numerical difficulties. We believe that both transformations will perform equally well in MCMC applications, however, the present transformation may be slightly preferred as Skilling's transformation has higher computation cost because evaluation of the incomplete gamma function is required for its use.

TABLE 1. Comparison between the present and Skilling's transformation for sample points that lie inside or on the edges of a unit square. Here, $EPS = 2^{-52}$.

Values of Parameters (u_1, u_2)	Present Transformation (x_1, x_2)	Skilling's Transformation (x_1, x_2)
(EPS, EPS)	(0.00, 0.00)	(0.50, 1.00)
(EPS, 0.50)	(0.00, 0.71)	(0.98, 1.00)
(EPS, 1.00)	(0.00, 1.00)	(1.00, 1.00)
(0.5, EPS)	(0.00, 0.00)	(0.02, 1.00)
(0.5, 0.5)	(0.35, 0.71)	(0.32, 0.64)
(0.5, 1.0)	(0.50, 1.00)	(0.39, 0.39)
(1.0, EPS)	(0.00, 0.00)	(0.00, 1.00)
(1.0, 0.5)	(0.71, 0.71)	(0.00, 0.39)
(1.0-EPS, 1.0-EPS)	(1.00, 1.00)	(0.00, 0.00)

APPLICATION

For MCMC codes that require uniform prior *pdfs* for the model parameters, the transformations described by (19) and (24) are particularly useful when the model has related sets of ordered parameters. As an example, consider the sound energy decay model used in [6], which can be simplified to

$$g(\alpha, \tau, t) = \alpha_1 \exp(-t/\tau_1) + \alpha_2 \exp(-t/\tau_2) + \alpha_3 \exp(-t/\tau_3) \qquad (26)$$

where from the prior information we know that

$$1 > \alpha_1 > \alpha_2 > \alpha_3 > 0 \text{ and } 0 < \tau_1 < \tau_2 < \tau_3 < 1. \qquad (27)$$

Equation (26) can be re-parameterized so that the prior *pdfs* for the new parameters $\mathbf{u} = \{u_1, u_2, u_3, u_4, u_5, u_6\}$ are independent and uniformly distributed using the following substitutions: $\alpha_1 = u_1^{1/3}$, $\tau_1 = u_2 u_4^{1/2} u_6^{1/3}$, $\alpha_2 = u_3^{1/2} u_1^{1/3}$, $\tau_2 = u_4^{1/2} u_6^{1/3}$, $\alpha_3 = u_5 u_3^{1/2} u_1^{1/3}$, and $\tau_3 = u_6^{1/3}$. Substituting the expressions above into (26) yields the re-parameterized model:

$$g(\mathbf{u}, t) = u_1^{1/3} \exp\left(-t/u_2 u_4^{1/2} u_6^{1/3}\right) + u_3^{1/2} u_1^{1/3} \exp\left(-t/u_4^{1/2} u_6^{1/3}\right) \\ + u_5 u_3^{1/2} u_1^{1/3} \exp\left(-t/u_6^{1/3}\right). \qquad (28)$$

SUMMARY

We have addressed the problem of assigning a prior *pdf* for model parameters that are known to be ordered and bounded. The joint prior *pdf* assigned is uniform in the region where the bounds and the ordering relationships are satisfied. The resulting marginal prior *pdfs* for the individual parameters are shown to be Beta distributions. As required for certain numerical Bayesian methods, we have derived transformation functions that can be use to re-parameterize models with ordered and bounded parameters so that the prior for the new parameter set is uniform over the unit hypercube. The derived transformation functions are an alternative to those given by Skilling. The present transformation has a lower computation cost than Skilling's transformation. For models with related sets of ordered and bounded parameters, the transformation functions presented here are particularly useful.

ACKNOWLEDGMENTS

For useful discussions and insights, we thank C. Ray Smith and John Skilling.

REFERENCES

1. J. Skilling, and D. J. C. Mackay, *Annals of Statistics* **31**, 753–755 (2003).
2. J. Skilling, "Using the Hilbert Curve," in *The 23rd Annual Conference on Bayesian Methods and Maximum Entropy in Science and Engineering*, Jackson Hole, WY., 2003, pp. 388–405.
3. D. J. C. Mackay, *Information Theory, Inference and Learning Algorithms*, Cambridge University Press, 2003.
4. R. M. Neal, *Annals of Statistics* **31**, 705–767 (2003).
5. J. Skilling, *BayeSys and MassInf*, Maximum Entropy Data Consultants Ltd (2004), page 24.
6. N. Xiang, and P. M. Goggans, *Journal of the Acoustical Society of America* **113**, 2685–2697 (2003).

Bayesian Inference Featuring Entropic Priors

Tilman Neumann

Tilman.Neumann@lycos.de

Abstract. The subject of this work is the parametric inference problem, i.e. how to infer from data on the parameters of the data likelihood of a random process whose parametric form is known a priori. The assumption that Bayes' theorem has to be used to add new data samples reduces the problem to the question of how to specify a prior before having seen any data. For this subproblem three theorems are stated. The first one is that Jaynes' Maximum Entropy Principle requires at least a constraint on the expected data likelihood entropy, which gives entropic priors without the need of further axioms. Second I show that maximizing Shannon entropy under an expected data likelihood entropy constraint is equivalent to maximizing relative entropy and therefore reparametrization invariant for continuous-valued data likelihoods. Third, I propose that in the state of absolute ignorance of the data likelihood entropy, one should choose the hyperparameter α of an entropic prior such that the change of expected data likelihood entropy is maximized. Among other beautiful properties, this principle is equivalent to the maximization of the mean-squared entropy error and invariant against any reparametrizations of the data likelihood. Altogether we get a Bayesian inference procedure that incorporates special prior knowledge if available but has also a sound solution if not, and leaves no hyperparameters unspecified.

Keywords: parametric inference, Bayesian inference, Maximum Entropy Principle, entropic prior, reparametrization invariance, non-informative prior, least-informative prior, expected entropy change maximization, stable inference
PACS: 02.50.Cw, 02.50.Tt, 05.20.-y

1. THE PARAMETRIC INFERENCE PROBLEM

If we know the parametric form of the data likelihood function $L(x|\theta)$ of a random process with random variable(s) X and have observed n data samples $\{x_1,...,x_n\} = x^n \in X^n$, then what can we say about the parameters θ after having seen the data?

2. FROM BAYES TO ENTROPIC PRIORS

Let's assume that the right way to update our degrees of belief in some θ's being the real parameters θ^* from new data samples is Bayes' theorem. Emphasizing its quality as an "update rule" this theorem can be stated as

$$P(\theta|x^n) \propto P(\theta|x^{n_1})L(x^{n_2}|\theta) \qquad (1)$$

with $n_1 + n_2 = n$ and $n_1, n_2 \geq 0$. Then, all our knowledge (and uncertainty) about the parameters θ after having seen data x^n is represented by the posterior density $P(\theta|x^n)$.

Adopting this point, the parametric inference problem reduces to the question of how to determine a prior $P(\theta) \equiv P(\theta|x^0)$ expressing the state of mind of the "reasoner" before having seen any data. In order to avoid confusions with priors that already incorporate knowledge derived from data, in the following I shall call these priors "first priors".

Note that a first prior might contain real information, although it represents a state of mind before having seen any data. Such information could stem from knowledge about the design of the random experiment, like that we know that a die is loaded and has its mass center near the six face. If we don't have such prior knowledge, a first prior will be equivalent to what is most often called a "non-informative" or "least-informative" prior.

The problem of how to assign priors dates back at least to the work of Laplace and Bernoulli [1] and is still an active research area. Founded in 1957 by Edwin T. Jaynes, the approach that has probably received most attention is the Principle of Maximum Entropy [2] [3] [4] [5]. Let's restrict the discussion in this section to the case of discrete-valued data likelihoods; then the ME principle claims that among all possible densities $P(\theta)$ satisfying a couple of constraints given as expectation values (and a normalization constraint), we should choose the one that maximizes the Shannon entropy

$$S_P = -\int_\Theta P(\theta)\ln P(\theta)d\theta \qquad (2)$$

Note that doubts on the uniqueness of the entropy measure raised e.g. by Uffink [6] have been countered by Caticha and Giffin [7].

A particular proposal for least-informative priors are the so-called "entropic priors" [8] [9] [10] [11] [12] [13], that have first been derived by Skilling [8] from the axioms of Maximum Entropy and an additional "quantification" argument. Entropic priors $P_\alpha(\theta) \equiv P(\theta)$ owe their name to the shape of their density

$$P_\alpha(\theta) \propto e^{\alpha S_L(\theta)} \qquad (3)$$

where

$$S_L(\theta) = -\int_X L(x|\theta)\ln L(x|\theta)dx \qquad (4)$$

is the entropy of the data likelihood function with parameters θ.

In 2004, Caticha and Preuss recognized that in order to solve problems with repeatable experiments, entropic priors need a constraint on the expected data likelihood entropy $\langle S_L \rangle$ (see [12], page 5), which is defined as

$$\langle S_L \rangle = \int_\Theta S_L(\theta)P(\theta)d\theta \qquad (5)$$

However, application of Jaynes' Maximum Entropy Principle with a constraint like $\langle S_L \rangle = S$ always gives an entropic prior! Thus, such a constraint has the same power as Skilling's quantification argument, and defines the assumption made in entropic priors.

Theorem 1 (Entropic Priors) *Application of the Maximum Entropy Principle requires at least a constraint on the expected data likelihood entropy like $\langle S_L \rangle = S$. The result is an entropic prior. The information contained in a "pure" entropic prior is an expectation about the data likelihood entropy and nothing else.*[1]

---

[1] I call an entropic prior "pure" if there are no more constraints than one on the expected data likelihood entropy and one for the normalization. Further constraints may lead to more complex expressions than equation 3; on the other hand, as we will see in section 5.1, the expressions may simplify as well if the data likelihood entropy takes on a logarithmic form.

3. CONTINUOUS-VALUED DATA LIKELIHOODS

It is a well-known fact that for continuous-valued data likelihoods, equation 2 is not invariant under reparametrizations of the data likelihood; consequently, mere reparametrizations might yield different inference results. To overcome this problem, we have to introduce a measure $m(\theta)$ in the log that transforms as $P(\theta)$ does. This gives the relative entropy:[2]

$$S_{P|m} = -\int_\Theta P(\theta) \ln \frac{P(\theta)}{m(\theta)} d\theta \tag{6}$$

Note that reparametrization invariance would be guaranteed by any $m(\theta)$, and that the ME principle doesn't give us a hint which one to use. However, we can formulate a couple of desirable properties:

- **Jaynes' Argument:** "Except for a constant factor, the measure $m(\theta)$ is also the prior describing 'complete ignorance' of θ." (Jaynes in [14], page 377)
- **Axiomatic Consistency:** Even for continuous-valued data likelihoods, the resulting prior still has to obey the restrictions imposed by Skillings axioms.
- **Limit Argument:** For continuous-valued data likelihoods that can be derived as some limit of a discrete-valued likelihood (e.g. hypergeometric and binomial), the following two solutions should be equivalent: First, solving the variational problem with the simple Shannon-entropy for the discrete-valued data likelihood and getting the limit of the solution; and second, solving the variational problem with the relative entropy for the continuous-valued likelihood.

First proposed by Rodriguez [9], a popular approach for entropic priors is that we should maximize the relative entropy under a normalization and an expected data likelihood entropy constraint, giving the solution

$$P_\alpha(\theta) \propto m(\theta) e^{\alpha S_L(\theta)} \tag{7}$$

where

$$m(\theta) \propto \sqrt{\det g_{ij}(\theta)} \tag{8}$$

is Jeffreys' prior which is based on the Fisher information matrix.

It is easy to see that the solution proposed by equations 7 and 8 doesn't satisfy the first two demands defined above. But did we already put in all the information we have to find the right entropy functional and/or $m(\theta)$? I think we can do better than just demanding reparametrization invariance. What we really want is: *Maximizing our entropy functional under a normalization constraint and an expected data likelihood entropy constraint should give the same result, no matter which parametrization we choose.* Let's have a look at the Lagrangian describing this using the simple Shannon entropy:

$$\mathcal{L} = (-\int_\Theta P(\theta) \ln P(\theta) d\theta) + \alpha(-\int_\Theta P(\theta) \int_X L(x|\theta) \ln L(x|\theta) dx d\theta - \bar{S})$$

---

[2] Relative entropy is often written with an opposite sign. I follow the notation of [12].

$$+ \lambda(\int_\Theta P(\theta)d\theta - 1)$$
$$= (-\int_\Theta P(\theta)(\ln P(\theta) + \alpha \int_X L(x|\theta)\ln L(x|\theta)dx)d\theta) - \alpha\bar{S} + \lambda(\int_\Theta P(\theta)d\theta - 1)$$
$$= (-\int_\Theta P(\theta)\ln \frac{P(\theta)}{m_\alpha(\theta)}d\theta) + \lambda(\int_\Theta P(\theta)d\theta - 1) - \alpha\bar{S} \qquad (9)$$

where
$$m_\alpha(\theta) \propto e^{-\alpha \int_X L(x|\theta)\ln L(x|\theta)dx} \equiv e^{\alpha S_L(\theta)} \qquad (10)$$

Since the $\alpha\bar{S}$-term cancels when expression 9 is maximized, we see:

Theorem 2 (Reparametrization Invariance) *Maximizing the simple Shannon entropy under an expected data likelihood entropy constraint is equivalent to maximizing relative entropy with the underlying measure $m(\theta)$ given by $m_\alpha(\theta)$, and therefore reparametrization invariant. Consequently, both the inference procedure and the resulting prior $P_\alpha(\theta)$ are exactly the same for discrete- and continuous-valued data likelihoods.*

The argument above shows in my opinion that it is a tautology to have both an $m(\theta)$ given by equation 8 and the entropic term as in equation 7. One of the two is enough! Nevertheless, my personal belief is that this solution is only slightly wrong, because Jeffreys' prior is quite similar to the entropic term, and therefore the two terms are almost linear dependent conditions in the solution of the variational problem.[3]

4. THE HYPERPARAMETER α

In this section we are going to see how we can deal with the hyperparameter α still present in the "generic" entropic priors derived so far. Note that like the internal energy of an ideal gas, the expected data likelihood entropy $\langle S_L \rangle$ of an entropic prior is a function of (inverse) temperature and nothing else, i.e. $\langle S_L \rangle \equiv \langle S_L \rangle(\alpha)$ (see figure 1 for a typical example). Therefore, if we have a concrete expectation like $\langle S_L \rangle = \bar{S} = 0.35$, then we just pick the α that realizes the expected $\langle S_L \rangle = \bar{S}$, and our prior is fixed, i.e. contains no more variables than the θs.

The more difficult case is when we have no idea at all about $\bar{S}$, which is when first priors become equivalent to least-informative priors. How shall we deal with α? The most popular approach currently is to treat α as a nuisance parameter and eliminate it via outmarginalization [12] [17] [18]. Nevertheless, I see a couple of problems with this:

- **Technical Problems:** The outmarginalization procedure requires a prior $P(\alpha)$. In my eyes, this just means that the problem of determining the least-informative prior is moved to another, less "visible" place. Furthermore, we have even less intuition on how to determine a prior on a hyperparameter than on θ. Consequently, the

---

[3] Actually, I think that Jeffreys' prior is a second-order approximation to the entropic prior, the small error being caused by merely asymptotically valid expansions, possibly those in [15], page 13, and [16], page 4.

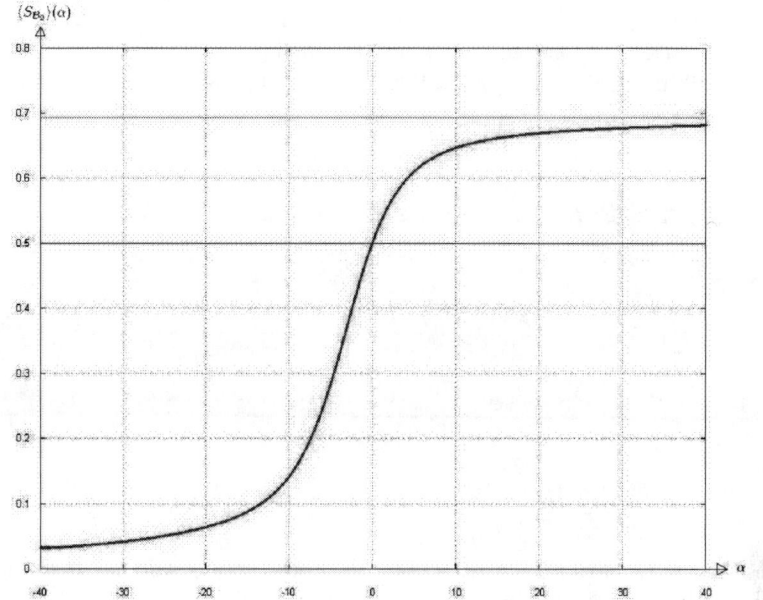

FIGURE 1. Expected entropy as a function of α given the entropic prior for 2-class discrete data

attempts to specify $P(\alpha)$ I've seen so far don't convince me very much: Strauss, Wolf and Wolpert (see [17], page 115) simply assumed a "reasonable" flat prior, which is obviously not a well-founded argument. Preuss and Caticha [12] advocate for an entropic prior on α, but then they get another hyperparameter (e.g. β) they have to deal with. Rodriguez [19] proposed an infinite progression of entropic priors, but how can we compute this?

- **Axiomatic Consistency:** The result of an outmarginalization will usually not match anymore the entropic form required by Skilling's axioms.
- **Epistemic Argument:** We are doing Bayesian inference, because we are interested in the uncertainties of all possible θs being the real θ^*s, based upon the state of knowledge we are in after having seen certain data. But we are not interested at all in the distribution of a hyperparameter. Therefore in this case I'm missing the motivation for a Bayesian treatment.

Which alternatives do we have? Skilling advanced the view that α could not be fixed a priori. (see [8], page 51). Nevertheless, my opinion is that the desiderata formulated above can only be satisfied by a point estimate, and that *though no particular value is correct for any possible data likelihood, there might exist a rule to determine α for a given data likelihood.*

In order to progress into that direction, let's again have a look at figure 1 and recall that any choice of α corresponds to a particular expectation $\langle S_L \rangle$. If we don't have a clue about the real value of the data likelihood entropy, we would surely not want to expect a zero ("the die always gives the same number") or maximal value ("the die is fair"), which would be examples of great prior knowledge. But which of the "moderate" α-values makes most sense, and what is this sense? My opinion is:

Theorem 3 (Least-Informative Priors) *If we are implicitly making an assumption on the data likelihood entropy anyway, although we don't know which one to expect, the least biased choice is the α^* that maximizes the assumption error, i.e. the entropy variance. This is the exact meaning of "least-informativity".*

This choice has amazing properties, for example the following equivalence:

$$\langle S_L^2 \rangle(\alpha) - \langle S_L \rangle(\alpha)^2 = \frac{\int_\Theta S_L(\theta)^2 e^{\alpha S_L(\theta)} d\theta}{\int_\Theta e^{\alpha S_L(\theta)} d\theta} - \left(\frac{\int_\Theta S_L(\theta) e^{\alpha S_L(\theta)} d\theta}{\int_\Theta e^{\alpha S_L(\theta)} d\theta}\right)^2$$

$$= \frac{(\frac{\partial}{\partial \alpha} \int_\Theta S_L(\theta) e^{\alpha S_L(\theta)} d\theta)(\int_\Theta e^{\alpha S_L(\theta)} d\theta) - (\int_\Theta S_L(\theta) e^{\alpha S_L(\theta)} d\theta)(\frac{\partial}{\partial \alpha} \int_\Theta e^{\alpha S_L(\theta)} d\theta)}{(\int_\Theta e^{\alpha S_L(\theta)} d\theta)^2}$$

$$= \frac{\partial}{\partial \alpha} \frac{\int_\Theta S_L(\theta) e^{\alpha S_L(\theta)} d\theta}{\int_\Theta e^{\alpha S_L(\theta)} d\theta} = \frac{\partial}{\partial \alpha} \langle S_L \rangle(\alpha) \quad (11)$$

Thus, choosing the α^* that maximizes the data likelihood entropy variance of an entropic prior is equivalent to choosing the turning point of $\langle S_L \rangle(\alpha)$,

$$\alpha^* = \arg \max_{\{\alpha\}} \left(\frac{\partial \langle S_L \rangle(\alpha)}{\partial \alpha}\right) \quad (12)$$

Therefore, theorem 3 could as well be called a "Maximum Entropy Change Principle". Further interesting properties of this principle are

- It is *absolutely invariant against reparametrizations* of the probabilistic model, because the principle itself is based solely on the data likelihood function. If the likelihood function is reparametrized, the functional form of the condition for the turning point of $\langle S_L \rangle(\alpha)$ changes correspondingly so that the resulting α^* always stays the same.
- It realizes a *stable inference* solution as proposed by Tikochinsky [20]: The α to choose is the one that maximizes the change in expected entropy; the other way round, it is the choice where a small error in our assumption (say, the expected entropy differs from the real one) has the least effect on the shape of the resulting density and estimators derived from it. Since the quantity whose stability is guaranteed is an entropy, we could call that solution "entropy-stable".
- That the proposed choice *maximizes the mean-squared entropy error* is in agreement with [21] that the quadratic loss is the unique loss function consistent with the entropy measure.
- The principle has many meaningful transformations; for example it may be rewritten as the demand to set the expected entropy skew to zero.

Note last not least that maximization of expected entropy change has been proposed in nonequilibrium thermodynamics by Q.A. Wang [22] [23] before, and that we could state the choice of α as well in a Bayesian style with the prior $P(\alpha)$ being a delta function.

5. APPLICATION OF THE MAXIMUM ENTROPY CHANGE PRINCIPLE

I applied the proposed rule to the normal data likelihood function as well as to data likelihoods for discrete-valued random processes. Computations usually consist of two components: First an approximation of the expected data likelihood entropy and/or some higher central moments for a given α, and second a kind of Newton-step procedure to find the α^* where the entropy change is maximized.

5.1. Normal data

For a normal data likelihood $\mathcal{N}(x|\mu,\sigma) \equiv L(x|\theta)$ with

$$\mathcal{N}(x|\mu,\sigma) = \frac{1}{\sqrt{2\pi}\sigma} e^{-\frac{(x-\mu)^2}{2\sigma^2}} \tag{13}$$

the data likelihood entropy $S_{\mathcal{N}}(\mu,\sigma) \equiv S_L(\theta)$ is given by

$$S_{\mathcal{N}}(\mu,\sigma) = \ln\sqrt{2\pi}\sigma \tag{14}$$

The entropic prior with hyperparameter α therefore has the form

$$P_\alpha(\mu,\sigma) \propto e^{\alpha \ln\sqrt{2\pi}\sigma} \tag{15}$$

Though $\langle S_{\mathcal{N}} \rangle(\alpha)$ gives ∞, its derivative towards α can be computed. Maximizing the change of $\langle S_{\mathcal{N}} \rangle(\alpha)$ with respect to α gives $\alpha^* = -1$ and with this, the entropic prior resolves to

$$P(\mu,\sigma) \propto \frac{1}{\sigma} \tag{16}$$

This is the result preferred by Jeffreys although it contradicts his "general rule" (see [24], page 1345).

5.2. Discrete random processes

Now we consider random processes that can only take on a finite number of values $X \in \{X_1,...,X_k\}$ with probabilities $p_1,...,p_k, \sum_{i=1}^k p_i = 1$. A single random experiment with such a likelihood function is called a Bernoulli trial for $k=2$, and a Bernoulli scheme for any $k \geq 2$; I will refer to the corresponding distributions in general as Bernoulli likelihoods $\mathcal{B}_k$. If we combine several Bernoulli experiments ignoring their order, we get binomial or multinomial data likelihoods. Inference from such data likelihoods has turned out to be a tough problem [25] [26], so let's see what the new approach delivers:

The entropy of a Bernoulli likelihood is given by the standard Shannon entropy

$$S_{\mathcal{B}_k}(p_1,...,p_{k-1}) = -\sum_{i=1}^k p_i \ln p_i \tag{17}$$

The entropic prior with hyperparameter α is

$$P_\alpha(p_1,...,p_{k-1}) \propto e^{\alpha S_{\mathcal{B}_k}(p_1,...,p_{k-1})} \quad (18)$$

and the expected data likelihood entropy

$$\langle S_{\mathcal{B}_k}\rangle(\alpha) = \int ... \int_\mathbf{P} S_{\mathcal{B}_k}(p_1,...,p_{k-1}) P_\alpha(p_1,...,p_{k-1}) d\mathbf{p} \quad (19)$$

where we have to integrate over the $(k-1)$-dimensional simplex
$\mathbf{P} = \{(p_1,...,p_{k-1}) \mid p_1 = 0..1, p_2 = 0..(1-p_1),..., p_{k-1} = 0..(1-p_1-...-p_{k-2})\}$.

The required computations are pretty time-consuming, because the expected entropy integrals have the tendency to converge very slowly. Already for $k = 3$, it was quite necessary to use special numerical integration techniques. The best method I applied is an Adaptive Quadrature algorithm [27] using Gauss-Legendre polynomials and some analytical simplifications. The results are given in table 1.

TABLE 1. α^*-choice for the entropic prior for k-class discrete data

k	α^*	$\langle S_{\mathcal{B}_k}\rangle(\alpha^*)$
2	-3.118356848554...	0.3685467...
3	-4.772026959...	0.5676038...
4	-6.0437688...	0.7127738...
5	-7.104524...	0.8308972...
6	-8.03223...	0.932265...
7	-8.9216...	1.01628...

The entropic prior for 2-class discrete data resembles much Jeffreys' prior $P_J(\theta) \propto \theta^{-1/2}(1-\theta)^{-1/2}$; in fact, we can develop the entropic prior into a power series around $\theta_0 = 1/2$ to which Jeffreys' prior is a second order approximation.

As indicated by table 1, the α^*-values keep the expected data likelihood entropies near to $\frac{\ln k}{2}$ for any k.

6. FINAL REMARKS

The major points of this work are expressed in the three theorems. The first claims that Jaynes' Maximum Entropy Principle requires necessarily a constraint $\langle S_L\rangle = \bar{S}$ on the expected entropy of the considered data likelihood $L(x|\theta)$. Practically, such a constraint means that we have an idea about the complexity of the random problem. If we apply the ME principle with such a constraint, we will always get an entropic prior $P_\alpha(\theta)$.

The second theorem shows that maximization of Shannon entropy under an expected data likelihood entropy constraint is equivalent to maximizing relative entropy without such a constraint, but with a particular choice of the underlying measure $m(\theta)$. Therefore, reparametrization invariance is guaranteed for continuous-valued data likelihoods by exactly the same formalism that was derived for discrete-valued data likelihoods.

The third and last theorem is a proposal for priors in the absence of any prior knowledge. It claims that in this case, we should choose the hyperparameter α such that the data likelihood entropy variance or equivalently, the change of expected data likelihood entropy against α is maximized. Among other interesting properties, this principle is completely invariant against reparametrizations of the data likelihood and could be called an *entropy-stable* inference solution. Furthermore, the principle shows us that Jose Bernardo's famous sentence "Non-informative priors do not exist" [28] is absolutely correct: Any entropic prior implies an assumption on the data likelihood entropy, and if we don't know which entropy to expect, all we can do is to minimize our assumption error by maximizing the entropy variance.

Putting all pieces together, we get a "universal" solution procedure (in the sense that it is applicable to any data likelihoods) for the parametric inference problem that leaves no hyperparameters unspecified. This procedure has the following components:

- You are given the parametric form of the data likelihood of a random problem.
- Determine the "first prior" using the ME principle with a constraint on the expected data likelihood entropy. The result is an entropic prior with hyperparameter α. If you really have an expectation about the data likelihood entropy, choose the α that realizes it; else take the value that maximizes the data likelihood entropy variance (or equivalently, the change of data likelihood entropy).
- Use Bayes' theorem to update your degrees of belief from the first prior with data.
- Compute the desired parameter estimates from the posterior.

Concerning statistical inference, I think that the whole of the theory developed here is very self-consistent, explains some things that have not been explained yet, and gives encouraging results. An aspect I want to emphasize is that it prefers special ("subjective") knowledge if present. In fact, if we understand the ME principle as a mere tool to cast prior knowledge into a nicely shaped prior, then the approach presented here could help to close the gap between the positions of "objectivists" and "subjectivists". I'm now looking forward to use the results of this paper as a building block to tackle more complex problems, like non-parametric inference for high-dimensional real-world problems. That's where the right choice of priors will have the biggest impact.

Concerning physics, like Wang I believe that the Maximum Entropy Change Principle is a kind of natural law, and I wonder if there are further problems and theories to which it might be applicable. Some potential candidates are already suggested by the near relationship between inference and thermodynamics through the ME principle, for example the problems of temperature fluctuations [10] [29] [30] or a thermodynamical uncertainty relation [31] [32]. Other interesting relationships are those with black hole thermodynamics [33] [34] and with the theories that can be derived by application of the ME principle with exotic probabilities [35].

ACKNOWLEDGMENTS

I would like to thank Ariel Caticha, John Skilling and Matthew Brand for valuable comments, my mother Karin for her support, and my son Max for just being there.

REFERENCES

1. E. T. Jaynes, *IEEE Transactions on System Science and Cybernetics* **4**, 227–241 (1968).
2. E. T. Jaynes, *Physical Review* **106**, 620–630 (1957).
3. E. T. Jaynes, *Physical Review* **108**, 171–190 (1957).
4. W. T. Grandy Jr., "The three phases of statistical mechanics," in *Maximum Entropy and Bayesian Methods*, edited by J. Skilling, Kluwer, Dordrecht, 1989, pp. 73–91.
5. M. Mihelcic, Maximum entropy method and Bayesian probability theory, Technical Report Jül-3395, Forschungszentrum Jülich (1995).
6. J. Uffink, *Studies in History and Philosophy of Modern Physics* **26**, 223–261 (1995), URL http://citeseer.ist.psu.edu/uffink97can.html.
7. A. Caticha, and A. Giffin, Updating Probabilities, *arXiv:physics/0608185* (2006).
8. J. Skilling, "Classic Maximum Entropy," in *Maximum Entropy and Bayesian Methods*, edited by J. Skilling, Kluwer, Dordrecht, 1989, pp. 45–52.
9. C. C. Rodriguez, "The metrics induced by the Kullback number," in *Maximum Entropy and Bayesian Methods*, edited by J. Skilling, Kluwer, Dordrecht, 1989, pp. 415–422.
10. A. Caticha, "Maximum entropy, fluctuations and priors," in *Maximum Entropy and Bayesian Methods in Science and Engineering*, edited by A. Mohammad-Djafari, AIP, Melville, 2001, pp. 94–105.
11. A. Caticha, and R. Preuss, Entropic priors, *arXiv:physics/0312131* (2003).
12. A. Caticha, and R. Preuss, *Physical Review E* **70** (2004).
13. C. G. Chakrabarti, N. C. Das, and K. De, *Czechoslovak Journal of Physics* **52**, 911–918 (2002).
14. E. T. Jaynes, *Probability Theory: The Logic of Science*, Cambridge University Press, Cambridge (UK), 2003.
15. S. Amari, and H. Nagaoka, *Methods of Information Geometry*, American Mathematical Society, 2000.
16. V. Balasubramanian, Statistical inference, occam's razor and statistical mechanics on the space of probability distributions, *arXiv:cond-mat/9601030* (1996).
17. C. M. E. Strauss, D. H. Wolpert, and D. R. Wolf, "Alpha, evidence, and the entropic prior," in *Maximum Entropy and Bayesian Methods*, edited by A. Mohammad-Djafari, Kluwer, Dordrecht, 1993, pp. 113–120.
18. R. Fischer, W. von der Linden, and V. Dose, "On the importance of α marginalization in maximum entropy," in *Maximum Entropy and Bayesian Methods*, edited by K. Hanson, and R. Silver, Kluwer, Dordrecht, 1996, pp. 229–236.
19. C. C. Rodriguez, Entropic Priors for Discrete Probabilistic Networks and for Mixtures of Gaussian Models, *arXiv:physics/0201016* (2002).
20. Y. Tikochinsky, N. Z. Tishby, and R. D. Levine, *Physical Review A* **30**, 2638–2644 (1984).
21. Y. Tikochinsky, and R. D. Levine, *J. Math. Phys.* **25**, 2160–2168 (1984).
22. Q. A. Wang, Maximizing entropy change and least action principle for nonequilibrium systems, *arXiv:cond-mat/0312329* (2003).
23. Q. A. Wang, Action principle and Jaynes' guess method, *arXiv:cond-mat/0407515* (2004).
24. R. E. Kass, and L. Wasserman, *Journal of the American Statistical Association* **91**, 1343–1370 (1996).
25. P. Walley, *Journal of the Royal Statistical Society B* **58**, 3–57 (1996).
26. M. Zhu, and A. Y. Lu, *Journal of Statistics Education* **12** (2004).
27. A. Genz, "An Adaptive Numerical Integration Algorithm for Simplices," in *Computing in the 90s*, edited by N. Sherwani, E. de Doncker, and J. Kapenda, Springer, New York, 1991, pp. 279–292.
28. J. M. Bernardo, T. Z. Irony, and N. D. Singpurwalla, *Journal of Statistical Planning and Inference* **65**, 159–189 (1997).
29. G. D. J. Phillies, *American Journal of Physics* **52**, 629–632 (1984).
30. H. B. Prosper, *American Journal of Physics* **61**, 54–58 (1993).
31. Y. Alhassid, and R. D. Levine, *Chemical Physical Letters* **73**, 16–20 (1980).
32. F. Schlögl, *Journal of Physics and Chemistry of Solids* **49**, 679–683 (1988).
33. J. Baez, This week's finds in mathematical physics – week 111 (1997), URL http://math.ucr.edu/home/baez/week111.html.
34. R. M. Wald, Black holes and thermodynamics, *arXiv:gr-qc/9702022* (1997).
35. S. Youssef, Physics with exotic probability theory, *arXiv:hep-th/0110253* (2001).

Strong Nonlinear Correlations, Conditional Entropy and Perfect Estimation

Christopher S. Jones*,†, John M. Finn* and Nicolas Hengartner**

*T-15, Plasma Theory, Los Alamos National Laboratory, Los Alamos, NM 87545
†Present address: ID Analytics, San Diego, CA
**CCS-3, Information Sciences, Los Alamos National Laboratory, Los Alamos, NM 87545

Abstract. This paper deals with parameter estimation in which measurements subjected to highly correlated noise allow for very accurate estimation. For linear regression with normally distributed noise, this generically occurs when the noise becomes highly linearly correlated. For a linear model with nonnormal noise distributions, there may exist nonlinear regressions that allow for accurate estimation if a conditional entropy is small (analogous to linear correlations in the normal case approaching ±1.) Nonlinear regression may also yield an accurate estimate if a nonlinear model is subjected to strongly linearly correlated noise.

Keywords: Nonlinear correlations, perfect estimation
PACS: <07.05.-t, 95.75.Mn>

1. INTRODUCTION

Authors in various fields have previously noted the unusual behaviors of linear regression in the presence of strong correlations [1, 2, 3, 4, 6, 7, 8, 9, 10]. Certain features related to strong correlations were remarked upon, often with puzzlement. These include (a) negative weighting, in which an estimate of a mean lies outside the range of the data – Peelle's Pertinent Puzzle (see Ref. [6] and references therein), and (b) the possibility of both beneficial and deleterious effects of strong correlations with respect to the quality of an estimate. In Ref. [1] the decrease of the estimate variance for strong linear correlations, and its vanishing for perfect correlation, was noted in the context of plasma reconstruction. (Plasma reconstruction is intrinsically nonlinear, but in the cases considered in Ref. [1] a linear approximation is valid if the noise level is small enough.) In these applications, large correlations can occur because the fluctuations treated as noise are in fact high frequency plasma fluctuations outside the scope of the model, which nevertheless can be strongly spatially correlated, either linearly or nonlinearly. In Ref. [11], the present authors explained these features and showed how, generically, perfect estimation is possible in the limit of full positive or negative correlation.

In this paper, we discuss via some simple examples how the results of Refs.[1, 11] regarding linear regression may be generalized to nonlinear estimation procedures. In Section 2, we briefly review the results of Ref. [11] and clarify the effects of strong linear correlations. We discuss two simple problems, namely the estimation of a single parameter with two data points and the estimation of the slope and intercept of a line using three data points. In Section 3, we discuss situations in which nonlinear regression may yield an exact estimate. For a linear model subjected to highly nonlinearly correlated noise, we show in Sec. 3.1 that the vanishing of the conditional entropy [5] between certain noise variables leads to the possibility of perfect estimation. We illustrate with three examples: estimation of a single parameter with either two or three measurements, and estimate of a slope and intercept with three measurements. In Sec. 3.2 we discuss the case of Sec. 3.1 by a completely different approach involving conditional expectation and expected conditional variance. Similarly, we show in Sec. 3.3 that nonlinear models in the presence of fully linearly correlated noise allow for perfect estimation if a particular linear combination of the model functions is invertible, leading to zero conditional entropy.

In Ref. [11] the behavior of the estimate variance for the limiting case $|\rho| \to 1$ was studie, as well as the more general case with large linear correlations $|\rho| \sim 1$. It was shown that in the latter case the estimate variance could be quite small. Similarly, the perfect nonlinear estimation studies in this paper for zero conditional entropy indicate the possibility of very good nonlinear estimation when the conditional entropy (or expected conditional variance) is small.

2. LINEAR ESTIMATION WITH HIGHLY CORRELATED NOISE

This section is a review of work presented in Ref. [11]. We begin with a brief discussion of the estimation of a single parameter using highly correlated errors, i. e.

$$\begin{pmatrix} y_1 \\ y_2 \end{pmatrix} = \begin{pmatrix} 1 \\ 1 \end{pmatrix} a + \begin{pmatrix} \eta_1 \\ \eta_2 \end{pmatrix} \quad (1)$$

or $\vec{y} = \mathsf{X} a + \vec{\eta}$, with design matrix $\mathsf{X} = (1,1)^t$. The mean of $\vec{\eta}$ is zero and its covariance matrix is of the form $\Sigma = \mathsf{SRS}$, where

$$\mathsf{S} = \begin{pmatrix} \sigma_1 & 0 \\ 0 & \sigma_2 \end{pmatrix}, \quad \mathsf{R} = \begin{pmatrix} 1 & \rho \\ \rho & 1 \end{pmatrix}.$$

The correlation coefficient must satisfy $-1 \le \rho \le 1$. The effect we discuss occurs for $|\rho| \sim 1$ but is easiest to understand for $\rho \to \pm 1$. Let us for now assume $\rho \to 1$ (perfect *positive* correlation). Then Σ has rank unity for $\rho = 1$, with one eigenvalue $\sigma_1^2 + \sigma_2^2 = \text{trace}(\Sigma)$ with eigenvector $\vec{v}_1 = (\sigma_1, \sigma_2)^t$. The other eigenvector $\vec{v}_2 = (\sigma_2, -\sigma_1)^t$ has eigenvalue zero. Perfect correlation $\rho = 1$ means that $\eta_2 = \eta_1 \sigma_2 / \sigma_1$. Multiplying equation (1) by $\vec{v}_2$, we find

$$(\sigma_1 - \sigma_2)\hat{a} = \sigma_1 y_2 - \sigma_2 y_1,$$

yielding the estimate $\hat{a}$ exactly, i.e. with no uncertainty, as long as $\sigma_1 \ne \sigma_2$. That is, the variance of the estimate, the *posterior variance*, is zero. In Ref. [11], this noise elimination process was described in terms of the projection of the design matrix X to a *reduced design matrix* $\tilde{\mathsf{X}}$ in a *noise-free subspace*. In this case, the reduced design matrix is simply the coefficient $(\sigma_1 - \sigma_2)$ and the noise-free subspace is the one-dimensional subspace spanned by the vector $\vec{v}_2$. The exception to the vanishing of the posterior variance is the special case $\sigma_1 = \sigma_2$. This situation is illustrated geometrically in the

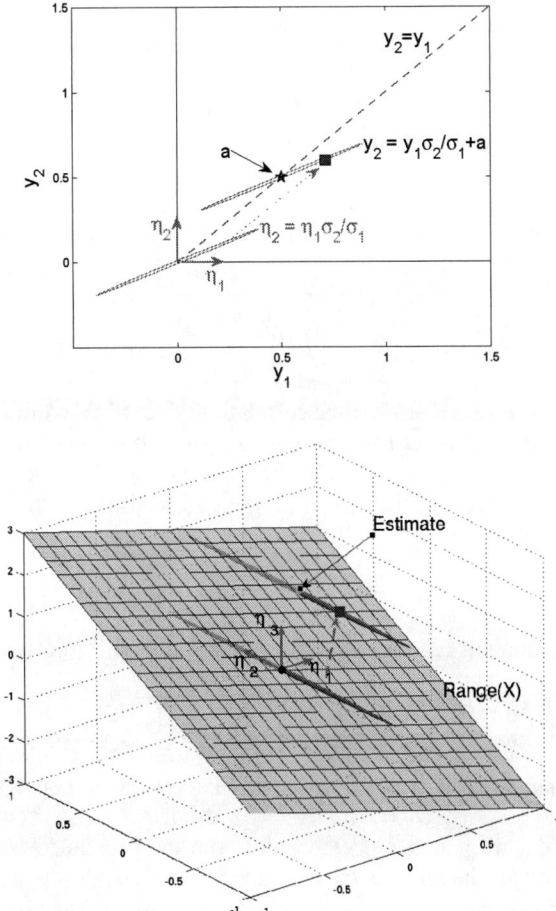

FIGURE 1. In the first panel, two measurements y_1, y_2 (the square) are subjected to highly correlated ($\rho \to 1$) errors with $\sigma_1 > \sigma_2$. The errors are distributed within a thin ellipse aligned along $\eta_2 = \eta_1 \sigma_2/\sigma_1$; a line fitting the data (y_1, y_2) crosses the line $y_2 = y_1$, giving an estimate with zero error, for $\rho \to 1$. Alternatively, the thin ellipse indicating the distribution of errors $\vec{\eta}$ is transported parallel to $y_2 = y_1$ to match the data (y_1, y_2), yielding an estimate with zero error. Clearly this is possible unless $\sigma_1 \neq \sigma_2$. In the second panel, the analogous picture is shown for estimation of the slope and intercept of a line using three data points. In this case, the thin ellipsoid is transported in two dimensions parallel to the plane, which is the span of the columns of the design matrix.

first panel of Fig. 1. Clearly the posterior variance remains zero if more measurements $y_3, \ldots, y_n$ are used.

We illustrate further with the example of estimating a line $y = a + bx$ using three data points

$$\begin{pmatrix} y_1 \\ y_2 \\ y_3 \end{pmatrix} = X \begin{pmatrix} a \\ b \end{pmatrix} + \begin{pmatrix} \eta_1 \\ \eta_2 \\ \eta_3 \end{pmatrix} = \begin{pmatrix} 1 & x_1 \\ 1 & x_2 \\ 1 & x_3 \end{pmatrix} \begin{pmatrix} a \\ b \end{pmatrix} + \begin{pmatrix} \eta_1 \\ \eta_2 \\ \eta_3 \end{pmatrix}. \quad (2)$$

Again the covariance matrix of the data is of the form $\Sigma = \text{SRS}$, where now

$$S = \begin{pmatrix} \sigma_1 & 0 & 0 \\ 0 & \sigma_2 & 0 \\ 0 & 0 & \sigma_3 \end{pmatrix}, \quad R = \begin{pmatrix} 1 & \rho_{12} & \rho_{13} \\ \rho_{12} & 1 & \rho_{23} \\ \rho_{13} & \rho_{23} & 1 \end{pmatrix},$$

with $|\rho_{ij}| \leq 1$. Full positive or negative correlations occur in the limit $\rho_{ij} \to \pm 1$ ($i \neq j$), although the signs must be consistent, e.g. $\rho_{12} \to 1$, $\rho_{23} \to -1$ imply $\rho_{13} \to -1$. In the full correlation limit, the rank of the correlation matrix becomes unity and we have $\rho_{ij} = e_i e_j$ where $e_i = \pm 1$. The limit $\rho_{12} \to 1$, $\rho_{13} \to 1$ leads to the errors being distributed along the line $\mathscr{L}$, for which

$$\eta_2 = \eta_1 \sigma_2 / \sigma_1 \quad \eta_3 = \eta_1 \sigma_3 / \sigma_1. \quad (3)$$

Substituting in equation (2) we obtain a reduced model, projected to the noise-free subspace,

$$\begin{pmatrix} \sigma_1 - \sigma_2 & \sigma_1 x_2 - \sigma_2 x_1 \\ \sigma_1 - \sigma_3 & \sigma_1 x_3 - \sigma_3 x_1 \end{pmatrix} \begin{pmatrix} \hat{a} \\ \hat{b} \end{pmatrix} = \begin{pmatrix} \sigma_1 y_2 - \sigma_2 y_1 \\ \sigma_1 y_3 - \sigma_3 y_1 \end{pmatrix}. \quad (4)$$

The noise-free subspace is the null space of the covariance matrix Σ and is spanned by the vectors $\vec{v}_2 = (-\sigma_2, \sigma_1, 0)^t$ and $\vec{v}_3 = (-\sigma_3, 0, \sigma_1)^t$. The vector $\vec{v}_1$ is the single eigenvector of Σ with a nonzero eigenvalue, the total variance trace(Σ), and is the direction in which all the noise is concentrated, orthogonal to the noise-free subspace. The 2×2 matrix on the left-hand side of (4) is the reduced design matrix $\tilde{X} = \begin{pmatrix} \vec{v}_2^t \\ \vec{v}_3^t \end{pmatrix} X$. Again, the limit of full correlation leads to an estimate with zero posterior variance. The exceptional cases, for which no zero variance estimate can be found, occur if the determinant $(\sigma_1 - \sigma_2)(\sigma_1 x_3 - \sigma_3 x_1) - (\sigma_1 - \sigma_3)(\sigma_1 x_2 - \sigma_2 x_1)$ vanishes. This is equivalent to

$$\begin{pmatrix} \sigma_1 \\ \sigma_2 \\ \sigma_3 \end{pmatrix} = \begin{pmatrix} 1 & x_1 \\ 1 & x_2 \\ 1 & x_3 \end{pmatrix} \begin{pmatrix} \alpha \\ \beta \end{pmatrix}$$

for some α, β, implying that $(\sigma_1, \sigma_2, \sigma_3)^t$ is in the range of the design matrix X, denoted R(X). [In the more general case allowing for the possibility of full negative correlations, the determinant is zero if $(e_1 \sigma_1, e_2 \sigma_2, e_3 \sigma_3)^t$ is in the range of the design matrix, R(X).]

The geometric picture of this situation is illustrated in the second panel of Fig. 1. In the three-dimensional space (y_1, y_2, y_3), the distribution of $\vec{\eta}$ is along the line $\mathscr{L}$ consistent with equation (3). If this line does not lie in the plane spanned by the columns $(1,1,1)^t$ and $(x_1, x_2, x_3)^t$ of X [if $(\sigma_1, \sigma_2, \sigma_3)$ is not in the range R(X)], then the line $\mathscr{L}$

can be transported parallel to R(X) to contain the data point (y_1, y_2, y_3), and the estimate is known precisely.

These results are special cases of our general result of Ref. [11]: As the correlations between noise variables become large in absolute value ($|\rho_{ij}| \to 1$), and if the vector of signed variances $\vec{v}_1 = (e_1\sigma_1, ..., e_n\sigma_n)^t$ is not in the range of the design matrix R(X), then there exists a noise-free subspace spanned by vectors orthogonal to $\vec{v}_1$. The projection of the measurements $\vec{y} = X\vec{a} + \vec{\eta}$ onto this noise-free subspace yields $n-1$ equations free of noise. Any m of these equations may be chosen to yield the estimate $\hat{\vec{a}}$ exactly. In the absence of model error, the measurements projected to the $(n-1)$-dimensional noise-free subspace are in the range of $\tilde{X}$ (the projection of the design matrix X into the noise-free subspace), which is m-dimensional. For the example of fitting a line with positive correlations, the estimate is obtained exactly by projecting onto the noise-free subspace, as in equation (4), as long $(\sigma_1, \sigma_2, \sigma_3)^t$ is not in the range R(X), where X is defined in equation (2).

It should be stressed that in Refs. [1, 11] it was found that the posterior variance could be quite small for realistically large correlations, and not only in the limit in which the correlation coefficients $|\rho_{ij}| \to 1$.

3. NONLINEAR ESTIMATION

In this section, we discuss situations analogous to those discussed in the previous section in which the estimate is instead a *nonlinear* function of the data, and the noise distribution is strongly correlated. This may occur for linear models when the noise is highly nonlinearly correlated, as discussed in Section 3.1, and when the model itself is nonlinear, as discussed in Section 3.3. In Sec. 3.2 we discuss an alternate formulation of the situation of Sec. 3.1, based on conditional expectation and expected conditional variance.

3.1. Linear model with fully nonlinearly correlated noise distribution

As with the first example of Section 2, let us consider the estimate of a constant with a model given by equation (1). Recall that with a normal distribution of noise, the limit of full positive linear correlation ($\rho \to 1$) leads to a linear relationship between the noise η_1 and η_2. In particular, the distribution of noise becomes

$$p_n(\eta_1, \eta_2) \propto \exp\left[-(\sigma_1\eta_1 + \sigma_2\eta_2)^2/2(\sigma_1^2 + \sigma_2^2)^2\right] \delta(\eta_2/\sigma_2 - \eta_1/\sigma_1). \quad (5)$$

Instead, suppose that the distribution is of the form

$$p_n(\eta_1, \eta_2) = \psi(\eta_1, \eta_2)\, \delta(\phi(\eta_1, \eta_2)), \quad (6)$$

where ψ and ϕ are chosen such that p_n, and therefore the marginal distributions of η_1 and η_2, are normalized probability density functions of zero mean. As in Section 2, knowledge of the relationship between η_1 and η_2 allows for the elimination of noise from the pair of equations (1). The distribution of estimates $\hat{a}$ for given measurements y_1 and y_2 is determined from

$$\phi(y_1 - \hat{a}, y_2 - \hat{a}) = 0, \qquad (7)$$

for those values of $\hat{a}$ for which $\psi(y_1 - \hat{a}, y_2 - \hat{a}) > 0$. When only a single value of $\hat{a}$ satisfies equation (7) and the corresponding constraint on ψ, the estimate is known exactly.

Clearly, the existence of a unique solution to equation (7) depends on the properties of the functions ψ and ϕ, or equivalently, the distribution of noise $p_n(\eta_1, \eta_2)$. The situation is shown in the first panel of Fig. 2. The noise distribution is transported parallel to $y_2 = y_1$ until it intersects the data, and if equation (7) has a unique solution for the estimate $\hat{a}$, the mean of the transported distribution yields the estimate with no error. (As illustrated in Fig. 2, the mean for a fully nonlinearly correlated distribution is typically not contained in the distribution.) If more than one intersection is possible, then equation (7) does not have a unique solution. This occurs if the tangent to the curve $\phi(\eta_1, \eta_2) = 0$ is parallel at some point to the range of X, i. e. parallel to $y_1 = y_2$. Note that the presence of a tangency is preserved under small perturbations of the distribution, unlike the linear case, in which $(e_1\sigma_1, ..., e_n\sigma_n)$ does not remain in the range of X under a small perturbation.

The uniqueness of the solution for $\hat{a}$ is related to the conditional entropy [5] of certain transformed noise variables. In particular, let $\xi_1 = (\eta_1 + \eta_2)/2$ and $\xi_2 = \eta_1 - \eta_2$, and

$$\tilde{p}_n(\xi_1, \xi_2) = p_n(\eta_1, \eta_2). \qquad (8)$$

Note that $\eta_1 = \xi_1 + \xi_2/2$ and $\eta_2 = \xi_1 - \xi_2/2$. The variable ξ_2 remains constant when the distribution is transported along $y_1 = y_2$, or $\eta_1 = \eta_2$. A unique solution of equation (7) exists if knowledge of ξ_2 yields definite knowledge of ξ_1, that is, if the noise distribution takes the form

$$\tilde{p}_n = \tilde{\psi}(\xi_1, \xi_2)\,\delta(\xi_1 - \tilde{\phi}(\xi_2)). \qquad (9)$$

In this case, the conditional entropy $H(\xi_1|\xi_2)$ vanishes. This condition holds for an arbitrary linear transformation of the noise variables, as long as $\xi_2 \propto (\eta_1 - \eta_2)$ and ξ_1 is a linearly independent combination of η_1 and η_2. (The map above from (η_1, η_2) to (ξ_1, ξ_2) is chosen to be area preserving because the joint entropy is unchanged under such a transformation. However, this restriction is not essential.)

For the case of estimating the intercept a and the slope b of a line from three measurements with fully nonlinearly correlated data, consider a noise distribution of the form

$$p_n(\eta_1, \eta_2, \eta_3) = \psi(\eta_1, \eta_2, \eta_3)\delta(\phi_1(\eta_1, \eta_2, \eta_3))\delta(\phi_2(\eta_1, \eta_2, \eta_3)). \qquad (10)$$

Again we eliminate the noise and find an implicit relation for the estimates $\hat{a}, \hat{b}$, namely

$$\phi_1(y_1 - \hat{a} - \hat{b}x_1, y_2 - \hat{a} - \hat{b}x_2, y_3 - \hat{a} - \hat{b}x_3) = \phi_2(y_1 - \hat{a} - \hat{b}x_1, y_2 - \hat{a} - \hat{b}x_2, y_3 - \hat{a} - \hat{b}x_3) = 0, \qquad (11)$$

and again the condition that $\hat{a}$ and $\hat{b}$ be known exactly depends on the solvability of equation (11) for $\hat{a}$ and $\hat{b}$ over the range where $\psi(y_1 - \hat{a} - \hat{b}x_1, y_2 - \hat{a} - \hat{b}x_2, y_3 - \hat{a} - \hat{b}x_3)$

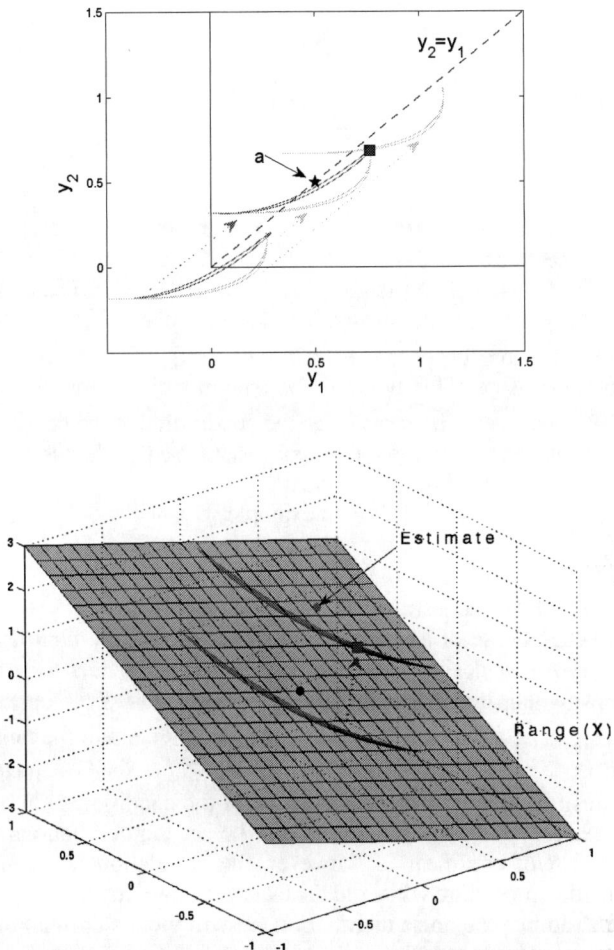

FIGURE 2. In the first panel, two measurements y_1, y_2 (the square) are subjected to an error distribution that becomes infinitely thin. If the error distribution (two dark curves) is such that transporting it along $y_1 = y_2$ intersects the data once, then the estimate is determined exactly. The three light curves display a failure of the estimation procedure to yield an exact result because this curve can be transported in two ways to intersect the data. This occurs because there exists a tangent to the light curves that is parallel to the range R(X) (i.e. the line $y_1 = y_2$). In the second panel, the analogous picture is shown for estimation of the slope and intercept of a line. Perfect estimation is possible as long as there is no tangent to the curve given by $\phi_1(\eta_1, \eta_2, \eta_3) = \phi_2(\eta_1, \eta_2, \eta_3) = 0$ that is parallel to the range $R(X) = \text{span}\{(1,1,1)^t, (x_1, x_2, x_3)^t\}$.

is positive. This situation is shown in the second panel of Fig. 2, and can again be described by a change of variables. For example, let $\vec{\xi} = T^t \vec{\eta}$, or

$$\xi_1 = \eta_1 = \vec{T}_1^t \vec{\eta},$$
$$\xi_2 = \eta_2 = \vec{T}_2^t \vec{\eta},$$
$$\xi_3 = (x_3 - x_2)\eta_1 + (x_1 - x_3)\eta_2 + (x_2 - x_1)\eta_3 = \vec{T}_3^t \vec{\eta},$$

with $\vec{T}_1 = (1,0,0)^t$, $\vec{T}_2 = (0,1,0)^t$ and $\vec{T}_3 = (x_3 - x_2, x_1 - x_3, x_2 - x_1)^t$. Note that $\vec{T}_3$ is orthogonal to the columns of the design matrix X, i.e. is in $R(X)^\perp$, the orthogonal complement of R(X). Thus, R(X) is given by the plane $\xi_3 = 0$. The condition that the distribution $p_n(\eta_1, \eta_2, \eta_3)$ can be transported along the plane R(X) to intersect the data uniquely is the condition [based on the relations $\phi_1(\vec{\eta}(\vec{\xi})) = \phi_2(\vec{\eta}(\vec{\xi})) = 0$] that ξ_1 and ξ_2 be functions of ξ_3. This holds if the conditional entropy $H(\xi_1, \xi_2 | \xi_3)$ of the distribution (10) equals zero. In terms of $\vec{\xi}$, the condition that the curve $\phi(\vec{\eta}) = 0$ does not possess a tangent parallel to the range space R(X) is equivalent to the condition that the derivatives $d\xi_1/d\xi_3$, $d\xi_2/d\xi_3$ be bounded.

These examples have $m = n - 1$. For an example with $m \neq n - 1$, consider estimating a mean with *three* measurements. If we let $\xi_1 = (\eta_1 + \eta_2 + \eta_3)/3, \xi_2 = \eta_1 - \eta_2, \xi_3 = \eta_1 - \eta_3$ [$\vec{T}_1 = (1/3, 1/3, 1/3)^t, \vec{T}_2 = (1, -1, 0)^t, \vec{T}_3 = (1, 0, -1)^t$], then the range of X is given by $\xi_2 = \xi_3 = 0$. Perfect estimation is possible if $H(\xi_1 | \xi_2, \xi_3) = 0$.

In the most general case of a linear model subjected to fully nonlinearly correlated data, we have an $n \times m$ design matrix X relating m parameters to n measurements. Consider a nonsingular, linear transformation T, with rows $\vec{T}_i^t$, such that the span of the vectors $\vec{T}_{m+1}, \ldots, \vec{T}_n$ is $R(X)^\perp$, the orthogonal complement to the range R(X). Then let $\vec{\xi} = T\vec{\eta}$. If the conditional entropy $H(\xi_1, \ldots, \xi_m | \xi_{m+1}, \ldots, \xi_n)$ vanishes, then there exists an exact estimate. Note that the vanishing of the conditional entropy implies the possibility of perfect estimation regardless of whether the correlations between noise variables are linear or nonlinear. If, however, the correlations are nonlinear, only a nonlinear estimation procedure will yield the exact estimate. In the case of the estimation of a mean, for example, the noise distribution (9) will yield a correlation coefficient ρ whose magnitude is less than unity for a nonlinear function $\tilde{\phi}$. Consequently, the best linear unbiased estimate (BLUE) will have a nonzero variance rather than the vanishing variance obtained from the solution of (7).

3.2. Alternate formulation in terms of conditional expectation and conditional variance

Let us return to the problem of Sec. 3.1, namely estimating a constant as in eq. (1) with highly nonlinearly correlated non-normal noise. Let us define

$$z_1 = y_1$$
$$z_2 = y_1 - y_2$$

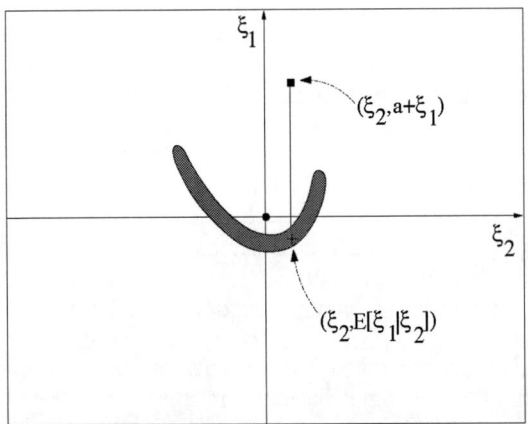

FIGURE 3. Illustration of the estimate based on the conditional expectation $E[\xi_1|\xi_2]$. This quantity is labelled $+$, and leads to an estimate improved relative to the estimate $z_1 = y_1$.

(again area preserving). In terms of the noise variables $\xi_1 = \eta_1$ and $\xi_2 = \eta_1 - \eta_2$ we have
$$z_1 = a + \xi_1$$
$$z_2 = \xi_2,$$
showing that there is no signal in z_2. Thus, z_2 is an ancillary statistic. Based on this, it is tempting to use z_1 as an estimator for the parameter a. However, a better estimate is given in terms of the conditional expectation $E[\xi_1|\xi_2]$, namely
$$\hat{a} = z_1 - E[\xi_1|\xi_2].$$

See Fig. 3. The conditional expectation $E[\xi_1|\xi_2]$ does not appear to be a function of the data, but it actually is because $E[\xi_1|\xi_2]$ is a function of $\xi_2 = z_2 = y_1 - y_2$, which is an observed quantity.

To see that $\hat{a}$ is a better estimate, note first that
$$E[\hat{a}] = a,$$
meaning that $\hat{a}$ is an unbiased estimate. Then the basic lemma on conditional expectation, namely $E[E(Y|X)] = E[Y]$, implies
$$V[\hat{a}] = E[(\xi_1 - E[\xi_1|\xi_2])^2]$$
$$= E\left[E[(\xi_1 - E[\xi_1|\xi_2])^2]|\xi_2\right]$$
$$= E[V(\xi_1|\xi_2)] \leq V[\xi_1] = V[z_1].$$

Here, $V[\xi_1|\xi_2]$ is the conditional variance of ξ_1 given ξ_2 and $E[V(\xi_1|\xi_2)]$ is the expected conditional variance. Clearly, $E[V(\xi_1|\xi_2)]$ is zero if and only if $V[\xi_1|\xi_2]$ is zero for all ξ_2.

We have several remarks concerning this inequality. First, it is clear that the estimator $\hat{a}$ improves on z_1 in the sense that its conditional variance is smaller. Second, suppose the conditional variance $V[\xi_1|\xi_2]$ is zero for all ξ_2, i. e. $E[V(\xi_1|\xi_2)] = 0$. This means that ξ_1 is a function of ξ_2 and the estimate $\hat{a}$ is exact. This is exactly the conclusion reached in the discussion of the first example in Sec. 3.1. Third, we can extend $z_1 = y_1$ to the more general area preserving transformation with $z_1 = \alpha y_1 + (1-\alpha)y_2$. If $E[V(\xi_1|\xi_2)] = 0$, the same conclusion $V[\hat{a}] = 0$ holds for all α. However, if $E[V(\xi_1|\xi_2)]$ is small but positive, then the posterior variance $V[\hat{a}]$ is positive but the parameter α can be used to minimize it.

Finally, we comment on the use of expected conditional variance in this subsection rather than conditional entropy as in Sec. 3.1. It is clear that the conditional entropy $H(\xi_1|\xi_2) = 0$ is equivalent to ξ_1 being a function of ξ_2. The same is true for the expected conditional variance $E[V(\xi_1|\xi_2)] = 0$. That is, for the limiting case in which the noise distribution is infinitely thin, these two formulations are equivalent, and the technical steps in this subsection are somewhat easier for the former. It remains to be seen which formulation is more useful when the noise distribution has small but nonzero thickness.

Consider the more general case where one observes n observations from the linear model $y_i = \vec{x}_i^t \vec{\beta} + \eta_i$, $i = 1, \ldots, n$, with both $\vec{x}_i$ and $\vec{\beta}$ vectors in $\mathscr{R}^m$. We write this compactly in matrix form as

$$\vec{y} = X\vec{\beta} + \vec{\eta}$$

where $X = [\vec{x}_1, \ldots, \vec{x}_n]^t$ is the design matrix. The data $\vec{y} = (y_1, \ldots, y_n)^t$ and the disturbances $\vec{\eta} = (\eta_1, \ldots, \eta_n)$ are vectors in $\mathscr{R}^n$, and the parameter $\vec{\beta}$ is as before an element of $\mathscr{R}^m$. We assume for ease of exposition that the design matrix X is of full rank m. Let $\vec{T}_1, \ldots, \vec{T}_m$ span the range R(X) and $\vec{T}_{m+1}, \ldots, \vec{T}_n$ spand the orthogonal complement R(X)$^\perp$. Define for $j = 1, \ldots, n$ the transformed variables

$$z_j = \vec{T}_j^t \vec{y} = \vec{T}_j^t X \vec{\beta} + \vec{T}_j^t \vec{\eta} = \gamma_j + \xi_j$$

By construction, $\gamma_j = 0$ for $j > m$, meaning that these γ_j are ancillary statistics. While z_j for $j \leq m$ is an unbiased estimator for γ_j, the estimator

$$\hat{\gamma}_j = z_j - E[\xi_j | \xi_{m+1}, \ldots, \xi_n]$$

has variance $E[V(z_j|z_{m+1}, \ldots, z_n)] \leq V(z_j)$. The parameter γ_j is estimated perfectly when $E[V(z_j|z_{m+1}, \ldots, z_n)] = 0$, which occurs when ξ_j is a function of $\xi_{m+1}, \ldots, \xi_n$ for $j = 1, \ldots, m$.

The treatments in this section and in Sec. 3.1 are alternate formulations which become equivalent in the limit of perfect nonlinear correlation. The findings are both generalizations of the linear, gaussian case of Ref. 11, summarized in Sec. 2.

3.3. Nonlinear models

Let us now consider estimation when we have a nonlinear model:

$$y_i = X_i(\vec{a}) + \eta_i, \quad i = 1, \ldots, n, \tag{12}$$

where X_i are n nonlinear functions of the m parameters $\vec{a}$. This is similar to the situations depicted in Figures 1 and 2, except that the m-dimensional hyperplanes representing the range of the linear model X are replaced by some general m-dimensional surface. Let us also restrict our attention to the simple case of normally distributed noise.

The discussion of perfect estimation in this case is very similar to the construction of the noise-free subspace discussed in Section 2 and in more detail in Ref. [11]. Let us first suppose that we have three measurements nonlinearly related to two parameters, a, b. In the limit of fully, positively linearly correlated noise $\rho_{12} = \rho_{13} = 1$, we again have the relation between the noise variables expressed in equation (3). We may eliminate the noise, yielding [cf. equation (4)]

$$F_{12}(\hat{a},\hat{b}) := \sigma_1 X_2(\hat{a},\hat{b}) - \sigma_2 X_1(\hat{a},\hat{b}) = \sigma_1 y_2 - \sigma_2 y_1$$
$$F_{13}(\hat{a},\hat{b}) := \sigma_1 X_3(\hat{a},\hat{b}) - \sigma_3 X_1(\hat{a},\hat{b}) = \sigma_1 y_3 - \sigma_3 y_1,$$

where the functions $F_{12}(\hat{a},\hat{b}), F_{13}(\hat{a},\hat{b})$ depend implicitly on the the deviations $\sigma_1,...,\sigma_n$. The possibility of perfect estimation arises if these equations are invertible for $\hat{a}$ and $\hat{b}$. Thus, for given deviations σ_1, σ_2, the determinant of the Jacobian of the functions F_{12} and F_{13} must be nonvanishing, i.e.

$$\det \begin{pmatrix} \partial F_{12}/\partial \hat{a} & \partial F_{12}/\partial \hat{b} \\ \partial F_{13}/\partial \hat{a} & \partial F_{13}/\partial \hat{b} \end{pmatrix} \neq 0, \quad \forall \hat{a}, \hat{b}.$$

As in the case of the linear estimation of a slope and intercept discussed in Section 2, there exists a 2-dimensional noise-free subspace, the null space of the covariance matrix Σ, spanned by the vectors $\vec{v}_2 = (-\sigma_2, \sigma_1, 0)^t$ and $\vec{v}_3 = (-\sigma_3, 0, \sigma_1)$. The reduced model in the noise-free subspace, determined by F_{12} and F_{13}, remains nonlinear.

For m parameters nonlinearly related to n measurements, as in equation (12), we consider the eigenvectors $\vec{v}_1,...,\vec{v}_n$ of the limiting, fully correlated covariance matrix ranked in order of decreasing eigenvalues. The eigenvector $\vec{v}_1$ is, in the limit, the vector of signed deviations $(e_1\sigma_1,...,e_n\sigma_n)^t$, and its eigenvalue is the total variance $\sigma_1^2 + ... + \sigma_n^2 = \text{trace}(\Sigma)$. The other eigenvalues vanish in the limit, and we may thus consider the reduced model defined by the m equations

$$F^{(i)}(\hat{\vec{a}}) := \vec{v}_i^t \vec{X}(\hat{\vec{a}}) = \vec{v}_i^t \vec{y}, \qquad (13)$$

where we again define the functions $F^{(i)}(\hat{\vec{a}})$ in terms of $\vec{X}$, the vector of functions $(X_1(\hat{\vec{a}}),...,X_n(\hat{\vec{a}}))$, and m vectors $\vec{v}_i$, chosen from among the eigenvectors $\vec{v}_2,...,\vec{v}_{n-1}$ that span the noise-free subspace. If this system of m nonlinear equations (from which the noise has been eliminated) admits a unique solution, i.e. if the determinant of the Jacobian of the functions $F^{(i)}$ with respect to the elements of $\hat{\vec{a}}$ is nonvanishing, then the estimate $\hat{\vec{a}}$ is determined exactly. In this case, the range of the functions $F^{(i)}(\hat{\vec{a}})$ is an m-dimensional surface within the noise-free subspace.

4. CONCLUSIONS

In this paper, we have shown how the results of Ref. [11] may be generalized for nonlinear estimation. That is, we have shown that for linear models subjected to fully nonlinearly correlated noise, there may exist a nonlinear estimator that exactly yields the model parameters. The same may be true for nonlinear models subjected to fully linearly correlated noise.

Unlike the general linear discussion of Ref. [11], we have focused In Secs. 2, 3.1, 3.3 on the limiting case of fully correlated noise and, in particular, those cases in which perfect estimation occurs. Less than perfect nonlinear correlation has been discussed in Sec. 3.2. The limit of full correlation is extreme and unlikely to be found in most applications. However, the existence of the limit suggests that estimation may be very faithful in the approach to the limit if strong correlations (nonlinearly, small conditional entropy or small expected conditional variance) are taken into account. Additional concerns may arise for nonlinear estimation in the approach to the limit, such as whether estimates derived from the posterior are affected by bias (although the specific example of Sec. 3.2 actually involves an unbiased estimator.) However, the benefits to be gained from nonlinear estimation may outweigh these concerns if the magnitude of the correlation coefficients are not close to unity but the appropriate conditional entropy or expected conditional variance is small.

REFERENCES

1. C.S. Jones and J.M. Finn, *Nucl. Fusion* **46**, 335 (2006).
2. P.L. Canner, *Am. Stat.* **23**, 239 (1969).
3. G. D'Agostini, *Nucl. Instr. Meth. in Phys. Research* **346**, 306 (1994).
4. G. D'Agostini, *Bayesian Reasoning in Data Analysis* (World Scientific, 2003), p. 202.
5. T. M. Cover and J. A. Thomas, *Elements of Information Theory* (Wiley, 1991), p. 15ff.
6. K.M. Hanson, T. Kawano and P. Talou, *AIP Conf. Proc.* **769**, 304 (2004).
7. D.S. Sivia, *AIP Conf. Proc.* **707**, 303 (2004).
8. R.L. Winkler, *Manage. Sci.* **27**, 479 (1981).
9. R.T. Clemen and R.L. Winkler, *Oper. Res.* **33**, 427 (1985).
10. D.G. Morrison and D.C. Schmittlein, *Oper. Res.* **39**, 519 (1991).
11. C.S. Jones, J.M. Finn and N. Hengartner, submitted to *J. Multivariate Anal.* (2007).

APPLICATIONS

A Bayesian re-analysis of HD 11964: evidence for three planets

P. C. Gregory

*Physics and Astronomy Department, University of British Columbia,
Vancouver, BC V6T 1Z1*

Abstract. Astronomers searching for the small signals induced by planets inevitably face significant statistical challenges. Bayesian inference has the potential of improving the interpretation of existing observations, the planning of future observations and ultimately inferences concerning the overall population of planets. This paper illustrates how a re-analysis of published radial velocity data sets with a Bayesian multi-planet Kepler periodogram is providing strong evidence for additional planetary candidates. The periodogram is implemented with a Markov chain Monte Carlo (MCMC) algorithm that employs an automated adaptive control system. For HD 11964, the data has been re-analyzed using 1, 2, 3 and 4 planet models. The most probable model exhibits three periods of $38.02^{+0.06}_{-0.05}$, 360^{+4}_{-4} and 1924^{+44}_{-43} d, and eccentricities of $0.22^{+0.11}_{-0.22}$, $0.63^{+0.34}_{-0.17}$ and $0.05^{+0.03}_{-0.05}$, respectively. Assuming the three signals (each one consistent with a Keplerian orbit) are caused by planets, the corresponding limits on planetary mass ($M \sin i$) and semi-major axis are $(0.090^{+0.15}_{-0.14}M_J, 0.253^{+0.009}_{-0.009}\text{au})$, $(0.21^{+0.06}_{-0.07}M_J, 1.13^{+0.04}_{-0.04}\text{au})$, $(0.77^{+0.08}_{-0.08}M_J, 3.46^{+0.13}_{-0.13}\text{au})$, respectively.

Keywords: Extrasolar planets, Bayesian methods, Markov chain Monte Carlo, time series analysis, periodogram, HD 11964

INTRODUCTION

Improvements in precision radial velocity measurements and continued monitoring are permitting the detection of lower amplitude planetary signatures. One example of the fruits of this work is the detection of a super earth in the habital zone surrounding Gliese 581 by Udry et al. [19]. This and other remarkable successes on the part of the observers is motivating a significant effort to improve the statistical tools for analyzing radial velocity data, e.g., [7, 6, 5, 9, 3, 15, 14]. Much of the recent work has highlighted a Bayesian MCMC approach as a way to better understand parameter uncertainties and degeneracies and to compute model probabilities.

Gregory [8, 9, 10, 11, 12] presented a Bayesian MCMC algorithm that makes use of parallel tempering to efficiently explore the full range of a large model parameter space starting from a random location. The prior information insures that any periodic signal detected satisfies Kepler's laws and thus the algorithm functions as a Kepler periodogram [1]. In addition, the Bayesian MCMC algorithm provides full marginal parameters distributions for all the orbital elements that can be determined from radial velocity data. The samples from the parallel chains can also be used to compute the marginal

---

[1] Following on from the pioneering work on Bayesian periodograms by Jaynes [13] and Bretthorst [1].

likelihood for a given model [8] for use in computing the Bayes factor that is needed to compare models with different numbers of planets. The parallel tempering MCMC algorithm employed in this work includes an innovative two stage adaptive control system that automates the selection of efficient Gaussian parameter proposal distributions through an annealing operation. This feature coupled with parallel tempering makes it practical to attempt a blind search for multiple planets simultaneously. This was done for the analysis of the current data set and for the analysis of the HD 208487 reported earlier [11]. Of course, there is no guarantee that the algorithm will discover all modes in a multiple mode problem. More discussion of the control system is given below.

This paper illustrates how a Bayesian re-analysis of the 87 precision radial velocity measurements for HD 11964 published by Butler et al. [2] is providing strong evidence for additional planetary candidates. A more detailed account of many aspects of this analysis can be found in Gregory [12]. Butler et al. [2] reported the detection of a single planet with a period of 2110 ± 270d after removing a trend in the data.

RE-ANALYSIS OF HD 11964

The Bayesian multi-planet Kepler periodogram utilizes a parallel tempering Markov chain Monte Carlo algorithm which yields samples of the joint probability density distribution of the model parameters and permits a direct comparison of the probabilities of models with differing numbers of planets. In parallel tempering, multiple MCMC chains are run in parallel with each chain corresponding to a different temperature. We parameterize the temperature by its reciprocal, $\beta = 1/T$ which varies from zero to 1. The joint probability density distribution for the parameters ($\vec{X}$) of model M_i for a particular chain is given by

$$p(\vec{X}|D,M_i,I,\beta) = P(\vec{X}|M_i,I) \times p(D|\vec{X}M_i,I)^\beta \quad (1)$$

For parameter estimation purposes 12 chains ($\beta = \{0.05, 0.1, 0.15, 0.25, 0.35, 0.45, 0.55, 0.65, 0.70, 0.80, 0.90, 1.0\}$) were employed. At intervals, a pair of adjacent chains on the tempering ladder are chosen at random and a proposal made to swap their parameter states. The mean number of iterations between swap proposals was set $= 8$. A Monte Carlo acceptance rule determines the probability for the proposed swap to occur. This swap allows for an exchange of information across the population of parallel simulations. In the higher temperature simulations, radically different configurations can arise, whereas in higher β (lower temperature) states, a configuration is given the chance to refine itself. The final samples are drawn from the $\beta = 1$ chain, which corresponds to the desired target probability distribution. For $\beta \ll 1$, the distribution is much flatter. The choice of β values can be checked by computing the swap acceptance rate. When they are too far apart the swap rate drops to very low values. For the β values employed the swap rate was $\sim 50\%$. The lowest β value was chosen to achieve a broad sampling of the prior parameter range. A more common strategy is to propose a swap after each iteration and use fewer more widely spaced chains that achieve a swap rate of $\sim 25\%$. During the early development phase of the algorithm, this latter strategy appeared not to be quite as satisfactory, but we plan to re-visit this issue.

The samples from hotter simulations can also used to evaluate the marginal (global) likelihood needed for model selection, following Section 12.7 of Gregory [8] and Gregory [12]. Marginal likelihoods estimated in this way require many more parallel simulations. For HD 11964, 40 β levels were used spanning the range $\beta = 10^{-8}$ to 1.0.

For a one planet model the predicted radial velocity is given

$$v(t_i) = V + K[\cos\{\theta(t_i + \chi P) + \omega\} + e\cos\omega], \qquad (2)$$

and involves the 6 unknown parameters

$V = $ a constant velocity.
$K = $ velocity semi-amplitude.
$P = $ the orbital period.
$e = $ the orbital eccentricity.
$\omega = $ the longitude of periastron.
$\chi = $ the fraction of an orbit, prior to the start of data taking, that periastron occurred at. Thus, $\chi P = $ the number of days prior to $t_i = 0$ that the star was at periastron, for an orbital period of P days.
$\theta(t_i + \chi P) = $ the angle of the star in its orbit relative to periastron at time t_i measured with the focus of the orbital ellipse as the origin, also called the true anomaly.

We utilize this form of the equation because we obtain the dependence of θ on t_i by solving the conservation of angular momentum equation. Gregory [11] describes the advantage of this approach.

In a Bayesian analysis we need to specify a suitable prior for each parameter. The priors used in the current analysis are given in Table 1 of Gregory [12]. Following Gregory [9], all of the models considered in this paper incorporate an extra additive noise whose probability distribution is Gaussian with zero mean and standard deviation s. Marginalizing s has the desirable effect of treating anything in the data that can't be explained by the model and known measurement errors (e.g., stellar jitter) as noise, leading to conservative estimates of orbital parameters. Following Gregory [11], we employed a modified Jeffrey's prior for s with a knee, $s_0 = 1\text{m s}^{-1}$.

MCMC ADAPTIVE CONTROL SYSTEM

The process of choosing a set of useful proposal σ's when dealing with a large number of different parameters can be very time consuming. In parallel tempering MCMC, the problem is compounded because of the need for a separate set of proposal σ's for each chain. We have automated this process using an innovative two stage statistical control system (CS) in which the error signal is proportional to the difference between the current joint parameter acceptance rate and a target acceptance rate, typically 25% [17].

In the first stage, an initial set of proposal σ's ($\approx 10\%$ of the prior range for each parameter) are used for each chain. During the major cycles, the joint acceptance rate is measured based on the current proposal σ's. During the minor cycles, each proposal σ is separately perturbed to determine an approximate gradient in the acceptance rate for that parameter. The σ's are then jointly modified by a small increment in the direction of this gradient. This is done for each of the parallel simulations or chains as they are sometimes called. Proposals to swap parameter values between tempering levels are allowed during major cycles but not within minor cycles.

Although the first stage CS achieves the desired joint acceptance rate, it often happens that a subset of the proposal σ's are too small leading to an excessive autocorrelation in the MCMC iterations for these parameters. Part of the second stage CS corrects for this as follows.

The goal of the second stage is to achieve a set of proposal σ's that equalizes the MCMC acceptance rates when new parameter values are proposed separately and achieves the desired acceptance rate when they are proposed jointly. Let $\text{acc}(1)$ equal the acceptance for single parameter proposals and $\text{acc}(m)$ the desired acceptance rate (typically 0.25) for m parameter joint proposals. We make use of the following relationship between $\text{acc}(1)$ and $\text{acc}(m)$

$$\text{acc}(1) = \text{acc}(m)^{1/m^{k\alpha}}, \tag{3}$$

where α is given by

$$\alpha = 0.8061 - 1.1205 \times 10^{-2} m + 3.1233 \times 10^{-4} m^2 - 3.0357 \times 10^{-6} m^3, \tag{4}$$

and $k = 0.85$ is an empirical determined quantity. Equ. (3) was arrived at in the following way. An MCMC simulation was run on an m parameter multivariate normal target probability distribution with a mean for each parameter of zero and a covariance matrix equal to an identity matrix. New parameters were proposed using another multivariate normal with mean zero and a covariance matrix equal to γ^2 times the identity matrix. Thus, γ is the ratio of the proposal σ to the target distribution σ for each parameter. For each choice of γ in the range 0.4 to 1.1, the MCMC acceptance rate for joint parameter proposals was determined as a function of m. For each γ the acceptance rate was well fit by a function of the form

$$\text{acc}(m) = \text{acc}(1)^{m^{\alpha_\gamma}}, \tag{5}$$

and the value of $m = m_\gamma$ at which $\text{acc}(m) = 0.25$ was noted. For γ ranging from 0.4 to 1.1, m_γ varied from 34 to 5.4 and α_γ from 0.667 to 0.755. A cubic polynomial was fit to the $(m_\gamma, \alpha_\gamma)$ pairs yielding Equ. (3) without the k value. Of course, the actual Kepler target distribution is not a multivariate normal but with the inclusion of the empirically determined k value, Equ. (3) provides a useful scaling relationship.

The next step is to adjust the individual parameter proposal σ's to achieve an acceptance of $\text{acc}(1)$ given by Equ. (3). Using the proposal σ's obtained in the first stage CS, each parameter is allowed to vary one at a time during a minor cycle and the acceptance rate measured. Let $acc_1 =$ the measured acceptance rate when the proposal σ for the parameter in question was σ_1. We then update the proposal σ for this parameter to σ_2 according to

$$\sigma_2 = \sigma_1 \sqrt{\frac{(acc_1 + \Delta)}{\text{acc}(1)} \frac{(1 - \text{acc}(1))}{(1 - acc_1 + \Delta)}}, \tag{6}$$

where we use a $\Delta = 0.01$.

If $acc_1 = acc(1)$, then Equ. (6) leaves the proposal σ unchanged except for the small effect of the Δ term. The Δ term is there to handle the extremes of $acc_1 = 0$ and 1 gracefully. If $acc_1 = 1$, then we want to increase the proposal σ for that parameter. From Equ. (6) and $m = 17$ parameters, $\sigma_2/\sigma_1 = 6.7$. If on the other hand acc_1 is too low, say $acc_1 = 0.25$, we want to decrease the size of the proposal distribution. In this case, Equ. (6) yields $\sigma_2/\sigma_1 = 0.39$. Equ. (6) can be iterated for each parameter to achieve a final set of proposal σ's that achieve equal acceptance rates and a final joint acceptance rate of $acc(m)$. In practice we iterate Equ. (6) twice for each parameter. Other forms of Equ. (6) could also achieve the same goal in an iterative fashion.

In general, the burn-in period occurs within the span of the first stage CS, i.e., the significant peaks in the joint parameter probability distribution are found, and the second stage improves the choice of proposal σ's for the highest probability parameter set. Occasionally, a new higher (by a user specified threshold) target probability parameter set emerges after the first two stages of the CS are completed. The control system has the ability to detect this and re-activating the second stage. In this sense the CS is adaptive. If this happens the iteration corresponding to the end of the control system is reset. The useful MCMC simulation data is obtained after the first two stages of the CS are switched off.

Although inclusion of the control system may result in a somewhat longer effective burn-in period, there is a huge saving in time because it eliminates many trial runs to manually establish a suitable set of proposal σ's. When the σ's are large all the MCMC chains explore broadly the prior distribution and locate significant probability peaks in the joint parameter space. As the proposal σ's are refined these peaks are more efficiently explored, especially in the higher β chains. This annealing of the proposal σ's typically takes place over the first 5,000 to 150,000 (unthinned) iterations for one planet and first 5,000 to 300,000 iterations for three planets. This may seem like an excessive number of iterations but keep in mind that (a) we are dealing with sparse data sets that can have multiple, widely separated probability peaks, (b) the typical start location in parameter space is far from the target posterior peak, and (c) we want the MCMC to locate the most significant probability peak before finalizing the choice of proposal σ's. Within each chain, the CS corresponds to an annealing operation. Taken together with the parallel tempering, the two operations enhance the chances of detecting peaks in the target posterior compared to just implementing either one.

RESULTS

Panel (a) of Figure 1 shows the precision radial velocity data for HD 11964 from Butler et al. [2] who reported a single planet with $M \sin i = 0.61 \pm 0.10$ in a 2110 ± 270 day orbit with an eccentricity of 0.06 ± 0.17. Panels (b) and (c) show our best fitting three planet velocity curve and residuals. The initial starting location in period parameter space that was used for the Kepler periodogram ($P_1 = 10, P_2 = 500$ and $P_3 = 2300$d) was significantly different from the best location the algorithm found. Similar results were obtained with other different starting positions.

Table 1 gives our Bayesian three planet orbital parameter values and their marginal uncertainties. The parameter values given for our analysis are the median of the marginal

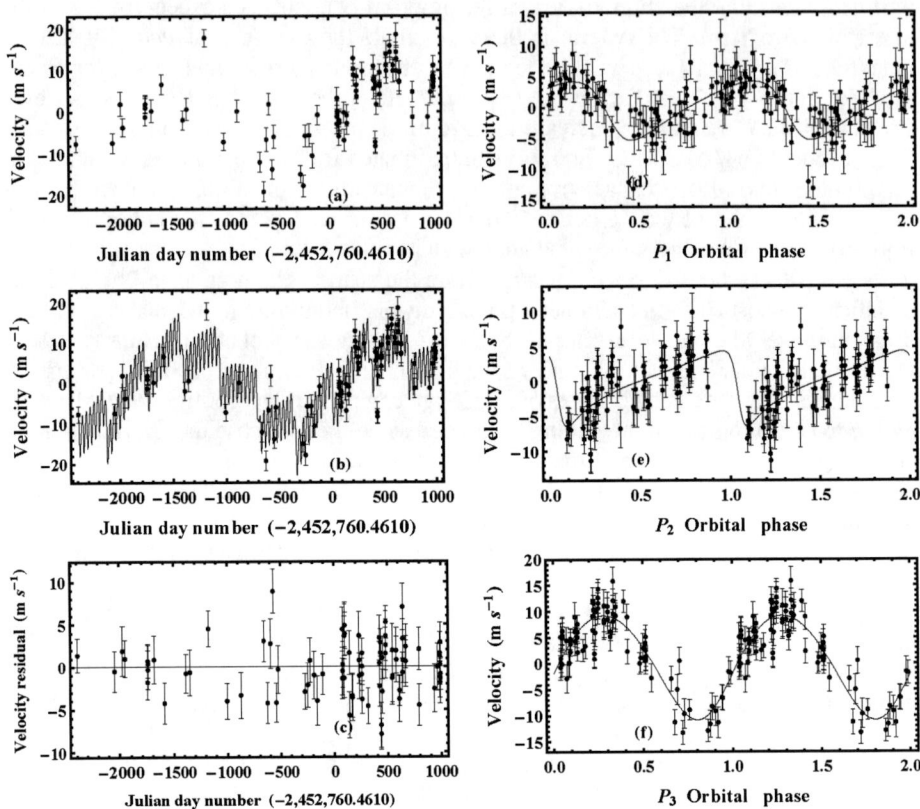

FIGURE 1. The data is shown in panel (a), the best fitting three planet ($P_1 = 38$, $P_2 = 357$, & $P_3 = 1928$ days) model versus time in (b), and the residuals in (c). Panel (d) shows the data plotted versus P_1 phase for two cycles, with the best fitting P_2 and P_3 orbits removed. Panel (e) shows the data plotted versus P_2 phase with the best fitting P_1 and P_3 orbits removed. Panel (f) shows the data plotted versus P_3 phase with the best fitting P_1 and P_2 orbits removed.

probability distribution for the parameter in question and the error bars identify the boundaries of the 68.3% marginal credible regions. The value immediately below in square brackets is the maximum *a posteriori* (MAP) value determined using the Nelder-Mead [16] downhill simplex method. Next to this, in parenthesis, is the MCMC parameter value corresponding to the largest joint posterior probability density, which is an approximation to the MAP value. The values derived for the semi-major axis and $M \sin i$, and their errors, are based on the assumed mass of the star $= 1.49 \pm 0.15$ $M_\odot$ [18]. Butler et al. [2] assumed a mass of $= 1.12$ $M_\odot$ but also quote Valenti & Fischer [18] as the reference. Panel (d) of Figure 1 shows the data with the best fitting P_2 and P_3 orbits subtracted, for two cycles of P_1 phase with the best fitting P_1 orbit overlaid. Panel

TABLE 1. Three planet model parameter estimates.

Parameter	planet 1	planet 2	planet 3
P (d)	$38.02^{+0.6}_{-0.5}$ [38.07](38.08)	360^{+4}_{-4} [357](356)	1925^{+44}_{-44} [1928](1914)
K (m s^{-1})	$4.3^{+0.7}_{-0.7}$ [4.8](5.4)	$6.1^{+3.0}_{-3.3}$ [5.4](5.8)	$9.7^{+0.8}_{-0.8}$ [10.0](9.3)
e	$0.23^{+.10}_{-.22}$ [0.31](0.34)	$0.63^{+.35}_{-.13}$ [0.63](0.60)	$0.05^{+.03}_{-.05}$ [0.09](0.07)
ω (deg)	123^{+41}_{-48} [111](108)	103^{+38}_{-34} [107](90)	195^{+80}_{-74} [205](208)
a (au)	$0.2527^{+.0085}_{-.0085}$ [0.253](0.253)	$1.132^{+.039}_{-.039}$ [1.124](1.123)	$3.46^{+.13}_{-.13}$ [3.46](3.45)
$M \sin i$ (M_J)	$0.090^{+.014}_{-.015}$ [0.098](0.098)	$0.213^{+.058}_{-.067}$ [0.191](0.209)	$0.77^{+.08}_{-.08}$ [0.795](0.735)
Periastron passage (JD - 2,440,000)	12737^{+6}_{-3} [12736](12736)	12397^{+35}_{-32} [12421](12370)	10535^{+401}_{-414} [10564](10598)

(e) shows the data plotted versus P_2 phase with the best fitting P_1 and P_3 orbits removed. Panel (f) shows the data plotted versus P_3 phase with the best fitting P_1 and P_2 orbits removed.

Following Gregory [11], the marginal likelihoods and their uncertainties for the 1,2,3 and 4 planet models were computed for the HD 11964 data set. Assuming that all the models are equally probable *a priori*, the three planet model was found to be ≥ 600 times more probable than the next most probable model which is a two planet model. A detailed comparison of the different marginal likelihood estimates is given in a Gregory [12].

For the most probable three planet model, the estimated stellar jitter based on the MAP value of the s parameter is 1.9m s^{-1}.

CONCLUSIONS

In this paper, we provided further details of the innovative adaptive control system employed by our automated parallel tempering MCMC nonlinear model fitting algorithm. This has been applied to the analysis of precision radial velocities for HD 11964 using 1, 2, 3 and 4 planet models. Assuming that all the models are equally probable *a priori*, the three planet model was found to be ≥ 600 times more probable than the next most probable model which is a two planet model. The most probable model exhibits three periods of $38.02^{+0.06}_{-0.05}$, 360^{+4}_{-4} and 1924^{+44}_{-43} d. The small difference (1.3σ) between the 360 day period and one year suggests that it might be worth investigating the barycentric correction for the HD 11964 data. Based on our three planet model results, the remaining unaccounted for stellar jitter parameter is ~ 1.9m s^{-1}.

This research was supported in part by grants from the Canadian Natural Sciences and Engineering Research Council at the University of British Columbia.

BIBLIOGRAPHY

REFERENCES

1. Bretthorst, G. L., 1988, Bayesian Spectrum Analysis and Parameter Estimation, New York: Springer-Verlag
2. Butler, R. P., Wright, J. T., Marcy, G. W., Fischer, D. A., Vogt, S. S., Tinney, C. G., Jones, H. R. A., Carter, B. D., Johnson, J. A., McCarthy, C., and Penny, A. J., 2006, ApJ, 646, 505
3. Cumming, A. 2004, MNRAS, 354, 1165
4. Clyde, M., in 'Statistical Challenges in Modern Astronomy IV,' G. J. Babu and E. D. Feigelson (eds.), San Francisco:Astron. Soc. Pacific (in press 2006)
5. Ford, E. B., 2005, AJ, 129, 1706
6. Ford, E. B., 2006, ApJ, 620, 481
7. Ford, E. B., & Gregory, P. C., in 'Statistical Challenges in Modern Astronomy IV', G. J. Babu and E. D. Feigelson, eds, San Francisco:Astron. Soc. Pacific (in press 2006) astro-ph 0608328
8. Gregory, P. C., 2005a, 'Bayesian Logical Data Analysis for the Physical Sciences: A Comparative approach with *Mathematica* Support', Cambridge University Press
9. Gregory, P. C., 2005b, ApJ, 631, 1198
10. Gregory, P. C.,2005c, in 'Bayesian Inference and Maximum Entropy Methods in Science and Engineering', San Jose, eds. A. E. Abbas, R. D. Morris, J. P.Castle, AIP Conference Proceedings, 803, 139
11. Gregory, P. C., 2007, MNRAS, 374, 1321
12. Gregory, P. C., MNRAS (in press 2007), astro-ph/0709.0970
13. Jaynes, E.T. (1987), 'Bayesian Spectrum & Chirp Analysis,' in *Maximum Entropy and Bayesian Spectral Analysis and Estimation Problems*, ed. C.R. Smith and G.L. Erickson, D. Reidel, Dordrecht, p. 1.
14. Loredo, T., 2003, in 'Bayesian Inference And Maximum Entropy Methods in Science and Engineering: 23rd International Workshop', G.J. Erickson & Y. Zhai, eds, AIP Conf. Proc. 707, 330 (astro-ph/0409386).
15. Loredo, T. L. and Chernoff, D., 2003, in 'Statistical Challenges in Modern Astronomy III', E. D. Feigelson and G. J. Babu, eds , p. 57
16. Nelder, J. A. and Mead, R., Comput. J. 7, 308
17. Roberts, G. O., Gelman, A. and Gilks, W. R., 1997, Annals of Applied Probability, 7, 110
18. Valenti, J. A., & Fischer, D. A. 2005, ApJS, 159, 141
19. Udry, S., Bonfils, X., Delfosse, X., Forveille, T., Mayor, M., Perrier, C., Bouchy, F., Lovis, C., Pepe, F., Queloz, D., and Bertaux, J.-L., A&A in press 2007.

Bayesian Analysis of RR Lyrae Luminosities and Kinematics

Thomas R. Jefferys*, Thomas G. Barnes†, Andrei Dambis** and William H. Jefferys‡

*University of Texas at Austin
†Department of Astronomy, University of Texas at Austin
**Sternberg Astronomical Institute, Universitetskii pr. 13, Moscow, 119992 Russia
‡University of Texas at Austin, and University of Vermont

Abstract. We are using a hierarchical Bayes model to analyze the distances, luminosities, and kinematics of RR Lyrae stars. Our model relates these characteristics to the raw data of proper motions, radial velocities, apparent luminosities and metallicities of each star. A combination of Gibbs and Metropolis-Hastings sampling, using latent variables for the actual velocity and luminosity of each star, is used to draw a sample from the full posterior distribution of these variables, with consideration to identifiability and the properness of the hierarchical model, and draw inferences on the quantities of interest in the usual way. We have applied our model to the large HIPPARCOS database, and we have attempted to include metallicity and period in our model, which has not been done previously.

Keywords: RR Lyrae, magnitude, luminosity, HIPPARCOS, Kinematics, Hierarchical Bayes, MCMC
PACS: 02.70.Uu, 98.10.+z, 97.30.Kn, 97.10.Ri

DESCRIPTION OF THE PROBLEM

RR Lyrae stars are a class of pulsating variable stars. They are readily recognizable from their periods (0.6 ± 0.2 days) and characteristic light curves [1]. These stars have two desirable properties that make them very important in astronomy. The first property is that they are fairly bright, about forty times the brightness of the Sun. Thus, an RR Lyrae star can be seen to fair distances in the Galaxy. The other important property is that the intrinsic visual-band luminosities of RR Lyraes are nearly constant from star to star. This second property is known from studies of RR Lyraes in clusters, where all the stars in the cluster are at the same distance.

These properties make these stars useful as "standard candles" for estimating distances in the universe: We observe an RR Lyrae's apparent luminosity, and from the intrinsic luminosity of the class, use the fact that the apparent luminosity falls off as the square of the distance to work out the distance to the star. Therefore, if an RR Lyrae is embedded in some structure, by finding the distance to it, we then know the distance to the structure in which it is embedded.

RR Lyrae stars have two other properties useful for our study. They have distinctive kinematics, a statistical description of their motions in the galaxy, that imply that they formed in an epoch in galactic history before star formation was confined to the galactic disk. Secondly, they formed over an interval during which the abundances of elements heavier than helium, the *metallicity,* changed from very low to near solar. Third, they pulsate with a period lasting about a day.

Our goals, therefore, are to determine the characteristic absolute magnitude (proportional to the negative logarithm of luminosity) of these stars as a group, investigate any variation of absolute magnitude with metallicity and period, investigate the *cosmic scatter* of the magnitude (i.e., variation about the mean unexplained by the covariates), and investigate the kinematics of the stars as a group.

Overview of our Method

Our raw data for each star include the proper motions $\vec{\mu}$ (vector of angular motion/time unit of the motions of the stars across the sky, in seconds of arc per century), the radial velocity ρ of the motions towards or away from the Sun, in kilometers/second, obtained by observing the doppler shift of spectral lines, and the apparent magnitudes m (proportional to the negative logarithm of the flux) of the star, which is measured with negligible error. For each star, we also have two components of error on the proper motion, σ_μ, and error on the radial velocity, σ_ρ.

We could infer the the distance to a star from the proper motion if we knew the transverse velocity. The proper motions are related to the transverse velocities (in km/sec) by multiplying the former by the distance s to the star:

$$s\vec{\mu} \propto \vec{V}^\perp. \tag{1}$$

Similarly, we could infer the distance to that star from the flux and the inverse square law if we knew the luminosity. The inverse square law, mapped to magnitudes is

$$s = 10^{0.2(m-M+5)}. \tag{2}$$

Extinction, the attenuation of light due to dust between us and the star, is included as a pre-analysis adjustment to m for each star.

We have neither luminosity nor transverse velocity for the stars. However, if we assume that the proper motions and radial velocities are characterized by the same kinematical parameters, we can statistically infer the distances and the magnitudes luminosities of the RR Lyraes through through (2), since we probe these stars in various directions.

THE HIERARCHICAL MODEL

If for a single star $\vec{\mu}^o = (\mu_\alpha^o, \mu_\delta^o)$ are the two components of the observed proper motion vector (in the plane of the sky), and ρ^o is the observed radial velocity, we assume

$$\begin{aligned} \mu_\alpha^o \mid V_\alpha^\perp, s, \sigma_{\mu_\alpha} &\sim N(V_\alpha^\perp/s, \sigma_{\mu_\alpha}^2) \\ \mu_\delta^o \mid V_\delta^\perp, s, \sigma_{\mu_\delta} &\sim N(\vec{V}_\delta^\perp/s, \sigma_{\mu_\delta}^2) \end{aligned} \tag{3}$$

$$\rho^o \mid V^{\|}, \sigma_\rho \sim N(V^{\|}, \sigma_\rho^2),$$

where $V_\alpha^\perp, V_\delta^\perp$ are the α and δ components of the transverse space velocity, $\vec{V}^\perp$, respectively. The variances are measured at the telescope and assumed known with negligible uncertainty. The variables without 'o' superscripts are the "true" values. The joint distribution over all stars for these three components is our likelihood.

The likelihood defined above constrains the hierarchical model proper, which makes predictions on the proper motions, radial velocities and fluxes of the stars from a set of latent variables. We define the individual absolute magnitude for the star i as $M_i = M + U_i$, where M is the characteristic absolute magnitude for RR Lyrae stars, and U_i is star i's individual deviation from M. Then,

$$s_i = s(M, U_i) = 10^{0.2(m_i - M - U_i + 5)}. \tag{4}$$

We note that the s_i's are not new parameters; they have no distribution that is not completely inherited from M and the U_i's. The reviewer suggested that it may be possible to include prior information on the space distribution of s_i in the anaylsis [2]. We are considering this for future research.

We choose a flat prior on M (we have also tried a somewhat informative prior based on known data, but the results were not significantly different). Evidence from other sources (e.g., studies of RR Lyrae stars in clusters) indicates a cosmic scatter of about 0.15 magnitudes in M_i. Thus a prior on U_i of the form

$$U_i \mid \sigma_U \sim N(0, \sigma_U^2 = (0.15)^2) \tag{5}$$

is appropriate; we explored the effect of this choice (see below).

The priors on the true space velocities, $\vec{V}_i$, are obtained by assuming that the velocities of the individual stars are drawn from a three-dimensional multivariate normal distribution with mean $\vec{V}_\odot$ (the solar motion) and covariance matrix W (the *velocity ellipsoid*):

$$\vec{V}_i \mid \vec{V}_\odot, W \sim N(\vec{V}_\odot, W). \tag{6}$$

We choose a flat prior on $\vec{V}_\odot$, and to avoid an improper posterior distribution, a "hierarchical independence Jeffreys prior" on W [3], which for a three-dimensional distribution implies

$$\pi(W) \propto \mid I + W \mid^{-2}. \tag{7}$$

SAMPLING STRATEGY

We can use Gibbs sampling to sample on $W, \vec{V}_i$ and $\vec{V}_\odot$. Sampling on $\vec{V}_i$ and $\vec{V}_\odot$ are straightforward normal distributions with appropriate parameters. Sampling on W is more tricky, as the full conditional is not a closed form. However, we can sample from the full conditional using an importance sampler, using the following proposal:

$$W^* \mid \{\vec{V}_i\}, \vec{V}_\odot \sim \text{InverseWishart}(T, df = N) \tag{8}$$

where $T = \sum(\vec{V}_i - \vec{V}_\odot)(\vec{V}_i - \vec{V}_\odot)'$, and an importance ratio $P = (|W^*|/|I+W^*|)^2$. Fortunately, in this case the dimension of W is small, N is large, and the proposal $|W^*|$ is typically large, so the acceptance probability is high. (For a more thorough treatment of this problem, see Berger et al. [3].)

In our model, the conditional distributions of M and U_i come from the following equality,

$$p_N(\vec{\xi}_i^o \mid \vec{\xi}_i, \Lambda_i) = s_i^2 p_N(\vec{V}_i^o \mid \vec{V}_i, S_i), \qquad (9)$$

where $p_N(\cdot)$ is a multivariate normal pdf, $\vec{\xi}_i^o = (\mu_{\alpha,i}^o, \mu_{\delta,i}^o, \rho_i^o)$ and $\vec{\xi}_i = (\mu_{\alpha,i}, \mu_{\delta,i}, V_i^{\parallel})$ are the observed and true state vector for the star i, $\vec{V}_i^o = s_i \vec{\mu}_i^o + \rho_i^o \hat{n}$ is the "observed" space velocity, $\Lambda_i = X_i + E_i$ is the observational covariance matrix and $S_i = s_i^2 X_i + E_i$, both formed from X_i, the covariance matrix for the observed proper motion vector $\vec{\mu}_i^o$, and E_i, the covariance matrix for the observed radial velocity ρ_i^o. The derivation is left as an exercise for the reader, where the identities $|\Lambda_i| = s_i^{-2} |S_i|$ and $S_i^{-1} = s_i^2 X_i^+ + E_i^+$ will be useful.[1] Both identities hold because X_i and E_i are not full rank and live in orthogonal subspaces.

The right-hand side of equation (9) give the conditional distribution of both M and U_i, products over all i's in the case of M. This conditional must be sampled with random-walk Metropolis-Hastings for both M and U_i.

For sampling M, we propose $M^* \sim N(M, w)$, with an appropriate standard deviation parameter w adjusted for good mixing.

For sampling U_i under the informative prior (5), our proposal for a U_i^* is a location-scale family of a t distribution centered on the origin with an appropriate choice of degrees of freedom and scale adjusted for good mixing. The conditional (9), because of independence in U_i, may be sampled in parallel.

RESULTS

Our key astrophysically interesting result is the characteristic absolute magnitude of RR Lyrae stars, M. In addition, we are also interested in the solar motion, $\vec{V}_\odot$, and the velocity ellipsoid of the galactocentric orbits of the observed RR Lyraes, W.

Hawley et al. [4] used a maximum likelihood technique to study this problem using proper motion data from the *Annals of the Shanghai Observatory* and radial velocity, apparent magnitude, and reddening data from Hemmingway and various sources (all reproduced in their paper, and referred to henceforth as "SHANG"). We compare this to our own results using the same data to demonstrate the improvement using our method, and to our results using the HIPPARCOS data set and improved velocity data [5] from our colleague Andrei Dambis (henceforth, "HIPP").

Hawley et al. ran cases with a number of subsets of stars, broken down by various criteria. We looked only at the "ab" stars (normal pulsators, $N_{ab} = 143$) and the "c" stars (overtone pulsators, $N_c = 17$), and compared the weighted average of these two groups

---

[1] A^+ denotes the Moore-Penrose matrix inverse of A.

in Hawley's study with our own.[2] Table 1 includes both results for the "ab" and "c" stars analyzed as segregated groups shown, and a weighted average of the two results, but also our own analysis of the HIPP data set as an integrated group ($N_{HIPP} = 349$). Our results are posterior marginal means and standard deviations.

TABLE 1. Characteristic Absolute Magnitude, M, by study

	Data Set	Star Group	M
This study	SHANG	"ab" stars	0.71 ± 0.11
		"c" stars	0.72 ± 0.26
		weighted average	0.71 ± 0.10
	HIPP	all stars	0.75 ± 0.07
Hawley et al.	SHANG	"ab" stars	0.68 ± 0.14
		"c" stars	1.01 ± 0.38
		weighted average	0.72 ± 0.13

The sample of "ab" stars agrees within the errors with the analysis of Hawley et al. [4] However, the "c" sample gives a discrepant value of M. Their value is anomalously high, compared to our study and to the general consensus on the value of M. The best direct measurement of the distance of an RR Lyrae star, by Hubble Space Telescope (Benedict et al. [6]), gives $M = 0.61 \pm 0.10$ magnitudes. The results of Skillen et al. [7], using the Surface Brightness method (similar to the method used by Jefferys et al. [8] and reported at ISBA 2000 [8]) give $M = 0.65 \pm 0.10$ magnitudes.

Other results of interest are the solar motion $\vec{V}_\odot$ relative to the sample and the velocity ellipsoid of the galactocentric orbits W.

The solar motion vector informs us of the relative motion of the Sun through the ensemble of RR Lyrae stars in the sample (units are km/sec). Marginal summaries appear in Table 2, with coordinates in galactic cylindrical coordinates (ϖ, θ, z); that is, cylindrical coordinates with the origin at the Galactic center.

TABLE 2. Solar motion, $\vec{V}_\odot$, by component

	Data Set	Star Group	V_ϖ	V_θ	V_z
This study	SHANG	"ab" stars	-12 ± 11	-136 ± 9	-8 ± 7
		"c" stars	-25 ± 24	-113 ± 26	-2 ± 11
		weighted average	-15 ± 10	-134 ± 9	-6 ± 5
	HIPP	all stars	-22.67 ± 7.97	-141 ± 5.58	-11.34 ± 3.96
Hawley et al.	SHANG	"ab" stars	-10 ± 13	-155 ± 12	-9 ± 8
		"c" stars	-26 ± 25	-124 ± 25	-6 ± 13
		weighted average	-13 ± 12	-149 ± 10	-8 ± 6

We believe this to be a significant result: using the HIPP data set, we have detected solar motion along both the V_ϖ and the V_z axes.

---

[2] The data of Hawley et al. had some incorrect apparent magnitudes. Reanalysis with corrected data shows that the value of M should be decreased by approximately 0.08 magnitudes. This correction has been applied to Hawley's figures. We have made similar corrections to our own result using SHANG—HIPP needed no such corrections.

The velocity ellipsoid tells us how the galactocentric orbits of the RR Lyrae stars are oriented in a statistical sense, as described by the covariance matrix of the velocities. In galactic cylindrical coordinates (ϖ, θ, z) the matrix is believed roughly diagonal, with the on-diagonal dispersions decreasing from ϖ to θ to z. (Units are km/sec).

TABLE 3. Velocity ellipsoid, W, for normal and overtone pulsators

	Data Set	Star group	σ_ϖ	σ_θ	σ_z
This study	SHANG	"ab" stars	132 ± 9	108 ± 7	75 ± 5
		"c" stars	97 ± 21	101 ± 21	44 ± 10
		weighted average	127 ± 8	107 ± 7	69 ± 4
	HIPP	all stars	145.60 ± 5.73	100.16 ± 4.04	71.10 ± 2.96
Hawley et al.	SHANG	"ab" stars	150 ± 59	120 ± 47	87 ± 33
	SHANG	"c" stars	101 ± 57	71	51

There is a significant difference in the estimated standard deviations between this study and Hawley et al. [4]. Hawley et al. have noted that their use of simplex optimization in the maximum likelihood method made estimating the uncertainties difficult and somewhat arbitrary [9].

Because of the small sample size, in their analysis of these data Hawley et al. [4] did not solve for the full covariance matrix of the velocity ellipsoid for the "c" stars, the overtone pulsator RR Lyraes. Instead, they set the off-diagonal terms to zero, and fixed the ratios for the on-diagonal terms to values given by the reduction of the larger "ab" data. This is why the σ_θ and σ_z do not have standard deviations attached, and no weighted average appears for these results.

DISCUSSION

Sampling on U_i

We find the marginal distribution of the U_i to be totally dominated by the prior; as we make changes to the prior, the samples drawn from the posterior change in step—for instance, varying σ_U in (5), earlier on, a sample drawn from the posterior has a variance of around this σ_U. From this evidence we conclude that the individual U_i's (and consequently, the cosmic scatter) are unidentified in this model.

This would suggest simply leaving U_i out of the model altogether, sampling only on M, which would result in a considerable increase in the efficiency of the sampler: We would have only one Metropolis-Hastings step, instead of two. However, J. Berger raised concerns about leaving out the sampling on these variables [10]. Leaving these latent variables in allows us to better estimate the variance on M. We find that omitting this set of latent variables did not seem to alter the resulting means significantly, however.

Unidentifiability of Metallicity and Log-Period

We had hoped to extend this investigation into the effects of metallicity and period on the individual luminosities of RR Lyraes. The absolute magnitude of a given RR Lyrae star is known to depend on the star's metallicity and period. A commonly used phenomenological relation is:

$$M_i = M + a[Fe/H]_i + b \log P_i, \tag{10}$$

where M_i is the absolute magnitude of a certain RR Lyrae star, $[Fe/H]_i$ is a logarithmic measure of its metallicity (with respect to solar), P_i is its period, and M is the characteristic absolute magnitude of RR Lyrae stars. The constants a and b are coefficients relating metallicity and period to deviations from the characteristic absolute magnitude.

We find that actually including metallicity effects in the model produces an unidentified model, as any change in the coefficient a is exactly offset by an appropriate change in the individual velocities of the stars, and so the model is unidentified with respect to a, and likely with respect to b as well, although we haven't tested this. Since modeling individual velocities is at the core of our model, we suspect that this is a fundamental limitation of our model.

ACKNOWLEDGMENTS

This material is based in part upon work by Thomas G. Barnes III while serving at the National Science Foundation. Any opinions, findings, and conclusions or recommendations expressed in this material are those of the authors and do not necessarily reflect the views of the National Science Foundation.

REFERENCES

1. H. A. Smith, *RR Lyrae Stars*, Cambridge University Press; Cambridge: New York, NY, 1995, chap. 1,2,4, ISBN 0-521-32180-8.
2. T. Loredo, *Private communication* (2007).
3. J. O. Berger, W. Strawderman, and D. Tang, *Annals of Statistics* **33**, 606–646 (2005).
4. S. L. Hawley, W. H. Jefferys, T. G. Barnes III, and W. Lai, *The Astrophysical Journal* **302**, 626–631 (1986).
5. T. C. Beers, M. Chiba, Y. Yoshii, I. Platais, R. B. Hanson, B. Fuchs, and S. Rossi, *The Astronomical Journal* **119**, 2866–2881 (2000).
6. G. F. Benedict, B. E. McArthur, L. W. Fredrick, T. E. Harrison, J. Lee, C. L. Slesnick, J. Rhee, R. J. Patterson, E. Nelan, W. H. Jefferys, W. van Altena, P. J. Shelus, O. G. Franz, L. H. Wasserman, P. D. Hemenway, R. L. Duncombe, D. Story, A. L. Whipple, and A. J. Bradley, *The Astronomical Journal* **123**, 473–484 (2002).
7. I. Skillen, J. Fernley, R. Stobie, and R. Jameson, *Montly Notices of the Royal Astronomical Society* **265**, 301 (1993).
8. W. Jefferys, T. Barnes III, R. Rodriguez, J. Berger, and P. Mueller, "Model selection for Cepheid star oscillations," in *Bayesian Methods With Applications to Science, Policy and Official Statistics*, edited by E. George, Luxemborg: Office for Official Publications of the European Communities, 2001, pp. 243–252.
9. S. L. Hawley, *Private communication* (2005).
10. J. O. Berger, *Private communication* (2005).

Estimating Background Spectra

M. K. Tse[a], J. Choinsky[a,b], D. F. Carbon[c], and K. H. Knuth[a,d]

a. Univ. at Albany, Dept of Physics, Albany NY USA
b. Univ. at Albany, Dept of Computer Science, Albany NY USA
c. NASA Ames Research Center, NAS, Moffett Field CA USA
d. Univ. at Albany, Dept of Informatics, Albany NY USA
http://knuthlab.rit.albany.edu/

Abstract. All measurements consist of a mixture of the signal of interest and additional signals called the background. Here we focus on the problem of measuring infrared spectra emitted by interstellar clouds. The signals of interest are infrared emissions from polycyclic aromatic hydrocarbons (PAHs), which are a class of complex organic molecules. The PAH emissions are characterized by emission bands near 3.3, 6.2, 7.7, 8.6, 11.2, and 15-20 microns. The background consists of a host of associated spectral signals which, in the simplest case, can include emissions from multiple Planck blackbodies as well as broadband and narrowband emissions. To analyze the PAH spectra we must accurately assess this background. To do this, we have developed a Bayesian algorithm based on nested sampling (Skilling 2005, Sivia & Skilling 2006). The spectral model consists of a mixture of Planck functions and Gaussians. We demonstrate this algorithm on both synthetic data and infrared spectra recorded from interstellar clouds. The result shows that the algorithm can accurately identify and remove simple backgrounds. In future work, we plan to incorporate mixtures of PAH spectra and more complex models for the background so that the algorithm will simultaneously estimate both the signals of interest and the background.

Keywords: background, spectrum, nested sampling, MCMC, Planck blackbody, mixture of Gaussians, astrophysics, astrobiology
PACS: 98.70.Lt, 98.70.Vc

1. INTRODUCTION

In astrophysics, all observed spectra consist of a mixture of the signal of interest and additional signals collectively called the background. The background can be decomposed into emission from unknown sources and known sources. Here we focus on the problem of analyzing infrared spectra emitted by interstellar clouds (Figure 1). Infrared spectra in regions of star formation have strong emission from a class of benzene-based molecules known as Polycyclic Aromatic Hydrocarbons (PAHs). The observed spectra are the combined emission of numerous PAH species, both neutral and ionized, in addition to emission from atomic species plus dust emission and absorption at multiple temperatures. While characterizing these factors promises to enhance the level of understanding of star-forming regions, identifying individual PAHs in a spectrum that consists of a mixture of a large number of PAH species is a difficult problem in itself.

The definition of background depends on both the application and the interests of the researcher; one person's background is another person's signal. In this paper, the

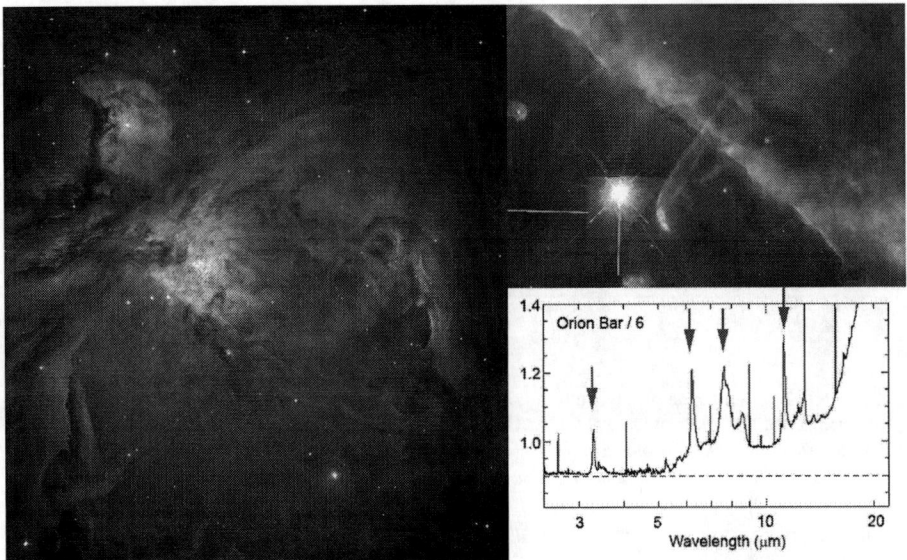

FIGURE 1. The Orion Nebula (left) is an active star-forming region. The Orion Bar magnified on the right is where PAHs in the nebula are being excited by ultraviolet light from young stars. The intensity of the spectral features are shown in the lower right (HST/ACS Mosaic from STScI).

signals of interest are the PAH emissions, which are characterized by bands near 3.3, 6.2, 7.7, 8.6, 11.2 and 15-20 microns. There are thousands of PAH species, each with their unique emission spectrum (Figure 2 shows some examples of PAHs and their spectra). However, there are additional signals present in the recorded infrared spectrum. There are dust clouds along the line-of-sight that both radiate and absorb radiation. Each of the clouds has distinct physical conditions. In addition, these are unknown sources which cannot be described phenomenologically. For this reason, estimating the background is a difficult problem.

In this paper, Nested Sampling (Skilling 2005; Sivia & Skilling 2007) will be used to characterize the background spectrum. We will demonstrate how the algorithm can isolate PAHs from the other sources, such as dust emission. Furthermore, we will use the evidence calculations to evaluate whether the contributions of what are interpreted as unknown sources are warranted by the data.

2. MODELING SPECTRA

2.1 Blackbody Radiation

Interstellar clouds of dust in star-forming regions are heated by the radiation emitted by young stars (Figure 1). The particles in the clouds re-radiate this energy as infrared radiation with a spectrum dictated by the physical conditions in the cloud. In the very

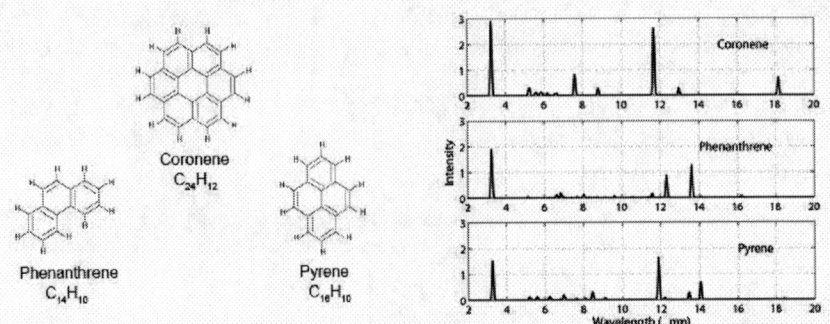

FIGURE 2. Three examples of common Polycyclic Aromatic Hydrocarbons (PAH). The infrared emissions of these molecules explain the UV-driven IR emissions in star-forming regions. Infrared spectra for three PAH species are shown on the right (data from the NASA Ames PAH library). While each PAH has unique spectral features, note that they also share commonalities.

simplest approximation, the cloud radiation can be represented by a Planck blackbody function at some temperature T:

$$Planck(\lambda, T_k) = \sqrt{\frac{\lambda_{max}}{\lambda}} \frac{\exp(\frac{hc}{\lambda_{max} kT}) - 1}{\exp(\frac{hc}{\lambda kT}) - 1} \qquad (1)$$

where k is Boltzmann's constant, h is Planck's constant, c is the speed of light, and λ_{max} is the wavelength where the blackbody spectral energy peaks.

We employ three distinct types of source spectra models: a dictionary of atomic and PAH spectra, blackbody radiators describing dust radiation and a 'non-parametric' mixture of Gaussians describing the unknown source radiation. To isolate the signals emitted from PAHs, the background must be first identified and separated from the linear superposition of the various PAH species. The mixture of Gaussians and the blackbody radiators are used to model the background.

2.2 Modeling Unknown Signals

The recorded infrared spectra usually contain a certain level of broad spectral features. These features cannot be attributed to any known source. To model these unknown sources, we utilize a mixture of Gaussians. Since unknown spectral sources are not a focus of this project, the mixture of Gaussians model is sufficient to represent these background features to a first approximation.

$$\sum_i N(\mu_i, \sigma_i) = \sum_i \frac{1}{\sqrt{2\pi\sigma_i^2}} \exp(\frac{x - \mu_i}{2\sigma_i^2}) \qquad (2)$$

where $\sum_i N(\mu_i,\sigma_i)$ is the mixture of Gaussian, i is the index of Gaussians, μ_i is the mean of the i-th Gaussian and σ_i is the standard deviation of the i-th Gaussian.

2.3 PAH Spectra

A Polycyclic Aromatic Hydrocarbon (PAH) is an assembly of hexagonally-shaped carbon rings of the simplest aromatic molecule. Three examples of PAHs are shown in Figure 2. Each PAH has distinct spectral features, which arise from the vibrational states unique to the structure of each particular molecule. Ionization and element substitution can modify PAH wavelengths and relative intensities. In order to identify the characterized PAH emissions, our partners at NASA Ames Research Center have compiled a library of over a thousand PAH spectra based on laboratory measurements and quantum mechanical simulations. Figure 2 (right) shows spectra for three PAH species from the NASA Ames PAH library.

2.4 Combined Spectral Model

The infrared spectra we are working with are the combined emission of the PAH molecular species, the Planck radiation and unknown sources.

$$F(\lambda) = \sum_k Planck(\lambda,T_k) + \sum_i N(\lambda;\mu_i,\sigma_i) + \sum_j PAH_j(\lambda) \qquad (3)$$

where $F(\lambda)$ is the modeled spectral flux at wavelength λ. The first term represents a sum of blackbodies, the second term is a mixture of Gaussians representing the unknown sources, and the third term represents the PAH contributions. With this parameterized model of the observed spectra, we can begin to make inferences about the contributions from blackbody radiation, unknown sources, or PAH spectra. Since we have constrained the σ_i to have values above some minimum value substantially larger than the widths of the atomic features, we have not explicitly modeled the atomic emission in the current study. As a consequence of this approximation, the atomic features are treated here as though they were "PAHs". In later work, we will add a mixture of very sharp Gaussians and a catalog of wavelengths for known atomic emission species.

3. NESTED SAMPLING

We employ a new technique called Nested Sampling (Skilling 2005, Sivia & Skilling 2006). The main goal of Nested Sampling is to numerically integrate the posterior and obtain the evidence. In future work it is expected that this will be invaluable as we test multiple spectral theories. Nested sampling works by sampling

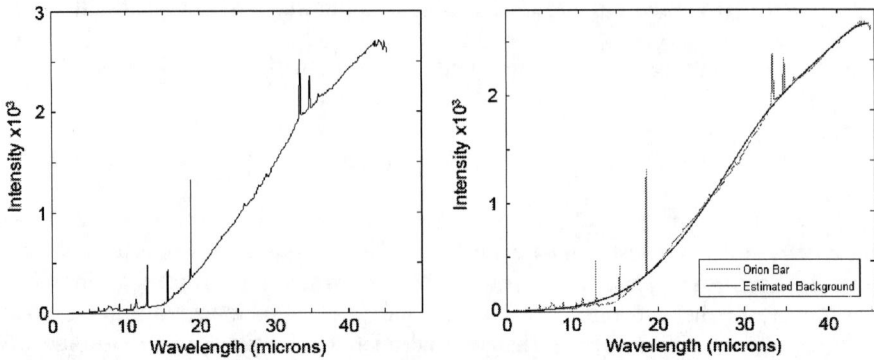

FIGURE 3. (Left) Measured Infrared spectra from the Orion Nebula. The unit of y-axis is intensity of the signal and the unit of x-axis is the wavelength (microns). (Right) The estimated background after running the Nested Sampling algorithm with 200 objects and 10000 iterations. The resulting curve describes the background well, except for a broad dip centered at 15 microns and systematic overshooting, then undershooting, between roughly 20 and 35 microns

from the prior probability within a likelihood constraint. The likelihood constraint defines an implicit boundary over which the algorithm shall not cross.

The algorithm maintains a set of samples, each of which describes a hypothesized spectral model. The likelihood of each sample is computed and the sample with the least likelihood is flagged and discarded. To maintain the set of samples, a new sample is generated, by duplicating one of the valid samples, and varied by changing its parameter values slightly several times until it is decorrelated from the initial sample. Each sample is assigned a weight based on the product of the estimated prior probability shrinkage ratio and the likelihood. The end result is a set of samples with weights that denote their contribution to the integral of the posterior probability.

We work with uniform priors, so that the algorithm samples uniformly from the hypothesis space. The likelihood of each sample is calculated by using the Student t-distribution, which is proportional to the sum of squared differences between observed spectrum $\Phi(\lambda)$ and model spectrum $F(\lambda)$

$$\log L = -(N/2)\log\left(\sum_N (\Phi(\lambda) - F(\lambda))^2\right) \qquad (4)$$

where N is the number of measured spectral fluxes.

There are several advantages to nested sampling. First, the algorithm naturally favors a minimal number of Gaussians and Planck blackbodies, which serves to prevent over-fitting the observed spectrum. More importantly, from the list of samples, one can compute mean quantities and uncertainties of the model parameter values as well as the evidence of the model itself.

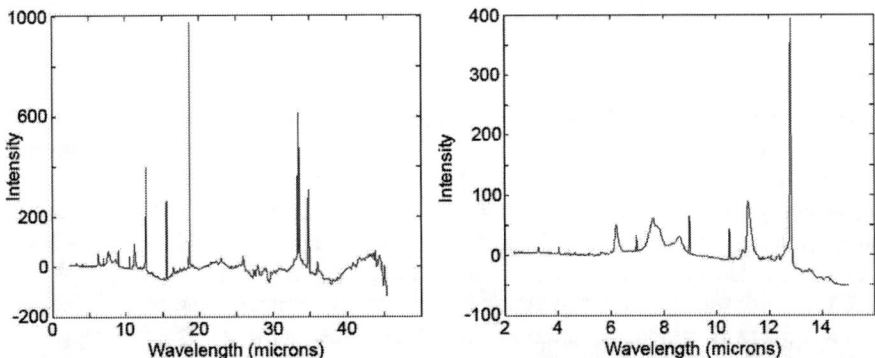

FIGURE 4. The left panel displays the signal of interest obtained from the Orion Nebula after removing the background. We observe several remaining difficulties due to the negative spectral intensities. The right panel focuses in on the PAH spectra in the range of 2.3-15 microns.

4. RESULTS

We illustrate the technique using sample infrared emissions recorded from the Orion Bar which is within the Orion Nebula (Figure 1). The data used here were collected by the Infrared Space Observatory and processed by Sloan et al. (2003). The nested sampling algorithm worked with 200 objects and ran for 10,000 iterations. The results show (Figures 3 and 4) that nested sampling can begin to isolate the background from the PAH plus atomic emission. However, there are some difficulties. For instance, the dip at 15 microns and the structure beyond 38 microns suggest that the actual dust emission is more complex than can be modeled by Gaussians and blackbody functions. In addition, the current procedure tends to give excessive weight to errors in the fit at the longer wavelengths. This leads to the unfortunate outcome that the background is not reasonably defined in the shorter wavelength regions where important PAH emission occurs. This can be readily seen in right panel of Figure 4 where we show an expanded view of the 2.3 – 15 micron region from Figure 4. Clearly work needs to be done to better fit the background in this crucial region. We are currently working on refining the fitting procedure so as to give more weight to the PAH regions.

Our modeling approach does provide additional information. For example, the Planck term indicates that there are possibly (36% chance) two blackbody radiators: one with a temperature of 61.043 +/- 0.004K and the other in the range of about 20K. This type of additional quantitative information may prove valuable in understanding star-forming regions.

5. CONCLUSION

Separating a mixed spectrum is a formidable problem, especially when the possibility exists for so many chemical species. This problem is made even more difficult by the fact that we can only record one spectrum, and that there may be multiple clouds each with different physical conditions along the line-of-sight.

Despite the difficulty of the task, we aim to infer as much as possible about the astrochemical environment. To this end, the nested sampling algorithm with its ability to estimate evidence, which is critical for model testing, is central to our methodology. Here prior information is our greatest asset and we can incorporate a great deal into the form of the likelihood function (4), which depends on the spectral model we have adopted (3). Our future work will be to improve the ability of the algorithm to identify the background signal, as well as to identify and characterize individual PAHs by relying on the NASA Ames PAH library of over a thousand PAH spectra.

ACKNOWLEDGEMENTS

This research was supported by NASA AISR NNH05ZDA001N. Analysis based on observations with the Infrared Space Observatory, a European Space Agency (ESA) project with instruments funded by ESA Member States (especially the Principal Investigator countries: France, Germany, the Netherlands, and the United Kingdom) and with the participation of the Institute of Space and Astronautical Science (ISAS) and the National Aeronautics and Space Administration (NASA).

REFERENCES

Knuth K.H. Tse M.K., Choinski J., Maunu H.A., Carbon D.F. "Bayesian source separation applied to finding complex organic molecules in space". Proceedings of the Statistical Signal Processing Meeting 2007, Madison Wisconsin, USA.

Sivia D.S. and Skilling J. 2006. "Data Analysis A Bayesian Tutorial 2nd Edition". Oxford Univ. Press.

Skilling J., "Turning ON and OFF," In: K.H. Knuth, A. Abbas, R. Morris (eds.), Bayesian inference and Maximum Entropy Methods in Science and Engineering, San Jose, California, USA, AIP Conference Proceedings 803, American Institute of Physics, Melville NY, pp.3-24.

Sloan, G.C., Kraemer, K.E., Price, S.D., Shipman, R.F. "A Uniform Database of 2.4-45.4 Micron Spectra from the Infrared Space Observatory Short Wavelength Spectrometer", 2003. ApJSupp, 147, 379.

PARAMETER ESTIMATION IN ULTRASONIC MEASUREMENTS ON TRABECULAR BONE

Karen R. Marutyan, PhD*, Christian C. Anderson, MA[†], Keith A. Wear PhD**, Mark R. Holland, PhD[†], James G. Miller, PhD[†] and G. Larry Bretthorst, PhD*

*Biomedical Magnetic Resonance Laboratory, Mallinckrodt Institute of Radiology, Washington University, St. Louis, MO 63110
[†]Laboratory For Ultrasonics, Department of Physics, Washington University, St. Louis, MO 63130
**U.S. Food and Drug Administration, Center for Devices and Radiological Health, Silver Spring, MD 20993

Abstract. Ultrasonic tissue characterization has shown promise for clinical diagnosis of diseased bone (e.g., osteoporosis) by establishing correlations between bone ultrasonic characteristics and the state of disease. Porous (trabecular) bone supports propagation of two compressional modes, a fast wave and a slow wave, each of which is characterized by an approximately linear-with-frequency attenuation coefficient and monotonically increasing with frequency phase velocity. Only a single wave, however, is generally apparent in the received signals. The ultrasonic parameters that govern propagation of this single wave appear to be causally inconsistent [1]. Specifically, the attenuation coefficient rises approximately linearly with frequency, but the phase velocity exhibits a *decrease* with frequency. These inconsistent results are obtained when the data are analyzed under the assumption that the received signal is composed of one wave. The inconsistency disappears if the data are analyzed under the assumption that the signal is composed of superposed fast and slow waves. In the current investigation, Bayesian probability theory is applied to estimate the ultrasonic characteristics underlying the propagation of the fast and slow wave from computer simulations. Our motivation is the assumption that identifying the intrinsic material properties of bone will provide more reliable estimates of bone quality and fracture risk than the apparent properties derived by analyzing the data using a one-mode model.

Keywords: Bayesian probability theory, tissue characterization, bone, ultrasound, osteoporosis
PACS: 43.35.Cg

INTRODUCTION

Osteoporosis is a disease that results in a decrease in the mineral density of trabecular bone, a porous material that fills the inner cavity of bones. By establishing correlations between ultrasonic parameters and mineral density of trabecular bone, ultrasonic tissue characterization has become an accepted method for clinical diagnosis of osteoporosis [2, 3]. The physics of ultrasound interaction with the composite structure of bone is not yet completely understood. In most measurements a single wave is apparent in the acquired signals [1, 3]. However, the propagation of two types of compressional waves, known as a fast and a slow wave, has been independently observed by several researches [1, 2, 4]. Furthermore, the ultrasonic characteristics of bone appear to violate the conditions imposed by causality. In particular, as illustrated in Fig. 1, for media

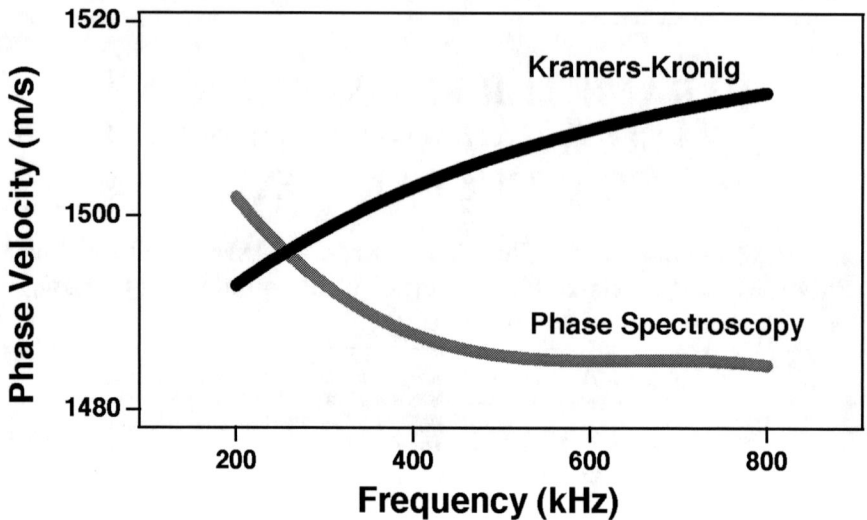

FIGURE 1. This figure illustrates the apparent inconsistency in the ultrasound measurements on bone. The phase velocity inferred from phase spectroscopy analysis exhibits negative dispersion. However, the phase velocity predicted from the attenuation coefficient using the nearly-local approximation of the Kramers-Kronig relations (Eq. 5) rises logarithmically with frequency

with an attenuation coefficient that rises approximately linearly with frequency, such as bone, the nearly-local approximation of the Kramers-Kronig relations [1] suggests that an increase of phase velocity with frequency or *positive* dispersion should be expected, as opposed to the decrease with frequency or *negative* dispersion that is often reported. These inconsistent results are obtained when the data are analyzed under the assumption that the received signal is composed of one wave. However, this inconsistency disappears if the data are analyzed under the assumption that the signal is composed of superposed fast and slow waves [5].

The motivation for the current study is the assumption that identifying the true ultrasonic properties of bone will provide a more reliable estimate of bone quality and fracture risk than the apparent properties derived from conventional techniques. Consequently, our objective is to make the best possible inferences about the ultrasonic characteristics of bone given that the data consist of two interfering fast and slow waves [1, 2]. Bayesian probability theory provides a powerful and well-developed apparatus tailored to solving problems of this kind. In what follows, we will first review the model for ultrasonic propagation in trabecular bone and apply Bayesian probability theory to compute the posterior probability for each parameter in the model. Finally, we apply the calculations to simulation data to determine how accurately the parameters for two interfering modes can be estimated.

MODEL FOR ULTRASONIC PROPAGATION

For linear ultrasound propagation, the real-valued time-series data are related to the complex spectrum of the propagated pulse via the discrete Fourier transform:

$$d_i = \text{Real}\left[\frac{1}{N}\sum_{j=1}^{N} P_j \exp\{-i\omega_j t_i/N\}\right] + n_i \qquad (1)$$

where d_i is a data value sampled at time t_i, P_j is a value of the complex spectrum at angular frequency ω_j, N is a number of points in the data, and n_i is an additive noise.

The data are modeled as the sum of the fast and slow modes. In the frequency domain this sum is given by

$$P_j = \gamma A_j H_{fast,j} + (1-\gamma) A_j H_{slow,j} \qquad (2)$$

where A_j is the complex spectrum of ultrasonic pulse prior to propagation through the bone. The fractional amplitude γ expresses what fraction of the signal amplitude is in the fast wave. Quantitatively γ varies from 0 (only a slow wave propagates) to 1 (only a fast wave propagates). The quantities $H_{fast,j}$ and $H_{slow,j}$ characterize the material properties of medium and are commonly called the medium transfer functions. The transfer functions govern the ultrasonic propagation of the fast and slow waves and have identical functional form for each mode. Therefore, only the equations for the fast mode are given below. Equivalent formulas for the slow mode are realized on the replacement of the label *fast* by *slow*.

For linear plane wave propagation $H_{fast,j}$ is given as

$$H_{fast,j} = \exp\{-\alpha_{fast,j} l\} \exp\{-i\omega_j l/v_{fast,j}\} \qquad (3)$$

in which l is the specimen thickness, $v_{fast,j}$ is the phase velocity and $\alpha_{fast,j}$ is the (linear-with-frequency) attenuation coefficient:

$$\alpha_{fast,j} = \beta_{fast}\frac{\omega_j}{2\pi}. \qquad (4)$$

The parameter β_{fast} is frequently referred to as the slope of attenuation.

The slope of attenuation, and therefore the attenuation coefficient, and phase velocity of the medium are not independent. The interrelation between these properties mathematically are expressed through the nearly-local approximation of the Kramers-Kronig relations [1]. For media characterized with an approximately linear-with-frequency attenuation coefficient, such as bone, the nearly-local approximation takes the form

$$v_{fast,j} \approx v_{fast} + \frac{\beta_{fast}}{\pi^2} v_{fast}^2 \ln\left(\frac{\omega_j}{\omega_r}\right) \qquad (5)$$

in which v_{fast} is a phase velocity at some arbitrarily chosen reference angular frequency ω_r. Following the earlier study [5] the reference frequency $\omega_r/2\pi$ of 300kHz is used in the calculations presented here.

BAYESIAN CALCULATIONS

The objective is to evaluate the posterior probability density functions for each parameter appearing in the model. These probability densities can be computed from the joint posterior probability for all of the parameters. Consequently, in what follows we will compute the joint posterior probability for all the parameters and then use a Markov chain Monte Carlo simulation with simulated annealing to approximate the respective posterior density functions.

If we denote $\Theta \equiv \{\beta_{fast}, \beta_{slow}, \nu_{fast}, \nu_{slow}, \gamma\}$, then joint posterior probability for Θ given all the data D and background information I, $P(\Theta|DI)$, is given by the Bayes' theorem

$$P(\Theta|DI) = \frac{P(\Theta|I)P(D|\Theta I)}{P(D|I)} \qquad (6)$$

where the prior probability for the data $P(D|I)$ is a normalization constant and may be dropped provided we normalize $P(\Theta|DI)$ at the end of the calculation, $P(\Theta|I)$ is the joint prior probability for the parameters, and $P(D|\Theta I)$ is the direct probability for the data given the parameters.

Assuming that the parameters are logically independent, the joint prior probability can be factored using the product rule

$$P(\Theta|I) = P(\nu_{slow}|I)P(\nu_{fast}|I)P(\beta_{slow}|I)P(\beta_{fast}|I)P(\gamma|I) \qquad (7)$$

where we have one prior for each parameter appearing in the model. Each prior probability in Eq. (7) is assigned using a bounded Gaussian to represent what is known about each of the parameters. Specifically, the prior probability for kth parameter in Θ is given by

$$P(\Theta_k|I) \propto \begin{cases} \exp\left\{-\frac{[\Theta_k - M_k]^2}{2\delta_k^2}\right\} & \text{if } L_k \leq \Theta_k \leq H_k \\ 0 & \text{otherwise} \end{cases} \qquad (8)$$

where M_k, δ_k, L_k and H_k are the mean, standard deviation and the low and high bounds for the respective parameters. The quantities M_k, δ_k, L_k and H_k are assumed known.

The direct probability for data given Θ, $P(D|\Theta I)$, cannot yet be assigned because it is a marginal probability. To assign this direct probability, the standard deviation of the noise prior probability, σ, must be introduced into the calculation. Applying the sum and product rules gives

$$P(D|\Theta I) = \int P(D|\Theta \sigma I)P(\sigma|I)d\sigma \qquad (9)$$

where $P(\sigma|I)$ is the prior probability for the standard deviation of the noise prior and $P(D|\Theta\sigma I)$ is the direct probability for the data given both Θ and σ or, in this case, the likelihood. The prior probability for σ is typically assigned using a Jeffreys' prior [6]

$$P(\sigma|I) \propto \frac{1}{\sigma}. \qquad (10)$$

The direct probability for the data given both Θ and σ will be assigned using a Gaussian noise prior probability

$$P(D|\Theta\sigma I) = (2\pi\sigma^2)^{-N/2} \exp\left\{-\frac{Q}{2\sigma^2}\right\} \qquad (11)$$

where Q is the sum of squared difference between data and the model

$$Q = \sum_{i=1}^{N} \left(d_i - \text{Real}\left[\frac{1}{N}\sum_{j=1}^{N} P_j \exp\{-i\omega_j t_i/N\}\right]\right)^2. \qquad (12)$$

Substituting Eq. (10) and (11) into Eq. (9) and performing integration over σ yields

$$P(D|\Theta I) \propto Q^{-\frac{N}{2}} \qquad (13)$$

which is of the form of Student's t-distribution.

This completes the assignment of the joint posterior probability for the parameters, $P(\Theta|DI)$. The program that implements this calculation uses a Markov chain Monte Carlo simulation with simulated annealing to draw samples from this joint posterior probability density function. At the end of the annealing phase samples are drawn from this joint posterior probability and Monte Carlo integration is then used to obtain samples from the marginal posterior probability for each parameter.

DISCUSSION

There is increasing evidence that ultrasonic signals acquired on bone are comprised of the overlapping fast and slow waves [2, 3, 5]. Because the conventional broadband spectroscopy analyses assume a single mode in the data, there is a need for an alternative analysis appropriate for the multi-modal signals.

In the current study the feasibility for accurate estimation of the bone parameters is investigated by applying Bayesian probability theory to the simulated bimodal data. The parameters used to generate this data were assigned from the empirical values as: $\beta_{fast} = 20$dB/cm/MHz, $\beta_{slow} = 6.9$dB/cm/MHz, $v_{fast} = 2100$m/s, $v_{slow} = 1500$m/s, [1, 2]. It is these parameter values that we seek to recover in the Bayesian analysis. The other model parameters, γ, l and σ, were systematically varied in the synthesized data to assess the performance of Bayesian methods under various conditions. First, data with signal-to-noise ratios of 500:1, 100:1 and 50:1 were generated. Next, data sets with varying overlaps between the fast and the slow wave were produced by changing the samples thickness l. The overlap was quantified as

$$\text{overlap }[\%] = \frac{t_{pulse} - l(1/v_{slow} - 1/v_{fast})}{t_{pulse}} \times 100\% \qquad (14)$$

in which t_{pulse} is the temporal length of the unpropagated pulse. Last, data sets were also generated by varying the relative fraction of fast and slow wave, γ.

A plot of some of the marginal posterior probability density functions are given in Fig. 2. These posterior probabilities were computed using three simulated data sets

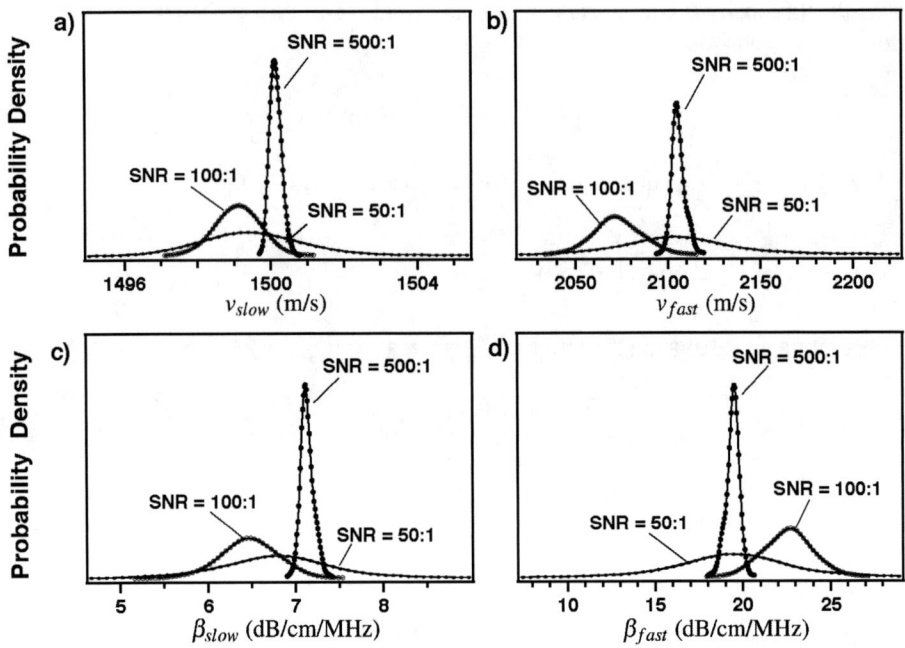

FIGURE 2. The marginal posterior probability density functions for v_{slow} (panel **a**), v_{fast} (panel **b**), β_{slow} (panel **c**) and β_{fast} (panel **d**). Posterior probabilities shown in each plot are computed from simulated data with peak signal-to-noise ratio set at 500:1, 100:1 and 50:1 level.

having signal-to-noise ratios of 500:1, 100:1 and 50:1 respectively. As illustrated in Fig. 2, the widths of the posterior probabilities scale inversely with the signal-to-noise ratio. However, close inspection of scales in panels **a** and **b**, reveals that the widths of the posterior for the fast waves parameters are roughly 15 times greater than the posterior probabilities associated with the slow waves parameters. This is in part due to the specific values of parameters used to simulate the data. The analyzed waveforms are composed of 70% slow wave and 30% fast wave. The fast wave is highly attenuated. Consequently, the signal-to-noise ratio is markedly lower for the fast wave than for the slow wave, which results in the large widths of the posterior probability density functions for the parameters of the fast wave.

The parameter estimates for all of the simulated data are summarized in Fig. 3. These summaries are mean ± standard deviation parameter estimates. The mean for a given data set is represented by the dark bar, the standard deviations by error bars and the true value of each parameter are represented by the line. The percents at the bottom of this figure are the percent fast and slow waves in the simulated data. The standard deviations are smallest for the 500:1 signal-to-noise ratio data and increase as the noise level increases. Signal-to-noise ratio of 100:1 are typical for data acquired *in vitro*, whereas a lower value is anticipated in measurements *in vivo*. Except for the first three cases, the parameter estimates shown in Fig. 3 are for 50:1 signal-to-noise ratio data. For almost all the data (with the only exception being the rightmost case in Fig. 3) the

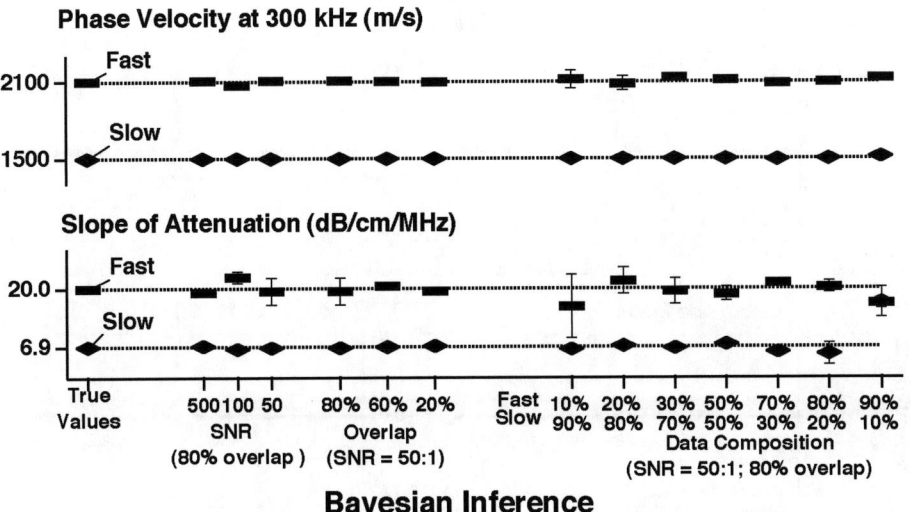

FIGURE 3. The true values for the phase velocity at 300kHz and the slope of attenuation for the fast and the slow wave are compared to the mean ± standard deviation estimates computed from marginal posterior probability for each parameter. Unless specified differently, the simulated data are comprised of 30% fast wave and 70% slow wave. In almost all cases the Bayesian posterior probabilities overlap the true value used in generating the data.

estimated parameter values lie within one or two standard deviations of the true values.

Although the complementary problem of model selection is beyond the scope of this paper, we briefly examine the results of analyzing simulated bimodal data using both a one-mode and a two-mode model. In Fig. 4 panels **a** and **b** are the simulated bimodal data. We have repeated the data in panel **b** for easier comparison. The simulated bimodal data are comprised of 30% fast wave and 70% slow wave with 80% overlap. The peak signal-to-noise ratio is 50:1. Panels **c** and **d** are models generated from the parameters that maximized the joint posterior probability for the parameters when the data are analyzed using a one-mode model (panel **c**) and a two-mode model (panel **d**). The residuals, the difference between the data and the model, are shown in panels **e** and **f**. Panel **e**, the residuals generated from the one-mode model, have a strong systematic artifact; while, panel **f**, the residuals generated from the two-mode model, are random and on the order of the noise. While panels **e** and **f** are suggestive that a two-mode model is needed to explain this data, it is not enough. One needs to go further and compute the posterior probability for the models before one can know for certain, see [7] for more details on how this is done.

SUMMARY

We have applied the Bayesian probability theory to simulated data containing both a fast and a slow wave that mimic those seen in trabecular bone. Conventional phase and power spectroscopy analyses have no mechanism to handle these bimodal data. However, Bayesian probability theory provides a rigorous approach to analyzing such data.

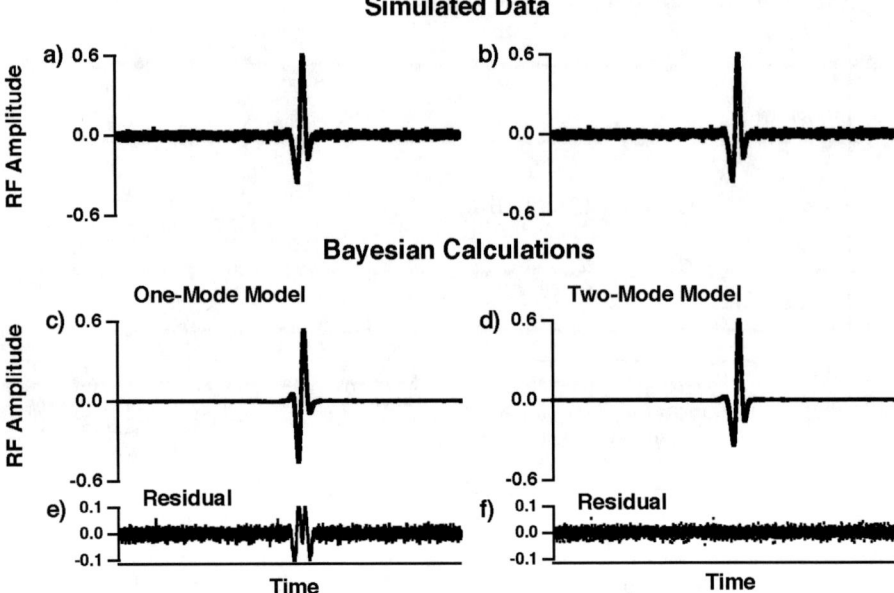

FIGURE 4. A comparison of the residuals generated when bimodal simulated data (panels **a** and repeated in **b**) are analyzed using a one-mode and a two-mode model. Panels **c** is the model generated from the parameters that maximized the joint posterior probability for the parameters given a one-mode model; while panel **d** is the model generated from the parameters that maximized the joint posterior probability given a two-mode model. The residuals for each models are shown on panels **e** and **f**, respectively. The simulated bimodal data are comprised of 30% fast wave and 70% slow wave with 80% overlap. The peak signal-to-noise ratio is 50:1.

In all of the investigated data sets, including low signal-to-noise ratio data, the Bayesian posterior probabilities cover the true value of the parameters. Thus the Bayesian approach is an effective method for extracting properties of bone both *in vitro* and *in vivo*.

REFERENCES

1. Waters, K.R. and Hoffmeister, B.K. (2005). Kramers-Kronig analysis of attenuation and dispersion in trabecular bone. J Acoust Soc Am, 118(6):3912–3920.
2. Hosokawa, A. and Otani, T. (1998). Acoustic anisotropy in bovine cancellous bone. J Acoust Soc Am, 103(5):2718–2722.
3. Padilla, F. and Laugier, P. (2000). Phase and group velocities of fast and slow compressional waves in trabecular bone. J Acoust Soc Am, 108(4):1949–1952.
4. Hughes, E.R., Leighton, T.G., Petley, G.W. and White, P.R. (1999). Ultrasonic propagation in cancellous bone: a new stratified model. Ultrasound Med Biol, 25(5):811–821.
5. Marutyan, K.R., Holland, M.R. and Miller, J.G. (2006). Anomalous negative dispersion in bone can result from the interference of fast and slow waves. J Acoust Soc Am, 120(5):EL55–61.
6. Jeffreys, H. (1961). *Theory of probability*. The International series of monographs on physics. Clarendon Press, Oxford, 3rd edition.
7. Anderson, C.A., Marutyan, K.R., Wear, K.A., Holland, M R amd Miller, J.G. and Bretthorst, G.L. (2007). Model selection in ultrasonic measurements on trabecular bone. K. Knuth, Maximum entropy and Bayesian methods in science and engineering, in this issue.

Model Selection in Ultrasonic Measurements on Trabecular Bone

Christian C. Anderson, M.A.*, Karen R. Marutyan, Ph.D.[†], Keith A. Wear Ph.D.**, Mark R. Holland, Ph.D.*, James G. Miller, Ph.D.* and G. Larry Bretthorst, Ph.D.[†]

Laboratory For Ultrasonics, Department of Physics, Washington University, St. Louis, MO 63130
[†]*Biomedical Magnetic Resonance Laboratory, Mallinckrodt Institute of Radiology, Washington University, St. Louis, MO 63110*
**U.S. Food and Drug Administration, Center for Devices and Radiological Health, Silver Spring, MD 20993*

Abstract. Previous work from our laboratory showed that the widely reported decrease in phase velocity with frequency (negative dispersion) for ultrasonic waves propagating through trabecular bone can arise from the interference of two compressional waves, each of which exhibits a positive dispersion. Previous simulations suggest that Bayesian probability theory can be employed to recover the material properties linked to these two interfering waves, even when the waves overlap sufficiently that visual inspection cannot distinguish two modes. In the present study, Bayesian probability theory is applied first to simulated data and then to representative experimental bone data to determine whether one or two compressional wave modes are present. Model selection is implemented by evaluating the posterior probability for each model. The calculation is implemented by defining a model indicator and then using Markov chain Monte Carlo with simulated annealing to draw samples from the joint posterior probability for the ultrasonic parameters and the model indicator. Monte Carlo integration is used to evaluate the marginal posterior probability for each parameter given the model indicator.

Keywords: Bayesian probability theory, model selection, tissue characterization, bone, ultrasound, osteoporosis
PACS: 43.35.Cg

INTRODUCTION

Quantitative ultrasound has become an accepted modality to aid in the assessment of osteoporotic fracture risk. However, as outlined in a companion article [1], the interaction of ultrasound with trabecular (cancellous) bone is not well understood. In particular, the observed decrease of phase velocity with frequency (negative dispersion) in human calcaneus appears to be inconsistent with the requirements of causality. More specifically, the causality-based Kramers-Kronig relations that relate attenuation to dispersion suggest that media with a linear-with-frequency increase in attenuation coefficient should show an increase in phase velocity with frequency (positive dispersion) [2]. Because bone has been demonstrated to have an approximately linear- or quasilinear-with-frequency increase in attenuation coefficient [2, 3], the observed negative dispersion is perplexing.

Our laboratory recently proposed a hypothesis to explain the observed negative dispersion in human calcaneus–namely, that received signals are not comprised of one propagating compressional wave, but are instead a superposition of two overlapping compressional waves, a fast wave and a slow wave [4]. In this view, the individual fast and slow waves exhibit positive dispersion in accordance with the Kramers-Kronig relations, but interfere in such a way as to produce a waveform with an apparent negative dispersion when analyzed conventionally under the assumption that one wave is present. The true ultrasonic properties of bone, as represented by the properties of the fast and slow waves, could provide more diagnostically relevant information than the apparent properties (e.g. negative dispersion) obtained through conventional means. The propagation of two compressional waves through trabecular bone has been observed experimentally [5] and predicted theoretically [6, 7].

Bayesian probability theory shows promise in recovering the properties of individual superposed waves [8]. The companion article [1] addresses the parameter estimation problem involving ultrasonic measurements on trabecular bone. In that work, Bayesian inference methods are used to recover the properties of simulated waveforms generated through the superposition of fast and slow waves. In the current article, we apply Bayesian probability theory to the model selection problem; that is, the problem of determining whether the received waveforms are composed of one wave or two interfering waves.

ULTRASONIC PROPAGATION MODELS

In a typical ultrasound experiment, two cylindrical ultrasonic transducers are immersed in water and coaxially aligned; one acts as a transmitter, and the other acts as a receiver. Data are acquired with and without a sample placed between the transducers. Since water is a well-characterized medium, analysis on the sample proceeds by referencing the signal that travels through the sample to the signal that travels only through water (the reference water-path signal).

Although acquired ultrasonic signals generally consist of time-domain data, we are concerned with properties that are conveniently analyzed in the frequency domain (i.e., frequency-dependent attenuation and dispersion). Accordingly, our ultrasonic propagation models are designed using a frequency-domain approach and act on the data by the discrete Fourier transform of the model. The propagated frequency spectra are converted back to temporal signals to provide an intuitive representation of the acquired waveform.

Following the convention in [1], we model the received data as

$$d_k = \text{Real}\left[\frac{1}{N}\sum_{j=1}^{N} P_j \exp(-i\omega_j t_k/N)\right] + n_k \qquad (1)$$

where d_k is a real data value sampled at time t_k, P_j represents a value of the propagated complex spectrum at angular frequency ω_j, N is the number of points in the frequency-domain representation of the data, and n_k is additive noise.

At this point, we depart from the notation used in [1], because that article assumes that two compressional ultrasonic waves propagate through bone. Here, we evaluate whether individual received signals are composed of one or two compressional waves. In this treatment, the propagated spectra, P_j in Eq. (1), depend on the respective models for ultrasonic propagation through bone. In the models considered here, the transmitted signals and the received propagated signals are related by mathematical representations known as transfer functions. The P_j are of the form

$$P_j = A_j \sum_{z=1}^{m} H_{z,j} \qquad (2)$$

where A_j is the complex spectrum of the ultrasonic pulse before it encounters the bone specimen, m is the number of wave modes present in the model, and $H_{z,j}$ is the zth transfer function for the propagation model. The one-mode model has one associated transfer function ($m = 1$), whereas the two-mode model involves two transfer functions ($m = 2$). For linear plane wave propagation, $H_{z,j}$ is given as

$$H_{z,j} = S_z \exp\left(-\alpha_{z,j} l\right) \exp\left(-i\omega_j \frac{l}{v_{z,j}}\right) \qquad (3)$$

where S_z is an amplitude scaling parameter representing frequency-independent loss, l is the specimen thickness, $\alpha_{z,j}$ is the attenuation coefficient, and $v_{z,j}$ is the phase velocity at frequency ω_j. If we assume that the attenuation coefficient rises linearly with frequency, $\alpha_{z,j}$ becomes

$$\alpha_{z,j} = \beta_z \frac{\omega_j}{2\pi}. \qquad (4)$$

The parameter β_z is commonly known in the tissue characterization community as the "slope of attenuation," while in the bone literature it is referred to as "normalized Broadband Ultrasound Attenuation (nBUA)."

The phase velocities $v_{z,j}$ are of the form

$$v_{z,j} \approx v_z + \frac{\beta_z}{\pi^2} v_z^2 \ln\left(\frac{\omega_j}{\omega_r}\right) \qquad (5)$$

where v_z is a phase velocity at some arbitrarily chosen reference angular frequency ω_r [4]. The value of the reference frequency is chosen based on the spectral characteristics of the interrogating ultrasonic pulse. Equation (5) is consistent with the causality-imposed Kramers-Kronig relations for media with an attenuation coefficient of the form in Eq. (4).

BAYESIAN CALCULATIONS

The goal of model selection problems is to calculate the posterior probability for the model, conditional on the input data and background information. To accomplish this calculation, a model indicator u is introduced. Two models are considered here, so we constrain the possible values of u to $\{1,2\}$, where $u = 1$ corresponds to the hypothesis, "the received data are described by a one-mode propagation model," and $u = 2$ represents the hypothesis, "the received data are described by a two-mode propagation model." The posterior probability for the model indicator can now be written $P(u|D,I)$, where D is the data and I is the background information. The posterior probability for the model indicator is computed by the application of Bayes' theorem

$$P(u|D,I) = \frac{P(u|I)P(D|u,I)}{P(D|I)}. \tag{6}$$

The term $P(u|I)$ in Eq. (6) is the prior probability for the model indicator, and $P(D|u,I)$ is the marginal direct probability for the data given the model indicator and background information. The probability for the data given the background information, $P(D|I)$, is a normalization constant. If we normalize Eq. (6) at the end of the calculation, then the equality becomes a proportionality

$$P(u|D,I) \propto P(u|I)P(D|u,I). \tag{7}$$

The direct probability for the data given the model indicator, $P(D|uI)$, is a marginal probability because it does not contain any model parameters. The marginal direct probability can be computed by reintroducing these parameters and applying the sum and product rules:

$$P(D|u,I) = \int d\Theta_u P(\Theta_u, D|u,I), \tag{8}$$

in which Θ_u represents the parameters for model u. In the models for ultrasonic wave propagation, there are three unknown parameters for each wave: S_z, v_z, and β_z in Eqs. (3-5). Thus, when $u = 1$, the integral in Eq. (8) is three-dimensional, and when $u = 2$, it is six-dimensional. The joint probability for the parameters and the data, $P(\Theta_u, D|u,I)$, can be factored using the product rule of probability theory; Eq. (8) becomes

$$P(D|u,I) = \int d\Theta_u P(\Theta_u|u,I) P(D|\Theta_u, u, I), \tag{9}$$

where $P(D|\Theta_u, u, I)$ is the probability for the data given the model indicator, parameters for the indicated model, background information and $P(\Theta_u|u,I)$ is the prior probability for the model parameters given the model indicator and background information. Substituting Eq. (9) into Eq. (7), we arrive at an expression for the posterior probability for the model indicator,

$$P(u|D,I) \propto P(u|I) \int d\Theta_u, P(\Theta_u|u,I) P(D|\Theta_u, u, I), \tag{10}$$

where terms of the form $P(\Theta_u|u,I)P(D|\Theta_u,u,I)$ are, up to a normalization constant, the joint posterior probability for the parameters given the model indicator. These calculations are given in [1], and we do not repeat them here.

We now turn our attention to assigning the prior probability for the model indicator, $P(u|I)$. Before performing any calculations, we have no reason to prefer the two-mode model over the one-mode model (or vice versa), so we assign a flat (uniform) distribution to reflect this state of knowledge.

Integrals such as the one in Eq. (10) rarely have analytic solutions, and numerical methods must therefore be employed to determine the posterior probability for the model indicator. The method used here is Markov chain Monte Carlo with simulated annealing. The relative proportions of Markov chain Monte Carlo simulations having model indicator equal to one and two serve as a model selection criterion [9].

DISCUSSION

A broadband ultrasonic pulse was experimentally acquired using paired transducers in a water bath. The nominal center frequency of the transducers was 500 kHz, and the -6 dB bandwidth limits of the pulse spanned the approximate frequency range 280-640 kHz. In this frequency range, propagation of the pulse through water was assumed to have a negligible effect on its attenuation and dispersion characteristics. The pulse was altered using the models described above to simulate one-mode propagation and two-mode propagation through a bone sample. The parameters used to generate the waveforms were selected so that the simulated data were qualitatively similar to experimentally acquired data. The simulated one-mode waveform was generated with $\{S_1, \beta_1, v_1\} = \{0.7, 30 \text{ Np/m/MHz}, 1500 \text{ m/s}\}$, and the simulated two-mode waveform was generated with $\{S_1, \beta_1, v_1; S_2, \beta_2, v_2\} = \{0.6, 30 \text{ Np/m/MHz}, 1500 \text{ m/s}; 0.3, 50 \text{ Np/m/MHz}, 1600 \text{ m/s}\}$. In both cases, the values of l and $\omega_r/2\pi$ were 1.85 cm and 500 kHz, respectively. Additive Gaussian noise with a signal-to-noise ratio of 50:1 was incorporated into the simulated waves, which were used as inputs to the Bayesian analysis program. The original, unaltered water-path pulse served as a reference signal.

Bayesian inference was also applied to experimental data obtained by interrogating 24 human calcaneus samples with ultrasound. These data were acquired with the same transducers used to generate the pulse described above. A more complete description of the experimental methodology can be found elsewhere [10]. Conventional phase spectroscopy analysis of the received signals revealed that a large majority of the samples exhibited negative dispersion.

In the program that implements this calculation multiple Markov chain Monte Carlo simulations are run simultaneously and in parallel. Each of these simulations has a model indicator that has value 1 or 2. One of the steps in the Monte Carlo simulation is to propose a different value for the model indicator. Consequently, the number of simulations having model indicator 1 or 2 changes as a function of the annealing parameter. When the simulations begin, the annealing parameter is zero; the data are being ignored. Hence, the number of simulations having model indicator 1 or 2 reflects the prior probability. The prior is uniform, so the number of simulations having a one-mode model is roughly equal to the number of simulations having a two-mode model. This effect is illustrated in Fig. 1, where the fractions of Markov chain Monte Carlo simulations having model indicator equal to 1 is plotted in gray as a function of the annealing parameter β. Similarly, the fractions of simulations having model indicator equal to 2 are plotted in black. In panel **a**, the synthetic data were generated with a model containing one mode. As the annealing parameter increases, the simulations transition to the one-mode model because, for all practical purposes, the one and two-mode models have equal likelihoods, but all the additional parameters in the two-mode model reduce its posterior probability. However, in panel **b**, the synthetic data contain two wave modes. It is illustrated in [1] that when data containing two modes is analyzed using a single mode model, the residuals generated from the parameters that maximize the posterior probability contain large systematic artifacts. Because the mean-square residual is essentially the negative logarithm of the likelihood, larger systematic variations in the residuals means a lower likelihood; and, consequently, the posterior probability for the one-mode model is much lower than the posterior probability for

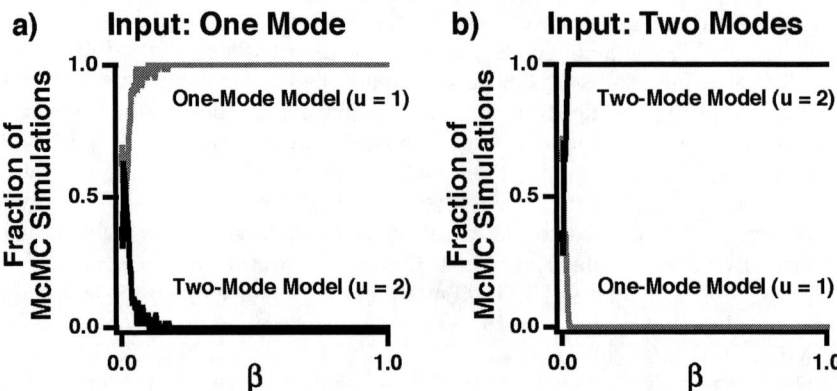

FIGURE 1. The fractions of Markov chain Monte Carlo simulations having model indicator equal to 1 (gray) and 2 (black) is plotted as a function of the annealing parameter β. As β increases, the simulations transition to the one-mode model when the data consist of one mode (panel **a**) and to the two-mode model when the data consist of two modes (panel **b**).

the two-mode model. This effect is illustrated in panel **b** where there is a very sharp transition to the two-mode model because a one-mode model cannot fit the data. Note that this transition is much sharper than the transition in panel **a** where it is the prior probabilities that force the transition to the one-mode model.

Figure 2 illustrates the results of the analysis when simulated one-mode, simulated two-mode, and actual experimental data are analyzed, panels **a**, **b** and **c** respectively. The simulated signals appear in panels **a** and **b**, and a single representative signal propagated through human calcaneus is shown in panel **c**. Panels **d**, **e**, and **f** are output waveforms constructed from the parameters that maximized the joint posterior probability for the parameters and the model indicator. When simulated one-mode data are input, the one-mode model is selected. When a two-mode waveform is the input, the two-mode model is selected. In both instances, the residuals (panels **g** and **h**) are random and on the order of the noise in the input data. When an experimentally acquired data set is analyzed, the two-mode model is selected. However, systematic variations are apparent in the residuals (panel **i**). Similar results were obtained for all experimentally acquired signals. The artifacts in the residuals imply that the wave propagation models presented earlier are only an idealized, approximate representation of actual ultrasound propagation through bone. Indeed, several factors related to the experimental design, including but not limited to beam diffraction, phase cancellation at the face of a phase sensitive receiver, and issues relating to transmission and reflection at the tissue-water interface, are not fully incorporated into the propagation models. Consequently, further study is necessary to determine conclusively if the two-mode hypothesis is a preferred alternative to other explanations of the observed negative dispersion in bone.

Despite the apparent incompleteness of the current model for two-mode propagation,

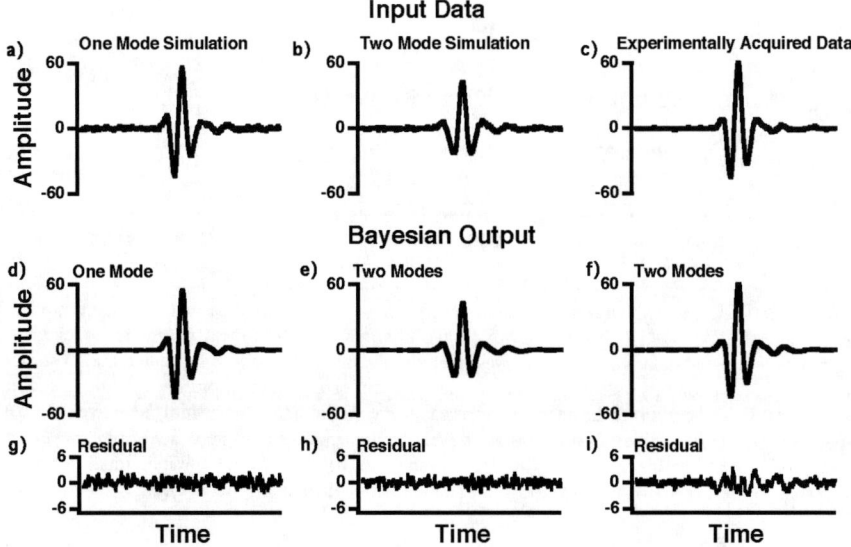

FIGURE 2. Panels **a** and **b** show simulated one and two-mode data, whereas panel **c** is experimental data. Panels **d**, **e**, and **f** are the models generated from the parameters that maximized the joint posterior probability for the parameters and the model indicator given each of the three data sets. For the one and two-model simulated data, the program correctly identifies the number of modes and generates residuals that model the data down to the noise, panels **g** and **h**. However, the residuals generated from the experimental data, panel **i**, show systematic artifacts even for the two-mode model.

it appears to be more capable of recovering the frequency dependence of phase velocity than a one-mode model. This is illustrated in Fig. 3 where the dispersion computed from experimental and simulated data are shown. These dispersion curves are computed from the following equation

$$v_{phase}(\omega) = \frac{v_{water}}{1 - \frac{v_{water}}{l}\frac{\Delta\phi(\omega)}{\omega}} \tag{11}$$

where $v_{phase}(\omega)$ is phase velocity at frequency ω, v_{water} is speed of sound in water, l is sample thickness, $\Delta\phi(\omega)$ is the difference in phase between the reference signal and through-sample signal. It is calculated by Fourier transforming the reference and sample signals, computing the (unwrapped) phase and then taking the difference. The dispersion curves for the experimental data are shown as the gray curves in panels **a** and **b**. These experimental data were then analyzed using a one and two-mode model. The parameters that maximized the joint posterior probability were then used to generate simulated data and these simulations were then use to compute the dispersion curves. The dispersion curves computed from the simulated one and two-mode models are shown as the black curves in panels **a** and **b** respectively. The dispersion curves for the experimental data show negative dispersion, whereas the dispersion curves computed from the one-mode model are positive. However, the dispersion curve computed from the two-mode model shows negative dispersion, even though the individual fast and slow waves each show

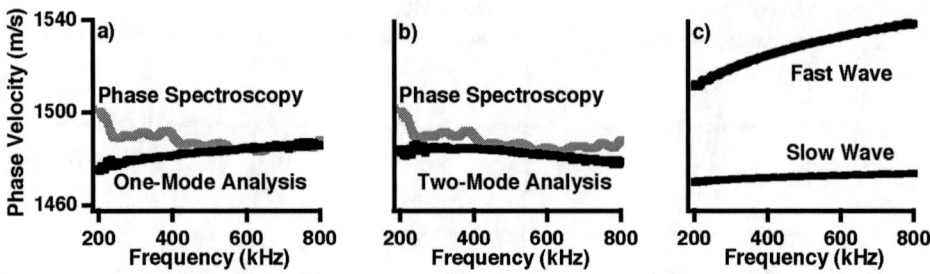

FIGURE 3. The experimental data was analyzed using both a one-mode and two-mode model. The parameters that maximized the joint posterior probability for a one-mode and a two-mode model were used to generate simulated data. Dispersion curves were then computed for these simulated and experimental data sets; see text for this calculation. The dispersion curves generated from the one-mode model are always positive (black curves in panel **a**), whereas the experimental data exhibit negative, decreasing, dispersion (gray curves in panels **a** and **b**). However, the dispersion curves generated from the two-model model produces a negative dispersion (black curve in panel **b**), even though the individual fast and slow waves that compose the two-mode waveform exhibit positive dispersion (panel **c**).

a positive dispersion (panel **c**). Throughout the usable bandwidth (approximately 280-640 kHz), the agreement between the phase spectroscopy analysis on the two-mode model and the experimental data is quite good, especially with regard to the frequency dependence of phase velocity.

SUMMARY

We have addressed the question of the number of wave modes present in simulated and experimental data from signals transmitted through trabecular bone using Bayesian probability theory. A program that implements the Bayesian calculations accurately determined the number of wave modes present in simulated data, and preferred the two-mode propagation model to describe experimentally obtained signals transmitted through human calcaneus. The models for ultrasonic propagation require further refinement if definitive conclusions about ultrasonic propagation through bone are to be drawn. Nevertheless, Bayesian probability theory shows promise as an effective method of evaluating different ultrasonic propagation models.

REFERENCES

1. Marutyan, K.R., Anderson, C.C., Wear, K.A., Holland, M.R., Miller, J.G. and Bretthorst, G.L. (2007). Parameter estimation in ultrasonic measurement on trabecular bone. Ed. K. Knuth, Maximum entropy and Bayesian methods in science and engineering, in this issue.
2. Waters, K.R. and Hoffmeister, B.K. (2005). Kramers-Kronig analysis of attenuation and dispersion in trabecular bone. J Acoust Soc Am, 118(6):3912–3920.
3. Droin, P., Berger, G. and Laugier, P. (1998). Velocity dispersion of acoustic waves in cancellous bone. IEEE Trans Ultrason Ferroelectr Freq Control, 45(3):581–592.

4. Marutyan, K.R., Holland, M.R. and Miller, J.G. (2006). Anomalous negative dispersion in bone can result from the interference of fast and slow waves. J Acoust Soc Am, 120(5):EL55–61.
5. Hosokawa, A. and Otani, T. (1997). Ultrasonic wave propagation in bovine cancellous bone. J Acoust Soc Am, 101(1):558–562.
6. Haire, T. and Langton, C. (1999). Biot theory: A review of its application to ultrasound propagation through cancellous bone. Bone, 24(4):291–295.
7. Hughes, E.R., Leighton, T.G., Petley, G.W. and White, P.R. (1999). Ultrasonic propagation in cancellous bone: a new stratified model. Ultrasound Med Biol, 25(5):811–821.
8. Marutyan, K.R., Bretthorst, G.L. and Miller, J.G. (2007). Bayesian estimation of the underlying bone properties from mixed fast and slow mode ultrasonic signals. J Acoust Soc Am, 121(1):EL8–15.
9. Bretthorst, G.L., Hutton, W.C., Garbow, J.R. and Ackerman, J.J. (2005). Exponential model selection (in NMR) using Bayesian probability theory. Concepts in Magnetic Resonance, 27A:64–72.
10. Wear, K.A. (2000). Measurements of phase velocity and group velocity in human calcaneus. Ultrasound Med Biol, 26(4):641–646.

Computing the Probability of Local Brain Connectivity using Diffusion Tensor Imaging

Joshua S. Shimony[1], Adrian A. Epstein[1], and G. Larry Bretthorst[2]

Mallinckrodt Institute of Radiology[1] and Department of Chemistry[2],
Washington University Medical School
521 S. Kingshighway Blvd., St. Louis, MO 63110, USA

Abstract. The diffusion tensor model has been used to analyze magnetic resonance diffusion data and has been successful in both neuroscientific and clinical applications. We propose an enhancement of this model with a local connectivity parameter that better accords with the known structure of white-matter. In addition to providing diffusion tensor parameter estimation the calculation provides the probability that a given pixel is connected to one of its nearest neighbors. These probabilities can be used in further calculations to determine the probability of connectivity between different brain regions. Implementation of the algorithm is discussed in addition to its usage in simulated and in-vivo magnetic resonance diffusion tensor data.

Keywords: Diffusion Tensor Imaging, DTI, MRI, Brain Connectivity
PACS: 02.50.Cw

INTRODUCTION

Diffusion, the random thermal motion of microscopic particles, has been of great scientific interest ever since its initial description in 1827 by Robert Brown (hence the term "Brownian motion"). He observed this motion while using a microscope to examine a collection of pollen particles floating in a water droplet. Stejskal and Tanner [1] first proposed a method to measure diffusion in liquid samples early in the development of nuclear magnetic resonance spectroscopy and this was later applied to magnetic resonance (MR) imaging by Le Bihan *et al.* [2, 3]. The clinical use of diffusion imaging expanded rapidly after its sensitivity to the early detection of stroke was demonstrated [4-7].

When diffusion occurs in an isotropic medium, it is uniform in all directions and well described by a single diffusion parameter. Visually, this can be understood as the spherical diffusive spread of a small drop of dye in a uniform media. In biological tissue, diffusion displacements are hindered unevenly in three-dimensional (3D) space by cell membranes and other sub-cellular constituents. This uneven diffusion can be quantified and is termed diffusion anisotropy. It is especially prominent in tissue with a regular structural organization, such as nerve or muscle fibers. Visually, the diffusive spread of a drop of dye in this medium will form a 3D ellipsoidal shape. The dye will spread a greater distance along the direction of the nerve fibers, but will be relatively restricted in the direction perpendicular to them [8-11].

The use of the diffusion tensor (DT) model to describe the anisotropic 3D diffusive spread of water in the brain as measured with MR was first proposed by Basser et al. [12]. This model relies on a symmetric 3x3 diffusion matrix which represents the size of the diffusion coefficient in different directions in space as a 3D ellipsoid. In practice seven parameters are estimated for each image pixel from numerous MR diffusion measurements, each sensitized to diffusion to a different degree and in a different direction. In MR imaging the sensitization to diffusion is achieved by turning on a pair of pulsed magnetic field gradients. The first gradient pulse tags the proton spins by causing them to precess rapidly and increase the phase shifts across the image. The second gradient pulse is turned on in the opposite direction to the first and unwinds the phase shifts caused by the first pulse. Any proton that diffuses away from its original position during the interval between the gradient pulses will not completely re-phase and will cause signal loss in the image. The amount of signal loss in the diffusion image as compared to the original image gives an estimate of the diffusion in the direction of the pulsed gradients.

The estimated parameters can be understood as three diffusion coefficients along the axes of the ellipsoid (diffusion eigenvalues) and three angles that represent the direction of the ellipsoid in space. In equation form, the signal S is a function of a gradient vector $\mathbf{q} = \gamma \mathbf{G} \delta$ ($\mathbf{G}$ the gradient vector, γ the gyromagnetic ratio for protons, δ the gradient width) and can be expressed as

$$S(\mathbf{q}) = S(0)\exp\{-\Delta \mathbf{q}\mathbf{D}\mathbf{q}^T\} + C \qquad (1)$$

in which $S(0)$ is the baseline signal with gradients turned off, Δ is the inter-gradient delay, and $\mathbf{D}$ is a symmetric 3x3 diffusion matrix. The diffusion matrix can be specified in term of its eigenvalues, λ_1, λ_2, λ_3, and three rotation angles, φ, θ, ψ, as seen in the relationship

$$\mathbf{D} = \mathbf{R}(\varphi,\theta,\psi)\begin{pmatrix} \lambda_1 & 0 & 0 \\ 0 & \lambda_2 & 0 \\ 0 & 0 & \lambda_3 \end{pmatrix}\mathbf{R}^T(\varphi,\theta,\psi) \qquad (2)$$

where $\mathbf{R}$ is a 3x3 rotation matrix. The constant C represents a component of the signal that arises from highly constrained water molecules, and has been found to improve the representation of the data [13, 14]. Estimation of $S(0)$, C, the three rotation angles, and the three eigenvalues provides for eight adjustable parameters in this model. Standard analysis of data using the DTI formalism makes the assumption that each pixel is an independent diffusion compartment and that the diffusion follows the Gaussian model, i.e. the solution of the diffusion equation represents the probability of a water molecule diffusing as a Gaussian function of the displacement distance.

Brain connectivity, defined as the anatomical and functional (white-matter) connections between the computational units (gray-matter) of the brain is of great interest to neuroscientists and physicians. In the past connectivity information was limited and only available from human pathological specimens and from invasive studies, usually in non-human primates. Given the sensitivity of DT imaging to the local direction of nerve fiber bundles it has the potential to provide information on brain connectivity in a non-invasive manner. Many approaches have been used to derive connectivity information from DT data [15-17] and some have made good use of Bayesian probability theory to estimate the model parameters [15]; however most of these attempts have solved each pixel independently of its neighbors. In this project we extend the standard DT model by including a local connectivity parameter, Λ, and then use Bayesian probability theory to compute the probability that a given pixel is connected to its neighbors. Although introducing this connectivity parameter complicates the parameter estimation, this provides a more accurate model of the structure of the brain since it is known that nerve fiber bundles can extend across long distances in the brain, well over the dimension of a single imaging pixel. By adding the connectivity parameter we include in the model our prior anatomic information of the structure of the brain, information that is ignored by models that treat each pixel independently.

METHODS

Bayesian Inference and Markov chain Monte Carlo

Unlike the simple case of estimating the DT parameters for each pixel separately, the addition of the connectivity parameter creates a global problem of estimation where each pixel depends on the parameters of its neighbors. Each pixel has nine parameters to be determined, six (3 eigenvalues and 3 angles) from the diffusion tensor (**D, in bold**), the baseline signal of the measurement, $S(0)$, the additive constant C, and the connectivity parameter, Λ. We express the DT parameters for pixel i as Ω_i:

$$\Omega_i = \{\lambda_{1i}, \lambda_{2i}, \lambda_{3i}, \varphi_i, \theta_i, \psi_i, S_i(0), C_i, \Lambda_{ij}\} \tag{3}$$

where Λ_{ij} is the connectivity parameter for pixel i, and j specifies its 26 nearest neighbors (the neighbors to a pixel are contained in a 3x3x3 cube surrounding it). The total parameter space (Ω) includes the entire N pixels in the image:

$$\Omega = \{\Omega_1, \Omega_2, \Omega_3, \ldots, \Omega_N\} \tag{4}$$

The probability for the DT parameters given the data (D, *italicized*, not to be confused with the diffusion tensor, **D**, represented in bold), and relevant background information I is expressed in terms of Bayes' theorem [18] as

$$P(\Omega|DI) \propto P(D|\Omega I) P(\Omega|I) \tag{5}$$

where $P(D|\Omega I)$ is the direct probability for the data given the model parameters, and $P(\Omega|I)$ is the prior probability for the model parameters. We factor this expression:

$$P(\Omega|DI) = \prod_{i=1}^{N} P(D_i|\Omega_i I) P(\Omega_i|I) \tag{6}$$

The prior probability for the parameters in a given pixel, $P(\Omega_i|I)$, can be factored using the product rule and logical independence into a series of eight prior probabilities:

$$P(\Omega_i|I) \propto P(\lambda_{1i}|I) P(\lambda_{2i}|I) P(\lambda_{3i}|I) P(\varphi_i|I) P(\theta_i|I) \ldots P(\Lambda_{ij}|I) \tag{7}$$

The rotation angles were assigned uniform prior probabilities. The remaining parameters (with the exception of the connectivity parameter) were assigned Gaussian prior probabilities bounded by an appropriate physiologic range.

The connectivity parameter does not appear in the DT model, consequently in this calculation it appears only in the prior probabilities. The preference for connectivity is indicated by the prior probability of Λ_{ij}, which we expressed as the probability that a water molecule will diffuse from the central pixel to a neighboring pixel along a line connecting the two. According to the DT model the probability for a water molecule to travel a distance $\mathbf{r}$, can be represented as a Gaussian function. An accurate calculation would require a path integral but we approximate the probability of diffusion across adjacent pixels as the product of two Gaussian distributions, one from the ith pixel, representing the first half of the motion of the water molecule, and the second from the neighboring pixel, representing the second half. If $\mathbf{r}_{ij}$ expresses the distance between the two pixels, Δ is the inter-gradient delay, and $\mathbf{D}_i$ and $\mathbf{D}_j$ the diffusion tensors of the two pixels, the prior probability can be patterned after the real probability as:

$$P(\Lambda_{ij}|I) \propto \exp\left\{-\frac{w\Delta}{\mathbf{r}_{ij}\mathbf{D}_i\mathbf{r}_{ij}^T}\right\}\exp\left\{-\frac{w\Delta}{\mathbf{r}_{ij}\mathbf{D}_j\mathbf{r}_{ij}^T}\right\}, \qquad (8)$$

with w being an adjustable weighting factor.

The direct probability for the data, given the DTI measurements in a given pixel, $P(D_i|\Omega_i I)$, is a marginal probability when the standard deviation, σ_i, was removed using the sum and product rules:

$$P(D_i|\Omega_i I) = \int P(\sigma_i D_i|\Omega_i I)d\sigma_i \propto \int P(\sigma_i|I)P(D_i|\sigma_i\Omega_i I)d\sigma_i. \qquad (9)$$

Although the standard deviation of the noise is fairly uniform in the central portion of the image it has a strong dependence on position toward the periphery. Consequently, we assume a different σ_i for each pixel in the image. Assigning a Jeffreys' prior probability for $P(\sigma_i|I)$ and assigning the direct probability for the data, $P(D_i|\sigma_i\Omega_i I)$, using a Gaussian, the marginal probability for the data, $P(D_i|\Omega_i I)$, may be written as:

$$P(D_i|\Omega_i I) = \frac{1}{2}\Gamma\left(\frac{M}{2}\right)\left(\frac{Q}{2}\right)^{-M/2} \qquad (10)$$

resulting in the Student's t-distribution, where M is the number of data points per pixel and Q_i is the total square residual given by:

$$Q_i = \sum_{k=1}^{M}(E_{ik} - S_i(\mathbf{q}_k))^2 \qquad (11)$$

with E_i being the experimental diffusion measurements.

Substituting Eq. [7] and Eq. [10] into Eq. [6] results in an expression for the posterior probability for all of the parameters with only the prior $P(\Lambda_{ij}|I)$ linking each pixel i to its neighbors. The joint posterior probability for all of the parameters was sampled using a Metropolis-Hastings Markov chain Monte Carlo simulation with simulated annealing [19, 20].

Introducing the connectivity parameter complicates the parameter estimation because each pixel is dependent, via the connectivity prior, on the parameter values of its neighbors. When performing the calculations on a parallel processor it is important to insure that the neighboring pixels do not change while the estimation of the ith pixel is underway. We implemented this calculation by an iterative scheme that is analogous in 1-dimension to doing all the even pixels first and then returning to do the odd pixels. In 3D this requires 8 separate passes through the data to cover all the pixels in the sample once.

Simulated and Experimental Data

The algorithm was tested using a small simulated data set that was 3x3x3 pixels in size. This provided us with a single pixel and its 26 immediate nearest neighbors. The initial data values in the 27 pixels were set to simple examples with known answers to test the algorithm under different circumstances. From the known preset values of the parameters we generated the results that would be measured in a typical MR experiment with the addition of Gaussian noise.

The algorithm was then applied to a single slice of DT measurements performed on a 1.5T Siemens Sonata scanner (Erlangen, Germany) in a normal volunteer. In addition to anatomic T1- and T2-weighted images the DT data was acquired using a locally modified echo planar imaging sequence in 37 directions with 3 separate gradients strengths (b-value = 400-1200s/mm^2, TE = 113ms, TR = 7s). All data processing was done using a locally written software package.

RESULTS

Figure 1A represents the probability distribution for the connectivity parameter in a simulated experiment with a single fiber passing through the sample volume. All the neighboring pixels in these cases have the same anisotropy and are pointing in the same direction as the central pixel. The highest connectivity is seen with the two neighboring pixels that are pointed to by the fiber direction (coded as directions 10 and 16 in this case). The change in the connectivity parameter with a change in anisotropy (=0.15, 0.30, and 0.7) is demonstrated in the figure, giving the expected increase in the probability along the preferred directions with increased anisotropy. These values of anisotropy were selected to represent the typical range of anisotropy values in the brain, from dense white matter (0.7) to just above gray matter (0.15).

Figure 1B represents the probability distribution for the connectivity parameter in a simulated experiment with two crossing fibers passing through the sample volume. Resolving crossing fibers represents one of the major challenges with current methods for estimating brain connectivity. In this case the highest connectivity is seen with four neighboring pixels (coded as directions 10, 12, 14, and 16), two for each one of the fiber directions. The change in the connectivity parameter is also demonstrated with a change in the anisotropy, again demonstrating the expected increase in the connection probability along the preferred directions with an increase in anisotropy.

Figure 2 presents the results from applying the model to a data set acquired from the brain of a healthy volunteer. The image on the left is an anisotropy image of a coronal section, approximately in the middle of the brain. The bright regions in the image are of areas of white matter with high anisotropy, and the darker regions are of gray matter with low anisotropy. The marked white square on the left has been magnified and is displayed on the right of the image. Superposed on the magnified view are red whiskers that give an indication of the value of the connectivity parameter in each pixel. A larger connectivity is indicated by a longer whisker. The full results, which are in 3D, are limited by the 2D representation in the image, and

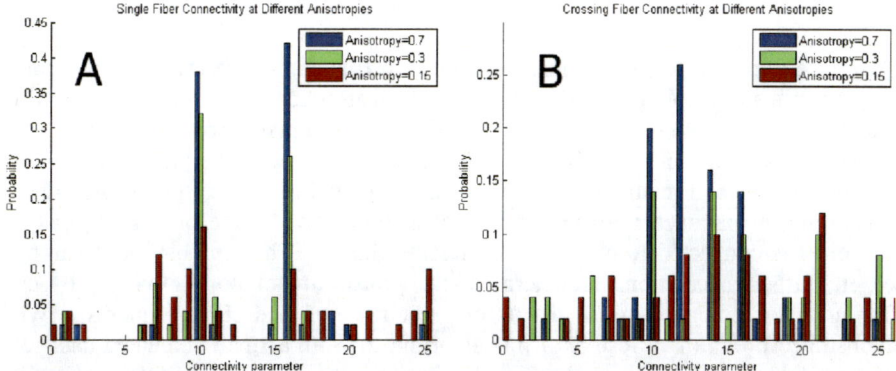

demonstrate a strong resemblance to the known anatomy in this area. Region 1 in this figure is of a patch of gray matter with low anisotropy, which is indicated by several approximately equal whiskers pointing in different directions without a dominant nerve fiber bundle. Region 2 represents the corpus callosum which is one of the

FIGURE 1. (A) The probability distribution of the connectivity parameter for the case of a single fiber. The preferred connectivity to two neighboring pixels is demonstrated in addition to changes in the distribution with a change in anisotropy. See text for further details. (B) The probability distribution of the connectivity parameter for the case of two crossing fibers. The preferred connectivity to four neighboring pixels is demonstrated in addition to changes in the distribution with a change in anisotropy. See text for further details.

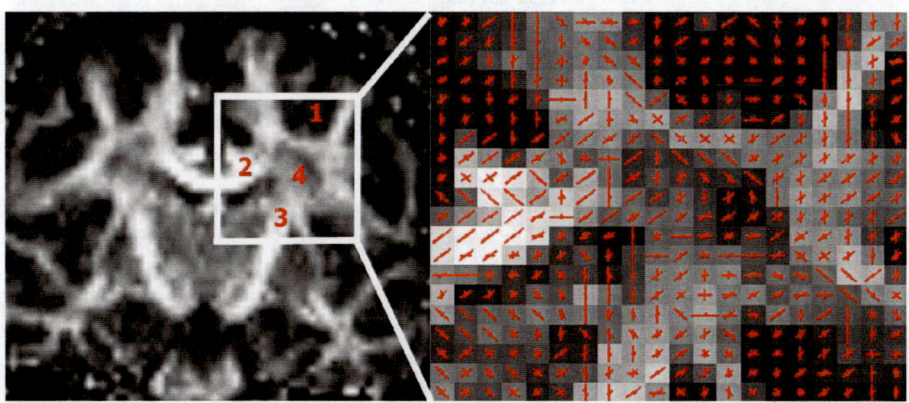

FIGURE 2. The connectivity parameter in a data set from the brain of a normal subject. The image on the left is an anisotropy image of a coronal section of the brain. The image on the right is a magnification of the region marked by the white square with superposed red whiskers indicating the value of the connectivity parameter in each pixel. See text for further details.

largest and most anisotropic fiber bundles in the brain. The whiskers in this region are very uniform and point in the dominant medio-lateral direction. Region 3 represents the internal capsule, which is also a dense fiber bundle that extends in the superior-inferior direction and is well represented by the whiskers. Region 4 contains many crossing fibers, which are represented in the whisker plot as small crosses.

SUMMARY

We have implemented an enhanced model for estimation of the DT from MR data of the brain using local connectivity information that better accords with the known structure of white-matter. Standard DT models treat each pixel separately and do not account for the fact that nerve fiber bundles in white matter of the brain can extend over many pixels and relatively large distances in the brain. In addition to providing diffusion tensor parameter estimation the calculation provides the probability that a given pixel is connected to one of its nearest neighbors. These probabilities can be used in further calculations to determine the probability of connectivity between different brain regions, a result that is of great research and clinical interest. We demonstrated preliminary results of this algorithm in both a small simulated data set with known expected results, and in a real diffusion data set obtained in the brain of a healthy volunteer, in which it demonstrated results that are in accord with the known anatomy. Future work will use the local connectivity results to calculate the probability that various regions in the brain are connected.

ACKNOWLEDGMENTS

We would like to thank NIH K23 HD053212 and National Multiple Sclerosis Society PP1262 for generous financial support.

REFERENCES

1. Stejskal, E.O. and J.E. Tanner, *Spin diffusion measurements: Spin echoes in the presence of time-dependent field gradients.* J. Chem. Phys., 1965. **42**: p. 288-292.
2. LeBihan, D., et al., *MR imaging of intravoxel incoherent motions: Application to diffusion and perfusion in neurologic disorders.* Radiology, 1986. **161**: p. 401-407.
3. LeBihan, D., et al., *Separation of diffusion and perfusion in intravoxel incoherent motion MR imaging.* Radiology, 1988. **168**: p. 497-505.
4. Moseley, M.E., et al., *Early detection of regional cerebral ischemia in cats: Comparison of diffusion- and T_2-weighted MRI and spectroscopy.* Magn. Reson. Med., 1990. **14**: p. 330-346.
5. Le Bihan, D., et al., *Diffusion MR imaging: clinical applications.* AJR Am J Roentgenol, 1992. **159**(3): p. 591-9.
6. Chien, D., et al., *MR diffusion imaging of cerebral infarction in humans.* AJNR Am J Neuroradiol, 1992. **13**(4): p. 1097-102; discussion 1103-5.
7. Warach, S., et al., *Fast magnetic resonance diffusion-weighted imaging of acute human stroke.* Neurology, 1992. **42**: p. 1717-1723.
8. Moseley, M.E., et al., *Diffusion-weighted MR imaging of anisotropic water diffusion in cat central nervous system.* Radiology, 1990. **176**: p. 439-445.
9. Hajnal, J.V., et al., *MR imaging of anisotropically restricted diffusion of water in the nervous system: technical, anatomic, and pathologic considerations.* J Comput Assist Tomogr, 1991. **15**(1): p. 1-18.
10. Chenevert, T.L., J.A. Brunberg, and J.G. Pipe, *Anisotropic diffusion in human white matter: demonstration with MR techniques in vivo.* Radiology, 1990. **177**: p. 401-405.
11. Beaulieu, C. and P.S. Allen, *Determinants of anisotropic water diffusion in nerves.* Magn. Reson. Med., 1994. **31**: p. 394-400.

12. Basser, P.J., J. Mattiello, and D. LeBihan, *MR diffusion tensor spectroscopy and imaging.* Biophys. J., 1994. **66**: p. 259-267.
13. Bretthorst, G.L., C.D. Kroenke, and J.J. Neil. *Characterizing water diffusion in fixed baboon brain.* in *24th International workshop on Bayesian inference and maximum entropy methods in science and engineering.* 2004. Garching, Germany.
14. Kroenke, C.D., et al., *Diffusion MR imaging characteristics of the developing primate brain.* Neuroimage, 2005. **25**(4): p. 1205-13.
15. Behrens, T.E.J., et al., *Characterization and propagation of uncertainty in diffusion-weighted MR imaging.* Magn Reson Med, 2003. **50**: p. 1077-1088.
16. Mangin, J., et al., *A framework based on spin glass models for the inference of anatomical connectivity from diffusion-weighted MR data - a technical review.* NMR Biomed, 2002. **15**: p. 481-492.
17. Parker, G.J.M., H.A. Haroon, and C.A.M. Wheeler-Kingshott, *A framework for a streamline-based probabilistic index of connectivity (PICO) using a structural interpretation of MRI diffusion measurements.* J Magn Reson Imaging, 2003. **18**: p. 242-254.
18. Jaynes, E.T., *Probability Theory: the logic of science*, ed. G.L. Bretthorst. 2003, Cambridge: Cambridge University Press.
19. Gilks, W.R., S. Richardson, and D.J. Spiegelhalter, *Markov Chain Monte Carlo in Practice.* 1996, London: Chapman and Hall.
20. Metropolis, N., et al., *Equation of state calculation by fast computing machines.* J Chem Phys, 1953. **21**: p. 1087-1092.

Generalised Skilling–Bryan Minimisation for Micro-Rotation Imaging in Light Microscopy

Danai Laksameethanasan and Sami S. Brandt

Helsinki University of Technology, Laboratory of Computational Engineering,
P.O. Box 9203, FI-02015 TKK, Finland

Abstract. The original Skilling–Bryan method, first introduced in the field of astronomy, minimises the statistical cost function assuming Gaussian noise and the entropy prior functional. In contrast to the conventional trust region methods, where the solution search in a high-dimensional space may be very expensive, the Skilling–Bryan scheme minimises the cost function in a subspace instead, yielding much more efficient computation. However, in low-photon-count processes, such as image formation in confocal microscopy, Poisson noise assumption is more suitable. In addition, it would be desirable to use the 2nd order, Skilling–Bryan optimisation framework in applications where general prior functionals with the positivity constraint is required. In this work, we hence generalise the Skilling–Bryan method to be used with the Poisson noise model as well as with various kinds of prior functions including the total variation prior. The proposed method is generic whereas here it is applied in the micro-rotation imaging, where we aim at 3D reconstruction of a rotating object from 2D projections taken by a light microscope. On the basis of our investigation, the extended Skilling–Bryan method is promising for computing estimates in high-dimensional problems efficiently.

Keywords: Maximum Entropy Method, Skilling–Bryan minimisation, Micro-Rotation imaging.

1. INTRODUCTION

A standard way for characterising a posterior distribution is computing the maximum a posteriori (MAP) estimate. Efficient numerical optimisation tools are essential for finding the MAP estimate, particularly in high-dimensional problems. Among numerous optimisation methods, line search and trust-region methods are conventional choices. The line search methods, in each iteration, search for a solution in a single direction whereas the trust-region methods compute the iterates in a high-dimensional space. In general, trust region methods tend to be more robust than line search methods, such as conjugate gradient [1]. However, the high-dimensional search may be expensive in the trust-region approach.

In contrast to the techniques above, Skilling and Bryan suggested to minimise a statistical cost function in a promising subspace, yielding more efficient computation [2, 3]. This approach is thus appropriate for solving nonlinear cost functions in high-dimensional spaces and, by their formulation, positivity of the solution is inherently enforced [2]. This method was first introduced in the field of astronomy and thereafter it has been widely used in other fields of science [4]. However, the original Skilling–Bryan method assumes the Gaussian noise model and entropy prior functional whereas in low-photon-count applications, such as image formation in confocal microscopy, Poisson noise assumption is more suitable. In addition, in many applications [5, 6], rather than the entropy prior, other priors may be needed.

Hence, we propose here a generalisation of the Skilling–Bryan method for minimising the statistical cost function with the Poisson noise model and general prior functions including the Gaussian and total variation (TV) priors. The method is generic and we apply it in solving a 3D reconstruction problem in the Micro-Rotation application in light microscopy [7, 8, 9]. Micro-Rotation imaging is an optical microscopic imaging technique which employs dielectric fields in rotating cells around a single axis parallel to the focal plane of the microscope [9, 10]. More precisely, the method involves manipulating dielectric fields to trap and control cells, aiming to achieve high resolution imaging of individual, intact live cells [10, 11]. Our ultimate goal is to obtain an efficient 3D reconstruction algorithm, based on Bayesian theory assuming Poisson noise with the general priors.

The paper begins by presenting the image formation model for micro-rotation imaging in Section 2 and introducing the statistical framework and the statistical cost functions in Section 3. The generalisation of the Skilling–Bryan minimisation of the cost functions is described in Section 4. The experimented results with simulated micro-rotation series are reported in Section 5. In Section 6, we discuss and conclude the paper.

2. IMAGE FORMATION MODEL

In this section, we describe the image formation model, used in the Micro-Rotation application. A typical microscope system can be characterised by its point spread function (PSF): if the microscope system is linear and shift invariant, the measurement image is

$$m_i(x,y) = h(x,y,z) * f_i(x,y,z)\Big|_{z=d}, \qquad (1)$$

where h is the 3D PSF, $f_i = R_i f$ is the rotated object density for the projection i, R_i is the rotation operator and $*$ denotes the 3D convolution operator. The measurement image is thus recorded as the plane corresponding to the focal plane $z = d$ and the optical axis is assumed to be parallel to the z-direction.

In minimisation problem, the projection model (1) requires discretisation of the 3D space and the images onto finite grids. Since the imaging operator (1) is linear, the total discretised model for the M projections is described as

$$\mathbf{m} = \mathbf{A}\mathbf{f}, \qquad (2)$$

where $\mathbf{f}$ is a vector of object density values, $\mathbf{m} = [\mathbf{m}_1^T, \cdots, \mathbf{m}_M^T]^T$ and $\mathbf{A} = [\mathbf{A}_1^T, \cdots, \mathbf{A}_M^T]^T$ are the joint measurement vector and the joint projection matrix, respectively, combining all the projections. The implementation details of the matrix $\mathbf{A}$ and its adjoint $\mathbf{A}^T$ can be found from [11, 12].

3. BAYESIAN INVERSION THEORY

By the Bayesian inversion approach, prior knowledge can be used in a systematic way. The solution is described via the posterior distribution which, in our application, takes the form

$$p(\mathbf{f}|\mathbf{m}) \propto p(\mathbf{f})p(\mathbf{m}|\mathbf{f}), \qquad (3)$$

where $p(\mathbf{m}|\mathbf{f})$ is the likelihood density and $p(\mathbf{f})$ is the prior density. The computation of the MAP estimate

$$\hat{\mathbf{f}} = \arg\max_{\mathbf{f}} p(\mathbf{f}|\mathbf{m}) \qquad (4)$$

corresponds to maximising the cost function

$$q(\mathbf{f}) = s(\mathbf{f}) - \lambda c(\mathbf{f}), \qquad (5)$$

where $c(\mathbf{f}) = -\ln p(\mathbf{m}|\mathbf{f})$ is the likelihood term and $s(\mathbf{f}) = \ln p(\mathbf{f})$ is the prior term. The auxiliary regularisation parameter λ balances the likelihood and the prior terms. Over regularisation (too large λ) leads the algorithm to diverge where as under-regularisation amplifies noise. The likelihood density is constructed from the measurement noise model, as follows.

In many applications, such as those of fluorescence microscopy, the amount of photons reaching the detector is well described by the Poisson process, i.e., the pixel measurement m_j is a sample of a Poisson distributed random variable with the expected value $(\mathbf{Af})_j$. Assuming independent measurements, the cost is

$$c_P(\mathbf{f}) = \mathbf{1}^T(\mathbf{Af}) - \mathbf{m}^T \log(\mathbf{Af}), \qquad (6)$$

where $\mathbf{1} = [1,1,...,1]^T$. If the amount of photons arrived at the detector is high, the Poisson noise is well approximated by Gaussian noise [13]. In the case of isotropic Gaussian noise with unit variance, the cost function is

$$c_G(\mathbf{f}) = \frac{1}{2}\|\mathbf{m} - \mathbf{Af}\|^2. \qquad (7)$$

To control the fitting, it is reasonable to set the bound c_{aim} to which we adjust the likihood c [2]. Typically, the constant c_{aim} is proportional to the noise level in the images. For the Gaussian noise model, c_{aim} is equal to $N\sigma^2$ where σ^2 is the noise variance and N is the number of measurement data. The Poisson case is more difficult, one possibility is to select the Cisiszár I-divergence [14] as c_{aim}.

In this work, we have experimented Gaussian, Entropy, and the TV prior models. The Gaussian, white noise prior, with the cost

$$s_G(\mathbf{f}) = -\frac{1}{2}\|\mathbf{f}\|^2, \qquad (8)$$

is a simple choice and is also known as Tikhonov regularisation. The entropy prior is obtained by setting

$$s_E(\mathbf{f}) = \mathbf{1}^T\mathbf{f} - \mathbf{f}^T \log(\frac{\mathbf{f}}{\mathbf{f}_0}), \qquad (9)$$

TABLE 1. Likelihood and prior terms with their gradients and Hessians used in the extended Skilling–Bryan minimisation method. The operations between two vectors are performed element by element.

	Functions	Gradient **g**	Hessian **H**
Gaussian noise c_G	$\frac{1}{2}\|\mathbf{m}-\mathbf{Af}\|^2$	$-\mathbf{A}^T(\mathbf{m}-\mathbf{Af})$	$\mathbf{A}^T\mathbf{A}$
Poisson noise c_P	$\mathbf{1}^T(\mathbf{Af}) - \mathbf{m}^T\log(\mathbf{Af})$	$-\mathbf{A}^T(\frac{\mathbf{m}}{\mathbf{Af}}-1)$	$\mathbf{A}^T\mathrm{diag}\left(\frac{\mathbf{m}}{(\mathbf{Af})(\mathbf{Af})}\right)\mathbf{A}$
Guassian prior s_G	$-\frac{1}{2}\|\mathbf{f}\|^2$	$-\mathbf{f}$	$-\mathbf{I}$
Entropy prior s_E	$\mathbf{1}^T\mathbf{f} - \mathbf{f}^T\log(\mathbf{f}/\mathbf{f}_0)$	$-\log(\mathbf{f}/\mathbf{f}_0)$	$-\mathrm{diag}(1/\mathbf{f})$
TV prior s_T	$-\mathbf{1}^T\beta^{-1}\log(\cosh(\beta\mathbf{Gf}))$	$-\mathbf{G}^T\tanh(\beta\mathbf{Gf})$	$-\mathbf{G}^T\mathrm{diag}\left(\beta\,\mathrm{sech}^2(\beta\mathbf{Gf})\right)\mathbf{G}$

where $\mathbf{f}_0$ is the initial object, for which the ideal choice would be the true object. As the true object is unknown, $\mathbf{f}_0$ is often selected to have a constant value.

For the total variation prior, we use the definition

$$s_T(\mathbf{f}) = -\mathbf{1}^T|\mathbf{Gf}|. \qquad (10)$$

where $\mathbf{G}$ is the Toeplitz matrix representing the 3D convolution with the Laplacian of Gaussian kernel. Since the absolute function is not differentiable, we use the smooth approximation $|t| \approx \beta^{-1}\cosh(\beta t)$, as suggested in [15], where β is a parameter. The advantage of the TV is that it preserves the edges of the object while it smooths out homogeneous regions [16, 6].

In minimising the cost, we need to compute the gradient **g** and Hessian **H** of the likelihood and the prior terms. Table 1 summaries all the functions with their gradients and Hessians, which will be used in the following section.

4. SKILLING–BRYAN MINIMISATION

In this section, we generalise the Skilling–Bryan algorithm to be applied with the Poisson noise model with the three priors, described in the previous section. The method aims at maximising the prior term s subject to the constraint $c = c_{\mathrm{aim}}$, by implicitly adjusting λ so that the constraint is satisfied. More precisely, in each iteration, s and c are projected onto a small dimensional subspace where s attains its maximum while the constraint is satisfied.

The cost function (5) can be approximated with the second-order Taylor expansion. Then the vector **p** is selected so that the posterior cost function q is maximised, i.e.,

$$\max_{\mathbf{p}} q(\mathbf{f}+\mathbf{p}) = q(\mathbf{f}) + \mathbf{g}_q^T\mathbf{p} + \frac{1}{2}\mathbf{p}^T\mathbf{H}_q\mathbf{p}, \qquad (11)$$

where $\mathbf{g}_q = \nabla q$ and $\mathbf{H}_q = \nabla\nabla q$. This implies that $\mathbf{g}_q = \mathbf{g}_s - \lambda\mathbf{g}_c$ and $\mathbf{H}_q = \mathbf{H}_s - \lambda\mathbf{H}_c$ where $\mathbf{g}_c = \nabla c$, $\mathbf{g}_s = \nabla s$, $\mathbf{H}_c = \nabla\nabla c$ and $\mathbf{H}_s = \nabla\nabla s$. Searching for the vector **p** in a high dimensional space is costly. To obtain an efficient computation, Skilling and Bryan suggested maximising the cost function in the trust subspace, as follows.

The solution for (11) is

$$\mathbf{p} = -(\mathbf{H}_q + \gamma \mathbf{I})^{-1}\mathbf{g}_q, \qquad (12)$$

$$\approx -(\mathbf{I} + \gamma^{-1}\mathbf{H}_q)\mathbf{g}_q, \qquad (13)$$

$$\approx -\mathbf{g}_s + \lambda \mathbf{g}_c + \gamma^{-1}[(\mathbf{H}_s - \lambda \mathbf{H}_c)(\mathbf{g}_s - \lambda \mathbf{g}_c)] \qquad (14)$$

where the inversion is approximated upto the second-order. Obviously, the search vector $\mathbf{p}$ is just a linear combination of the six vectors, $\mathbf{g}_c, \mathbf{g}_s, \mathbf{H}_c\mathbf{g}_c, \mathbf{H}_s\mathbf{g}_s, \mathbf{H}_s\mathbf{g}_c$ and $\mathbf{H}_c\mathbf{g}_s$ in the approximation. In practice, the linear combination of the last four terms can be used as one search direction. Thus, we obtain the three search directions

$$\mathbf{e}_1 = \mathbf{fg}_s, \qquad (15)$$

$$\mathbf{e}_2 = \mathbf{fg}_c, \qquad (16)$$

$$\mathbf{e}_3 = \mathbf{fH}_c\left(\frac{\mathbf{e}_1}{\|\mathbf{g}_s\|} - \frac{\mathbf{e}_2}{\|\mathbf{g}_c\|}\right) + \mathbf{fH}_s\left(\frac{\mathbf{e}_1}{\|\mathbf{g}_s\|} - \frac{\mathbf{e}_2}{\|\mathbf{g}_c\|}\right), \qquad (17)$$

which define a 3D subspace. The vector multiplications and divisions are performed element by element.

The gradient directions $\mathbf{g}_s$ and $\mathbf{g}_c$ are replaced by $\mathbf{fg}_s$ and $\mathbf{fg}_c$ to increase the weight for high values in order to achieve the positivity constraint. In the third direction, the normalisation by the length of gradient vector is performed before multiplying with the Hessian matrix. In the case of the entropy prior, the second term in (17) can be dropped since $\mathbf{fH}_s$ is then equal to unity matrix. Note that solving the search vector $\mathbf{p}$ from (12) corresponds to the Levenberg-Marquardt method whereas using the first-order Taylor approximation would coincide with optimisation with line search methods.

Within the 3D subspace, the search vector $\mathbf{p}$ is

$$\mathbf{p} = \mathbf{E}\mathbf{x} = x_1\mathbf{e}_1 + x_2\mathbf{e}_2 + x_3\mathbf{e}_3, \qquad (18)$$

where the matrix $\mathbf{E} = [\mathbf{e}_1\ \mathbf{e}_2\ \mathbf{e}_3]$ and the coefficient vector $\mathbf{x} = [x_1, x_2, x_3]^\mathrm{T}$. Now, we determine $\mathbf{x}$ in this subspace that gives the maximum of s subject to the constraint $c = \tilde{c}_{\mathrm{aim}} < c_{\mathrm{aim}}$. In order to estimate the maximum s in the 3D subspace, the three-element gradients and nine-element Hessian need to be computed. For details, see [2]. Finally, the current $\mathbf{f}$ is moved to the new location by

$$\mathbf{f}_{\mathrm{new}} = \mathbf{f} + \mathbf{E}\mathbf{x}, \qquad (19)$$

while the updated $\mathbf{f}$ needs to be protected against stray on negative values; if $\mathbf{f}_{\mathrm{new}}$ becomes negative, we scale the updated vector $\mathbf{E}\mathbf{x}$ down by the factors of two until it becomes positive. The iteration is repeated until the aim $c = c_{\mathrm{aim}}$ is achieved or the maximum number of iterations is reached. Compared to line search methods, the optimisation scheme requires much less number of iterations but each iteration consists of six projection and back-projection ($\mathbf{A}^\mathrm{T}\mathbf{A}$) pairs. So constructing the more sophisticated search directions is made on the expense of more frequent evaluation of the image formation model.

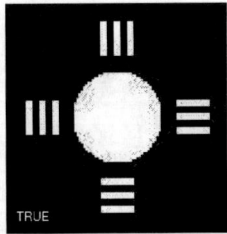 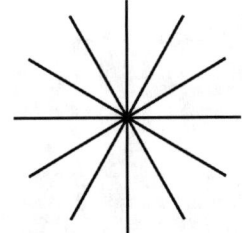

FIGURE 1. Left: The original 2D synthetic object. Right: Ideal geometries of the data acquisition of Micro-Rotation imaging where each line represents the focal plane in the object coordinate frame.

5. RESULTS

To evaluate the proposed method, we have experimented simulated wide-field microscope images. In the simulations, we created a synthetic 2D object in a volume of 100×100 samples. The object consists of a filled circle in the middle and three thin bars, arranged around the sphere in different directions, as illustrated in Figure 1. The object has only two intensity values, one for the background and the other for the object. Poisson noise was added to the object and it was then projected onto 50 one-dimensional views using the simulated micro-rotation imaging model.

In Figure 2, there are six reconstructions computed with different selections from the two noise models and the three prior functions. As the number of simulated photons was relatively large, the reconstructions with Poisson and Gaussian noise models have only a small difference when the same prior function is selected. Likewise, there is only a small difference between the Gaussian and entropy priors but the TV prior is the most competitive in smoothing the artifacts inside the homogeneous region inside the circle while the edges of the circle remain sharp. It can be additionally seen that the three bars oriented in the radial direction can not be reconstructed. We see that this is due to the rotation geometry and the elongation of the PSF function of the microscope which make the resolution weaker in the tangential direction of the rotation [11].

We also experimented with a 3D simulated cell. Figure 3 (left) shows some example images, obtained by the simulated image formation model with Poisson noise. The $100 \times 100 \times 100$ reconstruction using Poisson model and TV prior is illustrated in Figure 3 (right) which shows the rotated slice of the reconstruction corresponding to the original images. In the simulation, the algorithm converged in about 100 iterations for both Gaussian and Poisson noise models. In the early iterations, the algorithm typically has slow convergence but it becomes faster nearer the optimum. In addition, we found only slight differences in convergence curves between different prior models. However, the convergence of the algorithm highly depends on sharpness of the PSF; the narrower the PSF width, the faster the convergence seems to be.

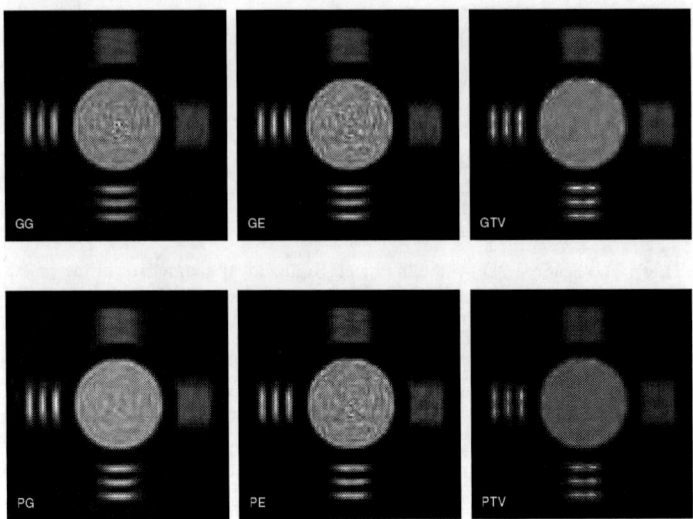

FIGURE 2. 2D reconstructions of simulated data using different priors and noise models. Upper row: Gaussian noise; lower row: Poisson noise; columns from left to right: Gaussian, Entropy, and Total Variation priors.

6. DISCUSSION

In this work, we have generalised the original Skilling–Bryan method to be applied with the Poisson noise model as well as the various kinds of prior structures, including the Gaussian, the Entropy and the TV priors. In general, our extended algorithm should be able to find the optimum solution for arbitrary convex, nonlinear cost functions with the positivity constraint. Only the gradients and Hessians of the cost functions is required for this method. The main features of the Skilling–Bryan method is that (1) a subspace of several search directions is used instead of a line search, (2) the algorithm directly controls the noise level constraint over the likelihood function so that the selection of regularisation parameters becomes implicit, and (3) the method inherently enforces the positivity constraint during each subspace search. These features make our extended algorithm attractive for solving high-dimensional, nonlinear problems. In addition, our experiments in the Micro-Rotation setting suggested that the extended algorithm is appropriate for solving the wide class of measurement noise and prior models within the Bayesian framework.

ACKNOWLEDGMENTS

This work is supported by the European Commission (NEST 2005 programme) in consortium AUTOMATION (Coordinator S. L. Shorte, Institut Pasteur, Paris). We additionally thank Scientific Volume Imaging, AUTOMATION project partner, for providing the 3D cell phantom and the PSF for testing.

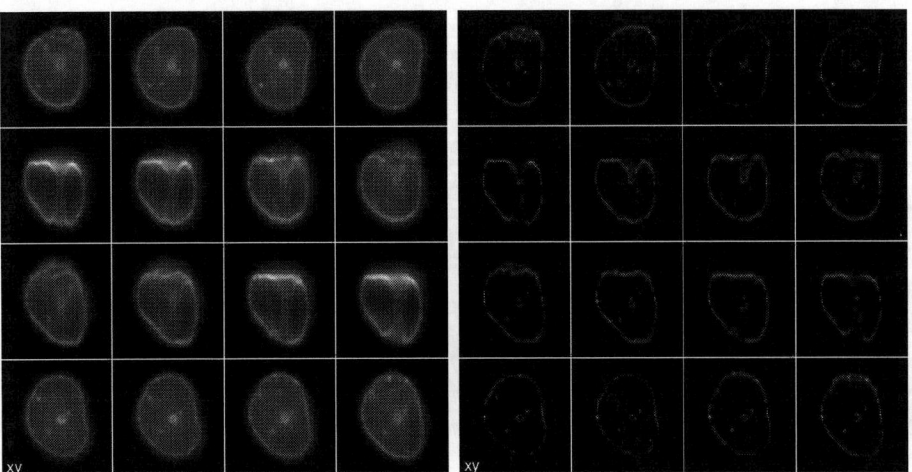

FIGURE 3. Left: Simulated, original wide-field micro-rotation images (SNR=20). Right: Deblurred images, i.e. the ideal slices of the reconstruction from the positions corresponding to the focal plane of the original images.

REFERENCES

1. J. Nocedal, and S. Wright, *Numerical Optimization*, Springer Verlag, 2006, 2nd edn.
2. J. Skilling, and R. K. Bryan, *Mon. Notices R. Astron. Soc.* **211**, 111–124 (1984).
3. J. Skilling, "Classic Maximum Entropy," in *Maximum Entropy and Bayesian Methods*, edited by J. Skilling, Kluwer Academic, 1989.
4. W. H. Press, S. A. Teukolsky, W. T. Vetterling, and B. P. Flannery, *Numerical Recipes in C: The Art of Scientific Computing*, Cambridge University Press, 1992.
5. P. J. Verveer, M. J. Gemkow, and T. M. Jovin, *Journal of Microscopy* **193**, 50–61 (1999).
6. N. Dey, L. Blanc-Féraud, C. Zimmer, Z. Kam, P. Roux, J. Olivo-Marin, and J. Zerubia, *Microscopy Research Technique* **69**, 260–266 (2006).
7. T. Schnelle, R. Hagedorn, G. Fuhr, S. Fiedler, and T. Müller, *Biochim. Biophys. Acta.* **1157**, 127–140 (1993).
8. G. Fuhr, T. Müller, T. Schnelle, R. Hagedorn, A. Voigt, S. Fiedler, W. M. Arnold, U. Zimmermann, B. Wagner, and A. Heuberger, *Naturwissenschaften* **81**, 528–535 (1994).
9. R. Lizundia, L. Sengmanivong, J. Guergnon, T. Müller, T. Schnelle, G. Langsley, and S. L. Shorte, *Parasitology* **130**, 1–7 (2005).
10. O. Renaud, R. Heintzmann, A. Saez-Cirion, T. Schnelle, T. Müller, and S. L. Shorte, *Current Opinion in Microbiology* (2007).
11. D. Laksameethanasan, S. S. Brandt, P. Engelhardt, O. Renaud, and S. L. Shorte (2007), submitted for publication.
12. D. Laksameethanasan, S. S. Brandt, and P. Engelhardt, "A Three-Dimensional Bayesian Reconstruction Method with the Point Spread Function for Micro-Rotation Sequences in Wide-Field Microscopy.," in *Proc. IEEE International Symposium on Biomedical Imaging*, Arlington, VA, USA, 2006, pp. 1276–1279.
13. J. Kaipio, and E. Somersalo, *Statistical and Computational Inverse Problems*, Springer Verlag, 2004.
14. I. Csiszár, *Ann. Stat.* **19**, 2032–2066 (1991).
15. V. Kolehmainen, S. Siltanen, S. Järvenpää, J. Kaipio, P. Koistinen, M. Lassas, J. Pirttilä, and E. Somersalo, *Phy. Med. Bio.* **48**, 1465–1490 (2003).
16. L. Rudin, S. Osher, and E. Fatemi, *Physica D* **60**, 259–268 (1992).

Bayesian Estimator for Angle Recovery: Event Classification and Reconstruction in Positron Emission Tomography

Angela M K Foudray*,† and Craig S Levin*

*Stanford University
Molecular Imaging Program at Stanford
Department of Radiology
Palo Alto, CA, USA
†University of California, San Diego
Department of Physics
La Jolla, CA, USA

Abstract. PET at the highest level is an inverse problem: reconstruct the location of the emission (which localize biological function) from detected photons. Ideally, one would like to directly measure an annihilation photon's incident direction on the detector. In the developed algorithm, Bayesian Estimation for Angle Recovery (BEAR), we utilized the increased information gathered from localizing photon interactions in the detector and developed a Bayesian estimator for a photon's incident direction. Probability distribution functions (PDFs) were filled using an interaction energy weighted mean or center of mass (COM) reference space, which had the following computational advantages: (1) a significant reduction in the size of the data in measurement space, making further manipulation and searches faster (2) the construction of COM space does not depend on measurement location, it takes advantage of measurement symmetries, and data can be added to the training set without knowledge and recalculation of prior training data, (3) calculation of posterior probability map is fully parallelizable, it can scale to any number of processors. These PDFs were used to estimate the point spread function (PSF) in incident angle space for (i) algorithm assessment and (ii) to provide probability selection criteria for classification. The algorithm calculates both the incident θ and ϕ angle, with $\sim$16 degrees RMS in both angles, limiting the incoming direction to a narrow cone. Feature size did not improve using the BEAR algorithm as an angle filter, but the contrast ratio improved 40% on average.

Keywords: 3D detectors, Compton scatter, event filtering, angular resolution, Bayes, PET, medical imaging, measurement estimation, event classification
PACS: 02.30.Zz, 02.50.Tt, 29.40.Gx, 87.58.Fg, 87.58.Pm, 87.58.Ce

INTRODUCTION

Positron emission tomography (PET) imaging has become a workhorse of disease detection and management. To resolve and detect smaller volumes of molecular function, requisite to image early stage cancer, higher resolution cameras are being developed. PET assumes the anti-co-linear (exactly opposite trajectory) production of two 511 keV photons from the annihilation of a positron emitted from an introduced isotope. Detecting both photons from one annihilation then localizes the possible emission positions to a line. All of the lines formed from the photon pairs, called lines of response (LORs), are used to tomographically reconstruct the emission's three-dimensional location(s) within the body. Traditional PET detectors have relatively large detection volumes over which the multiple interactions involved in high energy annihilation photon transport cannot be distinguished. The next generation PET detectors are beginning to detect these individual interactions, giving rise to more complicated, but potentially higher accuracy, event classification.

This work developed a Maximum Likelihood algorithm that includes a realistic model of the detector system (material interaction physics; detector energy, position and time resolution; device centroiding; and crystal binning and gaps) and incorporates Bayesian methods for extracting a single photon's incident angle with respect to the detector using the information from the detected multiple interactions.

Annihilation Photon Transport

The two major sources of background noise in a coincidence-collimated emission tomography system are present due to the finite energy and time resolution of the detection system, and will be discussed below. The finite position resolution of the detection system contributes to local intensity blurring in the reconstructed image and could also conceivably be addressed with a Bayesian estimation scheme, but wasn't developed here. Photons, at the energies used in emission imaging, are generally not fully absorbed upon interaction with matter. Annihilation photons interact most often via Compton interaction, so we will begin there.

Compton Scattering

Finite energy resolution in the detection system contributes to the inability to distinguish photons that have undergone small angle scatter before being detected. The photon's energy after a Compton Scatter is a function of its pre-interaction energy and the angle at which it scattered:

$$E'(E,\theta) = \frac{E}{1+\alpha(E)(1-cos(\theta))} \qquad (1)$$

where E and E' are the energies before and after scattering respectively, θ is the scattering angle, and $\alpha(E)$ is the relative energy of the photon to the rest mass of an electron. For an incoming annihilation (511 keV) photon, $\alpha = 1$. Detectors used for positron emission tomography have energy resolutions anywhere from 3 to 20%, but typically around 15% at 511 keV for the best clinical systems. Therefore by energy filtering, we are rejecting the photons that have undergone relatively large angle scatter. An event pair that contains one or two scattered photons, but are not filtered out by energy discrimination due to this energy blur, we call "Scatters". They cannot be labeled in a traditional system (if they could, those events would be eliminated!), but can be estimated. They can be directly quantified in a Monte Carlo simulation, and are described below (figure 1).

Random Coincidence

The other major contribution to large scale noise is due to the finite time resolution of a PET system. Events are constructed from pairing detected incident single photons. An annihilation event produces two photons at precisely the same time. A blur in timing information leads to possible pairing of detected incident photons from separate annihilations, which incurs contrast loss. Events that incorrectly pair photons from separate annihilation events are call "Randoms". One from a pair of generated photons can go undetected by (1) passing through the detectors without (completely) interacting, (2) traveling along a path that isn't directed toward detectors, (3) being absorbed in the body before getting to the detectors. "Singles" are events in which only one photon was detected. "Multiples" occur when multiple annihilation events have produced three or more detected photons within a chosen time window. Figure 1 shows the fractional rate of the "Trues", "Randoms" and "Scatters" events at typical activities used in laboratory studies, where "Trues" are coincidences formed from photons from the same emission event (neither photon has scattered). The events in figure 1 have survived the energy and time windows, and in a traditional system, would all be used in the reconstruction.

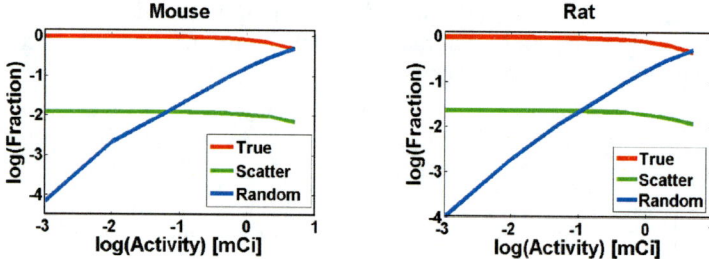

FIGURE 1. Fraction of events that are true, scattered and random coincidences for mouse and rat-sized objects, normalized to total events.

MULTIPLE INTERACTION BASED ELECTRONIC COLLIMATION (MIBEC)

Having information about the characteristics of an incoming photon could allow a higher degree of certainty about how that detected photon (and ultimately event) should be used, if at all. Benefits of having direction and energy information about a photon incident on the detectors include: more precise event typing (Singles, Randoms, Scatters, Multiples, Trues) for 1) filters, 2) reclamation of previously unusable events [1] and 3) for LOR assignment (event positioning). This paper studies the first benefit. The interactions occurring in the imaged object (Compton, Photoelectric, etc.), also occur in the detector. However, in the detector, the deposited energy is recorded, leaving information about the path of the photon. This group of detected interactions we will call clusters. This study undertakes using this information to estimate the incident direction of the photon and how this information affects spatial resolution and contrast.

Instrumentation Considerations

Often, tomographic emission imaging detectors are read-out in a multiplexed fashion, generally to reduce cost and complexity. Device multiplexing will inhibit the ability to distinguish individual interactions that occurred within the multiplexed arrangement during the finite acquisition integration time. In the position-sensitive avalanche photodiodes (PSAPDs) used in this study[2], multiple interactions within a single detector module during the charge integration period cannot be distinguished, but inter-detector interactions can be separately localized. Since the PSAPD photodetector has the highest quantum efficiency when detecting optical photons (~550-800 nm), the scintillation crystals are used to "convert" high energy photon interactions into optical wavelength photons. Optical transport in the scintillation crystal also contributes to spatial blurring, discretization and perceived energy loss from optical photon absorption. Photoelectric interactions and their characteristic x-rays, Bremsstrahlung, and Doppler broadening can shift the magnitude and location of energy deposited. The complex forward model coupled with device and electronic multiplexing make methods that have a less strict (or non-strict) requirement on the details of the forward model, such as Bayesian estimation, ideal candidates[3] for event processing.

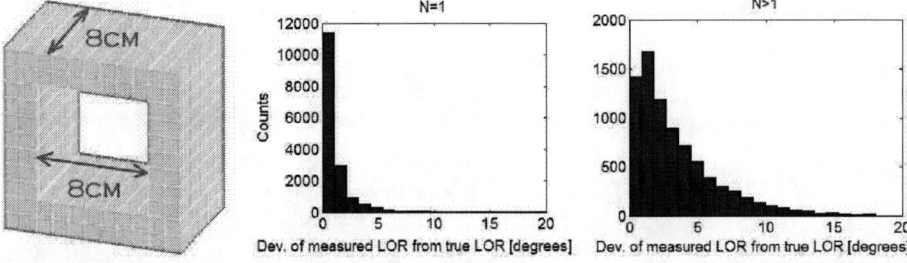

FIGURE 2. Left: an illustration of the high resolution PET detection system[2]. Center: histogram over all LORs of the angular difference between lines formed from the COM of a measured event cluster and 1) the COM of its coincident cluster, and 2) the annihilation photon emission point, for the N=1 case, Right: for the N>1 case. Events used in (center) and (right) were gathered for a source uniformly filling the interior of the detector.

Shown in figure 2b,c is the convergence of all of the particle transport and instrumentation blurring, quantified in angle space, knowing the emission position. This is, in effect, the intrinsic ability of the system (detection + clustering + positioning scheme) to localize the original photon emission (using the COM positioning algorithm), or the histogrammed LOR deviation due to intercrystal scatter and positioning. This deviation is histogrammed in figure 2b,c for N=1 and N>1, where N is the number of detected interactions in the cluster. Since this paper is assessing a MIBEC algorithm (N>1), for this instrumentation[2], figure 2c indicates the distribution of the LOR deviation we can expect inherently in the data to which the algorithm will be applied.

BEAR: A NAIVE BAYESIAN CLASSIFIER

Incident Angle Estimation

We consider a cluster of N interactions from a single photon entering the system, each interaction defined by their relative positions and energies (x_i, y_i, z_i, E_i) abbreviated X_i, and computing the probability of finding these interactions jointly for a given incident photon direction (θ, ϕ) abbreviated Θ (figure 3):

$$P(\Theta|X_1,...,X_N) = \frac{P(X_1,...,X_N|\Theta) \cdot P(\Theta)}{P(X_1,...,X_N)} \quad (2)$$

where the prior $P(\Theta)$, using the theoretical assumption that the singles are generated isotropically throughout the volume, was assumed to be one (all equally likely). Using the joint probability definitions, this can then be written as:

$$P(\Theta|X_1,...,X_N) = \prod_{i=1}^{N>1} \frac{P(X_i|X_j, \Theta)}{P(X_i|X_j)} \quad (3)$$

where X_j is $(X_{i-1}, X_{i-2}, ..., X_1)$. When i=1, X_j is $\oslash$. The decision rule for angle recovery is simply

$$\Theta_{cluster} = sup\{P(\Theta|X_1,...,X_N)\} \quad (4)$$

Event Space Representation

Separate methods were used to discretize the position and energy values for each interaction in the cluster[4]. An unsupervised static local discretization was used for the position dimensions, using equal width intervals in a space relative to the center of mass of the cluster. This COM reference space had a number of advantages: (1) a significant reduction in the size of the data in measurement space, making further manipulation and searches faster (2) the construction of COM space does not depend on measurement location, it takes advantage of measurement symmetries, and data can be added to the training set without knowledge and recalculation of prior training data, (3) calculation of posterior probability map is fully parallelizable, it can scale to any number of processors. A supervised static global discretization was used for the energy dimension. The energy bins near to large angle scatter and photoelectric interactions were sampled coarsely, whereas the small energy deposition interactions were sampled with more resolution to distinguish processes at low energies, while still keeping the number of bins low.

Probability Density Maps

The probabilities used in eq. 3 were calculated for each interaction cluster in event space for each incident (θ, ϕ). Two sets of interaction locations and energies for each location and angle studied were generated - one for the training set (likelihood and evidence calculations, right hand side of eq. 3) and one for the test set (posterior probabilities, left hand side of eq. 3).

Training sets were generated over incident angle space using the forward model of the annihilation photon transport physics in the Geant4 Application for Tomographic Emission (GATE) Monte Carlo package; the most validated simulation software for detailed PET physics [5]. A pencil beam with zero angular extent was directed into the detector system from a point source for every θ and ϕ angle value in the range detected by the system, which is nearly π radians in both θ and ϕ (see figure 3 for the trained locations).

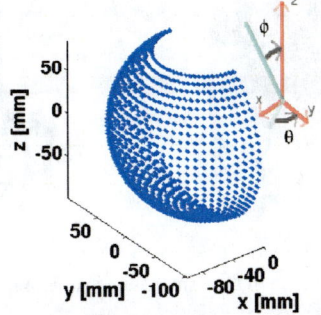

FIGURE 3. Illustration of coordinate system and the beam locations of the simulation annihilation photon source locations in (x,y,z). Each blue dot represents a training set of ~25,000 events. In the reference space used in the figure, the x-direction is the direction normal to the face of the detector.

Posterior probability maps were determined for each event by calculating the right hand side of eq. 3 at each (θ, ϕ), using the probabilities of the cluster interactions determined by the training set data. Figure 4 shows an example of these PDFs during the calculation of equation 4 for a typical cluster (one event) interacting in a central module of one sides of the system.

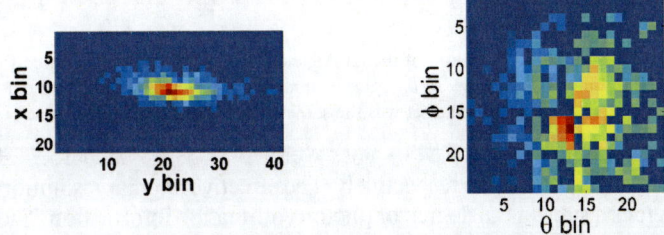

FIGURE 4. Log of the probability density maps for (a) likelihood in X_i space (an x-y plane slice), for $\theta=17$ and $\phi=13$, and (b) posterior probability in Θ space, for a particular interaction cluster.

Prediction Capabilities

The algorithm's ability to predict photon incident angle, only knowing the interaction locations and energies, was measured using test sets of known incident angle data, and calculating the point spread functions in (θ,ϕ) space (see figure 5). ϕ and θ are the angles that are made with the axial and radial direction (z and x in figure 3), respectively. The root mean squared (RMS) deviation from the true angle bin was used to calculate the angular resolution of the algorithm shown in figure 6a and b.

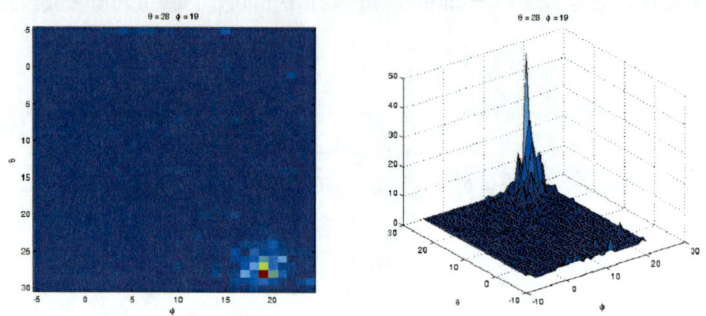

FIGURE 5. Point spread function for a single incident θ and ϕ value (θ=17 and ϕ=13).

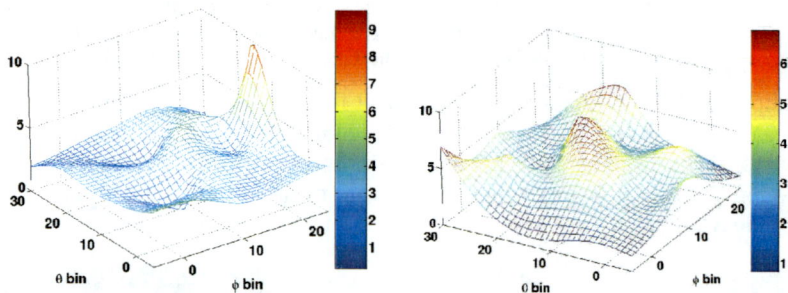

FIGURE 6. Shown is the RMS deviation of the BEAR-estimated angle from the angle of the line that goes through the annihilation in (a) ϕ and (b) θ angles in angle bins (vertical axis) where 1 bin = 4 degrees. The distribution of sources for the test data set was the same as that for the training data set.

The mean and variance of the RMS values were $15.2° \pm 5.1°$ and $15.7° \pm 6.9°$ for the theta and phi angle estimation respectively. The effective angular resolution from this algorithm provides a substantial degree of photon collimation information. The variation in the RMS values across all angular space in θ and ϕ would most likely go down as a result of including more event interaction data in the training set. With higher training set counts, the RMS values themselves will likely asymptotically approach a constant but non-zero value due to the finite energy resolution and position blurring effects (binning, scintillation light transport) in the detector modules.

Image Reconstruction in the Presence of Biologically Relevant Background

The BEAR algorithm was used in this study as a simple filter to reject lines whose estimated angles deviated more than a selected degree from the angle calculated from coincidence pairing. This deviation was chosen to be a 24 degree window (± 12 degrees), to enable a high number of accepted events for good statistics. Further, the PDF could be used as a weighted projector in reconstruction for a myriad of functions: multiples selection, single and scatter reconstruction, leading to new advances in tomographic reconstruction.

Detection in an emission tomography system is dependent on the system's ability to resolve sources of activity, in this paper quantified by the perceived spread of activity (FWHM) and relative intensity to background (contrast ratio (CR)). In this study, a plane of sphere sources; each quadrant with an array of a single diameter (1.25, 1.5, 2.5 and 3.5 mm in quadrant I, II, III, IV respectively), centers separated by 4 times their diameter; were simulated in a 6cm diameter, 8cm long, cylinder of scattering medium (water). The activity concentration in the spheres was 9 times that of the uniform background in the cylinder.

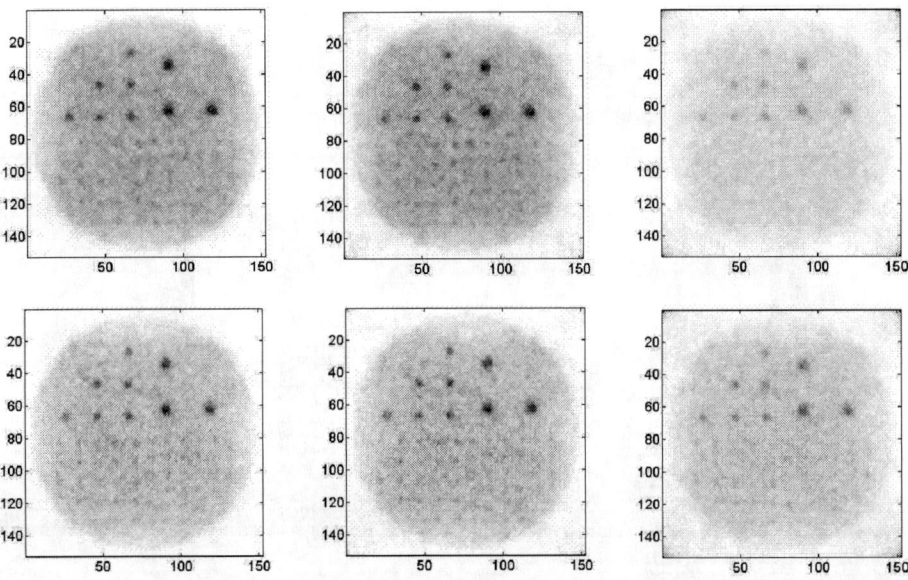

FIGURE 7. Shown are the unfiltered images (top) and filtering with the BEAR algorithm (bottom); for the 1% (left), 18% (middle), and 50% (right) acquisitions; each reconstructed by the iterative image reconstruction method Ordered Subset Expectation Maximization [6]: 12 subsets, 1 iteration, keeping the number of true coincidences the same in each study. Inverse grey scale: Darker represents higher intensity.

As the total activity imaged in the detector effects the fraction of random counts accepted by the system (figure 1), three studies, with different total activities were acquired: 100μCi, 1mCi, and 5mCi, leading to acquisitions with about 1%, 18% and

50% of the counts being random events (with volumetrically uniform background). The energy and arrival time were distributed according to Gaussian distributions with 12% FWHM, and 2 ns FWHM, respectively, and the window settings were twice those resolution values. The UF (unfiltered) data was filtered before reconstruction via BEAR to produce the BEAR data. The number of counts used in reconstruction were: 144 (UF) and 94 (BEAR) for the 1% case, 174 (UF) and 106 (BEAR) for the 18% case, and 348 (UF) and 197 (BEAR) for the 50% case, in millions of counts, respectively.

The features that are apparent in all images are the 2.5 and 3.5 mm spheres. In the 1% and 18% randoms fraction images (left and center in figure 7), some of the smaller spheres would potentially gain further contrast with more counts (increased SNR). A fit was made to each 2D image using a 2D Gaussian plus a contrast background (5 fit parameters) using MATLAB's fminsearch. The FWHM of the 2.5 and 3.5 mm features were extracted from the fit (table 1), and didn't improve significantly by using the BEAR algorithm.

TABLE 1. Fitted FWHM of the 3.5mm and 2.5mm spheres in all reconstructed images.

% Randoms	Filter	3.5mm FWHM	3.5mm StDev	2.5mm FWHM	2.5mm StDev
1	None	2.83	0.37	2.29	0.14
1	BEAR	2.80	0.22	2.11	0.26
18	None	2.80	0.47	2.11	0.27
18	BEAR	2.88	0.30	2.11	0.32
50	None	3.17	0.78	2.46	0.49
50	BEAR	3.06	0.24	2.35	0.12

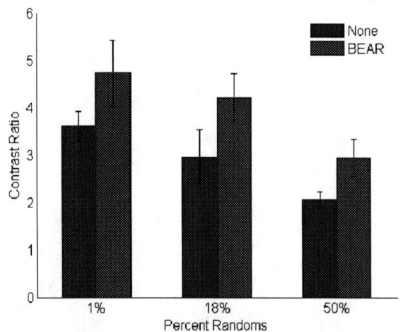
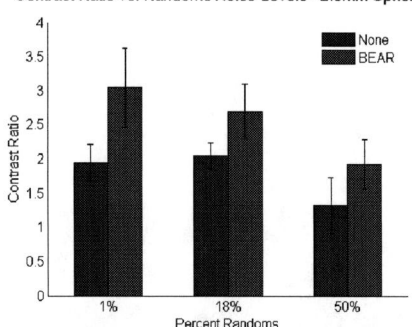

FIGURE 8. Shown are the contrast ratios for the 3.5 and 2.5 mm spheres in the three noise regimes for both unfiltered and BEAR filtered images.

Contrast ratio, on the other hand, improved markedly. The contrast ratio was defined in this paper by the peak height of the Gaussian fit divided by the constant background fit. Since a source concentration ratio of 9 used in the study, the reconstructed contrast ratio would ideally also be 9. However blurring effects in the detection and filtering schemes attenuate and spread the activity away from the point of origin. Using the BEAR algorithm, the 3.5mm spheres improved in CR by 38.4%±6.41%, and the 2.5mm spheres by 44.1%±12.4% across the images.

Removing object scatter and random counts alone does not account for the number of counts filtered or the contrast ratio improvement using this algorithm in the low activity data set. Mispositioning of the LOR due to intercrystal scatter is not activity dependent and would contribute to the background near the correct LOR (feature location) as seen from figure 2. These intercrystal scatter events, which have angular deviation from the true LOR, are also filtered by this algorithm. Further study will be performed using a contrast ratio phantom to investigate this issue.

ACKNOWLEDGMENTS

The authors would like to thank the Molecular Imaging Program at Stanford, and particularly the Sam Gambhir lab, for the use of the 32-node reconstruction cluster. We would also like to thank Garry Chinn for the use of the list-mode Ordered Subset Expectation Maximization reconstruction code used in this work.

Angela M K Foudray would like to thank PEO International for the Scholar Award, which partially funded this work. Angela M K Foudray would also like to thank Kevin Knuth and the MaxEnt organization for full sponsorship to attend and present this work at the 27th International MaxEnt Workshop.

This work was also supported in part by the following grants: NIH-National Cancer Institute R21 CA098691, NIH-NIBIB R21/R33 EB003283, and NIH-NCI CA119056.

REFERENCES

1. G. Chinn, A. M. K. Foudray, and C. S. Levin, *IEEE Nuclear Science Symposium and Medical Imaging Conference Record* pp. 1746–1751 (2006).
2. A. M. K. Foudray, G. Chinn, and C. S. Levin, *IEEE NSS-MIC Conference Record* **4**, 2108–2111 (2005).
3. D. J. C. MacKay, *Information Theory, Inference, and Learning Algorithms*, Cambridge University Press, 2005, version 7.2, 4th edn.
4. J. Dougherty, R. Kohavi, and M. Sahami, *Machine Learning: Proceedings* (1995).
5. S. Jan, and et. al., *Phys. Med. Bio.* **49**, 4543–4561 (2004).
6. H. M. Hudson, and R. S. Larking, *IEEE Transactions on Medical Imaging* **13**, 601–609 (1994).

Bayesian tomographic reconstruction of microsystems

Sofia Fekih Salem*, Alexandre Vabre* and Ali Mohammad-Djafari[†]

*CEA, LIST, Laboratoire Images et Dynamique, 91191 Gif-sur-Yvette, France
[†]Laboratoire des Signaux et Systèmes,
Unité mixte de recherche 8506 (CNRS-Supélec-UPS 11)
Supélec, Plateau de Moulon, 3 rue Joliot-Curie, 91191 Gif-sur-Yvette, France

Abstract. The microtomography by X ray transmission plays an increasingly dominating role in the study and the understanding of microsystems. Within this framework, an experimental setup of high resolution X ray microtomography was developed at CEA-List to quantify the physical parameters related to the fluids flow in microsystems. Several difficulties rise from the nature of experimental data collected on this setup: enhanced error measurements due to various physical phenomena occurring during the image formation (diffusion, beam hardening), and specificities of the setup (limited angle, partial view of the object, weak contrast).

To reconstruct the object we must solve an inverse problem. This inverse problem is known to be ill-posed. It therefore needs to be regularized by introducing prior information. The main prior information we account for is that the object is composed of a finite known number of different materials distributed in compact regions. This *a priori* information is introduced via a Gauss-Markov field for the contrast distributions with a hidden Potts-Markov field for the class materials in the Bayesian estimation framework. The computations are done by using an appropriate Markov Chain Monte Carlo (MCMC) technique .

In this paper, we present first the basic steps of the proposed algorithms. Then we focus on one of the main steps in any iterative reconstruction method which is the computation of forward and adjoint operators (projection and backprojection). A fast implementation of these two operators is crucial for the real application of the method. We give some details on the fast computation of these steps and show some preliminary results of simulations.

Key Words: Tomography, Image reconstruction, Bayesian inversion, MCMC, X ray, Projection, Backprojection.

Introduction

We consider the case of the microtomography by X ray transmission. Each part of the object attenuates the flux of photons according to its nature and thickness. The simplest model is the line integration model:

$$g(\mathbf{s}_i) = \int_{L_i} f(\mathbf{r}) dl_i \quad (1)$$

where $\mathbf{r} = (x,y,z) \in \mathfrak{R}$ is a voxel position, $\mathfrak{R}$ is the volume of the considered object, $\mathbf{s}_i$ is a detector position, L_i is a line connecting the X ray source position to the detector position $\mathbf{s}_i$ and dl_i is a unit element of this line. When discretized this equation becomes

$$\mathbf{g} = \mathbf{Hf} + \varepsilon \quad (2)$$

where $\mathbf{f} = (f(\mathbf{r}), f(\mathbf{r}) \in \Re^N, \mathbf{r} \in \Re)$ is a vector containing the discretized voxel values of the object, $\mathbf{g} = (g(\mathbf{s}_i), g(\mathbf{s}_i) \in \Re^M)$ is a vector containing the values of the projection data, ε is a vector representing the modeling and measurement noise and $\mathbf{H}$ is the matrix of the projection coefficients, also called "projection matrix".

The resolution of such a system has some specificities:

- The number of equations: if N indicates the number of voxels and M indicates the number of projection rays, the size of the matrix is equal to $M.N$. In real applications, M and N are of 10^6 in $2D$ and 10^9 in $3D$;
- The sparse matrix $\mathbf{H}$: the projection rays cross few voxels. Generally, less than 1% of matrix elements are significant.
- The existence and uniqueness of the solution: the acquired data are often noisy so the system can either be unconsistant or have an infinite number of possible solutions.

The problem is then severely ill-posed and prior knowledge is needed to obtain satisfactory reconstruction results. There are many works dealing with this inverse problem([1]).
Our resolution is based on the bayesian formulation which expresses the conditional probability $p(\mathbf{f} \mid \mathbf{g})$ as:

$$p(\mathbf{f} \mid \mathbf{g}) = \frac{p(\mathbf{g} \mid \mathbf{f}) p(\mathbf{f})}{p(\mathbf{g})} \quad (3)$$

where the likelihood $p(\mathbf{g} \mid \mathbf{f})$ represents the knowledge of the physical process of the measurements and $p(\mathbf{f})$ is the prior distribution of the object. When the Posterior probability $p(\mathbf{f} \mid \mathbf{g})$ is determined, one common solution is to find the Maximum A Posteriori which can be compared to the general regularization theory.

$$\hat{\mathbf{f}} = \arg\max_{\mathbf{f}}(p(\mathbf{f} \mid \mathbf{g})) = \arg\min_{\mathbf{f}}[-\ln p(\mathbf{g} \mid \mathbf{f}) - \ln p(\mathbf{f})] \quad (4)$$

The solution is defined as the minimizer of a compound criterion. This approach has been used with success in many applications([2], [3], [4], [5]). The main contribution of those works are in choosing appropriate regularization functional or equivalently appropriate prior probability laws for $\mathbf{f}$ to enforce some particular properties of the object such as smoothness, positivity or piecewise smoothness ([5], [6], [7]).

The studied objects consist of K known different materials. So, we know *a priori* that the reconstructed object is composed of a limited number of compact homogeneous regions with limited known type K of materials.

Proposed method

As we want to reconstruct an image with statistically homogeneous regions, we introduce a hidden variable $\mathbf{z} = (z(1), z(2), ..., z(N)) \in \{1, ..., K\}^N$ which represents the classification of the image voxels $\mathbf{f}$. The problem is now to estimate the set of variables $(\mathbf{f}, \mathbf{z})$ using the Bayesian approach

$$p(\mathbf{f},\mathbf{z}\mid\mathbf{g}) \propto p(\mathbf{g}\mid\mathbf{f})p(\mathbf{f}\mid\mathbf{z})p(\mathbf{z}) \quad (5)$$

Thus to be able to give an expression for $p(\mathbf{f},\mathbf{z}\mid\mathbf{g})$ using the Bayes formula, we need to define $p(\mathbf{g}\mid\mathbf{f})$, $p(\mathbf{f}\mid\mathbf{z})$ and $p(\mathbf{z})$. We assume the noise to be centered, white and Gaussian with the covariance matrix $\Sigma_\varepsilon = \sigma_\varepsilon^2 \mathbf{I}$

$$p(\mathbf{g}\mid\mathbf{f}) = N(\mathbf{Hf},\Sigma_\varepsilon) \propto \exp[\frac{-1}{2\sigma_\varepsilon^2}\|\mathbf{g}-\mathbf{Hf}\|^2] \quad (6)$$

If we note by $f_k = \{f(\mathbf{r}), r \in \Re_k\}$, the voxels of the image in class k are characterized by their mean m_k and their variance v_k. We can write:

$$p(f(\mathbf{r})\mid z(\mathbf{r})=k) = N(m_k,v_k), k=1,...,K \quad (7)$$

If we note by $p(z(\mathbf{r})=k) = \pi_k$, then this model can be recognized as a mixture of Gaussian model:

$$p(f(\mathbf{r})) = \sum_k \pi_k N(m_k,v_k) \quad (8)$$

We have now to assign $p(\mathbf{z})$. As we introduced the hidden variable $\mathbf{z}$ for finding statistically homogeneous regions in the images, it is natural to define a spatial dependency on these labels. The simplest model to account for this desired local spatial dependency is a Potts Markov Random Field Model [8]:

$$p(z(\mathbf{r})\mid z(\mathbf{r}'), \mathbf{r}' \in V(\mathbf{r})) \propto \exp\left[\alpha \sum_{\mathbf{r}' \in V(\mathbf{r})} \delta(z(\mathbf{r})-z(\mathbf{r}'))\right] \quad (9)$$

where the parameter α controls the mean size of the regions and $V(\mathbf{r})$ represents the set of voxel positions in the neighborhood of $\mathbf{r}$. In this work, we consider the 6-connex neighbours $V(\mathbf{r})$ shown in figure 1. If we note $\{\mathbf{z} = z(\mathbf{r}), \mathbf{r} \in \Re\}$, then using the Hammersley-Clifford theorem (1971)

$$p(\mathbf{z}) \propto \exp\left[\alpha \sum_{\mathbf{r}} \sum_{\mathbf{r}' \in V(\mathbf{r})} \delta(z(\mathbf{r})-z(\mathbf{r}'))\right] \quad (10)$$

In a first case, if we assume that the voxels belonging to different regions are independent, we can write:

$$p(\mathbf{f}\mid\mathbf{z}) \propto \prod_k \prod_{\mathbf{r} \in \Re_k} \exp\left[\frac{-1}{2v_k}|f(\mathbf{r})-m_k|^2\right]$$

$$\propto \prod_k \exp\left[\frac{-1}{2v_k}|f_k - m_k \mathbf{1}|^2\right]$$

$$\propto \prod_{\mathbf{r} \in \Re} \exp[\frac{-1}{2}(\frac{f(\mathbf{r})-m(\mathbf{r})}{\sqrt{v(\mathbf{r})}})^2]$$

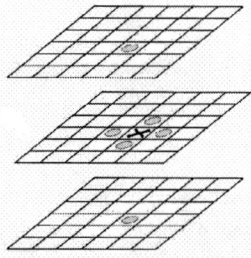

FIGURE 1. The 6-connex neighbours V(**r**)

where we used the notations $\Re_k = \{\mathbf{r} : z(\mathbf{r}) = k\}$ with $\bigcup_k \Re_k = \Re$, $m(\mathbf{r}) = \{m_k, \mathbf{r} \in \Re_k\}$ and $v(\mathbf{r}) = \{v_k, \mathbf{r} \in \Re_k\}$

In this model, all the voxels in a given class are assumed to have the same mean but they are not correlated. This hypothesis is not realistic. A better hypothesis is to account for a local Markovian relation between them (second case). To account for this property of such voxels, we introduce a contour hidden variable $q(\mathbf{r})$ which is related to the classification hidden variable $z(\mathbf{r})$ by a deterministic relation $q(\mathbf{r}, \mathbf{r}') = \delta(z(\mathbf{r}) - z(\mathbf{r}'))$ where $\mathbf{r}'$ represents a position in the neighborhood $V(\mathbf{r})$ of $\mathbf{r}$. With this new hidden variable, we can propose the following model:

$$p(\tilde{f}(\mathbf{r}) \mid q(\mathbf{r}, \mathbf{r}'), \tilde{f}(\mathbf{r}'), \mathbf{r}' \in V(\mathbf{r})) = N(\bar{\tilde{f}}(\mathbf{r}), \sigma_f^2) \quad (11)$$

where, again $V(\mathbf{r})$ represents the 6-connex neighbours of $\mathbf{r}$,

$$\tilde{f}(\mathbf{r}) = \frac{f(\mathbf{r}) - m(\mathbf{r})}{\sqrt{v(\mathbf{r})}} \quad (12)$$

$$\bar{\tilde{f}}(\mathbf{r}) = \beta(\mathbf{r}) \sum_{\mathbf{r}' \in V(\mathbf{r})} (1 - q(\mathbf{r}, \mathbf{r}')) \tilde{f}(\mathbf{r}) \quad (13)$$

$$\beta(\mathbf{r}) = \frac{1}{\sum_{\mathbf{r}' \in V(\mathbf{r})} (1 - q(\mathbf{r}, \mathbf{r}'))} \quad (14)$$

We have now all necessary prior laws and then we can give an expression for $p(\mathbf{f}, \mathbf{z}, \theta \mid \mathbf{g})$. However these probability laws have unknown parameters such as $\theta = (\theta_\varepsilon, \theta_f)$ where $\theta_\varepsilon = \sigma_\varepsilon^2$ and $\theta_f = (m_k, v_k)$. We have to assign prior laws to these hyperparameters. In this paper, we used conjugate priors for all of them, Gaussian for the means m_k and inverse Gamma for the variances v_k as well as for noise variance σ_ε^2. We can now use the expression of the posterior law

$$p(\mathbf{f}, \mathbf{z}, \theta \mid \mathbf{g}) \propto p(\mathbf{g} \mid \mathbf{f}, \theta_\varepsilon) p(\mathbf{f} \mid \mathbf{z}, \theta_f) p(\mathbf{z}) p(\theta) \quad (15)$$

either to find a point estimator such as Joint MAP

$$(\widehat{\mathbf{f}},\widehat{\mathbf{z}},\widehat{\boldsymbol{\theta}})_{MAP} = \arg\max_{(\mathbf{f},\mathbf{z},\boldsymbol{\theta})}\{p(\mathbf{f},\mathbf{z},\boldsymbol{\theta}|\mathbf{g})\} \quad (16)$$

or the posterior means or exploring it by generating samples from it using MCMC techniques.

The JMAP solution needs a global optimization algorithm, because $p(\mathbf{f},\mathbf{z},\boldsymbol{\theta}|\mathbf{g})$ may not be unimodal even if it is Gaussian when $\mathbf{z}$ and $\boldsymbol{\theta}$ are given. What we propose here is a combined solution using both optimizations (with respect to $\mathbf{f}$ when $\mathbf{z}$ and $\boldsymbol{\theta}$ is given and with respect to $\boldsymbol{\theta}$ when $\mathbf{f}$ and $\mathbf{z}$ are given) and sampling with respect to $\mathbf{z}$. Unfortunately, we can not give any proof of the convergence of the proposed algorithm or other properties of the obtained solution.

Proposed algorithm

Optimizing the joint *a posteriori* law (eq. 15) can be done in an iterative way:

- Estimate $\mathbf{f}$ using $p(\mathbf{f}\mid\widehat{\mathbf{z}},\widehat{\boldsymbol{\theta}},\mathbf{g})$ where

$$p(\mathbf{f}\mid\mathbf{z},\boldsymbol{\theta},\mathbf{g}) \propto p(\mathbf{g}\mid\mathbf{f},\boldsymbol{\theta}_\varepsilon)p(\mathbf{f}\mid\mathbf{z},\boldsymbol{\theta}_f) \quad (17)$$

$p(\mathbf{g}\mid\mathbf{f},\boldsymbol{\theta}_\varepsilon)$ and $p(\mathbf{f}|\mathbf{z},\boldsymbol{\theta}_f)$ are both Gaussians. Thus the conditional posterior law $p(\mathbf{f}\mid\mathbf{z},\boldsymbol{\theta},\mathbf{g})$ is also Gaussian $p(\mathbf{f}\mid\mathbf{z},\boldsymbol{\theta},\mathbf{g}) = N(\widehat{\mathbf{f}},\widehat{\Sigma})$. In this expression, $\widehat{\mathbf{f}}$ is the posterior mean and $\widehat{\Sigma}$ is the Posterior covariance. $\widehat{\mathbf{f}}$ is also the mode of the Posterior and can be obtained by maximizing $p(\mathbf{f}\mid\widehat{\mathbf{z}},\widehat{\boldsymbol{\theta}},\mathbf{g})$ or equivalently by minimizing $-\ln p(\mathbf{f}\mid\widehat{\mathbf{z}},\widehat{\boldsymbol{\theta}},\mathbf{g})$. It is an optimization of a quadratic criterion :

$$\widehat{\mathbf{f}} = \arg\min_{\mathbf{f}}\{J(\mathbf{f}\mid\widehat{\mathbf{z}},\widehat{\boldsymbol{\theta}},\mathbf{g}) = -\ln p(\mathbf{f}\mid\widehat{\mathbf{z}},\widehat{\boldsymbol{\theta}},\mathbf{g})\} \quad (18)$$

where:

$$J(\mathbf{f}\mid\mathbf{z},\boldsymbol{\theta},\mathbf{g}) = \|\mathbf{g}-\mathbf{H}\mathbf{f}\|^2 + \lambda\sum_k\sum_{\mathbf{r}\in\mathfrak{R}_k}|\frac{f(\mathbf{r})-m_k}{\sqrt{v_k}}|^2 \quad (19)$$

$$= \|\mathbf{g}-\mathbf{H}\mathbf{f}\|^2 + \lambda\sum_{\mathbf{r}\in\mathfrak{R}}|\bar{f}(\mathbf{r})|^2 \quad (20)$$

if we consider the first case, and:

$$J(\mathbf{f}) = \|\mathbf{g}-\mathbf{H}\mathbf{f}\|^2 + \lambda\sum_{\mathbf{r}\in\mathfrak{R}}|\bar{f}(\mathbf{r})-\bar{\bar{f}}(\mathbf{r})|^2 \quad (21)$$

if we consider the second case.
In both cases, $\lambda = 2\sigma_\varepsilon^2$ can be assimilated to a regularization parameter.

- Estimate **z** using $p(\mathbf{z} \mid \widehat{\mathbf{f}}, \widehat{\theta}, \mathbf{g}) \propto p(\mathbf{g} \mid \widehat{\mathbf{f}}, \theta_\varepsilon) p(\mathbf{z})$. The optimization of $p(\mathbf{z} \mid \widehat{\mathbf{f}}, \widehat{\theta}, \mathbf{g})$ is not an easy task. This is due to the fact that $p(\mathbf{z})$ has a markovian structure. We propose a Gibbs Sampling Scheme to generate samples from this Posterior and find the patterns with greatest posterior probabilities.
- Estimate θ using $p(\theta \mid \widehat{\mathbf{f}}, \widehat{\mathbf{z}}, \mathbf{g})$ where

$$p(\theta \mid \mathbf{f}, \mathbf{z}, \mathbf{g}) \propto p(\mathbf{g} \mid \mathbf{f}, \sigma_\varepsilon^2) p(\mathbf{f} \mid \mathbf{z}, \theta_f) p(\theta) \qquad (22)$$

Thanks to the conjugate priors (the Gaussian for the means m_k and inverse Gamma for the variances σ_ε^2 and v_k), the corresponding posteriors are also Gaussian and Inverse Gamma. The analytical expressions of these estimates can be obtained easily. Their detailed expressions can be found in ([9] [10] [11]).

When we initialize $z(\mathbf{r}) = 1$ and $q(\mathbf{r}) = 0, \forall \mathbf{r} \in \Re, \sigma_\varepsilon^2 = 1, m_1 = 0$ and $\sigma_1^2 = 1$, we could see that this step is equivalent to obtain a first solution of the reconstruction problem via a quadratic regularization, or equivalently, via a Gauss-Markov prior modeling of the image. This step is crucial for the success of the method. At this initializing step, obtaining a first estimate for z and then for the hyperparameters is also crucial by using our prior knowledge about K the number of materials, and their means and variances.

Implementation

In this part, we focus on one of the main steps in any iterative reconstruction method which is the computation of forward and adjoint operators (projection and backprojection). For example, if we focus on the first step of the proposed algorithm which needs the optimization of (eq. 20), we see that to compute the criterion J we need to compute **Hf** and to its gradient:

$$\nabla J = -2\mathbf{H}^t (\mathbf{g} - \mathbf{H}\mathbf{f}) + 2\lambda \sum_{\mathbf{r} \in \Re} \overline{f}(\mathbf{r}) \qquad (23)$$

We need also to compute $\mathbf{H}^t \Delta \mathbf{g}$ where $\Delta \mathbf{g} = \mathbf{g} - \mathbf{H}\mathbf{f}$. The first term **Hf** corresponds to the computation of the projections with a given object **f** and the second term $\mathbf{H}^t \Delta \mathbf{g}$ corresponds to the backprojection of the innovation $\Delta \mathbf{g}$.

The computation of those operators is the most costly computation time of our method. For this reason, we implemented it on C language. Those steps are used in the estimation of **f**. The other steps are implemented using Matlab.

Projection consists of calculating the values of the projection data $g(s_i)$ from the values of the object $\mathbf{f} = \{f_j, j = 1..N\}$ (N the number of voxels in the object) and the projection matrix H. This matrix is hollow because the projection rays intercept few voxels (see figure 2). Therefore we use a list of voxels for each projection ray i, noted I_i, and a list of ray projection which intercept each voxel j, noted J_j. The equation becomes:

$$g(s_i) = \sum_{j \in I_i} H_{ij} f_j \qquad (24)$$

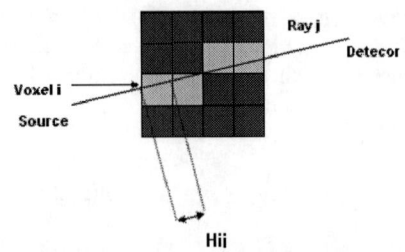

FIGURE 2. Basis of the projection and backprojection

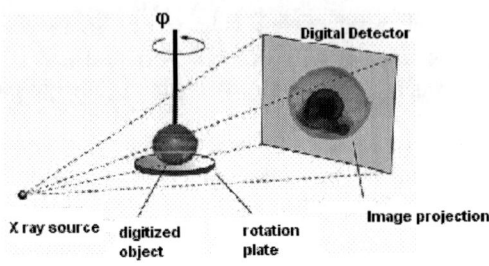

FIGURE 3. Sinogram creation

So we do not memorize all the matrix elements but only its non zero elements. $g(s_i)$ represents a 2D projection for an angle φ. From **g**, we can then create the sinogram (the object is imaged from several angles, producing a set of X-ray projections called sinogram. Each sinogram line corresponds to the X-ray projection at one angle) (see figure 3). In the same manner, backprojection consists on calculating $\mathbf{H}^t \Delta \mathbf{g}$ which is what is called "backprojected reconstruction".

It is expressed by

$$b_j = \sum_{i \in J_j} H_{ij} \Delta g_i \qquad (25)$$

$\underline{b} = b_j, j = 1,..,N$

We may note here that this does not corresponds to $\mathbf{H}^{-1} \Delta \mathbf{g}$. In fact, $\mathbf{H}^t \mathbf{g}$ corresponds to the reconstruction by backprojection which is not very far from $\mathbf{H}^{-1} \mathbf{g}$ if we have a great number of projections uniformly distributed around the object. We simulated the projection and the backprojection of an homogeneous cube using an image detector as used for the experiment (256 × 256 pixels). The dimensions of the simulated cube is 64 × 64 × 64. Figure 4 shows the sinogram construted from the computed projections of the cube. Figure 5 shows the result of the backprojection (on the left), and the comparaison with the ideal object (on the right).

FIGURE 4. Sinogram for 1 line of the 2D detector

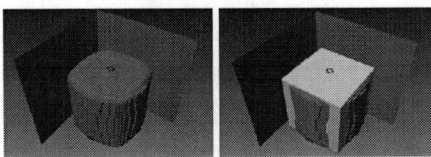

FIGURE 5. Results of the backprojection of the cube and comparison with the ideal object.

Conclusions and Perspectives

A new method for tomographic image reconstruction from a small number of limited angle projections is proposed. The originality of our method is mainly the use of the Bayesian estimation approach with an appropriate hierarchical Markov hidden variable which accounts for the specificity of our application (known number of materials).

The implementation of the projection and backprojection has been successfully tested in C. The next step of this work is to finish the interface between Matlab and C to use the results of the projection and the backprojection in the reconstruction method previously described. Then, we plan to apply this method to reconstruct experimental data of porous media. The method should then be compared with other classical reconstruction methods.

REFERENCES

1. Pierre Grangeat. *La tomographie : fondements mathématiques, imagerie microscopique et imagerie industrielle.* 2002.
2. Kenneth Lange and Richard Carson. EM reconstruction algorithms for emission and transmission tomography. Journal of Computer Assisted Tomography, 8(2):306–316, April 1984.
3. K.J Langenberg Gabor T.Herman, H.K Tuy and P.C. Sabatier. *Basic Methods of Tomography and Inverse Problems.* Adam Hilger, 13, 1987.
4. Stuart Geman and D.McClure. Statistical methods for tomografic image reconstruction. In *Proceedings of the 46th, Session of the ICI*, volume 52, 1987.
5. Charles A.Bouman and Ken D.Sauer. A generalized Gaussian image model for edge-preserving map estimation. *IEEE TRANS on Image Processing*, 2(3):296–310, July 1993.
6. Suhail S.Saquib Charles A.Bouman and Ken Sauer. Ml parameter estimation for markov random fields with applications to bayesian tomography. *IEEE TRANS on Image Processing*, pages 1029–1044, 1998.
7. Gilles Aubert Pierre Charbonnier, Laure Blanc-Féraud and Michel Barlaud . Deterministic edge-preserving regularization in computed imaging. *IEEE Transactions on image processing*, 6(2), February 1997.
8. Simon A.Barker. *Image Segmentation using Markov Random Field Models.* PhD thesis, University of Cambridge, July 1998.
9. Hichem Snoussi and Ali Mohammad-Djafari. Fast joint separation and segmentation of mixed images. *Journal of Electronic Imaging*, 13:349–361, April 2004.

10. Olivier Féron and Ali Mohammad-Djafari. Image fusion and joint segmentation using and MCMC algorithm. *Journal of Electronic Imaging*, 14(2), 4 April 2005.
11. Fabrice Humblot and Ali Mohammad-Djafari. Super-resolution using hidden markov model and bayesian detection estimation framework. *EURASIP Journal on Applied Signal Processing*, Special number on super-Resolution Imaging: Analysis, Algorithms, and Applications, 2005.

A Nonparametric Bayesian Approach For Emission Tomography Reconstruction

Éric Barat and Thomas Dautremer

CEA Saclay, Electronics and Signal Processing Laboratory, 91191 Gif sur Yvette, France

Abstract. We introduce a PET reconstruction algorithm following a nonparametric Bayesian (NPB) approach. In contrast with Expectation Maximization (EM), the proposed technique does not rely on any space discretization. Namely, the activity distribution – normalized emission intensity of the spatial poisson process – is considered as a spatial probability density and observations are the projections of random emissions whose distribution has to be estimated. This approach is nonparametric in the sense that the quantity of interest belongs to the set of probability measures on $\mathbb{R}^k$ (for reconstruction in k-dimensions) and it is Bayesian in the sense that we define a prior directly on this spatial measure. In this context, we propose to model the nonparametric probability density as an infinite mixture of multivariate normal distributions. As a prior for this mixture we consider a Dirichlet Process Mixture (DPM) with a Normal-Inverse Wishart ($\mathcal{NIW}$) model as base distribution of the Dirichlet Process. As in EM-family reconstruction, we use a data augmentation scheme where the set of hidden variables are the emission locations for each observed line of response in the continuous object space. Thanks to the data augmentation, we propose a Markov Chain Monte Carlo (MCMC) algorithm (Gibbs sampler) which is able to generate draws from the posterior distribution of the spatial intensity. A difference with EM is that one step of the Gibbs sampler corresponds to the generation of emission locations while only the expected number of emissions per pixel/voxel is used in EM. Another key difference is that the estimated spatial intensity is a continuous function such that there is no need to compute a projection matrix. Finally, draws from the intensity posterior distribution allow the estimation of posterior functionnals like the variance or confidence intervals. Results are presented for simulated data based on a 2D brain phantom and compared to Bayesian MAP-EM.

Keywords: Emission tomography, Bayesian nonparametrics, Dirichlet mixture, Poisson inverse problem.
PACS: 02.50.Tt, 02.50.-r, 02.50.Ng, 87.58.Ce, 87.58.Fg.

INTRODUCTION

Consider, on one hand, that the seminal filtered back-projection (FPB) algorithm is defined analytically on the continuous space but is not designed to take into account the Poisson statistics of observations. On the other hand, EM statistical approach makes use of the likelihood of the data but requires in its basis the discretization of the object space. We propose in this contribution an attempt to avoid this rather *ad hoc* space discretization while taking into account the Poisson nature of observations in the inverse problem. Our purpose is mainly to capture the structure of the activity distribution which can exhibit great complexity in PET imaging. In addition, since we are interested in the estimation of the posterior distribution and, particularly, in the quantification of the reconstruction error, we address the problem in the framework of Bayesian nonparametrics[1]. The normalized spatial intensity can be considered as a spatial probability density. A bayesian approach requires to construct a prior on this spatial measure. In the next section, we show how Dirichlet Process Mixtures define such prior and how they can provide a convenient tool for PET reconstruction.

METHODOLOGY

Dirichlet Process Mixture

The key idea is to represent the activity distribution as an infinite mixture of normal distributions. Dirichlet process mixtures (DPM), introduced by Antoniak [2] play an important role in nonparametric density estimation and clustering problems since they give a prior for these statistical mixtures. We first briefly recall the so-called stick-breaking representation of a Dirichlet Process (DP)[1]. Let us consider the independent and identically distributed samples $\mathbf{Z} = Z_1, Z_2, \ldots$ generated from any distribution G_0 and the independant and identically distributed samples $V_1, V_2, \ldots$ generated from a beta distribution with parameters 1 and α (Beta$(1, \alpha)$). Let construct $\mathbf{p} = p_1, p_2, \ldots$ as $p_1 = V_1$ and for all $k \geq 2$, $p_k = V_k \prod_{i=1}^{k-1}(1 - V_i)$. Then, the infinite sum

$$G(\cdot) = \sum_{k=1}^{\infty} p_k \delta_{Z_k}(\cdot) \tag{1}$$

is a random probability measure generated by a Dirichlet Process denoted as DP(αG_0) where δ_{Z_k} is the probability mass function localized in Z_k, α is the scaling parameter and G_0 the mean distribution of the DP. In the following, we denote by $S_\alpha(\cdot)$ the distribution of $\mathbf{p}$.

Due to discreteness of generated measures, DP cannot be used as a prior for estimating a probability density function. Nevertheless, the convolution of the DP by a parametric kernel produces continuous densities. We propose to use a multivariate normal kernel for tomographic inverse problems. We denote by $\mathcal{N}(\cdot|\mathbf{m}, \Sigma)$ the multivariate normal density parametrized by its mean $\mathbf{m}$ and covariance matrix Σ with $\mathbf{m}|\Sigma \sim \mathcal{N}(\mu_0, \Sigma/\rho)$ and $\Sigma^{-1} \sim \mathcal{W}(n_0, (n_0 \Sigma_0)^{-1})$, where $\mathcal{W}$ is the Wishart distribution. G_0 corresponds then to the commonly adopted Normal-Inverse Wishart model ($\mathcal{NIW}$) with ρ the precision parameter and n_0 (resp. Σ_0^{-1}) the degree of freedom (resp. mean) of the Wishart. The generated activity distribution is obtained as follows

$$f(\cdot) = \int \mathcal{N}(\cdot|\mathbf{m}, \Sigma) \, dG(\mathbf{m}, \Sigma)$$
$$G \sim \text{DP}(\alpha G_0) \tag{2}$$

Several techniques have been developped for sampling DPM (see [3]), among others we follow the approximation proposed in [4] for tractability and mixing properties. Using the stick-breaking representation, we get a convenient truncated Dirichlet Mixture representation

$$f(\cdot) = \sum_{k=1}^{N} p_k \mathcal{N}(\cdot|Z_k = (\mathbf{m}_k, \Sigma_k)) \tag{3}$$

with p_k defined as in (1).

Tomographic reconstruction

Suppose that we have n disintegrations localized in the object space at $\mathbf{X} = (X_1, X_2, \ldots, X_n)$. If all X_i were known, the estimation of the activity distribution would be straightforward (see [4]). Unfortunately, in the tomographic inverse problem, we do not observe emission locations but projections of these locations on detector pair units denoted $\mathbf{Y} = (Y_1, Y_2, \ldots, Y_n)$ where each Y_i corresponds to the index of the line of response of the ith coincidence. This notation is referred to as a list mode acquisition. Since the reconstruction algorithm proposed in this contribution is purely spatial, ordering of Y_i does not matter (data are *exchangeable*, see [1]). As a consequence, when working with sinograms, the jth line of response, containing m_j coincidences, may be decomposed in a sequence $\left(Y_{M_j+1} = j, Y_{M_j+2} = j, \ldots, Y_{M_j+m_j} = j\right)$ where $M_j = \sum_{l=1}^{j-1} m_l$. The method, though basically targetting list mode acquisition, is thus easily applicable to sinogram mode systems.

Annihilation locations $(X_1, X_2, \ldots, X_n)$ are then introduced as hidden variables in the reconstruction algorithm. As in EM approach, this data augmentation scheme will make explicit the expression of the likelihood. In our Bayesian approach, the ability to introduce conditionals on $\mathbf{X}$ let express a generative hierarchical model for PET data. We denote by $\mu(Y_i|X_i)$ the projection distribution. In addition, it is convenient, when infering with DP, to allocate each X_i to a component of the DP, see [4] for details. This is namely a clustering of the data. This is also an attempt to capture some data structure. We then denote by $\mathbf{K} = (K_1, K_2, \ldots, K_n)$ the classification variables defined such that $K_i = j$ if $(\mathbf{m}_i, \Sigma_i) = Z_j$. Here, $(\mathbf{m}_i, \Sigma_i)$ are the parameters of the normal component of the DP at which X_i is affected to. With these notations and $\mathbf{p}, \mathbf{Z}$ defined as in previous section, we may write the posterior distribution of the complete variables set

$$\Pr(\mathbf{X}, \mathbf{K}, \mathbf{p}, \mathbf{Z}|\mathbf{Y}) \propto \Pr(\mathbf{Y}|\mathbf{X}) \Pr(\mathbf{X}|\mathbf{Z}, \mathbf{K}) \\ \times \Pr(\mathbf{K}|\mathbf{p}) \Pr(\mathbf{p}, \mathbf{Z}) \tag{4}$$

with involved distributions

$$\begin{aligned} (Y_i|X_i) &\overset{\text{ind}}{\sim} \mu(Y_i|X_i) \\ (X_i|\mathbf{Z}, \mathbf{K}) &\overset{\text{ind}}{\sim} \mathcal{N}(X_i|Z_{K_i}) \\ (K_i|\mathbf{p}) &\overset{\text{i.i.d.}}{\sim} \sum_{k=1}^{N} p_k \delta_k(\cdot) \\ (\mathbf{p}, \mathbf{Z}) &\sim S_\alpha(\mathbf{p}) \times \mathcal{NIW}_{\rho, n_0, \mu_0, \Sigma_0}^N \end{aligned} \tag{5}$$

where $\mathcal{NIW}_{\rho, n_0, \mu_0, \Sigma_0}$ and S_α are defined in the previous section.

Based on this hierarchical model, we propose a Markov Chain Monte Carlo algorithm in order to generate draws $f(\cdot)|\mathbf{Y}$. The MCMC sampler will successively draw samples from the following conditional distributions

$$\begin{aligned}
\text{Annihilation location proposal}: & \quad (\mathbf{X}|\mathbf{Y},\mathbf{p},\mathbf{Z}) \\
\text{DPM component parameters}: & \quad (\mathbf{Z}|\mathbf{K},\mathbf{X}) \\
\text{Allocation of emission to DP}: & \quad (\mathbf{K}|\mathbf{p},\mathbf{Z},\mathbf{X}) \\
\text{DP weights (Stick breaking)}: & \quad (\mathbf{p}|\mathbf{K})
\end{aligned} \quad (6)$$

Except generation of $X_i|Y_i, \mathbf{p}, \mathbf{Z}$, all the conditional distributions allow direct sampling. The MCMC algorithm is then mainly a Gibbs sampler. Nevertheless, generation of $X_i|Y_i, \mathbf{p}, \mathbf{Z}$ requires a Metropolis-Hastings (MH) step and choice of a proposal distribution. Let express this distribution using the Bayes rule

$$\Pr(X_i|Y_i,\mathbf{p},\mathbf{Z}) \propto \Pr(Y_i|X_i)\Pr(X_i|\mathbf{p},\mathbf{Z}) \quad (7)$$

where $\Pr(Y_i|X_i)$ is the probability of projection on detector pair unit Y_i given emission location X_i and activity distribution characterized by $\mathbf{p}$ and $\mathbf{Z}$. Note that for an *idealized* PET system (*i.e.* if $\mu(Y_i|X_i)$ is uniform on an ideal line of response), $\Pr(X_i|Y_i,\mathbf{Z},\mathbf{p})$ is the intersection of a multivariate normal mixture with the line, which is always a mixture of univariate normals with rescaled weights on the ideal line. We use this property for the MH step and take this mixture of univariate normals as proposal distribution.

After convergence, the sampler generates draws from $(\mathbf{X},\mathbf{K},\mathbf{p},\mathbf{Z}|\mathbf{Y})$. Then from draws $(\mathbf{X}^*,\mathbf{K}^*,\mathbf{p}^*,\mathbf{Z}^*)$ we construct

$$f^*(\cdot) = \sum_{k=1}^{N} p_k^* \mathcal{N}(\cdot|Z_k^*) \quad (8)$$

Generating $f^*(\cdot)$, we can estimate the posterior distribution of $f|\mathbf{X}$ and its functionals. In particular, we compute the conditional mean, $\mathbb{E}(f|\mathbf{X})$, that we call the *predictive activity distribution*.

RESULTS AND DISCUSSION

We generated 3×10^5 decays in one slice of a voxelized brain phantom (see Figure 1 (a)). The simulation is 2D and all decays are detected by a single ring scanner made of 576 detectors. Attenuation and random coincidences were not simulated; detection probabilities were assumed purely geometrical. We compare our results to a Bayesian MAP EM algorithm for a 128×128 image, based on a Gibbs field with exponentially decreasing quadratic potentials.

Parameters for the Dirichlet mixture were chosen as follows : $\alpha = 30$, $\rho = 10^{-2}$, $\Sigma_0 = 3 \times \mathbb{I}_2$, $n_0 = 3$. Note that another hierarchical level allowing estimation of these hyperparameters may be implemented. Though potentially infinite the number of DP components affected with at least one data is of the order of 200. The nonparametric behaviour of the method is characterized by the fact that this number can increase with the amount of observations, and then automatically increase the activity distribution resolution. The MCMC sampler allows straightforward estimation of any posterior functionals as the posterior mean and standard deviations plotted on figure 1. Finally, note that the discretization of the object space steps in only for visualization purposes and may be chosen *a posteriori*.

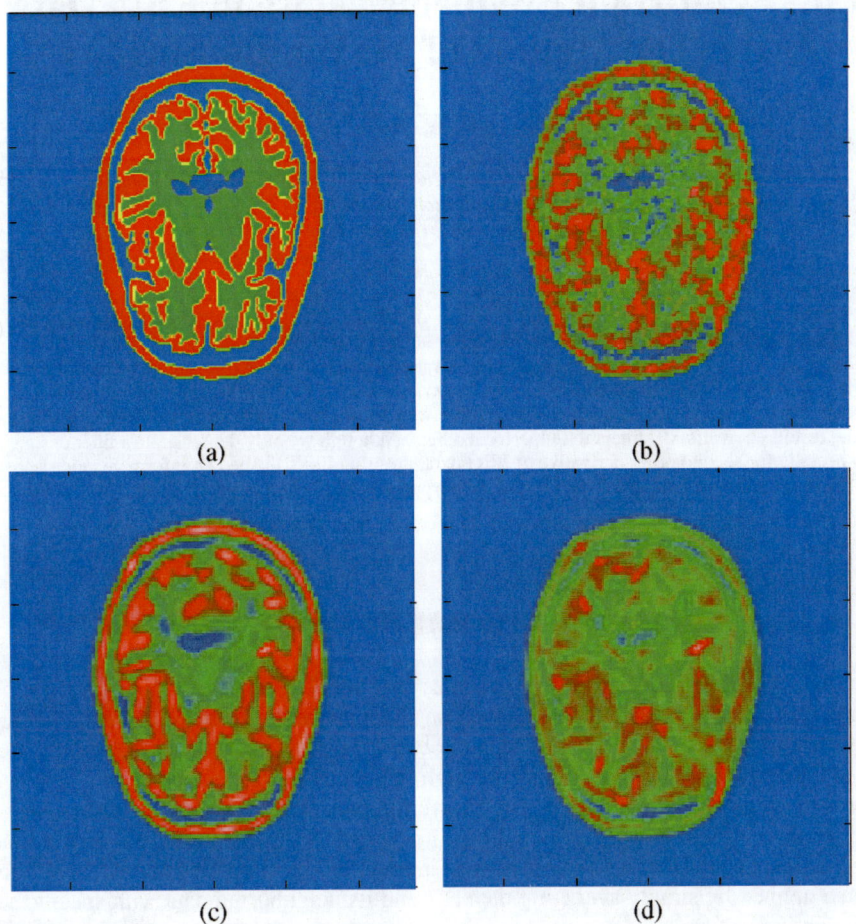

FIGURE 1. (a) 2D brain phantom, (b) MAP EM (100 iter), (c) NPB Conditional mean (10000 draws), (d) NPB Conditional standard deviation (10000 draws), scale in (d) is a quarter of scale in (a), (b), (c).

REFERENCES

1. J. K. Ghosh, and R. V. Ramamoorthi, *Bayesian Nonparametrics*, Springer, 2003.
2. C. E. Antoniak, *Ann. Statist.* **2**, 1152–1174 (1974).
3. M. I. Jordan, "Dirichlet processes, Chinese restaurant processes and all that," in *NIPS 2005*, 2005.
4. H. Ishwaran, and L. F. James, *J. Am. Stat. Assoc.* **96**, 161–173 (2001).

Filter Out High Frequency Noise in EEG Data Using The Method of Maximum Entropy

Chih-Yuan Tseng* and HC Lee[†*]

*Department of Physics and [†]Graduate Institute of Systems Biology and Bioinformatics
Computational Biology Laboratory
National Central University, Chungli, Taiwan 320, ROC

Abstract. We propose a maximum entropy (ME) based approach to smooth noise not only in data but also to noise amplified by second order derivative calculation of the data especially for electroencephalography (EEG) studies. The approach includes two steps, applying method of ME to generate a family of filters and minimizing noise variance after applying these filters on data selects the preferred one within the family. We examine performance of the ME filter through frequency and noise variance analysis and compare it with other well known filters developed in the EEG studies. The results show the ME filters to outperform others. Although we only demonstrate a filter design especially for second order derivative of EEG data, these studies still shed an informatic approach of systematically designing a filter for specific purposes.

Keywords: EEG, Current source density, Non-recursive filter, Maximum entropy
PACS: 87.10+e; 87.57.Gg

INTRODUCTION

Filtering out high frequency noise becomes extremely crucial in source localization problem of invasive EEG studies. Based on Freeman and Nicholson's works, source localization requires calculation of second order derivative of the EEG data [1, 2]. Because the derivative amplifies especially high frequency noise, Freeman and Nicholson proposed to introduce low-pass non-recursive filters to smooth it. [2]. Suppose a set of discrete data, $\Phi(r)$ labeled by spatial position r, which are distorted by noises ε_r and these data are equally spaced sampled with interval h. The simplest non-recursive filter is then defined by simply averaging data at r and its neighboring data with specific set of weighting values p_l,

$$\Phi_s(r) = \sum_{l=-z}^{z} p_l \Phi(r+lh), \qquad (1)$$

where z is an integer and $2z+1$ data points are considered. We will designate this $2z+1$-point spatial filter as a $S_{(2z+1)}$ filter. The mean filter, in which p_l equals to $1/(2z+1)$, is one filter widely used in many fields for data smoothing. Its applicability, however, is sometimes limited to data types [3].

Many methods have been developed to design low-pass filters in many fields. For example, there are adaptive beam-forming ([4] - [7]), wavelet de-noising [8], and entropic method ([9] - [14]) etc.. Yet no explicit discussions are given on smoothing noise in second order derivative of contaminated EEG data since Freeman and Nicholson's pioneering works. Freeman and Nicholson's approach will be briefly discussed in Sec. 2.

Two drawbacks, requirements of some empirical guidelines and low accuracy of second order derivative, in their approach will also be addressed.

Our goal is to provide a ME based approach to filter out high frequency noise especially in second order derivative of contaminated EEG data. This approach eliminates the two drawbacks in Freeman and Nicholson's approach and provides a systematic and robust method for smoothing noise. The ME approach and its application in the EEG studies will be presented in Sec. 3 and 4 respectively. We thereafter examine the proposed filter's performance and compare to some other filters such as the mean filter, Freeman and Nicholson's [2], and Rappelsberger's [15] filters through frequency and noise variance analysis in Sec. 5. A conclusion is given at last.

SOURCE LOCALIZATION FROM THE INVASIVE EEG DATA

In neurophysiology, electric field potentials are initiated during neural activity when ions flow across localized regions of cell membranes. These ionic flows establish a distribution of current sources and sinks in the extracellular space [1, 2, 16]. These field potentials can be measured in extracellular space discretely by the invasive EEG instrument, which usually has 16 to 32 detectors on its probe.

According to theory of electromagnetism, the current sources and sinks are associated with second order derivative of the field potentials. Freeman and Nicholson proposed the current source density (CSD) method for source localization [1, 2]. The CSD computes second order derivative approximately through substituting specific smoothed $\Phi_s(r)$, Eq. (1), into $\text{Ta}_3(r) = \sigma_c/h^2 (\Phi(r-h) - 2\Phi(r) + \Phi(r+h))$, where three data are considered, σ_c is conductivity, and h is separating distance between two sensors. A final expression of an approximated second order derivative is given by

$$D_2(r) = \sigma_c/h^2 \sum_{m=-M}^{M} a_m \Phi(r+mh), \qquad (2)$$

from 2M+1 datum with weightings $a_{\pm M} = p_z$, $a_{\pm M \mp 1} = p_{z-1} - 2p_z$, and $a_{\pm M \mp k} = p_{z-k} - 2p_{z-k+1} + p_{z-k+2}$, in which $k \geq 2$. The preferred smoothing function should be able to reduce high frequency noise and noise amplified by operation of the $\text{Ta}_3(r)$. Freeman and Nicholson devised several smoothing functions as shown in Table. 1, denoted by FNSn ([2]), where Sn stands for n-point spatial smoothing. Notes that the $\text{FNS}_3(2)$ is still used in Karwoski et al. [17] and Ulbert et al. [18]. These p_l are derived from the least square method [19, 20]. In addition, they also introduce some ad hoc rules to obtain 5 and 7-point filters. For example, the FNS_5 is obtained by smoothing data twice using set of (3/10, 4/10 ,3/10). Later, several successors have designed other smoothing functions according to their needs such as the Rappelsberger's RS_3 filter [15].

There are two drawbacks in Freeman and Nicholson's CSD method. The first is requirements of some ad hoc empirical guidelines for designing smoothing functions such as $\text{FNS}_3(1)$ or RS_3 shown in Table. 1. These empirical guidelines may limit the use of these smoothing functions to some specific EEG data. For example, Freeman and Nicholson's smoothing functions are suited for Anuran cerebellum studies [2]. The second is these smoothing functions are devised for second order derivative calculated via

Ta$_3(r)$, which is not an accurate calculation. In next section, we propose an informatic approach based on the ME method to eliminate these two drawbacks.

TABLE 1. Filters in EEG studies. The table lists five smoothing functions [2, 15]. MF: mean filter, FN: Freeman and Nicholson, R: Rappelsberger.

	p_0	$p_{\pm 1}$	$p_{\pm 2}$	$p_{\pm 3}$
MFS$_3$	0.33	0.33	0	0
FNS$_3$(1)	0.5	0.25	0	0
FNS$_3$(2)	0.43	0.29	0	0
RS$_3$	0.54	0.23	0	0
FNS5	0.34	0.24	0.09	0
FNS7	0.26	0.21	0.12	0.04

A MAXIMUM ENTROPY LOW-PASS FILTER

Rationale. The method of ME is developed as a tool for assigning a probability distribution of observing a system to be at a certain state according to information in hand, which is in the form of constraints [21]. Based on conventional studies on properties of the noise filter, Eq. (1), in the EEG source localization problems, which will be illustrated later, one can treat the weightings p_l as a probability distribution. Thus the ME provides a robust and objective approach for determining a preferred set of p_l once constraints relevant to noise reduction in the EEG problems are given.

The constraints. Conventional studies in EEG data analysis have shown some properties that are relevant to smooth noise using Eq. (1). First, the p_l are positive decimal values and symmetric around $l = 0$. These values obey a normalization condition $\sum_{l=-z}^{z} p_l = 1$. Second, the preferred set of the p_l should have noise in data being mostly removed through the $\Phi_s(r)$. We expand right hand side of Eq.(1) with respect to r according to the Taylor expansion and use $\sum_{l=-z}^{z} p_l = 1$ and $p_l = p_{-l}$, which results in the odd order derivative terms in the expansion vanished. The final result is $\Phi_s(r) = \Phi(r) + \Phi^{(2)}(r) h^2 \delta_2 / 2 + \cdots$, where $\Phi^{(2)}(r)$ denotes exact second order derivative of clean data and so on and $\delta_2 = \sum_{l=-z}^{z} p_l l^2$. It indicates that smoothing ability is actually related to the δ_2 value. Summarizing the above results, suggests the weightings p_l to be a probability distribution of having the Eq.(1) to reduce noise by the δ_2 approximately. Thus the ME can be applied to determine the preferred set of p_l.

The preferred low-pass filter for data smoothing. The ME states that maximizing entropy, $S = -\sum_{l=-z}^{z} p_l \ln p_l$, subject to the two constraints, the normalization condition of p_l and $\delta_2 = \sum_{l=-z}^{z} p_l l^2$ gives a preferred form of the p_l,

$$p_l = Z^{-1} \exp(-\alpha l^2), \qquad (3)$$

where α is a Lagrangian multiplier and partition function $Z = \sum_{l=-z}^{z} \exp(-\alpha l^2)$. The preferred set of p_l is a function of α. Namely, the ME method generates a family of $p_l(\alpha)$ distribution. In principle, the Lagrangian multiplier α can be determined by substituting Eq. (3) back into $\delta_2 = \sum_{l=-z}^{z} p_l l^2$, which gives $\delta_2 = -Z^{-1} \partial Z / \partial \alpha$, if the δ_2

value is given. Unfortunately, the δ_2 value is unknown at this point. Thus, we propose an alternative method to determine the α.

Since the preferred α value should result in corresponding set of the p_l to mostly reduce noise, if one can quantify noise the preferred α value can be determined. Although it is difficult to quantify noise, noise variance can be quantified by the autocorrelation function [3, 6]. Suppose field potentials $\Phi(r) = \Phi_0(r) + \varepsilon_r$ is repeatedly measured several times, in which a Gaussian white noise ε_r with variance σ is considered. For white noise, we have $\langle \varepsilon_r \rangle = 0$. Besides, noises are uncorrelated at different data points r and r', $\langle \varepsilon_r \varepsilon_{r'} \rangle = \sigma^2$ for $r = r'$ and 0 for $r \neq r'$ [3]. Because noise has zero mean, mean $\langle \Phi_0(r) \rangle$ is identical to $\Phi_0(r)$, noise-free data. Thus one can find autocorrelation of $\Phi(r)$ is $C_o = \langle (\Phi(r) - \langle \Phi(r) \rangle)^2 \rangle = \langle (\varepsilon_r - \langle \varepsilon_r \rangle)^2 \rangle = \sigma^2$, the noise variance. Similarly, one can compute the autocorrelation function of $\Phi_s(r)$, $C_{\text{smooth}} = \langle (\Phi_s(r) - \langle \Phi_s(r) \rangle)^2 \rangle = \sigma^2 \sum_{l=-z}^{z} p_l^2$. It indicates noise variance being amplified by $\sum_{l=-z}^{z} p_l^2$ after data is processed by the Eq.(1). Thus the preferred α value should minimize the noise variance C_{smooth}, which is found to be zero and $p_l(\alpha = 0) = 1/(2z+1)$. This is exactly the mean filter obtained from the least square method [3, 19]. Yet the ME approach roots in reducing noise variance C_{smooth} mostly, which is not clear in the least square method.

FILTER OUT NOISE IN SOURCE LOCALIZATION

Basic features. Freeman and Nicholson suggest to identify current sources and sinks through calculating second order derivative of the field potential via the $\text{Ta}_3(r)$ with specific smoothed $\Phi_s(r)$. However, the $\text{Ta}_3(r)$ does not provide accurate calculation. One straightforward way to raise the accuracy is to incorporate more data points into for calculation,

$$\text{Ta}_{2M+1}(r) = \sum_{m=-M}^{M} a_m^T \Phi(r+mh), \tag{4}$$

where 2M+1 data points are considered. To determine coefficients a_m^T, one first expand $\Phi(r+mh)$ with respect to r based on the Taylor expansion, $\text{Ta}_{2M+1}(r) = \sum_{m=-M}^{M} a_m^T \Phi(r) + \sum_{m=-M}^{M} a_m^T mh \Phi^{(1)}(r) + \sum_{m=-M}^{M} a_m^T (mh)^2/2 \Phi^{(2)}(r) + \cdots$. By requesting coefficient of second order term, $\sum_{m=-M}^{M} a_m^T (mh)^2/2 = 1$, and rests are zero, we have approximated $\text{Ta}_{2M+1}(r)$ to be identical to exact second order derivative $\Phi^{(2)}(r)$. One then can solve for coefficients a_m^T given these criteria. For three point calculation, coefficients are given in the Ta_3. Coefficients $a_{\pm 2}^T = -1/12$, $a_{\pm 1}^T = 16/12$, and $a_0^T = -30/12$ for Ta_5 and $a_{\pm 3}^T = 1/90$, $a_{\pm 2}^T = -13.5/90$, $a_{\pm 1}^T = 135/90$, and $a_0^T = -245/90$ for Ta_7.

Given these coefficients, one can immediately find the noise variance to be amplified by $C_{D_2} = \langle (D_2(r) - \langle D_2(r) \rangle)^2 \rangle = \sigma^2 \sum_{m=-M}^{M} (a_m^T)^2$ after operation of the $\text{Ta}_{2M+1}(r)$. For example, it is 6, 8.03, 9.75 times of the original noise variance for using $\text{Ta}_3(r)$, $\text{Ta}_5(r)$, and $\text{Ta}_7(r)$ respectively. The noise amplification is proportional to the number of data utilized in the calculation. This result suggests a trade-off of raising the accuracy of

TABLE 2. The preferred ME filters, denoted by $MES_n Ta_m$.

	$C_{D_{s2}} (\times \sigma^2)$	α	p_0	$p_{\pm 1}$	$p_{\pm 2}$	$p_{\pm 3}$
$MES_3 Ta_3$	0.28	0.031	0.42	0.28	0	0
$MES_3 Ta_5$	0.38	0.033	0.43	0.28	0	0
$MES_3 Ta_7$	0.42	0.035	0.44	0.27	0	0
$MES_5 Ta_3$	0.048	0.017	0.29	0.23	0.12	0
$MES_5 Ta_5$	0.061	0.018	0.29	0.23	0.11	0
$MES_5 Ta_7$	0.065	0.019	0.30	0.23	0.11	0
$MES_7 Ta_3$	0.014	0.011	0.22	0.19	0.12	0.06
$MES_7 Ta_5$	0.017	0.012	0.23	0.2	0.12	0.05
$MES_7 Ta_7$	0.019	0.012	0.23	0.2	0.12	0.05

second order derivative calculation with more data is amplification of the noise variance. Thus it is unlikely that the Freeman and Nicholson's and others filters can smooth this noise amplification because those filters are designed for $Ta_3(r)$ calculation only. One needs new smoothing functions.

The preferred ME filter. The new smoothing functions can be obtained through the following procedure. Substituting smoothed data, Eq. (1) into second order derivative $Ta_{2M+1}(r)$ first gives a generic expression

$$D_{s2}(r) = \sum_{n=-(M+z)}^{M+z} a'_n \Phi(r+nh) , \qquad (5)$$

where a'_n is a function of a_m^T and p_l. The preferred set of p_l is then determined by minimizing variance $C_{D_{s2}} = \left\langle (D_{s2}(r) - \langle D_{s2}(r)\rangle)^2 \right\rangle = \sigma^2 \sum_{n=-(M+z)}^{M+z} a'^2_n$. It should have high accuracy of second order derivative calculation and minimum noise variance simultaneously. Table 2 gives the results. We define operation based on the ME that utilizes n data points for noise smoothing, S_n, and m data for calculating second order derivative, Ta_m, as $MES_n Ta_m$ in the first column of Table. 2. Second column records the minimum $C_{D_{s2}}$ value. Next five columns list the corresponding α and the p_l value consecutively. The table first shows that when there is more data, S_3, S_5, and S_7, to be considered for smoothing given Ta_m, the noise variance $C_{D_{s2}}$ will be decreased dramatically. However, when one smoothes data with S_n and calculate second order derivative using more number of smoothed data points, Ta_3, Ta_5, and Ta_7, the noise variance $C_{D_{s2}}$ will then be amplified as expected.

PERFORMANCE ANALYSES

Instead of applying the proposed ME filter directly to study real EEG data for examinations, we will only analyze the filters's performance through frequency and noise variance analysis in this work. In addition, we will compare it to other filters, mean filter (MF), Freeman and Nicholson's (FN), Rappelsberger's (R) filters from Table 1.

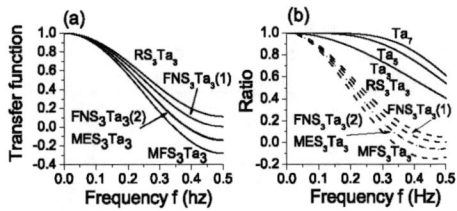

FIGURE 1. Frequency analysis of the five different 3-point S_3 filters in panel (a) and different S_3,Ta_3 in panel (b).

Frequency analysis. The frequency analysis applies the transfer function $H(\omega)$ for examining changes of signal's amplitudes after signal being processed by the filters and second order derivative calculation [3]. The transfer function $H(\omega) = \exp(-i\omega r)\Phi_s(r) = p_0 + 2\sum_{l=1}^{z} p_l \cos l\omega$ is derived by substituting $\Phi(r) = \exp(i\omega r)$, which is shown to be an eigenfunction of a linear time-invariant system, into $\Phi_s(r)$ of the Eq.(5), where angular frequency ω equals to $2\pi f$ rotational frequency [3]. When the transfer function is less then one, it indicates signal amplitude being attenuated by the filter. Panel (a) of Fig. 1 plots the results of using the different filters from Table 1. This figure demonstrates the mean filter has the largest amplitude attenuation and is followed by $FNS_3(2)$ and MES_3, $FNS_3(1)$, and RS_3 consecutively.

Similarly, one can compute the transfer function of second order derivative of smoothed data. Afterward, we compute the ratio of calculated and real second order derivative value through $R(\omega) = -D_{s2}(r)/\omega^2 = -1/\omega^2 \left(a_0 + 2\sum_{l=1}^{M+z} a'_l \cos l\omega\right)$ for explicit comparison of effects of the smoothing. When the ratio $R(\omega)$ equals to one, the calculated second order derivative of the smoothed data is identical to real value. Panel (b) of Fig. 1 plots not only the calculation through five filters, but also the three second order derivative calculations Ta_3, Ta_5, and Ta_7 respectively. First of all, frequency attenuation is the smallest in Ta_7, and is followed by Ta_5 and Ta_3. That explains again, why the second order derivative is more accurate when more data is comprehended for calculation. Given the second order derivative calculation with fixed number of smoothed data, for example, all S_3Ta_3 in this figure, it suggests that the mean filter MFS_3Ta_3 again has signal with high frequency attenuated the most among these filters. These attenuation patterns are similar to the panel (a), which is not surprising because smoothing data first and then calculating second order derivative is identical to calculating second order derivative first and then smooth the calculation.

Noise variance analysis. It seems the mean filter to be superior to all other choices from the previous analysis. However, the following noise variance analysis shows different results. We substitute the filters from Table 1 into Eq. (4) to compute new coefficient a'_m, and evaluate noise variance $C_{D_{s2}}$. The results are listed in Table 3. Because we design the ME filter to mostly reduce noise variance, the analysis shows the MES_3Ta_3 filter to exactly has the best performance among all filters.

Besides, Freeman and Nicholson have found noise in field potential measurements to be proportional to a K-value, $K = \sum_{l=-(M+z)}^{M+z} |a'_l|$ [2]. Thus we calculate the K-value

TABLE 3. Comparisons of filters. The table lists a'_n value and noise variance $C_{D_{s2}}(\times\sigma^2)$ in the third column and K value derived by Freeman and Nicholson for noise estimate in the fourth column.

	a'_0	$a'_{\pm 1}$	$a'_{\pm 2}$	$a'_{\pm 3}$	$a'_{\pm 4}$	$C_{D_{s2}}(\times\sigma^2)$	K
MFS$_3$Ta$_3$	0	-0.333	0.333	0	0	0.44	1.33
FNS$_3$Ta$_3$(1)	-0.5	0	0.25	0	0	0.38	1.01
FNS$_3$Ta$_3$(2)	-0.286	0.143	0.286	0	0	0.28	1.13
RS$_3$Ta$_3$	-0.623	0.082	0.229	0	0	0.57	1.34
MES$_3$Ta$_3$	-0.284	-0.144	0.286	0	0	0.28	1.14
FNS$_5$Ta$_3$	-0.2	-0.05	0.06	0.09	0	0.073	0.63
FNS$_7$Ta$_3$	-0.1	-0.04	0.01	0.04	0.04	0.019	0.39
MES$_5$Ta$_3$	-0.116	-0.055	-0.007	0.12	0	0.048	0.48
MES$_7$Ta$_3$	-0.06	-0.038	0.003	0.003	-0.06	0.014	0.27

after using different filters and list in the fourth column of Table 3. The K-value shows although MES$_3$Ta$_3$ is slightly outperformed by the FNS$_3$T$_3$(1), which has the same performance as the FNS$_3$T$_3$(2), the ME filter is superior than all other filters when more data points are included for smoothing.

Discussions. The frequency analysis has shown the mean filter MFS$_3$ and the MES$_3$ and FNS$_3$(2) mostly attenuate high frequency part in the signal. Yet the noise variance analysis, comparison of the $C_{D_{s2}}$ and K value, shows the MES$_3$Ta$_3$ and the FNS$_3$Ta$_3$(2) can mostly reduce noise variance among all three-point filters. These three analyses suggests the ME filter not only mostly reduce noise variance in second order derivative but also has the ability to mostly attenuate high frequency signal.

CONCLUSIONS

The filter design delicately depends on one's purpose and properties of a target system. It sometimes involves some ad hoc empirical guidelines that suit ones need. For example, Freeman and Nicholson have designed several filters for their needs [2]. We propose and demonstrate an informatic approach to design a low-pass filter based on the method of ME. In addition, we examine and compare the preferred ME filter with other well known filters in EEG studies through frequency and noise variance analysis. The performance of the preferred ME filter MES$_3$Ta$_3$ and the FNS$_3$Ta$_3$(2) filter are found to be almost identical. Even though the ME filter is designed for second order derivative calculation especially for the EEG data in this work only, the approach's systematics and flexibility ensures one can design filters that suit other purposes robustly without introducing ad hoc empirical guidelines.

ACKNOWLEDGMENT

This work is partially supported by NSC-95-2811-M008-016 to CYT and NSC-95-2112-M-008-004 to HCL from the National Science Council, Taiwan ROC.

REFERENCES

1. J. Freeman and C. Nicholson, "Theory of current source-density analysis and determination of conductivity tensor for anuran cerebellum," *J. Neurophysiol.* **38**, 356–368 (1975)
2. J. Freeman and C. Nicholson, "Experimental optimization of current source density technique for Anuran cerebellum," *J. Neurophysiol.* **38**, 369–382 (1975).
3. R. Hammings, "Digital filters," 3rd Ed., London: Prentice-Hall International, Inc., 1989.
4. G. Borgiotti and L. J. Kaplan, "Superresolution of uncorrelated interference sources by using adaptive array technique," *IEEE Trans. Antennas Propagat.*, **27**, 842–845 (1979).
5. A. Restrepo, A. C. Bovik, "Adaptive trimmed mean filters for image restoration," *IEEE Trans. Acous, Speech, and Signal Process.*, **36**, 1326–1337 (1988).
6. D. D. Feldman and L. J. Grifiths, "A constrained projection approach for robust adaptive beamforming," in *Proc. Int. Conf. Acoust., Speech, and Signal Processing*, 1991, pp. 1357–1360.
7. K. Sekihara, S. Nagarajan, D. Poeppel, A. Marantz, and Y. Miyashita, "Reconstructing spatio-temporal activities of neural sources using an MEG vector beamformer technique," *IEEE Trans. Biomed. Eng.*, **48**, 760–771 (2001).
8. S. Sardy, P. Tseng, and A. Bruce, "Robust wavelet de-noising," *IEEE Trans. Signal Process.*, **40**, 1146–1152 (2001).
9. N. Avesta and T. Aboulnasr, "Maximum entropy Kalman filter for image reconstruction and compression," *J. Electronic Imaging*, **13**, 738–755 (2004).
10. A. Beghdadi and A. Khellaf, "A noise-filtering method using a local information measure," *IEEE Trans. Image Process.*, **6**, 879–882 (1997).
11. G. A. Kivman, A. L. Kurapov, and A. V. Guessen, "An entropy approach to tuning weights and smoothing in the generalized inversion," *J. Atmospheric and Oceanic Tech.* **18**, 266–276 (2000).
12. J.-W. Wu, D. Erdogmus, and J. C. Principe, "Minimizing Fisher information of the error in supervised adaptive filter training," in *Proc. ICASSP*, 2004, pp.513–516.
13. S. Avitente and W. J. Williams, "Minimum entropy approach to de-nosing time frequency distributions," in *Proc. SPIE*, 2001, vol.4474, pp. 57–67.
14. V. E. DeBrunner, L. S. DeBrunner, S. Coone, and X. Hu "Using entropy to build efficient FIT digital filters," in *2004 IEEE 11th Digital Signal Processing Workshop and IEEE Signal Processing Education Workshop*, 2004, pp. 97–101.
15. P. Rappelsberger, H. Pockberger, and H. Petsche, "Current cource density analysis: Methods and application to simultaneously recorded field potentials of the rabbit's visual cortex," *Pflügers Arch.*, **389**, 159–170 (1981).
16. U. Mitzdorf, "Current source-density method and application in cat cerebral cortex: Investigation of evoked potentials and EEG phenomena," *Physiological Rev.* vol.**63**, pp. 37–100 (1985).
17. C. J. Karwoski, X. Xu, and H. Yu, "Current-source density analysis of the electroretinogram of the frog: methodological issues and origin of components," *J. Opt. Soc. Am. A*, **13**, 549–556 (1996).
18. I. Ulbert, E. Halgren, G. Heit, and G. Karmos, "Multiple microelectrode-recording system for human intracortical applications," *J. Neuroscience Meth.* **106**, 69–79 (2001).
19. C. Lanczos, "Applied analysis," Englewood Cliffs., N. J.: Prentice Hall, 1956.
20. R. Hammings, "Numerical methods for science and engineers," New York: McGraw-Hill, 1971, pp. 276–322.
21. E. T. Jaynes, "Information theory and statistical mechanics," *Phy. Rev.* **106**, 620–630 (1957).

Bayesian estimation of the learning effects of repeated pointing tasks

Koki Kyo

*Obihiro University of Agriculture and Veterinary Medicine
Inada-cho, Obihiro, Hokkaido 080-8555, Japan*

Abstract. Recently, in the field of human-computer interaction, the SH-model was developed for evaluating the performance of the input devices of a computer. This model was then modified by introducing a learning effect factor, which is expressed by using two deterministic functions. A remarkable merit of the use of deterministic functions is that parameters can be estimated easily by using the method of least squares. However, since the deterministic functions used to express the learning effect can only be fitted to some special patterns, performance of the modified SH-model may be somewhat restricted. In this paper, we apply a Bayesian modeling method for estimating the learning effect. We consider the parameters describing the learning effect as random variables and introduce smoothness priors for them. Results show that the learning effect can be estimated satisfactorily, thus providing proof of the validity of our model.

INTRODUCTION

Recently, with the rapid spread of information technology, it has become more and more important to evaluate the performance of the input devices of a computer and test the feasibility and efficiency of pointing tasks. A standard physical model for one-direction pointing task experiments is described by ISO 9241-9. A pointing device, controlled by a person, touches two targets on a screen one after the other. Since both targets have the same width, the pointing device must be moved between the targets continuously. The time taken between repetition of the movements is called the movement time and is considered as one of the indicators of the performance of the pointing task. One of the mathematical models used for pointing task evaluation and prediction is Fitts' law, which was proposed by Fitts (1954) based on Shannon's information capacity theory (see Shannon (1948)).

A version of the model for Fitts' law is as follows:

$$MT = a_0 + b_0(ID), \qquad (1)$$

where MT denotes the movement time, ID is the index of difficulty which is defined as

$$ID = \log_2(\frac{A}{W} + 1), \qquad (2)$$

and a_0 and b_0 are unknown parameters. In Eq.(2), A is the amplitude, indicating the distance between the centers of the targets and W is the width of the targets.

Fitts' law is flawed in that the performance in accuracy of pointing tasks and the distribution for MT are not taken into account. This prompted the studies of Ren, Kong and Jiang (2005) and Kong, Ren and Kyo (2006). They introduced the new SH-model and applied it to performance evaluation and comparison of pointing tasks. The SH-model is a linear model for the logarithm of MT instead of for MT with two factors being the explanatory variables, namely the systematic factor and the human factor. The SH-model is considered for the experiments which are designed for I types of amplitude, J types of target width, K subjects, and T times of repetition for each combination of the previous three indices. The model was constructed as follows:

$$y_{ij}^{(k)}(t) = a_0 + b_0 x_{ij} + c_0 z_{ij}(t) + e_{ij}^{(k)}(t) \quad (3)$$
$$(i=1,2,\ldots,I;\ j=1,2,\ldots,J;\ k=1,2,\ldots,K;\ t=1,2,\ldots,T),$$

where

$$y_{ij}^{(k)}(t) = \ln(MT_{ij}^{(k)}(t)), \quad x_{ij} = \ln(IS_{ij}), \quad z_{ij}(t) = \ln(IH_{ij}(t)), \quad (4)$$

a_0, b_0 and c_0 are unknown regression coefficients, and $e_{ij}^{(k)}(t)$'s denote the error terms, which are regarded as independent random variables normally distributed with unknown variance σ_0^2, i.e.,

$$e_{ij}^{(k)}(t) \sim N(0, \sigma_0^2). \quad (5)$$

Note that the necessity of the logarithmic transformations in Eq.(4) originates from the assumption in Eq.(5).

Moreover, in Eq.(4), IS_{ij} and $IH_{ij}(t)$ are defined respectively as

$$IS_{ij} = \log_2\left(\frac{A_i}{W_j} + \lambda + 1\right), \quad (6)$$

$$IH_{ij}(t) = \log_2\left(\frac{1}{p_{ij}(t)}\right). \quad (7)$$

Here, IS_{ij}, which is a systematic factor, denotes the self-information for the index of difficulty of each combination of i and j with λ being a parameter controlling the effective screen width. It is obvious from Eq. (6) that it may be more difficult to hit the target with a smaller value of W/A (the value of $x = \ln(IS)$ is larger), so the value of $y = \ln(MT)$ may be larger. As a result, we can surmise that there is a probable positive correlation between x and y, i.e., the value of b_0 may be positive. On the other hand, $IH_{ij}(t)$, which is a human factor, denotes the self-information for the operative accuracy with $p_{ij}(t)$ being the probability of direct hit for each combination of i, j and t. From Eq. (7) it can be seen that if a subject aims to hit the target with a larger value of p (the value of $z = \ln(IH)$ is smaller), then the value of $y = \ln(MT)$ may be larger. So, there is a probable negative correlation between z and y, i.e., the value of c_0 may be negative. Therefore, the larger the value of b_0 and the smaller the absolute value of c_0, the better is the performance of a pointing task. Obviously, the smaller the value of $a_0 > 0$ the better is the performance of a task.

It can logically be considered that for the same values of x and z, the value of $y = \ln(MT)$ may become smaller with repetition of a pointing task. This phenomenon is termed the learning effect. The learning effect can be observed on the values of a_0, b_0 and c_0 in Eq.(3). The learning effect observed on the parameters a_0, b_0 and c_0 respectively are named the general learning effect (GLE), the learning effect on systematic factor (LESF) and the learning effect on human factor (LEHF). In Zhang, Ren and Kyo (2006), the SH-model was modified by considering GLE as follows:

$$y_{ij}^{(k)}(t) = a(t) + b_0 x_{ij} + c_0 z_{ij}(t) + \varepsilon_{ij}^{(k)}(t), \quad \varepsilon_{ij}^{(k)}(t) \sim N(0, \sigma^2) \qquad (8)$$
$$(i = 1, 2, \ldots, I;\ j = 1, 2, \ldots, J;\ k = 1, 2, \ldots, K;\ t = 1, 2, \ldots, T),$$

where $a(t)$ denotes a function of t which indicates repetition times. Two formulae were introduced: the linear formula, $a(t) = \alpha_0 + \alpha_1 t$, and the inverse formula, $a(t) = \beta_0 + \beta_1 t^{-1}$ ($t = 1, 2, \ldots, T$). In the above formulae, α_0, α_1, β_0 and β_1 are considered as unknown parameters. We call this kind of approach the deterministic function approach (DFA). A merit of DFA is that the unknown parameters can be estimated by using the method of least squares. However, DFA can only be applied to estimate the learning effect as described by some special functions, so performance of a DFA-based model may be somewhat restricted, and the estimates of parameters may have bias due to lack of flexibility. In this paper, in order to negate the disadvantages of DFA, a Bayesian modeling method is introduced by applying the smoothness priors approach introduced by Kitagawa and Gersch (1996).

MODELS AND PARAMETER ESTIMATION

Model involving GLE

In this subsection, we consider a Bayesian model containing GLE, i.e., the learning effect is specified in the constant term, a_0. Formally, the model is the same as that in Eq. (8), but the Bayesian modeling method is used instead of DFA for estimating the parameters $a(t)$ ($t = 1, 2, \ldots, T$). The parameters $a(t)$ ($t = 1, 2, \ldots, T$) are considered as random variables and it is assumed that $a(t)$ varies over t smoothly. Then the smoothness priors can be applied to set up a prior distribution for $a(t)$ ($t = 1, 2, \ldots, T$). A second order stochastic difference equation is used:

$$a(t) - 2a(t+1) + a(t+2) = v(t), \quad v(t) \sim N(0, \tau^2) \quad (t = 1, 2, \ldots, T) \qquad (9)$$

with $v(t)$ indicating a stochastic disturbance. In Eq.(9), $a(T+1)$, $a(T+2)$ and τ^2 are hyperparameters. We also have the following assumptions: (a) $v(t_1)$ and $v(t_2)$ are independent of each other for $t_1 \neq t_2$. (b) $\varepsilon_{ij}^{(k)}(t)$ and $v(t)$ are independent of each other for any index.

The model in Eq.(8) can be expressed in the following matrix-vector form:

$$y_{ij}^{(k)} = a + b_0 x_{ij} 1_T + c_0 z_{ij} + \varepsilon_{ij}^{(k)}, \quad \varepsilon_{ij}^{(k)} \sim N(0_T,\ \sigma^2 E_T) \qquad (10)$$
$$(i = 1, 2, \ldots, I;\ j = 1, 2, \ldots, J;\ k = 1, 2, \ldots, K),$$

where E_T denotes a unit matrix with dimension T, and the vectors (all of dimension T) used here are defined respectively as follows: $y_{ij}^{(k)} = (y_{ij}^{(k)}(1), y_{ij}^{(k)}(2), \cdots, y_{ij}^{(k)}(T))^{\mathrm{t}}$,

$a = (a(1), a(2), \ldots, a(T))^t$, $z_{ij} = (z_{ij}(1), z_{ij}(2), \cdots, z_{ij}(T))^t$, $1_T = (1, 1, \ldots, 1)^t$, $\varepsilon_{ij}^{(k)} = (\varepsilon_{ij}^{(k)}(1), \varepsilon_{ij}^{(k)}(2), \ldots, \varepsilon_{ij}^{(k)}(T))^t$, $0_T = (0, 0, \ldots, 0)^t$.

Similarly, the prior for a in Eq.(9) can be rewritten as:

$$Da + H \begin{bmatrix} a(T+1) \\ a(T+2) \end{bmatrix} = v, \quad v \sim N(0_n, \tau^2 E_T), \tag{11}$$

where $v = (v(1), v(2), \ldots, v(n))^t$, $D = (d_{ij})$ is a $T \times T$ matrix defined as $d_{ij} = 1$ for $j = i$ or $j = i+2$, $d_{ij} = -2$ for $j = i+1$, and $d_{ij} = 0$ otherwise. $H = (h_{ij})$ is a $T \times 2$ matrix with $h_{(T-1)1} = h_{T2} = 1$, $h_{T1} = -2$, the other elements being zero.

Parameter estimation

At first, λ is estimated for the original SH-model in Eq.(3) before the learning effect is introduced. Thus the estimate can be obtained for a basic tendency as a standard expression of the system. $A_i/W_j + \lambda + 1 > 0$ is assumed in order to keep the values of IS_{ij} in Eq.(6) always finite for any values of i and j. For estimation of $p_{ij}(t)$, it is considered as "probability of success" in a Bernoulli trial, and a Bayesian method is used by introducing a uniform prior to it. So, the estimate of $p_{ij}(t)$ is given by the posterior mean as $p_{ij}(t) = (r_{ij}(t) + 1)/(K + 2)$ with $r_{ij}(t)$ being the number of direct hits on the target in K trials by all of the subjects for each combination of i, j and t.

Then the estimates for the other parameters are obtained based on the estimates of λ and $p_{ij}(t)$. A set of Bayesian linear models for a can be constructed based on the models in Eqs.(10) and (11). Jiang (1995) has introduced a procedure for estimating parameters using a Bayesian linear model based on the methodology proposed by Akaike (1980). This procedure is used in the present case as follows.

For given values of $\eta = \sigma/\tau$ and $d = (a(T+1), a(T+2), b_0, c_0)^t$ the posterior distribution is a normal distribution with

$$\hat{a}(\eta, d) = (W(\eta)^t W(\eta))^{-1} W(\eta)^t (y - G(\eta)d) \tag{12}$$

being the mode. The maximum likelihood estimate of σ^2 is

$$\hat{\sigma}^2(\eta, d) = \frac{1}{n} \|y - G(\eta)d - W(\eta)\hat{a}(\eta, d)\|^2, \tag{13}$$

and the log-likelihood of η and d is given by

$$\ell(\eta, d) = -\frac{n}{2}(\ln\{2\pi\hat{\sigma}^2(\eta, d)\} + 1) - \frac{1}{2}\ln\{\det(W(\eta)^t W(\eta))\} + \frac{T}{2}\ln\eta^2. \tag{14}$$

In the above equations, $\|x\| = \sqrt{x^t x}$ defines the norm of a vector x, y is an $(n+T)$-dimensional vector where the first n elements are defined by $y_{ij}^{(k)}$, for $i = 1, 2, \ldots, I$, $j = 1, 2, \ldots, J$ and $k = 1, 2, \ldots, K$, so that $n = T \times I \times J \times K$, while the last T elements are of zero value. $W(\eta)$ is an $(n+T) \times T$ matrix that is constructed with y by using

I_T n times and ηD. $G(\eta)$ is an $(n+T) \times 4$ matrix that is constructed correspondingly with y by using x_{ij}, z_{ij}, for $i = 1,2,\ldots,I$, $j = 1,2,\ldots,J$, and ηH. So, for a given value of η the estimate $\hat{d}$ of d can be obtained together with $\hat{a}$ by using the method of least squares (see Jiang (1995)), and the partial maximum log-likelihood is $\ell^*(\eta) = \ell(\eta, \hat{d})$. Finally, the estimate $\hat{\eta}$ of η can be obtained by maximizing the value $\ell^*(\eta)$. Thus, we can use $\hat{\sigma}^2(\hat{\eta}, \hat{d})$ as the estimate of σ^2, and then the value of τ^2 can be calculated as $\hat{\tau}^2 = \hat{\sigma}^2/\hat{\eta}^2$.

Incidentally, the value of AIC (Akaike information criterion) is given by $AIC = -2\ell^*(\hat{\eta}) + 2 \times N$, where N is the number of parameters to be estimated by the method of maximum likelihood. In this case we have parameters $a(T+1)$, $a(T+2)$, b_0, c_0, σ^2 and η or τ^2, i.e., $N = 6$. So, we can compare the model with other models by using the minimum AIC method (Akaike (1974)). Note that in Akaike (1980) AIC defined here is also called ABIC (Akaike's Bayesian information criterion) which is considered as an information criterion for evaluating priors. Schwarz (1978) proposed BIC (Bayesian information criterion) as a variation of AIC based on Bayesian consideration.

Models involving LESF and LEHF

We also consider separately the following two models where the learning effect is involved in b_0 or c_0 respectively as follows:

$$y_{ij}^{(k)}(t) = a_0 + b(t)x_{ij} + c_0 z_{ij}(t) + \zeta_{ij}^{(k)}(t), \quad \zeta_{ij}^{(k)}(t) \sim N(0, \phi^2) \quad (15)$$

$$y_{ij}^{(k)}(t) = a_0 + b_0 x_{ij} + c(t)z_{ij}(t) + \xi_{ij}^{(k)}(t) \quad \xi_{ij}^{(k)}(t) \sim N(0, \psi^2) \quad (16)$$

$(i = 1,2,\ldots,I;\ j = 1,2,\ldots,J;\ k = 1,2,\ldots,K;\ t = 1,2,\ldots,T).$

Similar to the Bayesian model for $a(t)$ $(t = 1,2,\ldots,T)$, we use the smoothness priors for $b(t)$ and $c(t)$ $(t = 1,2,\ldots,T)$ respectively as follows:

$$b(t) - 2b(t+1) + b(t+2) \sim N(0, \gamma^2) \quad (t = 1,2,\ldots,T), \quad (17)$$

$$c(t) - 2c(t+1) + c(t+2) \sim N(0, \omega^2) \quad (t = 1,2,\ldots,T), \quad (18)$$

with $b(T+1)$, $b(T+2)$, γ^2, $c(T+1)$, $c(T+2)$ and ω^2 being the hyperparameters.

TABLE 1. A summary of contending models

Name of model	Model for likelihood	Model for prior	Parameters as random variables	Parameters as unknown constants *
M0	Eq.(3)	Null	Null	$a_0, b_0, c_0, \sigma_0^2$
M1	Eq.(8)	Eq.(9)	$a(1), a(2),\ldots,a(T)$	$a(T+1), a(T+2), b_0, c_0, \sigma^2, \tau^2$
M2	Eq.(15)	Eq.(17)	$b(1), b(2),\ldots,b(T)$	$a_0, b(T+1), b(T+2), c_0, \phi^2, \gamma^2$
M3	Eq.(16)	Eq.(18)	$c(1), c(2),\ldots,c(T)$	$a_0, b_0, c(T+1), c(T+2), \psi^2, \omega^2$

* Here λ and $p_{ij}(t)$'s are omitted.

Then, two other Bayesian linear models are obtained and the parameters estimated by using the procedure in the previous subsection. Summarizing the above two sections, we have four contending models which are compared in Table 1.

AN EXAMPLE

As an example, we perform an analysis of the data from a pointing experiment on a PDA (see Ren, Kong and Jiang (2005) for the experiment in detail). The experiment was carried out by using three kinds of input methods, i.e., forefinger of subject (method F), a long stylus pen (method L) and a short stylus pen (method S). The number of observations for each input method is 3132.

First, we give the results of the estimated λ value. Table 2 shows the values of log-likelihood of the model M0 using different values of λ for the three input methods. From Table 2 we can see that the maximum likelihood estimate of λ for the method F is $\hat{\lambda} \approx 0.5$, that for the method L and the method S are $\hat{\lambda} \approx 1.5$ and $\hat{\lambda} \approx 6.0$ respectively.

TABLE 2. Values of log-likelihood at each λ value and characteristics of the model M0

Method	Log-likelihood (Value of λ)					Estimate of λ	Maximum log-likelihood	Value of AIC
F	-566.87 (-0.5)	-565.79 (0.0)	-565.76 (0.5)	-566.28 (1.0)	-567.13 (1.5)	0.5	-565.76	1139.52
L	-770.88 (0.5)	-770.72 (1.0)	-769.98 (1.5)	-770.35 (2.0)	-771.02 (2.5)	1.5	-769.98	1547.96
S	-755.85 (5.0)	-755.73 (5.5)	-755.69 (6.0)	-755.73 (6.5)	-755.83 (7.0)	6.0	-755.69	1519.38

The estimates of parameters, the values of maximum log-likelihood and the values of AIC for the models M1, M2 and M3 with each method are shown in Table 3, Table 4 and Table 5 respectively. By comparing the values of AIC for the models M1, M2 and M3 derived from the three different methods, we can see that performances of the models M1 and M2 are nearly the same and are better than for the model M3. It implies that models containing GLE or LESF perform better than those containing LEHF.

TABLE 3. Estimates of parameters and characteristics of the model M1

Method	$\hat{b}_0$	$\hat{c}_0$	$\hat{\sigma}^2$	$\hat{\tau}^2$	Maximum log-likelihood	Value of AIC
F	0.4873	-0.0629	0.0835	2.12×10^{-12}	-556.26	1124.52
L	0.6908	-0.0074	0.0952	2.29×10^{-6}	-765.53	1543.06
S	1.0998	-0.0108	0.0942	1.26×10^{-6}	-747.64	1507.28

TABLE 4. Estimates of parameters and characteristics of the model M2

Method	$\hat{a}_0$	$\hat{c}_0$	$\hat{\phi}^2$	$\hat{\gamma}^2$	Maximum log-likelihood	Value of AIC
F	5.3104	-0.0626	0.0835	2.12×10^{-12}	-556.15	1124.30
L	5.1698	-0.0083	0.0952	1.27×10^{-6}	-765.63	1543.26
S	4.5469	-0.0110	0.0942	5.24×10^{-7}	-747.64	1507.28

Line graphs for the estimates of $a(t)$, $b(t)$ and $c(t)$ ($t = 1, 2, \ldots, 29$), are shown in Figure 1, Figure 2 and Figure 3 respectively. From Figure 1 it can be seen that the mean level of $\hat{a}(t)$ ($t = 1, 2, \ldots, T$) for the method S is lower and has decreased rather more steeply than the others. This implies that the subjects perform better when using the

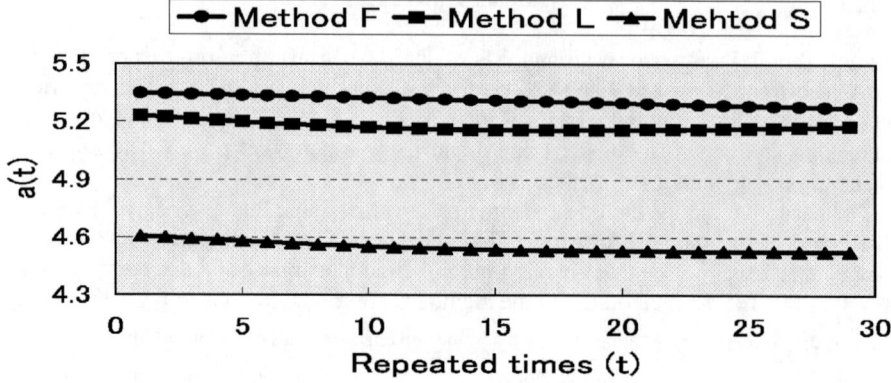

FIGURE 1. Line graphs for the estimates of $a(t)$ ($t = 1, 2, \ldots, 29$)

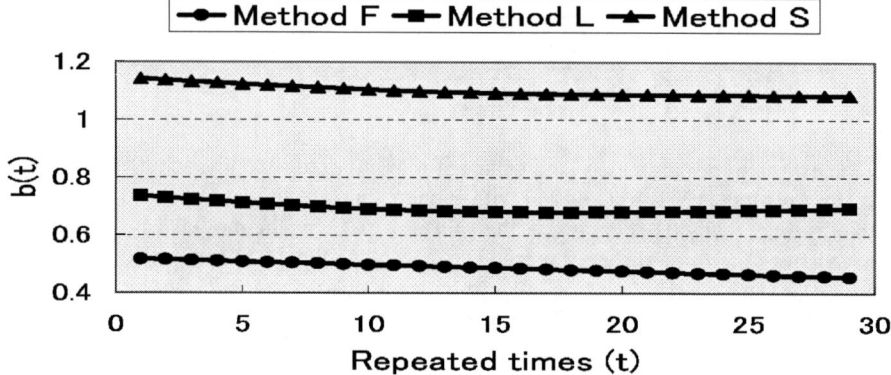

FIGURE 2. Line graphs for the estimates of $b(t)$ ($t = 1, 2, \ldots, 29$)

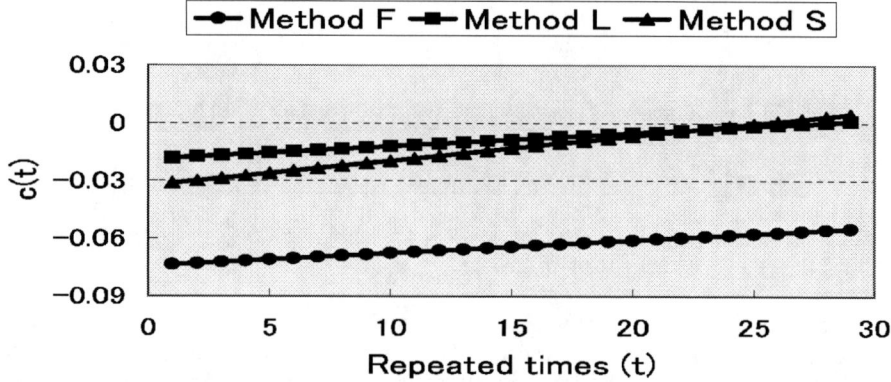

FIGURE 3. Line graphs for the estimates of $c(t)$ ($t = 1, 2, \ldots, 29$)

TABLE 5. Estimates of parameters and characteristics of the model M3

Method	$\hat{a}_0$	$\hat{b}_0$	$\hat{\psi}^2$	$\hat{\omega}^2$	Maximum log-likelihood	Value of AIC
F	5.3062	0.4896	0.0840	2.14×10^{-12}	-565.23	1142.64
L	5.1644	0.6968	0.0956	2.43×10^{-12}	-769.33	1550.66
S	4.5314	1.1097	0.0947	2.41×10^{-12}	-753.36	1518.72

method S and containing GLE. Figure 2 shows that the subjects perform better when using the method F containing LESF. It implies that the method F is shaped to a system with a higher degree of index of difficulty than the others. An interesting result is that in the method L $\hat{b}(t)$ drops to around $t = 15$ then increases slowly. It implies that in the beginning there is a learning effect, gradually the learning effect becomes weak to the point of exhaustion. Figure 3 shows that the subjects perform better in the methods L and S than in the method F using LEHF.

ACKNOWLEDGMENTS

I wish to thank Dr. Xiangshi Ren of Kochi University of Technology, Japan for allowing me to use his data. I also gratefully acknowledge the valuable comments of Professor Genshiro Kitagawa at the Institute of Statistical Mathematics, Japan. This study was carried out under the auspices of the ISM Cooperative Research Program (2007-ISM·CRP-2016).

REFERENCES

1. Akaike, H., "A New Look at the Statistical Model Identification," *IEEE Trans. Auto. Control*, AC-19, pp.716-723, 1974.
2. Akaike, H., "Likelihood and the Bayes Procedure," in: *Beyasian Statistics*, Bernardo, J.M. et al., eds., University Press, pp.143-166, 1980.
3. Fitts, P.M., "The Information Capacity of the Human Motor System in Controlling the Amplitude of Movement," *Journal of Experimental Psychology*, 47(6), pp.381-391, 1954.
4. Jiang, X.Q., "A Bayesian Method for the Dynamic Regression Analysis," *Transactions of the Institute of Systems, Control and Information Engineers*, 8(1), pp.8-16, 1995.
5. Kitagawa, G. and Gersch, W., *Smoothness Priors Analysis of Times Series*, Springer-Verlag, 1996.
6. Kong, J., Ren, X. and Kyo, K., "Comparison of Input Devices in Pointing Tasks through the Observation of the Human Effects: An Application of the SH-Model," *Transaction of Human Interface Society*, 8(2), pp.311-320, 2006.
7. Ren, X., Kong, J. and Jiang, X.Q., "SH-Model: A Model Based on both System and Human Effects for Pointing Task Evaluation," *IPSJ Journal*, 46(5), pp.1343-1353, 2005.
8. Schwarz, G., "Estimating the dimension of a model," *Annals of Statistics*, 6(2), pp.461-464, 1978.
9. Shannon, C.E., "A mathematical theory of communication," *The Bell System Technical Journal*, 27, pp.379-423 (Part I), pp.623-656 (Part II), 1948.
10. Zhang, X., Ren, X. and Kyo, K., "Modifying SH-model with Consideration of Learning Effect for Pointing Task Evaluation," in: *Proceedings of APCHI2006* (CD-ROM), 2006.

Classification of Maize and Weeds by Bayesian Networks

Michel Chapron[1], Alina Oprea[1], Bogdan Sultana[1], Louis Assemat[2]

[1]ETIS, ENSEA, UCP, CNRS, 14 avenue du ponceau, 95014 Cergy-Pontoise, France
[2]INRA, UMR Biologie et Gestion des Adventices, BP 86510, 21065 Dijon, France
chapron@ensea.fr

Abstract. Precision Agriculture is concerned with all sorts of within-field variability, spatially and temporally, that reduces the efficacy of agronomic practices applied in a uniform way all over the field. Because of these sources of heterogeneity, uniform management actions strongly reduce the efficiency of the resource input to the crop (i.e. fertilization, water) or for the agrochemicals use for pest control (i.e. herbicide). Moreover, this low efficacy means high environmental cost (pollution) and reduced economic return for the farmer. Weed plants are one of these sources of variability for the crop, as they occur in patches in the field. Detecting the location, size and internal density of these patches, along with identification of main weed species involved, open the way to a site-specific weed control strategy, where only patches of weeds would receive the appropriate herbicide (type and dose). Herein, an automatic recognition method of vegetal species is described. First, the pixels of soil and vegetation are classified in two classes, then the vegetation part of the input image is segmented from the distance image by using the watershed method and finally the leaves of the vegetation are partitioned in two parts maize and weeds thanks to the two Bayesian networks.

Key words: Bayesian networks, watershed segmentation, skeleton, plant classification, herbicide and pollution reduction, precision agriculture

INTRODUCTION

The pattern recognition or classification of vegetal species is one of the most relevant and difficult problems of precision agriculture. Instead of weed pulverization everywhere on the field, the farmer wants to pulverize only bad-infested sectors. It has a big impact on the chemical pollution due to herbicides especially in the under ground water which contains toxic products such as atrazine sprayed in big quantities everywhere on cereal fields. The fast technological developments in electronic devices such as cameras and computers permit us to design real time techniques for recognizing maize and weeds. Therefore, herbicides will only be sprayed on patches of weeds in the field and not uniformly as performed in conventional agriculture. A lot of work has been carried out in this domain, see for instance [1], [2] or [3]. In order to tackle this complex problem, we first eliminate the meaningless pixels of soil with the vegetation/soil classification. Then, a geometrical segmentation step founded upon the contour object distance image is performed concurrently with the skeleton computation of vegetal objects. These processings provide geometrical information about all parts of vegetation to the bayesian networks which classify them in maize or weeds. In the following, we will present the probability principal component analysis based classification of color pixels in two classes: soil and vegetation. Then the different leaves of vegetation will be classified into maize and weeds by using watershed algorithm and Bayesian networks and finally results and conclusions will be described and proposed.

VEGETATION/SOIL CLASSIFICATION

The classification of color pixels in two classes i.e. soil and vegetation is performed by Probabilistic Principal Component Analysis (PPCA) [4, 5]. The principal component analysis (PCA) is a well established technique for dimensionality reduction and can be improved in a probabilistic framework in order to get better results. The common Principal Component Analysis (PCA) is a linear projection which maximizes the variance in the projected space. For the color images [6], the pixels p_n, $n \in \{1,...,N\}$ are vectors, the dimension d of data is 3 (red, green, blue), the principal axes w_j, $j \in \{1,...,q\}$ are the ortho-normal dominant axes attached to the eigenvectors of the covariance matrix $S = \sum_n (p_n - \bar{p})(p_n - \bar{p})^T / N$ where $\bar{p}$ is the mean value of the pixels of the image, the w_j are such that $Sw_j = \lambda_j w_j$. The q dominant principal components of pixels p_n are given by the vector $x_n = W^T(p_n - \bar{p})$ where $W^T = (w_1,..,w_q)$ has its columns composed of dominant vectors which are linked to the highest eigenvalues of S. The variables x_i are uncorrelated and the covariance matrix $\sum_n x_n x_n^T / N$ is diagonal with elements λ_j. All orthogonal linear projections are $x_n = W^T(p_n - \bar{p})$. The principal component projection minimizes the squared reconstruction error $\sum_n \| p_n - \hat{p}_n \|^2$ with $\hat{p}_n$ as the optimal linear reconstruction of p_n given by $\hat{p}_n = W x_n + \bar{p}$. A latent variable model relates a 3-dimensional vector (pixel) to a q dimensional latent (unobserved) variable x. The relation between the variables is

$$p = Wx + \bar{p} + \varepsilon \qquad (1)$$

where W is the dxq matrix, the components ε_i represent the variability of pixel values. The latent variables offer a more parsimonious explanation of dependencies between the pixels. The latent variables x are independent and Gaussian N(0, I) with unit variance and the error ε is also gaussian N(0, ψ). Equation (1) provides a Gaussian distribution for the pixels $p \sim N(\bar{p}, WW^T + \psi)$. The model parameters may thus be determined by maximum likelihood. W and ψ are obtained by an iterative procedure. ψ is supposed to be diagonal with the same diagonal value σ^2 (isotropic error model). The observation covariance model then becomes $WW^T + \sigma^2 I$, where the parameters W and σ^2 are estimated iteratively by EM algorithm and I is the identity matrix. The probabilistic model takes into account the isotropic Gaussian noise N(0, $\sigma^2 I$) for ε in conjunction with equation (1) which implies that the x conditional probability distribution over p space is given by : $p|x \sim N(Wx + \bar{p}, \sigma^2 I)$.

The marginal distribution over the latent variables x is also Gaussian and follows $x \sim N(0, I)$. The marginal distribution for the observed pixel p is readily obtained by integrating out the latent variables and is also gaussian $p \sim N(\bar{p}, C)$ where the observation covariance model is $C = WW^T + \sigma^2 I$. The corresponding log-likelihood is then $L = (-N/2)(3\ln(2\pi) + \ln|C| + tr(C^{-1}S))$.

The maximum likelihood estimator for $\bar{p}$ is given by the mean of pixels, S by the covariance matrix of pixel p_n. The estimations of W and σ^2 are computed by an iterative maximization of L, herein by EM algorithm. The latent variables x, given the observed pixels p, can be calculated from Bayes' rule and is again Gaussian: $x|p \sim N(M^{-1}W^T(p-\bar{p}), \sigma^2 M^{-1})$ where $M=W^TW+\sigma^2 I$ has size qxq while C is 3x3.

The EM algorithm iteratively computes the estimates of $\hat{W}$ and $\hat{\sigma}^2$ until they converge.

$$\hat{W} = [\sum_n (p_n - \bar{p}) < x_n >^T][\sum_n < x_n x_n^T >]^{-1} \qquad (2)$$

$$\hat{\sigma}^2 = \frac{1}{3N} \sum_{n=1}^{N} (\|p_n - \bar{p}\|^2 - 2 < x_n >^T \hat{W}^T (p_n - \bar{p}) + tr(< x_n x_n^T > \hat{W}^T \hat{W})) \qquad (3)$$

With: $<x_n>=M^{-1}W^T(p_n-\bar{p})$ and $<x_n x_n^T> = \sigma^2 M^{-1} + <x_n><x_n>^T$ and $M=W^TW+\sigma^2 I$.

$\hat{W}$ is initialized by W computed from PCA and $\hat{\sigma}^2$ is initialized by computing the variance of the input image. Actually, the number of dominant axes has been chosen as 1 or 2. Then, all the pixels of the input image are projected by $x_n=W^T(p_n-\bar{p})$. These projections are approximated by a mixture of two gaussian distributions associated to the two classes: soil and vegetation.

WATERSHED ALGORITHM

The watershed algorithm transformation [7,8] appears to be a very powerful segmentation tool. Provided that the input image has been transformed to get an image whose minima mark relevant image objects and whose crest lines correspond to boundaries of image objects, the watershed transformation will partition the input image in meaningful regions. Let us consider the grey tone image as a topographic surface and assume that holes have been punched in each regional minimum of the surface. The surface is then slowly immersed into water. Starting from the minima at the lowest altitude, the water will progressively flood the catchment basins of the image. In addition, dams are raised at the places where the water coming from two different minima would merge. At the end of this flooding procedure, each minimum is surrounded by dams delineating its associated catchment basin. The whole set of dams corresponding to the watersheds provide a partition of the input image into its different catchment basins. We now formalize this flooding process. The smallest value taken by the gray scale image f on its domain Df is denoted by h_{min}, and the largest by h_{max}. The catchment basin associated with a minimum M is denoted by CB(M). The points p of this catchment basin which have an altitude f(p) less than or equal to h are denoted by CBh(M):

$$CB_h\{M\} = \{p \in CB(M) \mid f(p) \leq h\} \qquad (4)$$

X_h, the subset of all catchment basins which have a grey scale value less than or equal to h:

$$X_h = \cup_i CB_h(M_i) \qquad (5)$$

Finally, the set of points belonging to the regional minima of elevation h are denoted by $RMIN_h(f(p))$. The catchment basins are now progressively built by simulating the flooding process. The first image points that are reached by water are the points of lowest grey scale value. These points belong to the regional minima of the image at level h_{min}. They are also equivalent to X_{hmin}:

$$X_{hmin+1} = RMIN_{hmin+1}(f(p)) \cup IZ_{Tt \leq hmin+1(f(p))}(X_{hmin}) \qquad (6)$$

Let us give some definitions in order to introduce the Influence Zones (IZ) which are defined as follows :

$$z_X(Y_i) = \{x \in X / d_X(x, Y_i), \forall j \neq i \ d(x, Y_i) < d(x, Y_i)\} \qquad (7)$$

which is valid for finite distance between x and Y_i included in X.

The boundaries between the various IZs give the geodesic skeleton by zones of influence of Y in X. Hence, we can write :

$$IZ_X(Y) = \cup_i z_X(Y_i) \qquad (8)$$

The definition of X_{hmin+1} is based on the analysis of the flooding process up to the elevation h_{min} +1. The water either expands the regions of the catchment basins already reached by it or starts to flood the catchment basins whose minima have an altitude equal to h_{min} + 1. This recursion formula holds for all levels h.

The set of catchment basins of a grey scale image f is equal to the set X_{hmax}, i.e., once all gray levels have been flooded:

$$X_{hmin} = T_{hmin}(f) \qquad (9)$$

$$\forall \ h \in [h_{min},h_{max}-1], \ X_{h+1} = RMIN_{h+1}(f) \cup IZ_{Tt<=h+1(f)}(X_h) \qquad (10)$$

The catchment basin (CB) of a grey tone image is represented as a label image. Each labeled region corresponds to the catchment basin of a regional minimum of the input image. The WaterSheds WS of f correspond to the boundaries of the catchment basins of the image f. Problems arise when we try to extend the above approach to images which are not lower complete. In such images, non-minima plateaus with nonempty interior occur. When we directly apply the above definitions, the topographical distance between interior pixels of a plateau turns out to be identically zero. Hence, an additional ordering relation between such pixels is required. The usual solution is to compute geodesic distances to the lower boundary of the plateau. This can be formalized by first transforming the image to a lower complete image, to which the above definitions can then be applied. There are several advantages to apply this method: it calculates the widths (which we need for the classifying part) of each point of the skeleton. While calculating them, the numbers of pixels on the branches of the skeleton are calculated, thus giving us the lengths of the leaves. We can smooth the contours of the input image before computing the skeletons. It is rather important because the initial skeleton may contain many small loops and other small distortions due to the complexity of images with high-occluded leaves. It is also a very robust way to calculate the entire morphological description of the leaves – the surface, the perimeter and other derived measures. Like watershed algorithm, this method also has given good results. Our method is based on the width of leaves - distance from every pixel of the leaf to the nearest contour. Watershed methods give good segmentation of leaves and the next section will describe how to take into account the labels of leaves got by this segmentation and lengths and widths of leaves measured from the skeletons.

BAYESIAN NETWORKS

Over the last decade, the Bayesian networks [9, 10], which are graphical models for probabilistic relationships between variables, have become a popular representation for encoding uncertain expert knowledge. More recently, researchers have developed methods for constructing Bayesian networks from data. These techniques that have been developed are new and still evolving, but they have been shown to be remarkably effective for some data-analysis problem techniques for extracting and encoding knowledge from data. Bayesian networks can even process incomplete data sets, they permit us to learn about causal relationships. Learning about causal relationships is relevant for at least two reasons: the studied domain and the prediction from the knowledge of causal relationships with data. Bayesian networks in conjunction with Bayesian statistical techniques facilitate the combination of domain knowledge and data. They have a causal semantics that makes the encoding of causal prior knowledge. Pearl developed the Message Passing Algorithm for performing the inference in Bayesian networks. For a set A of instantiated values 'a' it determines the probabilities P(x|a) for all values x of each variable X. It permits the probabilities to propagate on a direct acyclic graph (DAG) from its neighbors. These neighbors, in return, pass recursively messages to their neighbors. The probability updating does not depend on the order of computing probabilities. In the following, the algorithm is applied on specific rooted trees which are direct acyclic graphs in which there is a unique root. Let E be a set of instantiated variables and $E = N_x \cup D_x$ where N_X is a set of a priori variables and D_X is a set of a posteriori variables. In this algorithm, there are two kinds of messages: λ and π messages:

$$\lambda(X) \propto P(D_x|X) \tag{11}$$

$$\pi(X) \propto P(X|N_X) \tag{12}$$

$$P(X|E=e) \propto \lambda(X)\pi(X) \tag{13}$$

The messages λ are computed as below:
For each child Y of X,

$$\lambda_Y(X = x) = \sum_y P(Y = y|X = x)\lambda(Y = y) \tag{14}$$

The value λ is calculated in every node by:
If X is instantiated, $\lambda(X) = [0\ 0\ 1....0]$
where the position of 1 corresponds to the value of X.
else
for Z the only parent of X:
If X is a leaf then $\lambda(X = x) = \prod_{Y \in Childf(Z) - \{X\}} \lambda_U(X = x)$ (15)

The messages π are calculated by the subsequent loops:

$$\pi_X(Z = z) = \pi(Z = z)\prod_{U \in Childf(Z) - \{X\}} \lambda_U(Z = z) \tag{16}$$

π is computed in every node as:
If X is instanced, $\lambda(X) = [001....0]$
else
If X is a leaf then, $\pi(x) = P(X)$
else $\pi(X = x) = \sum_z P(X = x|Z = z)\pi_X(Z = z)$ (17)

Now, we will provide some proofs of the above calculations. Let (G,P) be a Bayesian network whose DAG is a tree. Herein, G = (V, E) represents the graph with vertices V and edges E. P represents the probabilities attached to the edges E, specifying the causal relationship between nodes and A be a set of values which is a subset of V. For each variable X, let us define λ messages, λ values, π messages, and π values as follows:

Let us define λ messages:

For each child Y of X, for all values of x in X, we get :

$$\lambda_y(x) \equiv \sum_y P(y|x)\lambda(y) \quad (18)$$

and let us define λ values as follows:
If $X \in A$ and X's value is $\hat{x}$,
$$\lambda(\hat{x}) \equiv 1 \text{ and } \lambda(x) \equiv 0 \text{ for } x \neq \hat{x}$$

If $X \notin A$ and X is a leaf, for all values of x, $\lambda(x) = 1$.
If $X \notin A$ and X is a non-leaf for all values of x,

$$\lambda(x) = \prod_{U \in child(x)} \lambda_U(x) \quad (19)$$

The messages π are defined as:

If Z is the parent of X, then for all values of $z \in Z$ are: $\pi_X(z) = \pi(z) \prod_{U \in CH(x)-\{X\}} \lambda_U(z)$,

where CH(x) means the children of x, and the values π :

If $X \in A$ and X's value is $\hat{x}$,

$$\pi(\hat{x}) = 1 \text{ and } \pi(x) = 0 \text{ for } x \neq \hat{x}.$$

If $X \notin A$ and X is the root, for all values of x, $\pi(X) = P(x)$.
If $X \notin A$, X is not the root, and Z is the parent of X, for all values of x:

$$\pi(x) = \sum_z P(x/z)\pi_X(z) \quad (20)$$

From previous definitions, for each variable X, for all values $x \in X$, $P(x|a) = \alpha\lambda(x)\pi(x)$, where α is a normalized constant. Let D_x be the subset of A containing all members of A that are in the sub-tree rooted at X and N be the subset of A containing all members of A that are non-descendents of X. The demonstration holds for the case in which each node has two children and can be easily extended for more children.

$$P(x|a) = P(x|d_X, n_X) = \frac{P(d_X, n_X|x)P(x)}{P(d_X, n_X)}$$

$$= \frac{P(d_X|x)P(n_X|x)P(x)}{P(d_X, n_X)} = \frac{P(d_X|x)P(x|n_X)P(n_X)P(x)}{P(x)P(d_X, n_X)} = \beta P(d_X|x)P(x|n_X) \quad (21)$$

with β is a constant that does not depend on the value of x.

We will develop functions $\lambda(x)$ and $\pi(x)$ such that:

$$\lambda(x) \propto P(d_x|x) \text{ and } \pi(x) \propto P(x|n_x) \tag{22}$$

Then we have : $\qquad P(x|a) = \beta \lambda(x) \pi(x)$ (23)

Finally, this algorithm uses these equations in order to propagate the probabilities on the overall DAG corresponding to the application.

RESULTS

Image 1 shows the input image. The classification of pixels in two classes soil and vegetation is performed by using the probabilistic principal component analysis and the mixture of Gaussian distributions modeling the projections of pixels on the latent variables on the histogram of the image 1 from the color representation Lab of the input image. The utilization of one or two latent variables has been tested. The use of one latent variable is faster and provides good classification of soil and vegetation of color pixels. Image 2 presents the skeleton of the vegetation image, it is useful computing the leaf lengths of vegetal species. The watershed algorithm of segmentation has been applied on the distance image computed from the vegetation contour pixels of image 3; it provides the different regions of the segmentation of the vegetation. This distance image is quickly obtained through two scans of the vegetation image with recursive algorithms. Finally, the results of the proposed method appear in image 4. The Bayesian networks have been implemented on two levels with Pearl's algorithm, the first one takes into account the length, the width and the ratio length/width measured on the regions of leaves, the second one takes into account the results of the first Bayesian network, the type of neighboring leaves and their orientations. This second Bayesian network allows fusions of parts of leaves which are neighbors and in the same direction.

CONCLUSION

The above proposed method works well for all kinds of images and the results are consistent for the different images of the same field. The Bayesian networks used in our method are also well suited to the complexity of nature because they can be adapted to different steps of the evolution of vegetation. Further works will have to be done in order to consider in depth the occlusions of leaves and improve the learning stage of Bayesian networks by making it as automatic as possible. These Bayesian networks can easily incorporate another relevant parameter which is the height of leaves of different vegetal species.

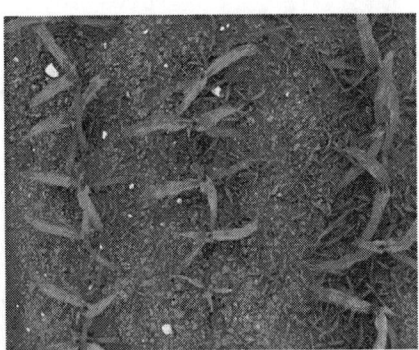

IMAGE 1 Input image

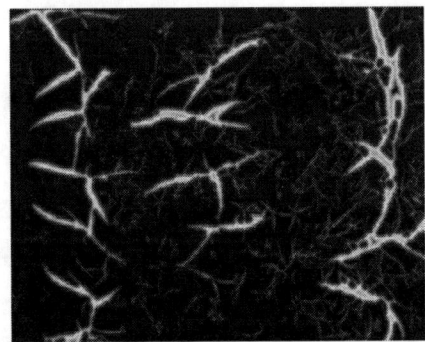

IMAGE 3 Distance map from contours

IMAGE 2 Skeleton of input image

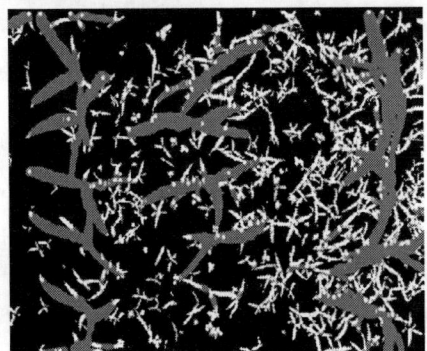

IMAGE 4 Classification maize/weeds

REFERENCES

1. Chapron M., Assemat L., Boissard P., Huet P., 2001. Weed and maize recognition using 2D and 3D data fusion. Third ECPA.
2. Manh A.G., Rabatel G., Assemat L., Aldon M.J., 2001. In field classification of weed leaves by machine vision using deformable templates. Third ECPA.
3. L. Tian , D. C. Slaughter, R. F. Norris, Machine Vision Identification of Tomato Seedlings for automated weed control, 2000
4. H. Hotelling, Analysis of a complex of statistical variables into principal components, journal of Educational Psychology 24, pp 417-441, 1933.
5. M. Tippling, C. M. Bishop, probabilistic principal component analysis, Journal of the Royal Statistical Society, Series B, vol 61, Part 3, pp 611-622.
6. Alain Trémeau, Image numérique couleur, Dunod, 2004
7. P. Soille , Morphological Image processing; Principles and Applications, Springer Verlag, 2003
8. Edward Dougherty, Roberto A. Lotufo, Hands-on Morphological Image Processing, SPIE PRESS Vol. TT59,July 2003
9. Richard Neapolitan, Learning Bayesian Networks, Prentice Hall, 2004
10. P. Naim, P.H. Wuillemin, P. Leray, O. Pourret, Les réseaux bayésiens, Eyrolles.

Separation of Stochastic and Deterministic Information from Seismological Time Series with Nonlinear Dynamics and Maximum Entropy Methods

Rafael M. Gutiérrez, Gina M. Useche and Elias Buitrago

Centro de Investigaciones, Universidad Antonio Nariño, Carrera 3 Este No. 47A – 15
Bogotá, Colombia
(rafael.gutierrez@uan.edu.co , gina.useche@uan.edu.co and elias.buitrago@uan.edu.co)

Abstract. We present a procedure developed to detect stochastic and deterministic information contained in empirical time series, useful to characterize and make models of different aspects of complex phenomena represented by such data. This procedure is applied to a seismological time series to obtain new information to study and understand geological phenomena. We use concepts and methods from nonlinear dynamics and maximum entropy. The mentioned method allows an optimal analysis of the available information.

Keywords: Time Series, Separation Problem, Nonlinear Dynamics, Maximum Entropy, Seismology.
PACS: 05.45.Tp, 05.45.-a, 05.70.-a, 91.30.Iv

INTRODUCTION

In this work we address the problem of constructing models from empirical information with very limited or no theoretical information, by identifying and separating the deterministic and stochastic information content of the empirical information. We consider empirical information as a set of measures of a certain observable of phenomena that cannot be reproduced or controlled with an experiment. Seismological information is an example. In general, empirical information corresponds to one observable of a complex system with many degrees of freedom or variables related in unknown ways. In this work we apply digital filters, nonlinear time series analysis and the maximum entropy method MEM, to organize, but does not yet self-organize, several methods to detect deterministic and stochastic information from empirical information. Applied to empirical seismological data, this method generates new information related to geological phenomena which cannot be obtained otherwise.

EMPIRICAL SEISMOLOGICAL INFORMATION

Empirical information is often irreproducible and has diverse sources of contamination that makes their analysis very challenging. In addition, we have limited theoretical knowledge of the phenomena involved and in particular of the signal to noise relation.

Seismological information

The seismological information that we use in this work corresponds to three active seismological nests in the region of Colombia where three tectonic plates converge: *Nazca*, *Caribean* and *Southamerican* plates.

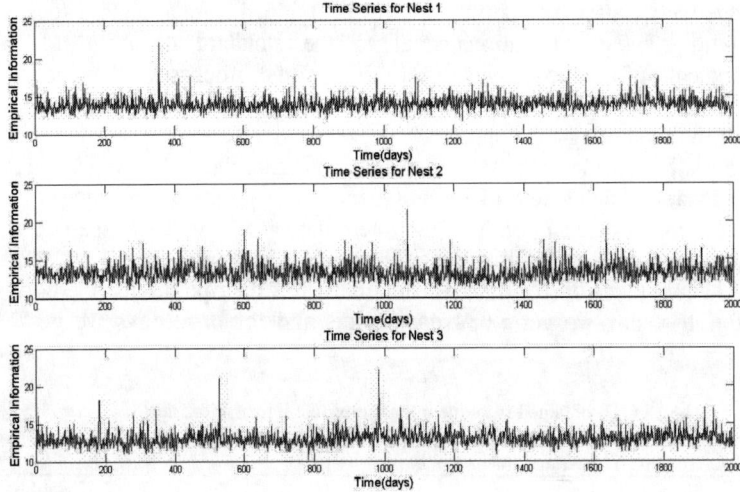

FIGURE 1. The empirical seismological information of the three Colombian nests: nest 1 *Bucaramanga*, nest 2 *Urabá* and nest 3 *Nariño*.

The initial seismological information is obtained from sensors which measure the local displacement of the earth's surface in the three possible spatial degrees of freedom. This initial information is processed with standard methods to obtain the magnitude, time and location of the seismological event: nest 1 named *Bucaramanga* with 37706 data points, nest 2 named *Urabá* with 5660 data points, and nest 3 named *Nariño* with 6606 data points [1]. In this context, we consider as contamination any characteristic or features of the time series which represent limitations to our capabilities to detect deterministic or stochastic information. Aside of noise and nonstationary effects, other sources of contamination may be artifacts introduced by the preprocessing of the seismological information, temporal and local effects on the measurements generated by climate changes or any other non-seismological phenomena.

Seismological activity is characterized by long periods of very low and sporadic activity interrupted by very strong events, making this kind of empirical information very irregular with large and isolated peaks [2]. Seismological information can be prepared in several ways in order to obtain seismological time series. We first assign the initial disperse seismological information to local events localized in one of the three mentioned nests. Then, we construct 18 original time series OTS (3 nests, times three time lags, times two different interpolation methods).

The interpolation methods

In order to generate time series with constant time lag Δt, between data points (the OTS), from a series of data points with different Δt between them (the initial seismological information), we use two different methods of interpolation. The first method is a standard linear interpolation where the closer neighbor has more weight to define the new data point. The second method is the same linear interpolation with an error added to the new data point; the error is obtained from a random source of white noise with standard deviation equal to the standard deviation of the initial seismological information. This error gives some stochastic characteristic to the created data points making the reconstructed information more consistent with the nature of the seismological events.

We initially consider three different Δt to construct the OTS: $\Delta t=1/2$ day, $\Delta t=1$ day and $\Delta t=2$ days. Independent of the number of data points of the three seismological nests, the number of data points of the interpolated OTS are the same depending only on the constant value of Δt. Since the information goes from June 4 of 1993 to May 6 of 2007, if we take $\Delta t=1/2$ day, we obtain 10.168 data points for the corresponding OTS, for $\Delta t=1$ day we get 5.084 data points and for $\Delta t=2$ days we get 2.542 data points, see table 1.

TABLE 1. The 18 OTS obtained from the initial empirical information: nest 1 *Bucaramanga*, nest 2 *Urabá* and nest 3 *Nariño*; the three time lags: $\Delta t = 1/2$ day, $\Delta t = 1$ day and $\Delta t = 2$ days; and the two interpolation methods: linear and linear-stochastic.

OTS	Seismological nest	Time lag, Δt	Interpolation method	Number of data points
1	1	½	Linear	10.156
2	1	½	Linear-stochastic	10.156
3	1	1	Linear	5.078
4	1	1	Linear-stochastic	5.078
5	1	2	Linear	2.538
6	1	2	Linear-stochastic	2.538
7	2	½	Linear	10.156
8	2	½	Linear-stochastic	10.156
9	2	1	Linear	5.078
10	2	1	Linear-stochastic	5.078
11	2	2	Linear	2.538
12	2	2	Linear-stochastic	2.538
13	3	½	Linear	10.156
14	3	½	Linear-stochastic	10.156
15	3	1	Linear	5.078
16	3	1	Linear-stochastic	5.078
17	3	2	Linear	2.538
18	3	2	Linear-stochastic	2.538

THE METHOD

The program has two fundamental levels: processing and analysis. In the first level of processing, we use a blind but systematic procedure to filter different regions of the power spectrum of the OTS. This procedure generates a set of processed time series PTS, that may be large or very large. We can generate a very large number of PTS as the raw information material for the following analysis procedure. In the second level of analysis, we apply three nonlinear deterministic time series analysis tools to all the generated PTS, in order to estimate the deterministic and stochastic information content of each one of the PTS. Then we use the MEM as a selection method.

In the block diagram of figure 2 we present a schematic representation of the whole process.

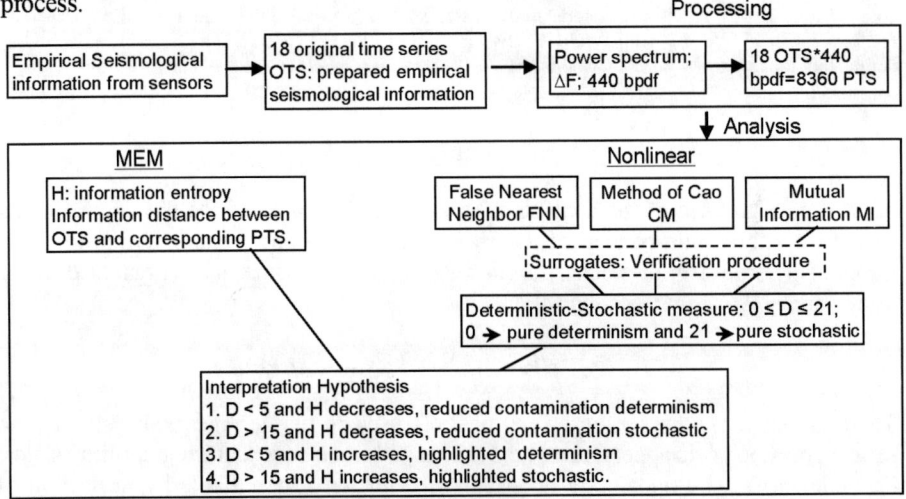

FIGURE 2. Block diagram of the general procedure.

Processing

Processing consists of the systematic application of different digital-band-pass filters to the OTS in order to generate a large set of PTS. The application of a filter to a OTS generates one PTS. The set of digital-band-pass filters is defined as follows: first, we obtain the power spectrum of the OTS and determine the frequencies F_1 and F_2 as the initial and final frequencies where 95% of the power is contained, see Fig. 3. The range of frequencies defined by F_1 and F_2 is divided in p equal parts of width ΔF. The set of digital-band-pass filters that generates the corresponding set of PTS is then defined by all the possible combinations of F_L and F_R, where F_L is the left frequency and F_R is the right frequency of the filter. F_L goes from F_1 to $F_1+(p-1)\Delta F$ and F_R goes from $F_1+\Delta F$ to F_2. The operation of adding different numbers of ΔF to F_1 and F_2 in all the possible combinations, corresponds to the translation and expansion of steps ΔF as shown in Fig. 2. We also vary the order of the digital-band-pass filter, n. The order of the filter defines the difference between the ideal and the real filter. We have fixed $p=10$ and we vary the order n from 1 to 8. We use the digital implementation of the IIR Butterworth filter [3]. The values of p and n can be changed at will considering the relation cost-benefit. With these considerations we obtain 440 digital-band-pass filters which generate the corresponding 440 PTS for each OTS.

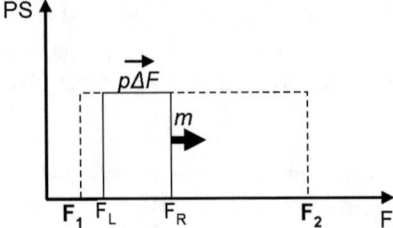

FIGURE 3. Digital-band-pass filters applied to the OTS to generate the corresponding PTS: all possible translations, p, and expansions, m, with step size ΔF, of the digital-band-pass filter defined by the frequencies F_L and F_R taking values between the frequencies F_1 and F_2. The frequencies F_1 and F_2 define the range of frequencies where 95% of the power spectrum of the corresponding OTS is contained.

Analysis

The main purpose of the analysis is to generate estimates of the deterministic and stochastic information content of the PTS, and to determine which PTS have clearer traces of determinism or stochastic information. These estimates are organized in the array named ***D***, we then use the MEM.

Deterministic and Stochastic Estimates

There are several compilation efforts of tools used in the nonlinear analysis of time series with stochastic appearance [4,5,6]. Their importance is that irregular time series, i.e. time series with apparently no structure or patterns identifiable by inspection or traditional methods in the time series itself nor in its power spectrum, cannot be modeled with traditional linear methods. However, the line between nonlinear deterministic time series with irregular behavior and stochastic time series is very fuzzy, in particular for empirical time series. The diverse methods to analyze nonlinear time series have different strengths and limitations. Therefore, it is important to combine more than one of them to maximize the strengths and minimize the limitations of each one individually. In this work we use three methods of nonlinear time series analysis: False Nearest Neighbors, FNN [7], Method of Cao, CM [8], and the method of mutual information, MI [9]. These three methods are in general and to some extent very robust to the natural limitations of empirical time series and very efficient to detect different traces of determinism. We also use the method of surrogates in two forms to strengthen the possible results [10]. The outputs of these three nonlinear tools are qualitative in the characteristics of the plots they generate. We transform the qualitative results into quantitative ones based on known results of well known time series with controlled contamination and characteristics. This quantification gives numerical values for each PTS between a minimum of 0, which corresponds to clear determinism, and a maximum of 21 corresponding to a clear stochastic time series.

Application of the MEM

We use the MEM similarly to the way it is used in image processing to improve a noisy image. We define the probabilities to be proportional to the difference between each data point of a PTS and the corresponding data point of the corresponding OTS. Since the information entropy H, is a measure of the "information distance" [11], it represents the information distance between the PTS to the corresponding OTS, being the first time series a filtered version of the second. Therefore, H helps to identify which PTS represents the information content of the corresponding OTS better, independently, whether the PTS has deterministic and/or stochastic information.

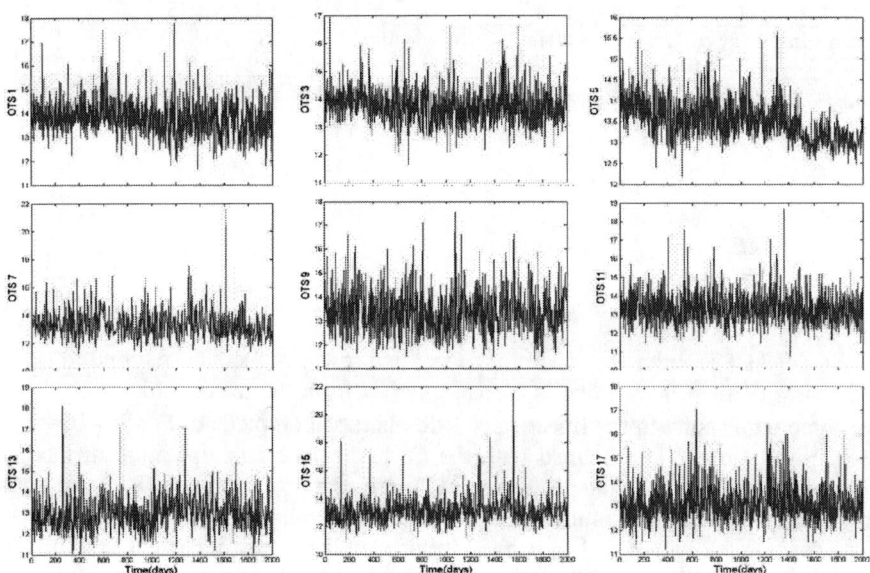

FIGURE 4. Segments of 9 of the 18 OTS, according to table 1, obtained from the empirical information at three time scales (½ day, 1 day and 2 days), for the three Colombian nests (nest 1 *Bucaramanga*, nest 2 *Urabá* and nest 3 *Nariño*), and two methods of interpolation (linear and linear-stochastic).

RESULTS AND DISCUSSION

The 18 OTS: three different time scales (½ day, 1 day and 2 days), two different methods of interpolation (linear and linear-stochastic), and three seismological nests. The 440 PTS for each one of the 18 OTS by the automatic application of the 440 digital-band-pass filters. The deterministic and stochastic estimation, array D, with the quantified values for each PTS. The application of the MEM by estimating the entropy $H = -\sum_{i}^{M} p_i \ln p_i$, where M is the number of data points and the probabilities p_i are proportional to the differences between data points of the PTS and the corresponding data points of the corresponding OTS.

Inspired by the application of the MEM in image processing to determine the best image from a set of "smoothed images" using a "default image" [12], we consider that the "smoothed images" correspond to the PTS and the "default image" is the corresponding OTS. In this context, the application of the MEM helps to detect which PTS are structurally closer to the corresponding OTS, despite the weak or strong modification that the corresponding filter may signify. With structurally closer we do not mean the goodness-of-fit statistic, chi-square, the structural closeness is related with the information content and is what the maximization of H indicates. In table 2 we present the most relevant PTS with the corresponding results.

TABLE 2. Results of D, H and the interpretation case for the 9 most relevant PTS.

Interesting PTS	The corresponding filter: F_L	F_R	n	Corresponding OTS	H : case			D
1	F_1	$F_1+\Delta F$	4	5	↓	:	1	3
2	$F_1+3\Delta F$	$F_1+4\Delta F$	4	6	↑	:	3	3
3	F_1	$F_1+8\Delta F$	5	6	↑	:	3	4
4	$F_1+\Delta F$	$F_1+10\Delta F$	5	2	↑	:	4	17
5	F_1	$F_1+2\Delta F$	3	2	↓	:	2	16
6	F_1	$F_1+3\Delta F$	6	2	↑	:	4	16
7	F_1	$F_1+7\Delta F$	5	2	↑	:	4	16
8	F_1	$F_1+7\Delta F$	2	10	↓	:	2	16
9	$F_1+3\Delta F$	$F_1+10\Delta F$	6	10	↑	:	4	16

We get 11 PTS with $D \leq 4$, all for $\Delta t = 2$ and seven of them correspond to nest 1. Only three of them have important changes of H from the mean value, two of these last three correspond to the linear-stochastic interpolation method. For $D \geq 16$ we get 36 PTS, all from OTS obtained with the linear-stochastic interpolation method, 24 from nest 1 with $\Delta t = ½$ and 12 from nest 2 with $\Delta t = 1$. Of these 36 PTS, we obtain only 6 with significant variations of H from the mean value: 4 from nest 1 with $\Delta t = ½$ and 2 from nest 2 with $\Delta t = 1$. In table 2 we present the 9 most relevant PTS with their corresponding details.

Using figure 2 to understand the filter specifications of table 2, we can make four observations from the results: 1) deterministic information is decontaminated in the first tenth of the power spectrum of OTS number 5 (nest 1, $\Delta t = 2$ and linear interpolation); 2) stochastic information is highlighted in the first half of the PS of the OTS 5 and 8; 3) the deterministic information is highlighted in two ways from OTS 6: in the fourth tenth of the PS and in the first eight tenths of the PS, though, in the first case the results are less clear; 4) the stochastic information is highlighted in several ways from the OTS 2 and 10: in the first half of OTS 2 and in the second half of the OTS 2 and 10.

Discussion and Conclusions

In very general terms, most of the results are consistent with a phenomenological fact. In general, but with some exceptions the results are simpler and with more similarities for many of the PTS from the nests 2 and 3 (*Urabá* and *Nariño*), located in the Colombian Pacific and aligned with the subduction line of the *Southamerican* plate under the *Nazca* plate. In the same general terms and exceptions, the results are more complex and richer for the nest 3 (*Bucaramanga*). In the geological region of nest 3 the mentioned subduction goes down about 150 km under the earth surface and into the continent from the Pacific. In addition, the closeness of the *Caribean* plate may also contribute to make the seismological sources richer and more complex.

We introduce an interpretation hypothesis to define four possible cases: Case 1: when D indicates determinism ($D < 5$), and H decreases from the mean value, the digital-band-pass filter generating the corresponding PTS reduces contamination in the deterministic information. Case 2: when D indicates a stochastic time series ($D > 15$), and H decreases from the mean value, the digital-band-pass filter reduces contamination of the stochastic information. Case 3: when D indicates determinism, and H increases from the mean value, the digital-band-pass filter highlights the deterministic information. Case 4: when D indicates stochastic time series, and H increases from the mean value, the digital-band-pass filter highlights the stochastic information. As a PTS becomes more structurally close to the corresponding OTS, the data become more featured: deterministic or stochastic patterns are highlighted, considering that empirical information is always incomplete and inexact information. The consistency of these results with the phenomenology of the tectonic dynamics of the region from where the empirical seismological information was collected, indicates that these results can be used to model and identify different aspects of the complex processes involved.

ACKNOWLEDGMENTS

We would like to thank the partial funding of the "Instituto Colombiano para el Fomento de la Ciencia y la Tecnología Francisco José de Caldas", COLCIENCIAS, under grant No. 257-2005, the support of the research group "Sistemas Complejos" and to all the personnel of the DNI-UAN.

REFERENCES

1. http://www.ingeominas.gov.co.
2. C. M. R. Fowler, *The Solid Earth: An Introduction to Global Geophysics*, Cambridge: Cambridge, 2005, pp. 100-157.
3. Matlab, Mathworks: http://www.mathworks.com/access/helpdesk/help/techdoc/matlab.html
4. H. Kantz and T. Schreiber, *Nonlinear Time Series Analysis*, Cambridge: Cambridge, 2005.
5. R. Hegger and H. Kantz, *Practical implementation of nonlinear time series methods: The TISEAN package*. arXiv:chao-dyn/9810005 v1 30 Sep 1998. http://www.mpipks-dresden.mpg.de/~tisean/
6. Ch. Merkwirth, U. Parlitz and W. Lauterborn, *Nonlinear time series analysis using TSTOOL.*, 2000, pp.256-260. http://www.physik3.gwdg.de/tstool/
7. H. D. I. Abarbanel, *Analysis of observed chaotic data*, New York: Springer-Verlag, 1996.
8. L. Cao, *Physica D* **110**, 43-50 (1997).
9. J. C. Sprott, *Chaos and Time-Series Analysis*, Oxford: Oxford University Press, 2003. pp. 314.
10. T. Schreiber and A. Schmitz, *Surrogate Time Series*. arXiv:chao-dyn/9909037 27 Sep 1999.
11. H. S. Robertson, *Statisthical Thermophysics*, Englewood Cliffs: PTR Prentice Hall, 1993, pp. 35-45.
12. J. Skilling, "Classic Maximum Entropy" in *Maximum Entropy and Bayesian Methods-1989*, edited by J. Skilling, Kluwer Academic, Norwell, MA, 1989, pp. 45-52.

Using Prior Information in Bayesian Inference – with Application to Fault Diagnosis

Anna Pernestål and Mattias Nyberg

Dept. of Electrical Engineering, Linköping University, Linköping, Sweden
{annap, matny}@isy.liu.se

Abstract. In this paper we consider Bayesian inference using training data combined with prior information. The prior information considered is response and causality information which gives constraints on the posterior distribution. It is shown how these constraints can be expressed in terms of the prior probability distribution, and how to perform the computations. Further, it is discussed how this prior information improves the inference.

Keywords: Bayesian Inference, Bayesian Classification, Prior Information, Fault Diagnosis

INTRODUCTION

In this paper we study the problem of making inference about a state, given an observed feature vector. Traditionally, inference methods rely either on prior information only or on training data consisting of simultaneous observations of the state variable and the feature vector [1], [2], [3]. However, in many inference problems there are both training data and prior information available. Inspired by the problem of fault diagnosis, where the feature vector typically is a set of diagnostic tests, and the states are the possible faults, we recognize two types of prior information. First, there may be information that some values of the features are impossible under certain states. In the present paper this information is referred to as *response information*, which for example can be that it is known that a test never alarms when there is no fault present. Second, it may be known that certain elements of the feature vector are equally distributed under several states, here referred to as *causality information*. In the fault diagnosis context this means that a diagnostic test is not affected by a certain fault.

The type of prior information studied in the present work typically appears in previous works on fault diagnosis. The response information is used for example in [4], [5], and [6]. The causality information is an interpretation of the Fault Signature Matrix (FSM) used for example in [7] and [8]. The main difference between these previous works and the present is that here we combine the prior information with training data instead of relying on prior information only.

How to compute the posterior probability for the states in the case of training data only is, although previously well studied, a nontrivial problem, see e.g. [9], [10], and [11]. In these previous works the computations are based on training data only. In the present work we go one step further, and discuss how the prior information in terms of response and causality information can be integrated into the Bayesian framework.

INFERENCE USING TRAINING DATA

We begin by introducing the notation used, and summarizing previous results on Bayesian inference using training data alone. Let $\mathbf{Z} = (\mathbf{X}, C)$ be a discrete variable, where the feature vector $\mathbf{X} = (X_1, \ldots X_R)$ is R-dimensional and the state variable C is scalar. The variables $\mathbf{X}$ and C can take K and L different values respectively, and hence $\mathbf{Z}$ can take $M = KL$ values. Use $\mathbf{z} = (\mathbf{x}, c) = ((x_1, \ldots, x_R), c)$ to denote a sample of $\mathbf{Z}$. Let $\mathbb{X}_i$, $\mathbb{X}$, $\mathbb{C}$, and $\mathbb{Z} = \mathbb{C} \times \mathbb{X}$ be the domains of X_i, $\mathbf{X}$, C and $\mathbf{Z}$ respectively. Enumerate the elements in $\mathbb{Z}$, and use $\zeta_i, i = 1, \ldots, M$, to denote the ith element.

Let $\mathscr{D}$ be the training data, i.e. a set of simultaneous samples of the feature vector and the state variable. The samples in training data are denoted by d_i, $i = 1, \ldots, N$.

We use $p(\mathbf{X} = \mathbf{x}|\mathbf{I})$, or simply $p(\mathbf{x}|\mathbf{I})$, to denote the discrete probability distribution for $\mathbf{X}$ given the current state of knowledge $\mathbf{I}$. For continuous probability density functions we use the notation $f(\mathbf{x}|\mathbf{I})$.

In the inference problem, the probability distribution $p(c|\mathbf{x}, \mathscr{D}, \mathbf{I})$ is to be determined. Note that for a given feature vector $\mathbf{x}$, the posterior probability for a state is proportional to the joint distribution of c and $\mathbf{x}$, $p(c|\mathbf{x}, \mathscr{D}, \mathbf{I}) = p(c, \mathbf{x}|\mathscr{D}, \mathbf{I})/p(\mathbf{x}|\mathscr{D}, \mathbf{I}) \propto p(c, \mathbf{x}|\mathscr{D}, \mathbf{I}) = p(\mathbf{z}|\mathscr{D}, \mathbf{I})$. Therefore we can study the probability distribution $p(\mathbf{z}|\mathscr{D}, \mathbf{I})$.

The computations of $p(\mathbf{z}|\mathscr{D}, \mathbf{I})$ when only training data is available are, under certain assumptions, given in detail for example in [9], [10], and [11]. In these references the arguments for the underlying assumptions are also discussed. Here we are content to present the most important assumptions briefly, and then we summarize the results in a theorem.

First, assume that there are parameters $\Theta = (\theta_1, \ldots, \theta_M)^T$ such that

$$p(\mathbf{Z} = \zeta_i | \Theta, \mathbf{I}) = \theta_i, \quad i = 1, \ldots, M, \tag{1a}$$

$$\theta_i > 0, \quad \sum_{\zeta_i \in \mathbb{Z}} \theta_i = 1. \tag{1b}$$

This is the most general parameterization of a discrete probability distribution. Furthermore, assume that the prior probability density $f(\Theta|\mathbf{I})$ for the parameters is Dirichlet distributed, i.e.

$$f(\Theta|\mathbf{I}) = \frac{\Gamma(\sum_{i=1}^{M} \alpha_i)}{\prod_{i=1}^{M} \Gamma(\alpha_i)} \prod_{i=1}^{M} \theta_i^{\alpha_i - 1}, \quad \alpha_i > 0, \tag{2}$$

where $\Gamma(\cdot)$ is the gamma function[1] and the parameters $\alpha = (\alpha_1, \ldots, \alpha_M)$ are given. The choice of the Dirichlet distribution as prior density for the parameters has several convenient properties. One is that the parameters α can be interpreted as *hypothetical samples* representing the prior state of knowledge about the different values $\mathbf{z}_i$. The properties of the Dirichlet distribution is further discussed in [9], [10], and [11]. We can now state the following theorem on the posterior probability for $\mathbf{Z}$.

Theorem 1 *Let $p(\mathbf{z}|\mathscr{D}, \mathbf{I})$ be discrete, and assume that (1) and (2) holds. Furthermore, assume that the samples in the training data are independent, let n_i be the count of samples in $\mathscr{D}$ where $\mathbf{Z} = \zeta_i$, and let $N = \sum_{i=1}^{M} n_i$ and $A = \sum_{i=1}^{M} \alpha_i$.*

---

[1] i.e. fulfills $\Gamma(n+1) = n\Gamma(n)$ and $\Gamma(1) = 1$

TABLE 1. Example of response information, where "•" means that the value of the feature is possible.

	$C = c_1$	$C = c_2$	$C = c_3$
$x_1 = 0$	•	•	•
$x_1 = 1$		•	•
$x_1 = 2$		•	

Then it holds that

$$p(\mathbf{Z} = \zeta_i | \mathcal{D}, \mathbf{I}) = \frac{n_i + \alpha_i}{N + A}. \tag{3}$$

In the following sections we will now discuss how the results from Theorem 1 can be extended to take response and causality information into account.

INFERENCE USING RESPONSE INFORMATION

Consider now the case where some values of the feature vector are known to be impossible in certain states of the system. We refer to this kind of information as *response information*. Formally, it means that there are sets $\gamma_{i,c} \subset \mathbb{X}_i$ representing "forbidden values" under state c, i.e.

$$p(x_i | c, \mathcal{D}, \mathbf{I}_\mathcal{R}) = 0, \text{ for } x_i \in \gamma_{i,c},$$

where we have used $\mathbf{I}_\mathcal{R}$ to denote that $\mathbf{I}$ includes response information.

To exemplify how the sets $\gamma_{i,c}$ can be determined, consider the following example with a three-valued feature X_1 with domain $\mathbb{X}_1 = \{0, 1, 2\}$. Assume that the information is given that in state c_1, the feature X_1 can only take the value 0. In state c_2 all values are possible, while in state c_3 all values except 2 are possible. This information is summarized in Table 1, where "•" means that the value of the feature is possible. The given response information is translated to the sets $\gamma_{1,c_1} = \{1, 2\}$, $\gamma_{1,c_2} = \{\emptyset\}$, $\gamma_{1,c_3} = \{2\}$.

One interpretation of the response information represented by Table 1 in terms of fault diagnosis is that X_1 is a diagnostic test where $X_1 = 0$ means that no alarm is generated, and $X_1 = 1$ and $X_1 = 2$ are two different types of alarms. Let c_1 represent the fault free state. The response information then means that no alarms are possible under the fault free state. The state c_2 represents a fault where both types of alarms are possible, and for c_3 only alarm type 1 is possible. In Table 1 the bullets in the first row in the columns for c_2 and c_3 are interpreted as that there may be missed detections.

Let $\gamma \subset \mathbb{Z}$ be the set of values such that if $x_i \in \gamma_{i,c}$, then $\mathbf{z} \in \gamma$. In our example we have $\gamma = \{(1, c_1), (2, c_1), (2, c_3)\}$. Assume that $p(\mathbf{z} | \Theta, \mathbf{I}_\mathcal{R})$ is parameterized by Θ as in (1a). By $\mathbf{I}_\mathcal{R}$ we have the following requirements on the parameters

$$\theta_i = 0 \quad \forall \zeta_i \in \gamma, \qquad \theta_i > 0 \quad \forall \zeta_i \in \mathbb{Z} \setminus \gamma, \qquad \sum_{\zeta_i \in \mathbb{Z} \setminus \gamma} \theta_i = 1. \tag{4}$$

Now, consider the prior probability density $f(\Theta | \mathbf{I}_\mathcal{R})$ for the parameters. Before adding the response information, we adopted a Dirichlet distribution for the parameters. The only new information that is added by the response information is the decrease of the number of possible values of $\mathbf{Z}$ to $\mathbb{Z}\setminus\gamma$. Thus, as prior density for the parameters we use the Dirichlet distribution over this set

$$f(\Theta|\mathbf{I}_\mathcal{R}) = \begin{cases} \frac{\Gamma(\Sigma_{\zeta_i \in \mathbb{Z}\setminus\gamma}\alpha_i)}{\Pi_{\zeta_i \in \mathbb{Z}\setminus\gamma}\Gamma(\alpha_i)} \Pi_{\zeta_i \in \mathbb{Z}\setminus\gamma} \theta_i^{\alpha_i-1}, \ \alpha_i > 0 & \text{if } \Theta \in \Omega_\mathcal{R} \\ 0 & \text{otherwise.} \end{cases} \quad (5)$$

The use of (5) is further discussed in [12].

We can now state the following theorem for the joint probability distribution when response information is available.

Theorem 2 *Assume that $p(\mathbf{Z}|\Theta, \mathbf{I}_\mathcal{R})$ is discrete and given by (1a) and (4), and that the prior probability $f(\Theta|\mathbf{I}_\mathcal{R})$ is given by (5). Furthermore, assume that the samples in the training data $\mathcal{D}$ are idependent. Let n_i be the count of samples in $\mathcal{D}$ where $\mathbf{Z} = \zeta_i$, and let $N = \sum_{i=1}^{M} n_i$ and $A = \sum_{i=1}^{M} \alpha_i$. Then it holds that*

$$p(\mathbf{Z} = \zeta_i | \mathcal{D}, \mathbf{I}_\mathcal{R}) = \begin{cases} 0, & \text{if } \mathbf{z} \in \gamma \\ \frac{n_i + \alpha_i}{N+A} & \text{otherwise.} \end{cases} \quad (6)$$

Proof: Apply Theorem 1 when $z \in \mathbb{Z}\setminus\gamma$, and use that (5) gives probability 0 for all $z \in \gamma_c$. A complete proof is given in [12]. □

INFERENCE USING CAUSALITY INFORMATION

Let us now turn to the case when there is information available that a certain feature is equally distributed in two states. We call this kind of information *causality information*. In this section we show how this information can be integrated in the problem formulation, and we also discuss a method for solving the problem.

Computing the Posterior Using Causality Information

The causality information is formally represented by

$$p(x_i|c_j, \Theta, \mathbf{I}_\mathcal{C}) = p(x_i|c_k, \Theta, \mathbf{I}_\mathcal{C}), \ i = 1, \ldots, K_i, \quad (7)$$

where $\mathbf{I}_\mathcal{C}$ is used to denote that causality information is present. Applying the product rule of probabilities on (8) we have

$$\frac{p(x_i, c_j|\Theta, \mathbf{I}_\mathcal{C})}{p(c_j|\mathbf{I}_\mathcal{C})} = p(x_i|c_j, \Theta, \mathbf{I}_\mathcal{C}) = p(x_i|c_k, \Theta, \mathbf{I}_\mathcal{C}) = \frac{p(x_i, c_k|\Theta, \mathbf{I}_\mathcal{C})}{p(c_k|\mathbf{I}_\mathcal{C})},$$

where $p(c_j|\mathbf{I}_\mathcal{C})$ and $p(c_k|\mathbf{I}_\mathcal{C})$ are the prior probabilities for the states c_j and c_k, and are assumed to be given by the background information $\mathbf{I}_\mathcal{C}$. Since the prior probabilities are known we can write $p(c_j|\mathbf{I}_\mathcal{C}) = \rho_{jk} p(c_k|\mathbf{I}_\mathcal{C})$ for a known constant ρ_{jk}. Thus, for each i we can transform the relation in (7) to the following relation on the joint probability of C and X_i,

$$p(c_j, x_i|\Theta, \mathbf{I}_\mathcal{C}) = \rho_{jk} p(c_k, x_i|\Theta, \mathbf{I}_\mathcal{C}). \quad (8)$$

For a certain value $x_i = \xi_i$ we have that

$$p(c_j, \xi_i | \Theta, \mathbf{I}_\mathscr{C}) = \sum_{\zeta_l \in \mathbb{Z}_{\xi_i, c_j}} p(\zeta_l | \Theta, \mathbf{I}_\mathscr{C}) = \sum_{\zeta_l \in \mathbb{Z}_{\xi_i, c_j}} \theta_l, \qquad (9)$$

where $\mathbb{Z}_{\xi_i, c_j} = \{\zeta_l \in \mathbb{Z} : \zeta_l = ((x_1, \ldots, \xi_i, \ldots, x_R), c_j)\}$, i.e. the set of all possible values ζ_l of $\mathbf{Z}$ in which $x_i = \xi_i$ and $C = c_j$. Equations (8) and (9) give requirements in the form

$$\sum_{\zeta_l \in \mathbb{Z}_{\xi_i, c_j}} \theta_l = \rho_{jk} \sum_{\zeta_l \in \mathbb{Z}_{\xi_i, c_k}} \theta_l. \qquad (10)$$

To exemplify, consider the following case with two states, $C \in \{c_1, c_2\}$, and a scalar feature $\mathbf{X} \in \{0, 1\}$. Define $\Theta = (\theta_1, \theta_2, \theta_3, \theta_4)$ by

$$p(\mathbf{X} = 0, C = c_1 | \Theta, \mathbf{I}) = \theta_1, \; p(\mathbf{X} = 0, C = c_2 | \Theta, \mathbf{I}) = \theta_2, \qquad (11a)$$
$$p(\mathbf{X} = 1, C = c_1 | \Theta, \mathbf{I}) = \theta_3, \; p(\mathbf{X} = 1, C = c_2 | \Theta, \mathbf{I}) = \theta_4. \qquad (11b)$$

Assume that the causality information $p(\mathbf{X}, C = c_1 | \mathbf{I}_\mathscr{C}) = p(\mathbf{X}, C = c_2 | \mathbf{I}_\mathscr{C})$ is given. Expressed in terms of the parameters this causality information means that $\theta_1 = \rho_{12} \theta_2$ and $\theta_3 = \rho_{12} \theta_4$.

Let $r \geq 0$ be the number of constraints in the form (7) given by the causality information. Note that there may be several constraints for the same feature X_i and for different classes. Each constraint gives one equation in Θ of the type (10), for each possible value of the X_i. Furthermore, Θ should fulfill the requirement (1b). All in all, this gives $1 + \sum_{i=1}^r K_i = R$ equations that Θ are required to fulfill. In matrix form we write

$$E\Theta = F, \qquad (12)$$

where $E \in \mathbb{R}^{R \times M}$ and $F \in \mathbb{R}^R$. In the example with parameters as in (11), and with $\rho_{12} = 1$, the matrices becomes

$$E = \begin{bmatrix} 0 & 0 & -1 & 1 \\ 1 & -1 & 0 & 0 \\ 1 & 1 & 1 & 1 \end{bmatrix}, \quad F = \begin{bmatrix} 0 \\ 0 \\ 1 \end{bmatrix}. \qquad (13)$$

Note that the requirement (1b) always is represented by one row in E consisting of ones only, and that the corresponding row in F is also a one.

To compute $p(\mathbf{Z} | \mathscr{D}, \mathbf{I}_\mathscr{C})$ marginalize over the set of parameters Ω that fulfill (1)

$$p(\mathbf{Z} | \mathscr{D}, \mathbf{I}_\mathscr{C}) = \int_\Omega p(\mathbf{Z} | \Theta, \mathscr{D}, \mathbf{I}_\mathscr{C}) f(\Theta | \mathscr{D}, \mathbf{I}_\mathscr{C}) d\Theta. \qquad (14)$$

The first factor in the integral (14) is independent of $\mathscr{D}$ since Θ is known. Thus, we have $p(\mathbf{Z} | \Theta, \mathscr{D}, \mathbf{I}_\mathscr{C}) = p(\mathbf{Z} | \Theta, \mathbf{I}_\mathscr{C})$, which is given by (1). To determine the second factor in the integral (14), apply Bayes' theorem

$$f(\Theta | \mathscr{D}, \mathbf{I}_\mathscr{C}) = \frac{p(\mathscr{D} | \Theta, \mathbf{I}_\mathscr{C}) f(\Theta | \mathbf{I}_\mathscr{C})}{\int_\Omega p(\mathscr{D} | \Theta, \mathbf{I}_\mathscr{C}) f(\Theta | \mathbf{I}_\mathscr{C}) d\Theta}.$$

Since the N samples in training data are assumed to be independent, and by using (1) we can write the likelihood as $p(\mathscr{D}|\Theta,\mathbf{I}_{\mathscr{C}}) = \prod_{i=1}^{N} p(d_i|\Theta,\mathbf{I}_{\mathscr{C}}) = \theta_1^{n_1} \ldots \theta_M^{n_M}$, where $\sum_{i=1}^{M} n_i = N$.

To determine the prior probability density $f(\Theta|\mathbf{I}_{\mathscr{C}})$ given causality information, we investigate the state of knowledge $\mathbf{I}_{\mathscr{C}}$. It consists of two parts, $\mathbf{I}_{\mathscr{C}} = \{\mathbf{I}, \mathbf{I}_E\}$. The first part, $\mathbf{I}$, is the basic prior information, stating that the probability is parameterized by Θ, that Θ is Dirichlet distributed, and knowledge about the prior probabilities for the states. The second part, $\mathbf{I}_E$, includes the information that Θ satisfies (12), as well as the values of E and F. By using Bayes' theorem we have that $f(\Theta|\mathbf{I}_{\mathscr{C}}) = f(\Theta|\mathbf{I},\mathbf{I}_E) \propto f(\Theta|\mathbf{I})f(\mathbf{I}_E|\Theta,\mathbf{I})$, where $f(\Theta|\mathbf{I})$ is given by (2), and $f(\mathbf{I}_E|\Theta,\mathbf{I}) = f_{E\Theta=F}(\Theta)$ is the density function where all probability mass is uniformly distributed over the set $\Omega_E = \{\Theta : \Theta \in \Omega, E\Theta = F\}$, i.e. over the parameters Θ that fulfills (12). Thus, we have

$$p(\mathbf{Z}=\mathbf{z}_i|\mathscr{D},\mathbf{I}_{\mathscr{C}}) = \frac{\int_{\Omega_E} \theta_1^{n_1+\alpha_1-1} \ldots \theta_i^{n_i+\alpha_i} \ldots \theta_M^{n_M+\alpha_M-1} f_{E\Theta=F}(\Theta) d\Theta}{\int_{\Omega_E} \theta_1^{n_1+\alpha_1-1} \ldots \theta_i^{n_i+\alpha_i-1} \ldots \theta_M^{n_M+\alpha_M-1} f_{E\Theta=F}(\Theta) d\Theta}. \quad (15)$$

We will now give one example of how this integral can be solved using variable substitution.

A Solution Method Based on Variable Substitution

To solve the integrals in (15) substitute variables $\Theta = H\Phi + G$, where Φ are new variables parameterizing the set Ω_E of Θ fulfilling $E\Theta - F = 0$. The matrix E has full row rank (otherwise there would be redundant information about the parameters Θ, and rows could be removed from E until full row rank is obtained). Thus, we can find a permutation matrix P such that $EP = \tilde{E} = [\tilde{E}_R \quad \tilde{E}_{M-R}]$ where $\tilde{E}_R \in \mathbb{R}^{R\times R}$ has full rank. By using the permutation matrix P the requirement (12) is transformed to

$$\tilde{E}\tilde{\Theta} = F, \quad (16)$$

where $P^T\Theta = \tilde{\Theta} = (\tilde{\theta}_1, \ldots, \tilde{\theta}_M)^T$. Similarly for the counts of training data $n = (n_1, \ldots, n_M)$ and the hypothetical samples we have

$$P^T n = \tilde{n} = (\tilde{n}_1, \ldots, \tilde{n}_M), \quad P^T \alpha = \tilde{\alpha} = (\tilde{\alpha}_1, \ldots, \tilde{\alpha}_M).$$

Multiply (16) by $\tilde{E}_R^{-1}$ to obtain

$$[I_R \quad \tilde{E}_R^{-1}\tilde{E}_{M-R}]\tilde{\Theta} = \tilde{E}_R^{-1}F \quad \Leftrightarrow \quad \tilde{\Theta}_{1:R} + \tilde{E}_R^{-1}\tilde{E}_{M-R}\tilde{\Theta}_{R+1:M} = \tilde{E}_R^{-1}F, \quad (17)$$

where $\tilde{\Theta}_{1:R}$ are the first R rows of $\tilde{\Theta}$ and $\tilde{\Theta}_{R+1:M}$ are the last $M-R$ rows. In in (17), augment $\tilde{\Theta}_{1:R}$ with $\tilde{\Theta}_{R+1:M}$ and let $\Phi = \tilde{\Theta}_{R+1:M}$. Then, rearranging the terms gives

$$\tilde{\Theta} = \underbrace{\begin{bmatrix} -\tilde{E}_R^{-1}\tilde{E}_{M-R} \\ I_{M-R} \end{bmatrix}}_{H} \Phi + \underbrace{\begin{bmatrix} \tilde{E}_R^{-1}b \\ 0_{M-R\times 1} \end{bmatrix}}_{G}. \quad (18)$$

Let H_i and G_i be the i:th rows in H and G respectively. Then $\theta_i = H_i\Phi + G_i$, and we can write the integrals in (15) as

$$\int_\Omega \tilde{\theta}_1^{\tilde{k}_1} \ldots \tilde{\theta}_M^{\tilde{k}_M} \prod_{i=1}^R \delta(\tilde{\theta}_i - \tilde{\theta}_i^0(\Phi)) d\tilde{\theta} = \int_{\Omega_\Phi} (H_1\Phi + G_1)^{\tilde{k}_1} \ldots (H_M\Phi + G_M)^{\tilde{k}_M} d\Phi, \quad (19)$$

where $\delta(\cdot)$ is the dirac delta function, $\theta_i^0(\Phi)$ is the solution to the equation $H_i\Phi + G_i = 0$, $\Omega_\Phi = \{\Phi : H\Phi + G > 0\}$, and $\tilde{k}_j = \tilde{k}_j(\tilde{n}_j, \tilde{\alpha}_j)$.

The area of integration for the left hand side of (19) is determined by, for each ϕ_i in $\Phi = (\phi_i, \ldots, \phi_{M-R})$, finding the lower boundary by solving the optimization problems

$$\min_{\Sigma=(\sigma_1,\ldots,\sigma_{M-R})} \sigma_i \quad (20)$$
$$\text{subject to} \quad H\Sigma > 0$$
$$\sigma_k = \phi_k, \quad k = 1, \ldots, i-1.$$

For the upper boundary, min is replaced by max in (20).

To investigate the computations in detail, return to the example with E and F given by (13). Here we use the identity matrix for P. Then the integral (19) becomes

$$\int_0^{0.5} (0.5 - \phi_1)^{\tilde{k}_1} (0.5 - \phi_1)^{\tilde{k}_2} \phi_1^{\tilde{k}_3} \phi_1^{\tilde{k}_4} d\phi_1 = \frac{1}{2^{1+\sum_{i=1}^4 \tilde{k}_i}} \frac{\Gamma(\tilde{k}_1 + \tilde{k}_2 + 1)\Gamma(\tilde{k}_3 + \tilde{k}_4 + 1)}{\Gamma(2 + \sum_{i=1}^4 \tilde{k}_i)}.$$

Although an analytical solution was easily found in the example considered here, this is generally not the case. To the authors knowledge, there is no closed formula for solving the integral on the right hand side in (19) in general. One possibility is to use Laplace approximation [13], where the integrand is approximated by an unnormalized Gaussian density function. See [12] for more details on the Laplace approximation applied to the current problem.

FAULT DIAGNOSIS EXAMPLE

To illustrate the methods, consider the following fault classification example with two-dimensional feature vector $\mathbf{X} = (X_1, X_2)$, where $x_i \in \{0, 1\}$, and the two faults (states) $C \in \{c_1, c_2\}$. To simplify notation, assume that the classes have equal prior probability. Enumerate the parameters as

C	1	2	1	2	1	2	1	2
X_1	0	0	1	1	0	0	1	1
X_2	0	0	0	0	1	1	1	1
$p(\mathbf{z}|\Theta, \mathbf{I}_\mathscr{C})$	θ_1	θ_2	θ_3	θ_4	θ_5	θ_6	θ_7	θ_8

and assume that we are given the causality information

$$p(x_1|\Theta, c_1, \mathbf{I}_\mathscr{C}) = p(x_1|\Theta, c_2, \mathbf{I}_\mathscr{C}).$$

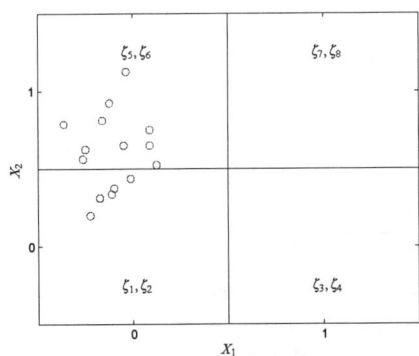

FIGURE 1. Example of training data from state c_2.

For this particular example, the integrals in (15) have the form

$$\int_{\Omega_E} (0.5 - \phi_1 - \phi_4 - \phi_5)^{\tilde{k}_1} (\phi_1 + \phi_4 - \phi_3)^{\tilde{k}_2} (0.5 - \phi_1 - \phi_4 - \phi_2)^{\tilde{k}_3} \phi_1^{\tilde{k}_4} \phi_2^{\tilde{k}_5} \phi_3^{\tilde{k}_6} \phi_4^{\tilde{k}_7} \phi_5^{\tilde{k}_8} d\Phi,$$

where we have used the permutation $\tilde{U} = [U_4\ U_1\ U_7\ U_2\ U_3\ U_5\ U_6\ U_8]$, where $U = n, \alpha, E, \Theta$. Let $\alpha_i = 1$, $i = 1, \ldots 8$ and consider for example the case when there is no data available from class c_1, i.e. $n_i = 0$, $i = 1, 3, 5, 7$, while there is training data $n_2 = 5$, $n_6 = 10$, $n_4 = n_8 = 0$ available. This example is plotted in Figure 1 and means that under class c_2 the observation $X_1 = 0$ is more likely than $X_1 = 1$. Since we have the causality information that X_1 is equally distributed under both classes we expect the observation $X_1 = 0$ to be more likely under class c_2 as well. This is verified by the computations

$$p(X_1 = 0, X_2 = 1, c = c_1 | \mathscr{D}, \mathbf{I}_\mathscr{C}) = p(\mathbf{Z} = \zeta_5 | \mathscr{D}, \mathbf{I}_\mathscr{C}) =$$
$$= \frac{\int_{\Omega_E} \phi_1^{n_2} \phi_3^{n_5} \phi_4^{n_6} d\Phi}{\int_{\Omega_E} \phi_1^{n_2} \phi_4^{n_6} d\Phi} \approx 0.41,$$

$$p(X_1 = 1, X_2 = 1, c = c_1 | \mathscr{D}, \mathbf{I}_\mathscr{C}) = p(\mathbf{Z} = \zeta_7 | \mathscr{D}, \mathbf{I}_\mathscr{C}) =$$
$$= \frac{\int_{\Omega_E} \phi_1^{n_2} (0.5 - \phi_1 - \phi_4 - \phi_2)^{n_7} \phi_4^{n_6} d\Phi}{\int_{\Omega_E} \phi_1^{n_2} \phi_4^{n_6} d\Phi} \approx 0.035,$$

and similar for the case where $X_2 = 0$. If causality information is not used, the probabilities becomes $p(X_1 = 0, X_2 = 1, c = c_1 | \mathscr{D}, \mathbf{I}) = p(X_1 = 1, X_2 = 1, c = c_1 | \mathscr{D}, \mathbf{I}) = 1/23 \approx 0.043$ by Theorem 1.

CONCLUSION

In the present work, it has been shown how the probabilistic inference problem can be formulated using training data combined with prior information given in terms of response and causality information. This type of prior information appears for example in traditional fault diagnosis problems. It has been shown how this prior information can be expressed as requirements on the parameters in the distributions.

A theorem for using response information in the inference problem has been given. Furthermore, it has been shown how the causality information can be introduced in the computations, and it is discussed how to solve the computations conceptually.

In the present work response and causality information alone has been considered one at a time, but they can also be used together to improve the inference further.

Introducing the prior information to the fault inference problem can, as shown in an example, improve the results significantly. It has been shown that the causality information makes it possible to reuse training data from one state when considering other states. This is particularly helpful when there is only a limited amount of training data available as is often the case in fault diagnosis.

ACKNOWLEDGMENTS

We acknowledge Udo von Toussaint for interesting discussions, in particular on methods for solving the integrals.

REFERENCES

1. R. O. Duda, P. E. Hart, and D. G. Storch, *Pattern Classification, 2nd Edition*, Wiley and Sons, 2001.
2. L. Devroye, L. Györfi, and G. Lugosi, *A Probabilistic Theory of Pattern Recognition*, Springer, 1996.
3. A. O'Hagan, and J. Forster, *Kendall's Advanced Theory of Statistics*, Arnold, 2004.
4. J. de Kleer, and B. C. Williams, "Diagnosis with Behavioral Modes," in *Readings in Model-based Diagnosis*, Morgan Kaufmann Publishers Inc., San Francisco, CA, USA, 1992, pp. 124–130, ISBN 1-55860-249-6.
5. J. M. Koscielny, M. Bartys, and M. Syfert, "The Practical Problems of Fault Isolation in Large Scale Industrial Systems," in *proceedings IFAC SAFEPROCESS*, 2006.
6. S. N. G. Biswas, *IEEE Trans. on Systems, Man And Cybernetics. Part A* **37**, 348–361 (2007).
7. M. Blanke, M. Kinnaert, J. Lunze, M. Staroswiecki, and J. Schröder, *Diagnosis and Fault Tolerant Control*, Springer, 2003.
8. J. J. Gertler, *Fault Detection and Diagnosis in Engineering Systems*, Marcel Decker, 1998.
9. P. Kontkanen, P. Myllymaki, T. Silander, H. Tirri, and P. Grunwald, "Comparing predictive inference methods for discrete domains," in *Proceedings of the Sixth International Workshop on Artificial Intelligence and Statistics, Ft. Lauderdale, Florida.*, 1997, pp. 311–318.
10. D. Heckerman, D. Geiger, and D. M. Chickering, *Machine Learning* **20**, 197–243 (1995).
11. A. Pernestål, and M. Nyberg, "Probabilistic Fault Diagnosis Based on Incomplete Data with Application to an Automotive Engine," in *Proceedings of European Control Conference*, 2007.
12. A. Pernestål, Using Data and Prior Information in Bayesian Classification, Tech. Rep. LiTH-ISY-R-2811, ISY, Linköping University (2007).
13. D. J. MacKay, *Information Theory, Inference and Learning Algorithms*, Cambridge University Press, 2005.

Application of the Maximum Entropy method to estimating parameter distributions for sonar signal processing

R. L. Culver, H. J. Camin, J. A. Ballard, C.W. Jemmott, and L.H. Sibul

Graduate Program in Acoustics and Applied Research Laboratory
The Pennsylvania State University, PO Box 30, State College, PA 16804

Abstract. The Maximum Entropy (MaxEnt) method and Bayesian inference have been employed to incorporate environmental knowledge into the signal processor for a sonar detection application. The sonar receiver is a new Estimator-Correlator structure that requires only that the probability density function (pdf) of the observation conditioned on the signal belongs to the exponential class, a requirement met by application of the MaxEnt method. Random but statistically correct realizations of the environment are constructed from the pdfs, and an acoustic propagation code is used to propagate acoustic energy through each realization of the environment in a Monte Carlo simulation. From the ensemble of received signals, statistical moments of the signal parameters are estimated and the MaxEnt method is again used to construct signal parameter pdfs. Using Bayesian inference, the predicted parameter pdfs are incorporated into the detection algorithm as *a priori* information. To evaluate the fidelity of the approach, the statistics of acoustic measurements made during a 1996 experiment in the Strait of Gibraltar are compared with MaxEnt pdfs and simulation predictions.

Keywords: Maximum entropy, Bayesian inference, sonar signal processing, uncertainty processing.
PACS: 30.Re, 60.Cg,

INTRODUCTION

We consider the following problem. A passive sonar receiver operating in the ocean has detected a signal, and we wish to estimate the location of signal source based upon the signal statistics. We have an acoustic propagation model (computer program) that can predict the characteristics of signals which have propagated through the ocean from a source to the receiver, and we can compute (or predict) the characteristics of the received signal for hypothetical source locations. The general approach to solving the problem is to compare the predicted signal characteristics with those of the actual received signal and identify the "best match" thereby obtaining the estimate of source location.

One solution to this problem, called matched field processing, consists of computing the pressure and phase at the receiver for sources at various ranges and depths, and correlating the predicted and actual acoustic fields (c.f. Sha and Nolte, 2005). Matched field processing works well when the receiver possesses vertical extent such that it can measure the vertical arrival angle structure of the signal caused by multipath. However, our interest is in receivers that have horizontal extent but little vertical extent, and for which matched field processing does not work well.

Signal amplitude, phase and frequency are perturbed by propagation through the ocean. We hypothesize that an acoustic propagation code can be used to predict the dependence of signal parameter statistics such as amplitude, frequency and phase on source location. For example, suppose the signal is sinusoidal with slowly varying amplitude caused by propagation through the ocean. Our hypothesis would be that the amplitudes of signals originating from sources at different ranges and depths in the ocean will have different statistics. If we can predict how the amplitude statistics depend upon source location, we can match the predicted and measure amplitude statistics and select the "most likely" source location.

THE SONAR RECEIVER

The receiver must provide a means of choosing between the following two hypotheses:

$$H_1 : r(t) = A_1 \cos(\omega t + \varphi) + n(t) \quad \text{source is at location 1}$$
$$H_2 : r(t) = A_2 \cos(\omega t + \varphi) + n(t) \quad \text{source is at location 2}$$
(1)

The observation is $r(t)$ and the noise $n(t)$ is taken to be zero mean Gaussian with power spectral density (PSD) equal to $N_0/2$. If we can obtain the probability density functions (pdfs) of the observed signal under H_1 and H_2, which are denoted $p_1(r)$ and $p_2(r)$, then we can chose the most likely hypothesis by calculating the *likelihood ratio* (Kay, 1998, Ch. 3):

$$\Lambda(r(t)) = \frac{p_1(r(t))}{p_2(r(t))}$$
(2)

If $\Lambda(r) \geq \gamma$, where γ is a threshold, we choose hypothesis H_1; otherwise we choose H_2. The threshold value can take into account the penalties or costs associated with making wrong decisions.

We further assume that the frequency and phase of the signal are constant and known. The signal amplitude A, however, is a random, slowly-varying function of time.[1] The randomness of the signal amplitude is due to propagation through sound speed inhomogenieties in the ocean and scattering by the ocean boundaries. We can calculate the pdfs of A under either hypothesis using the acoustic propagation code, but we cannot predict the exact value of A at a particular time without complete knowledge. The pdf of the observation conditioned on A is $p_1(r(t)|A)$. We can calculate the pdf of $r(t)$ under either hypothesis by integrating the conditional pdf times the prior pdf $p(A)$ over all possible values of A. In other words, if $p_{1,2}(A)$

---

[1] That is, slowly varying with respect to the period of the signal $(2\pi/\omega)$.

indicates the prior pdf of A under H_1 and H_2, respectively, calculated using the acoustic propagation code, then

$$p_{1,2}(r(t)) = \int_A p_{1,2}(r(t)|A) p_{1,2}(A) dA \qquad (3)$$

Ballard (2006; 2007) has derived an estimator-correlator solution to (2) and (3) by generalizing a development due to Schwartz (1977). The key requirement is that the conditional pdfs $p_{1,2}(r(t)|A)$ belong to the exponential class (Lehman, 1986):

$$p_{1,2}(r(t)|A) = K(A)\exp\{g_{1,2}(r(t),A) + B_{1,2}(r(t))\} \qquad (4)$$

In eqn. (4), $g_{1,2}(r(t),A)$ and $B_{1,2}(r(t))$ are arbitrary functions of $r(t)$ and A. Ballard (2007) shows that the log likelihood ratio can be expressed as

$$\ln \Lambda(r(t)) = G_1(r(t)) + B_1(r(t)) - G_2(r(t)) + B_2(r(t)) \qquad (5)$$

where

$$G_{1,2}(r(t)) = \int_{-\infty}^{r(t)} h_{1,2}(\xi) d\xi \qquad (6)$$

and

$$h_{1,2}(r(t)) = \int_{A_{1,2}} \frac{\partial}{\partial r(t)}\{g_{1,2}(r(t),A)\} p_1(A|r(t)) dA \qquad (7)$$

Equation (5) is a new result in that eqn (4) is not a stringent requirement. Whalen (1971, Ch. 7) provides the following expression for $p_{1,2}(r(t)|A)$ when the noise $n(t)$ is Gaussian (which is more restrictive than (4)):

$$p(r(t)|A) = k \exp\left\{-\frac{1}{N_0}\int_{t-T}^{t}[r(\tau) - A\cos(\omega\tau + \varphi)]^2 d\tau\right\}$$

$$= k \exp\left[-\frac{1}{N_0}\int_{t-T}^{t} r^2(\tau) d\tau\right] \exp\left[\frac{2A}{N_0} V(r(t)) - \frac{A^2 T}{2N_0}\right] \qquad (8)$$

Here T is the integration time and $V(r(t))$ is the coherent matched filter output at time t:

$$V(r(t)) \equiv \int_{t-T}^{t} r(\tau)\cos(\omega\tau + \varphi) d\tau \qquad (9)$$

Equation (8) applies to both hypotheses. Substituting (8) into (3) yields expressions for $p_{1,2}(r(t))$ that can be used with (2) to compute the likelihood ratio. Ballard has obtained an analytical solution for $\ln \Lambda(r(t))$ when $p_1(A)$ and $p_2(A)$ are Gaussian distributions with equal means but with different variances. The statistic $V(r(t))$ is found to be sufficient, and the maximum likelihood receiver compares the estimate of the variance of V to a threshold γ that depends upon the variances of the signals and the noise PSD (m is the signal mean and T is the integration time):

$$\left\{V(r(t)) - \frac{Tm}{2}\right\}^2 \overset{H_1}{\underset{H_2}{\gtrless}} \gamma \qquad (10)$$

Ballard has computed receiver operating characteristic (ROC) curves and examined receiver performance as a function of the signal amplitude variances and the signal-to-noise ratio (SNR).

APPLICATION OF THE MAXIMUM ENTROPY METHOD

An important use of the MaxEnt method is obtaining an exponential class pdf for the noise. In the case of passive sonar, noise samples can be obtained from adjacent beams or frequency bins. The form of the pdf in (4) is the solution to the standard MaxEnt problem in which we have the moments of a distribution and wish to obtain the pdf $p(r|A)$ which maximizes the entropy[2]

$$H = -\int p(r|A) \ln\{p(r|A)\} dA \qquad (11)$$

subject to the moment constraints

$$E(\phi_n(r)) = \int \phi_n(r) p(r|A) dr = \mu_n \qquad (12)$$

In eqn. (12) the $\phi_n(r)$, n = 0, 1, 2, ..., N are known functions of r, e.g. $r, r^2, \ln r$ etc., and μ_n is the value of the moment corresponding to the function $\phi_n(r)$.

We have implemented the numerical method developed by Mohammad-Djarari (1991) to obtain $p(A)$. Monte Carlo simulation is used to generate samples of r under H$_1$ and H$_2$. The moment functions are $\phi_0(A) = 1$, $\phi_1(A) = \ln A$, and $\phi_2(A) = A$, and the moments $\mu_1, \mu_2,$ and μ_3 are computed from the samples. Mohammad-Djarari's method is to linearize the set of N+1 non-linear equations

$$G_n(\lambda_0, \lambda_1, ...\lambda_N) = \int \phi_n(A) \exp\left(\sum_{k=0}^{N} \lambda_k \phi_k(A)\right) dA = \mu_n, \ n = 0, 1, ... N \qquad (13)$$

about trial values $\lambda^0 = \lambda_0^0, \lambda_1^0, ...\lambda_N^0$, compute the gradient of $G_n(\lambda_0, \lambda_1, ...\lambda_N)$, and iteratively search for the solution. The result is the "best fit" values for $\lambda_0, \lambda_1,$ and λ_2 in the resulting pdf

$$p_{1,2}(A) = \exp[-\lambda_0 - \lambda_1 \ln A - \lambda_2 A] \qquad (14)$$

An important question relative to our Monte Carlo – MaxEnt - Estimator-Correlator receiver is the fidelity of both the signal amplitude samples generated using simulation and of the corresponding pdfs estimated using the MaxEnt method. Accordingly, the method is being evaluated using acoustic and environmental measurements made in the Strait of Gibraltar in 1996 (Tiemann, et. al., 2001). The Strait of Gibraltar Acoustic Monitoring Experiment (SGAME) was an acoustic propagation

---

[2] From now on the dependence of r on t is suppressed in order to simplify the equations.

measurement consisted of acoustic transmissions across the strait accompanied by comprehensive sound velocity profile measurements.

The oceanography in the Strait is dominated by tidally driven interaction between the relatively warm, salty Mediterranean and the cooler, fresher Atlantic resulting in substantial SVP variability over the tidal cycle. A mean SVP model and comprehensive set of SVP drops spanning the full tidal cycle and propagation range provide the means to estimate environmental variability. In addition, there are significant internal waves that add to the complexity of the SVP. The acoustic measurements span several days and thus provide a measurement of acoustic variability over many tidal cycles.

Acoustic signals were transmitted across the Strait from a 100 m deep source to 10 hydrophones spaced from 81 m to 387 m in depth. Signals were transmitted hourly over a three day period yielding approximately 80 transmissions spread over 7 tidal cycles. Each transmission consisted of two identical 8 second pseudorandom, phase coded pulses centered at 250 Hz with approximately 75 Hz bandwidth.

Figure (1) compares measured and simulated pdfs for the 250 Hz SGAME acoustic data. Each panel corresponds to a single hydrophone depth, which is given beside the panel. The root mean square (rms) pressure of each received signal was calculated for each hydrophone, and the black asterisks in Figure (1) denote the histograms of the values. In addition, the sample moments of the ensemble of received signals were calculated and the MaxEnt method applied to obtain a pdf for the values at each hydrophone (shown in red).

The majority the rms pressure values are 120-126 dB re 1 μPa, but there is a long tail out to 140 dB. The source level was 192 dB and a rough estimate of transmission loss due to initially spherical and then cylindrical spreading is 69-72 dB [14]. The larger values in Figure (1) are due to noise from passing ships and are not present when the acoustic data are matched filtered.

The MaxEnt pdfs match the histograms quite well, despite some multi-modal behavior. Not all peaks are matched. For example, the 253 m deep hydrophone histogram has peaks at 123, 131 and 141 dB, while the MaxEnt pdf has peaks at 124 and 141 dB. It is possible to vary the bin spacing and cause the histograms to change their shape, and the λ's used in the MaxEnt pdf can be changed to allow more degrees of freedom in fitting the pdf to the data. This second option may be implemented in the sonar receiver.

The dashed blue lines in Figure (1) are the MaxEnt pdfs calculated from the moments of the simulation output, which is now discussed briefly. Further details may be found in (Camin et. al., 2006). During SGAME, a total of 129 conductivity and temperature vs depth (CTD) measurements (or drops) were made over the course of about two weeks during a period that included both spring and neap tides. Sound speed was calculated from the CTD drops approximately every meter in depth at seven locations along the vertical plane containing the source and receiver. The first two moments (mean and variance) were calculated for each depth and location, and the MaxEnt method used to generate pdfs. Realizations of the sound speed field were created using the pdfs, and the Range-dependent Acoustic Model (RAM) (Collins, 1993) was used to generate a received signal for each sound speed realization.

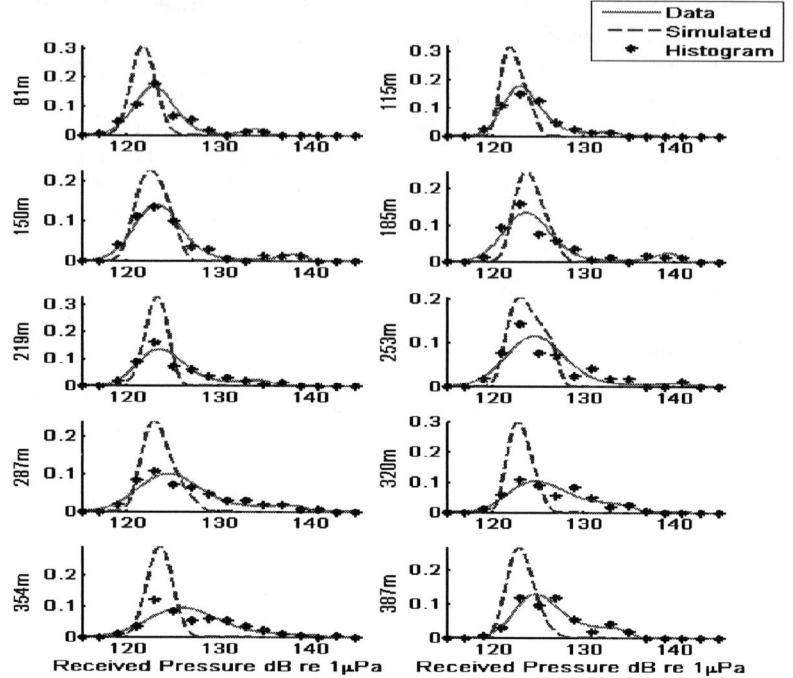

FIGURE 1. Distributions of rms received signal level in dB at the 10 receive hydrophones for the 250 Hz SGAME acoustic data. The vertical axis in each panel is frequency of occurrence. The hydrophone depth is given. Black asterisks denote the histogram of the measured received levels. The red line is an exponential pdf obtained using the MaxEnt method and the estimated moments. The blue curve is the pdf obtained using Monte Carlo simulation and the MaxEnt method.

Figure (2) shows transmission loss, in dB re 1 μPa, calculated using RAM for a single realization of the sound speed field. Several hundred realizations of the received signal were computed in Monte Carlo fashion. Rms pressure was calculated using the entire received signal, producing an ensemble of rms pressures. Sample moments were calculated and the MaxEnt method used to produce pdf's of rms pressure at each hydrophone location. The pdfs are shown in Figure 1 as dashed blue lines.

SUMMARY

An application of the Maximum Entropy (MaxEnt) method to incorporate knowledge of the environment into a passive sonar signal processor has been presented. The receiver is a new Estimator-Correlator structure that requires only that the conditional pdf of the observation conditioned on knowledge of the signal belong to the exponential class. This is less restrictive than previous derivations, which are based upon Gaussian assumptions. It also benefits from application of the MaxEnt method, which produces pdfs belonging to the exponential class.

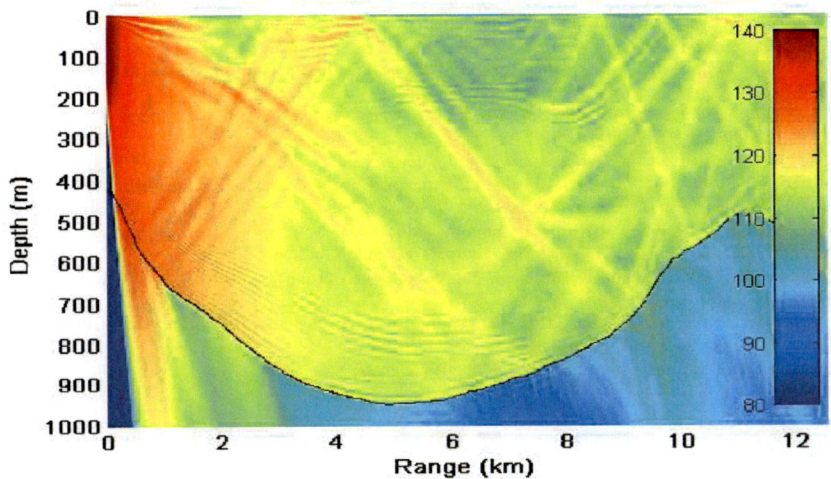

FIGURE 2. A single realization of received pressure (in dB re 1 μPa) computed for the Strait of Gibraltar Acoustic Monitoring Experiment (SGAME) 250 Hz source at 100 m using RAM [15]. The bottom is indicated by a black line

Our implementation of the MaxEnt method is derived from work by Mohammad-Diafari (1991) and is not novel. However, the effectiveness of our receiver depends upon the fidelity of an acoustic propagation simulation, validity of realizations of the environment, and the ability of the MaxEnt method to provide pdfs for multi-modal distributions. These issues are evaluated using data from a 1996 acoustic measurement in the Strait of Gibraltar. Comparison of pdfs produced from measured and simulated acoustic data is not complete, but clearly the MaxEnt method is capable of capturing the statistics of the data.

ACKNOWLEDGMENTS

This work was supported by ONR Undersea Signal Processing. The authors gratefully acknowledge assistance from Dr. Peter Worcester of Scripps Institution of Oceanography and Dr. Chris Tiemann of the Applied Research Lab, the University of Texas at Austin in providing data from the Strait of Gibraltar Acoustic Monitoring Experiment.

REFERENCES

Sha, L. and L. W. Nolte (**2005**). "Effects of environmental uncertainties on sonar detection performance prediction," J. Acoust. Soc. Am. **117** (4), pp. 1942-1953.

Ballard, J. A., Jemmott, C. W., Sibul, L. H., Culver, R. L., and Camin, H. J. (**2006**). "The Estimated Ocean Detector: Derivation and Predicted Performance under Gaussian Assumptions," *Proceedings of IEEE OCEANS 2006, 18-21 September 2006, Boston, MA.*

Ballard, J. A. (**2007**). The estimated signal parameter detector: Incorporating acoustic variability into the signal processing to classify between stochastic signals, MS thesis (The Pennsylvania State University, State College, PA).

Camin, H. J., Culver, R. L., Sibul, L. H., Ballard, J. A., Jemmott, C. W., Holland, C. W., and Bradley, D. L. (**2006**). "Received signal parameter statistics in random/uncertain oceans," *Proceedings of IEEE OCEANS 2006, 18-21 September 2006, Boston, MA*.

Collins, M. (**1993**). "A split-step Padá solution for the parabolic equation method," J. Acoust. Soc. Am, **93**, pp. 1736-1742.

Kay, S. M. (**1998**). *Fundamentals of Statistical Signal Processing: Volume II: Detection Theory* (Prentice Hall PTR, Upper Saddle River, NJ).

Lehmann, E. L. (**1986**). *Testing of Statistical Hypotheses* (Wiley and Sons, NY).

Mohammad-Djafari, A. (**1991**). "Maximum Likelihood Estimation of the Lagrange Parameters of the Maximum Entropy Distributions," in Maximum Entropy and Bayesian Methods, Seattle, 1991 – Proceedings of the Elevent International Workshop on Maximum Entropy and Bayesian Methods of Statistical Analysis, edited by C. R. Smith, G. J. Erickson and P. O Neudorfer (Kluwer Academic, Dordrecht, Netherlands).

Schwartz, S. C. (**1977**). "The estimator-correlator for discrete-time problems," IEEE Trans. On Information Theory, vol. **IT-23**, pp. 93-100.

Tiemann, C., P. Worcester, and B. Cornuelle (**2001**). "Acoustic scattering by internal solitary waves in the Strait of Gibralter," J. Acoust. Soc. Am. **109**, pp. 143-154.

Whalen, A. D. (**1971**). *Detection of Signals in Noise* (Academic Press, NY).

Modeling the Multiple-Antenna Wireless Channel Using Maximum Entropy Methods

M. Guillaud*, M. Debbah† and A.L. Moustakas**

*ftw., Vienna, Austria – guillaud@ftw.at
†Supelec, Gif-sur-Yvette, France – merouane.debbah@supelec.fr
**University of Athens, Athens, Greece – arislm@phys.uoa.gr

Abstract. Analytical descriptions of the statistics of wireless channel models are desirable tools for communication systems engineering. When multiple antennas are available at the transmit and/or the receive side (the Multiple-Input Multiple-Output, or MIMO, case), the statistics of the matrix H representing the gains between the antennas of a transmit and a receive antenna array, and in particular the correlation between its coefficients, are known to be of paramount importance for the design of such systems. However these characteristics depend on the operating environment, since the electromagnetic propagation paths are dictated by the surroundings of the antenna arrays, and little knowledge about these is available at the time of system design.

An approach using the Maximum Entropy principle to derive probability density functions for the channel matrix, based on various degrees of knowledge about the environment, is presented. The general idea is to apply the maximum entropy principle to obtain the distribution of each parameter of interest (e.g. correlation), and then to marginalize them out to obtain the full channel distribution. It was shown in previous works, using sophisticated integrals from statistical physics, that by using the full spatial correlation matrix $E\{\text{vec}(\mathbf{H})\text{vec}(\mathbf{H})^H\}$ as the intermediate modeling parameter, this method can yield surprisingly concise channel descriptions. In this case, the joint probability density function is shown to be merely a function of the Frobenius norm of the channel matrix $|\mathbf{H}|_F$.

In the present paper, we investigate the case where information about the average covariance matrix is available (e.g. through measurements). The maximum entropy distribution of the covariance is derived under this constraint. Furthermore, we consider also the doubly correlated case, where the intermediate modeling parameters are chosen as the transmit- and receive-side channel covariance matrices (respectively $E\{\mathbf{H}^H\mathbf{H}\}$ and $E\{\mathbf{H}\mathbf{H}^H\}$). We compare the maximum-entropy result obtained in this case with the well-known Kronecker model, and derive the channel probability distribution function in the case of single-side correlation constraint.

Keywords: Wireless MIMO Channel Model, Spatial Correlation, Maximum Entropy
PACS: 42.25.-p

INTRODUCTION

In wireless communications, truthful modeling of the electromagnetic propagation channel is critical for the optimization and performance evaluation of the channel codes and protocols. Since the introduction, at the end of the 90's, of the concept of Multiple-Input Multiple-Output (MIMO) communications [1], whereby multiple spatially separated antennas are used at both the transmitting and receiving side, a large number of models for MIMO wireless channels have been developed (see e.g. [2] for an overview). Recently, random matrix theory has proven to be an extremely successful tool in the analysis and design of wireless communications systems [3]. However, random matrix tools require that the channel properties be completely described analytically, which is not the case in general for the most refined (and most accurate) models used nowadays.

In this article, we focus on the derivation of analytical models for the channel fading coefficients through the use of the maximum entropy principle, with a focus on properly modeling the spatial correlation structure. Spatial correlation is a critical parameter for MIMO channels, through its influence on the channel capacity [4] and on the design of the channel code [5]. However, spatial correlation usually depends on the environment of the transmitter and receiver (through the position and electromagnetic properties of the various items or persons hampering the propagation of radio waves, referred to as *scatterers*). Therefore, the spatial correlation conditions are usually not known at the time of the design of the communication system, and a versatile design must be sought.

In this context, the capability of the maximum entropy principle [6] to represent the ignorance of certain properties is used as a modeling tool. In order to make the problem analytically amenable, the parameters of interest (e.g. the spatial correlation matrix) are first modeled independently, and later marginalized out to obtain the full channel probability density function (pdf), according to the method introduced in [7]. More specifically, we consider the channel as being described by the statistics of a multivariate fading process $\mathbf{H}$. We obtain its pdf (denoted by $P_\mathbf{H}(\mathbf{H})$) for the case where a parameter (denoted by $\mathbf{B}$ in the general case) is known to be particularly relevant for the design and engineering of communication systems. In this case, it is desirable that the range of $\mathbf{B}$ be maximally explored, which we ensure by application of the maximum entropy principle. Our method can be summarized as follows:

1. derive $P_\mathbf{B}(\mathbf{B})$ through entropy maximization,
2. derive $P_{\mathbf{H}|\mathbf{B}}(\mathbf{H}, \mathbf{B})$ through entropy maximization,
3. marginalize over $\mathbf{B}$ to obtain $P_\mathbf{H}(\mathbf{H}) = \int P_{\mathbf{H}|\mathbf{B}}(\mathbf{H}, \mathbf{B}) P_\mathbf{B}(\mathbf{B}) d\mathbf{B}$.

Note that the obtained $P_\mathbf{H}(\mathbf{H})$ is not, in general, the entropy maximizing distribution of $\mathbf{H}$. In the sequel, we present several results where $\mathbf{B}$ describes various correlations between the elements of $\mathbf{H}$.

NOTATIONS AND CHANNEL MODEL

We consider a wireless MIMO link with n_t transmit and n_r receive antennas, represented by the $n_r \times n_t$ matrix $\mathbf{H}$. We are only concerned with frequency-flat channels, i.e. the attenuation between transmit antenna j and receive antenna i (the (i,j)-th coefficient of $\mathbf{H}$) is a complex scalar that we denote $h_{i,j}$. Such a communication link is commonly represented by the equation

$$\mathbf{r}_n = \mathbf{H}_n \mathbf{s}_n + \mathbf{n}_n, \tag{1}$$

where $\mathbf{H}_n$ represents the realization of $\mathbf{H}$ at instant n, vectors $\mathbf{s}_n$, $\mathbf{r}_n$ and $\mathbf{n}_n$ represent the transmitted signal, the received signal, and an additive noise sample, respectively. In this article, we are only concerned with modeling the pdf of $\mathbf{H}_n$, assumed to be stationary, and therefore we drop the index n in the sequel.

In this article, we focus on the derivation of the fading characteristics of $\mathbf{H}$ in the form of the joint pdf of its coefficients, denoted by $P_\mathbf{H}(\mathbf{H})$. We are not concerned with the time-related properties of the channel, i.e. we assume that the process under study is stationary. We refer equivalently to the channel realization $\mathbf{H}$ or to its vectorized notation $\mathbf{h} = \text{vec}(\mathbf{H}) = [h_{1,1} \ldots h_{n_r,1}, h_{1,2} \ldots h_{n_r,n_t}]^T$. The notation $N = n_r n_t$ will also be used to denote the total channel dimension. We will sometimes use the alternative notation where the antenna indices are mapped into $[1 \ldots N]$, i.e. denoting $\mathbf{h} = [h_1 \ldots h_N]^T$. Finally, let $\mathcal{U}(N)$ denote the set of $N \times N$ unitary matrices, endowed with the Haar measure.

FULL CORRELATION MATRIX CASE

We first consider the simplest case, which consists in deriving the pdf of $\mathbf{H}$ with only information about the dimensions of the problem (n_t and n_r), and an energy constraint. This is achieved by maximizing the entropy of $P_\mathbf{H}(\mathbf{H})$. The energy constraint always exists, since in communications applications, the received energy is always limited by the (finite) transmit power. Here, we assume that we know the average energy of the channel, and that it is equal to NE_0. In [8], Debbah et al. show that the probability distribution that maximizes the entropy $\int_{\mathbb{C}^N} -\log(P(\mathbf{H}))P(\mathbf{H})d\mathbf{H}$, where $d\mathbf{H} = \prod_{i=1}^{N} d\text{Re}(h_i) d\text{Im}(h_i)$ is the Lebesgue measure on $\mathbb{C}^N$, under the cited constraints, is the Gaussian i.i.d. distribution

$$P_{\mathbf{H}|E_0}(\mathbf{H}) = \frac{1}{(\pi E_0)^N} \exp\left(-\sum_{i=1}^{N} \frac{|h_i|^2}{E_0}\right). \quad (2)$$

This distribution has been used classically to model MIMO channels when little was known about their fading characteristics. The Gaussian property has been confirmed experimentally, and can be explained by application of the central limit theorem, since the received signal is generally the sum of a large number of propagation paths that experience independent fading. The (spatial) independence of the coefficients of $\mathbf{H}$, on the other hand, is clearly the fruit of a lack of information from the modeler: measurements have shown that they are in general correlated, however a large variability in the correlation properties has been observed. Next, we shall incorporate some knowledge about the spatial correlation characteristics of $\mathbf{H}$. We first study the case where the correlation matrix is known deterministically, and subsequently extend the result to an unknown covariance matrix.

First, we review the case of the maximum entropy distribution of $\mathbf{H}$ under the assumption that the full covariance matrix $\mathbf{Q} = \int_{\mathbb{C}^N} \mathbf{h}\mathbf{h}^H P_{\mathbf{H}|\mathbf{Q}}(\mathbf{H})d\mathbf{H}$ is known, where $\mathbf{Q}$ is a $N \times N$ complex positive definite Hermitian matrix. Entropy maximization of $P_{\mathbf{H}|\mathbf{Q}}$ under these covariance constraints (the energy constraint being implicitly set by $\text{tr}(\mathbf{Q})$), through the Lagrange multipliers method, and elimination of the Lagrange coefficients through proper normalization, yields

$$P_{\mathbf{H}|\mathbf{Q}}(\mathbf{H}, \mathbf{Q}) = \frac{1}{\det(\pi \mathbf{Q})} \exp\left(-(\mathbf{h}^H \mathbf{Q}^{-1} \mathbf{h})\right). \quad (3)$$

Therefore, with the extra constraint of a deterministic correlation matrix, the maximum entropy principle yields a complex correlated Gaussian distribution.

If the covariance matrix $\mathbf{Q}$ is not known, we model it under two different assumptions: no knowledge about $\mathbf{Q}$ (except its domain, the set of positive semi-definite matrices), or knowledge about its mean.

No knowledge about the correlation matrix

Let $P_\mathbf{Q}$ be the probability density function of $\mathbf{Q}$. The entropy of $P_\mathbf{Q}$ is defined as

$$H(P_\mathbf{Q}) = \int_{\mathbf{Q} \geq 0} -\log(P_\mathbf{Q}(\mathbf{Q})) P_\mathbf{Q}(\mathbf{Q}) d\mathbf{Q} \quad (4)$$

where $d\mathbf{Q}$ is a $\mathscr{U}(N)$-invariant measure. The entropy-maximizing distribution of $\mathbf{Q}$ is obtained by solving

$$P_\mathbf{Q} = \arg \max_{E_\mathbf{Q}[\text{tr}(\mathbf{Q})] = NE_0} H(P_\mathbf{Q}), \quad (5)$$

where the constraint on the expectation of the trace, representing a constraint on the average channel energy of E_0 per coefficient. The optimization problem is solved using the Lagrange multipliers method, by maximizing the functional

$$L(P_\mathbf{Q}) = H(P_\mathbf{Q}) + \beta \left[\int_{\mathbf{Q} \geq 0} P_\mathbf{Q}(\mathbf{Q}) d\mathbf{Q} - 1 \right] + \gamma \left[\int_{\mathbf{Q} \geq 0} \text{tr}(\mathbf{Q}) P_\mathbf{Q}(\mathbf{Q}) d\mathbf{Q} - NE_0 \right]. \quad (6)$$

Let us perform the variable change to the eigenvalues/eigenvectors space, and denote $\Lambda = \text{diag}(\lambda_1 \ldots \lambda_N) \in \mathbb{R}^{+N}$ the diagonal matrix containing the eigenvalues of $\mathbf{Q}$, and $\mathbf{U} \in \mathscr{U}(N)$ the unitary matrix containing the corresponding eigenvectors. Therefore, $\mathbf{Q} = \mathbf{U} \Lambda \mathbf{U}^H$. As demonstrated in [9], maximization of eq. (6) in this new space yields a product distribution, where $\mathbf{U}$ is uniformly distributed over $\mathscr{U}(N)$, and the unordered eigenvalues are distributed according to

$$P'_\Lambda(\Lambda) = \left(\frac{N}{E_0}\right)^{N^2} \prod_{n=1}^{N} \frac{1}{n!(n-1)!} e^{-\frac{N}{E_0} \sum_{i=1}^{N} \lambda_i} \prod_{i<j} (\lambda_i - \lambda_j)^2. \quad (7)$$

Note [3, 10] that eq. (7) describes the unordered eigenvalue density of a complex $N \times N$ Wishart matrix with N degrees of freedom and covariance $\frac{E_0}{N} \mathbf{I}_N$ (denoted as $\tilde{\mathscr{W}}_N(N, \frac{E_0}{N} \mathbf{I}_N)$). Since the eigenvectors of $\mathbf{Q}$ are isotropically distributed, we can conclude that $\mathbf{Q}$ is itself a $\tilde{\mathscr{W}}_N(N, \frac{E_0}{N} \mathbf{I}_N)$ matrix. Note that the isotropic property of the obtained Wishart distribution is a consequence of the fact that no spatial constraints were imposed on the correlation. The energy constraint (imposed through the trace) only affects the distribution of the eigenvalues of $\mathbf{Q}$.

We have shown so far that the entropy-maximizing distribution for a $N \times N$ positive semidefinite matrix under an average trace constraint is a Wishart distribution with N degrees of freedom. The complete distribution of the correlated channel $\mathbf{H}$ is obtained by marginalizing over $\mathbf{Q}$:

$$P_\mathbf{H}(\mathbf{H}) = \int_{\mathbf{Q} \geq 0} P_{\mathbf{H}|\mathbf{Q}}(\mathbf{H}, \mathbf{Q}) P_\mathbf{Q}(\mathbf{Q}) d\mathbf{Q}. \quad (8)$$

It was shown in [9] that $P_H(\mathbf{H})$ can be described by a scalar function of the Frobenius norm of $\mathbf{H}$: using

$$P_x^{(L)}(x) = \frac{2}{x}\sum_{i=1}^{L}\left(-L\sqrt{\frac{x}{NE_0}}\right)^{L+i}\frac{K_{i+L-2}\left(2L\sqrt{\frac{x}{NE_0}}\right)}{[(i-1)!]^2(L-i)!}, \qquad (9)$$

the integral in eq. (8) is shown to yield $P_H(\mathbf{H}) = \frac{1}{S_N(\|\mathbf{H}\|_F^2)}P_x^{(N)}(\|\mathbf{H}\|_F^2)$, where $S_N(x) = \frac{\pi^N x^{N-1}}{(N-1)!}$ is the surface of the zero-centered complex hypersphere of radius $\sqrt{x}$, defined by $\mathbf{h}^H\mathbf{h} = x$. This indicates that $\mathbf{h}$ is isotropically distributed (uniformly on the sphere), and that the probability density of its squared Frobenius norm is described by $P_x^{(N)}$. Furthermore, if the rank of $\mathbf{Q}$ is constrained to be $L \leq N$ (the rank-limited covariance case is commonly used to model propagation situations with a limited number of scatterers), $P_H(\mathbf{H}) = \frac{1}{S_N(\|\mathbf{H}\|_F^2)}P_x^{(L)}(\|\mathbf{H}\|_F^2)$.

Note that these expressions provide a convenient method to numerically generate samples of $\mathbf{H}$ according to the pdf P_H, since $\mathbf{h}$ can be obtained by generating separately a normalized vector process uniformly distributed over the sphere of radius 1, and a scalar process representing the norm according to eq. (9) (e.g. by numerical inversion of the corresponding cumulative density function).

Mean correlation matrix knowledge

Adding the constraint that the mean of the covariance matrix $\mathbf{Q}$ must be an arbitrary positive definite matrix $\mathbf{M}$, we now consider the entropy-maximizing distribution

$$P_{\mathbf{Q}|\mathbf{M}} = \arg\max_{E_{\mathbf{Q}}[\mathbf{Q}]=\mathbf{M}} H(P_{\mathbf{Q}}). \qquad (10)$$

Let us seek the expression of $P_{\mathbf{Q}|\mathbf{M}}$, by considering the functional

$$L(P_{\mathbf{Q}}) = H(P_{\mathbf{Q}}) + \beta\left[\int_{\mathbf{Q}\geq 0}P_{\mathbf{Q}}(\mathbf{Q}) - 1\right] + \text{tr}\left(\Omega\left[\mathbf{M} - \int_{\mathbf{Q}\geq 0}\mathbf{Q}P_{\mathbf{Q}}(\mathbf{Q})\right]\right) \qquad (11)$$

where β and the $N\times N$ matrix Ω are Lagrange multipliers. Equating the functional derivative $\frac{\delta L(P_{\mathbf{Q}})}{\delta P_{\mathbf{Q}}}$ to zero yields the pdf $P_{\mathbf{Q}}(\mathbf{Q}) = \exp(\beta - 1)\exp(-\text{tr}(\Omega\mathbf{Q}))$. The density normalization constraint imposes that $\exp(\beta - 1) = \frac{\det(\Omega)^N}{\pi^{N(N-1)/2}\prod_{i=1}^{N}(i-1)!}$, and the constraint on the average lets us identify $\Omega = N\mathbf{M}^{-1}$. Therefore,

$$P_{\mathbf{Q}|\mathbf{M}}(\mathbf{Q},\mathbf{M}) = \frac{1}{\det(\frac{\mathbf{M}}{N})^N\pi^{N(N-1)/2}\prod_{i=1}^{N}(i-1)!}\exp\left(-\text{tr}\left\{\left(\frac{\mathbf{M}}{N}\right)^{-1}\mathbf{Q}\right\}\right). \qquad (12)$$

We recognize that in this case, $\mathbf{Q}$ is a complex Wishart matrix with N degrees of freedom and covariance $\frac{\mathbf{M}}{N}$, or $\widetilde{\mathscr{W}}_N(N,\frac{\mathbf{M}}{N})$.

The distribution of $\mathbf{H}$ is obtained by integration over all positive definite matrices $\mathbf{Q}$:

$$P_{\mathbf{H}|\mathbf{M}}(\mathbf{H},\mathbf{M}) = \int_{\mathbf{Q}>0} P_{\mathbf{H}|\mathbf{Q}}(\mathbf{H},\mathbf{Q}) P_{\mathbf{Q}|\mathbf{M}}(\mathbf{Q},\mathbf{M}) \tag{13}$$

$$= \int_{\mathbf{Q}>0} \frac{e^{-\mathbf{h}^H \mathbf{Q}^{-1} \mathbf{h}} e^{-\mathrm{tr}\left\{\left(\frac{\mathbf{M}}{N}\right)^{-1}\mathbf{Q}\right\}}}{\det(\pi\mathbf{Q})\det(\frac{\mathbf{M}}{N})^N \pi^{N(N-1)/2} \prod_{i=1}^{N}(i-1)!}. \tag{14}$$

Let us denote $\left(\frac{\mathbf{M}}{N}\right)^{-1} = \mathbf{U}\mathbf{D}\mathbf{U}^H$ the eigendecomposition of the mean of the Wishart distribution. Furthermore, let $\mathbf{Q}' = \mathbf{D}^{1/2}\mathbf{U}^H\mathbf{Q}\mathbf{U}\mathbf{D}^{1/2}$ become the integration variable (with the introduction of the corresponding Jacobian $\det(\mathbf{D})^{-N}$):

$$P_{\mathbf{H}|\mathbf{M}}(\mathbf{H},\mathbf{M}) = \int_{\mathbf{Q}'>0} \frac{\det(\mathbf{D})^{-N} e^{-\mathbf{h}^H \mathbf{U}\mathbf{D}^{1/2}\mathbf{Q}'^{-1}\mathbf{D}^{1/2}\mathbf{U}^H \mathbf{h}} e^{-\mathrm{tr}\{\mathbf{Q}'\}}}{\pi^{N(N-1)/2} \prod_{i=1}^{N}(i-1)! \det(\frac{\mathbf{M}}{N})^N \pi^N \det(\mathbf{D})^{-1} \det(\mathbf{Q}')}. \tag{15}$$

We note that if we consider the vector $\mathbf{g} = \mathbf{D}^{1/2}\mathbf{U}^H\mathbf{h}$ instead of $\mathbf{h}$, we obtain its pdf

$$P_{\mathbf{g}|\mathbf{M}}(\mathbf{g},\mathbf{M}) = \int_{\mathbf{Q}'>0} \frac{e^{-\mathbf{g}^H \mathbf{Q}'^{-1} \mathbf{g}} e^{-\mathrm{tr}\{\mathbf{Q}'\}}}{\pi^N \det(\mathbf{Q}') \pi^{N(N-1)/2} \prod_{i=1}^{N}(i-1)!}, \tag{16}$$

where we introduced the Jacobian $\det(\mathbf{D})^{-1}$, and used the fact that $\mathbf{D}$ and $\frac{\mathbf{M}}{N}$ have reciprocal determinants. Note that eq. (16) is in fact independent of $\mathbf{M}$. Furthermore, since the term under the integral is the product of the density of a complex Gaussian random variable $\mathbf{g}$ with covariance $\mathbf{Q}'$, and of the density of a Wishart $\mathscr{W}_N(N,\mathbf{I})$ matrix $\mathbf{Q}'$, we recognize that we are in the case described by eq. (8) (with $E_0 = N$). Therefore, we conclude that the pdf of $\mathbf{g}$ is given by $P_{\mathbf{g}}(\mathbf{g}) = \frac{1}{S_N(\|\mathbf{G}\|_F^2)} P_x^{(N)}(\|\mathbf{G}\|_F^2)$. Again, samples of $\mathbf{H}$ can be generated easily by generating $\mathbf{G}$ using the procedure outlined in the previous section, and by applying the transformation $\mathbf{h} = \mathbf{D}^{-1/2}\mathbf{U}\mathbf{g}$.

TRANSMIT AND RECEIVE COVARIANCES $\mathbf{Q}_T, \mathbf{Q}_R$

In this section, we consider the covariances at the transmitter side and/or at the receiver side, defined respectively as

$$\mathbf{Q}_T = \int_{\mathbb{C}^N} \mathbf{H}^H \mathbf{H} P_{\mathbf{H}}(\mathbf{H}) d\mathbf{H} \quad \text{and} \quad \mathbf{Q}_R = \int_{\mathbb{C}^N} \mathbf{H}\mathbf{H}^H P_{\mathbf{H}}(\mathbf{H}) d\mathbf{H}. \tag{17}$$

This "separable correlation" case arises frequently in wireless communications applications where $\mathbf{Q}_T$ accounts for the scatterers situated in the immediate vicinity of the transmitter, and $\mathbf{Q}_R$ for the scatterers situated near the receiver.

The Kronecker model, whereby $\mathbf{h}$ is a complex Gaussian random variable with a Kronecker covariance matrix (i.e. taking $\mathbf{Q} = \mathbf{Q}_T \otimes \mathbf{Q}_R$ in eq. (3)), is commonly used, and is indeed a sufficient condition for (17). However, it is not necessary, and indeed is has been shown in [11] that the entropy-maximizing pdf $P_{\mathbf{H}}$ under the constraints (17) is a correlated Gaussian distribution with a covariance matrix whose eigenvectors are the Kronecker products of the eigenvectors of $\mathbf{Q}_T$ and $\mathbf{Q}_R$, but whose eigenvalues are *not* the pairwise products of the eigenvalues of the transmit and receive covariances. There is no known closed-form analytical expression for the eigenvalues in this case.

Single-side correlation case

Let us consider the case where only one side of the MIMO wireless link exhibits correlation. This is a common occurrence, since the base station antennas are typically elevated above all likely scatterers, and are placed sufficiently apart from each other, so that no noticeable correlation is present. The space and usage constraints on the handset, on the other hand, typically impose a non-identity receive covariance matrix. Note that here, without loss of generality, we take the example of a downlink transmission, where the base station transmits and a mobile handset acts as the receiver. In this case, the distinction between the previous two cases vanishes: the entropy maximization under the constraint $E\left[\mathbf{HH}^H\right] = \mathbf{R}_R$, yields

$$P_{\mathbf{H}|\mathbf{Q}_R}(\mathbf{H},\mathbf{Q}_R) = \frac{1}{\pi^N \det(\mathbf{Q}_R)^{N_t}} e^{-\mathrm{tr}(\mathbf{HH}^H \mathbf{Q}_R^{-1})}, \qquad (18)$$

which is also a degenerate case of the Kronecker model, obtained for $\mathbf{Q}_T = \mathbf{I}$.

Using the complex Wishart distribution with n_r degrees of freedom for the receive covariance (shown before to be the entropy maximizing distribution), we have $\mathbf{Q}_R = \mathbf{U}_R \Lambda_R \mathbf{U}_R^H$ where $\mathbf{U}_R$ is Haar-distributed and $P'_{\Lambda_R}(\Lambda_R)$ is given as follows:

$$P'_{\Lambda_R}(\Lambda_R) = \left(\frac{n_r}{n_t E_0}\right)^{n_r^2} \prod_{n=1}^{n_r} \frac{1}{n!(n-1)!} \exp\left(-\frac{n_r}{n_t E_0} \sum_{i=1\ldots n_r} \lambda_i\right) \Delta(\Lambda_R)^2. \qquad (19)$$

This is similar to the expression for the full-covariance case of eq. (7) but scaled so that $E[\mathrm{tr}(\Lambda_R)] = n_t n_r E_0$. Marginalizing over $\mathbf{Q}_R$ yields

$$P_{\mathbf{H}}(\mathbf{H}) = \int_{\mathcal{U}(n_r) \times \mathbb{R}^{+n_r}} P_{\mathbf{H}|\mathbf{Q}_R}(\mathbf{H}, \mathbf{U_R}, \Lambda_R) P'_{\Lambda_R}(\Lambda_R) d\mathbf{U_R} d\Lambda_R \qquad (20)$$

$$= \int_{\Lambda_R \in \mathbb{R}^{+n_r}} \frac{e^{-\frac{n_r}{n_t E_0}\mathrm{tr}(\Lambda_R)}}{\pi^N \prod_{n=1}^{n_r} n!} \left(\frac{n_r}{n_t E_0}\right)^{n_r^2} \frac{\det(e^{-A_i/\lambda_j}) \det(\Lambda_R)^{n_r - n_t - 1} \Delta(\Lambda_R)}{\Delta(\mathbf{A})} \qquad (21)$$

where we used the Harish-Chandra-Itzykson-Zuber (HCIZ) integral [12] with $\mathbf{A} = \mathbf{HH}^H$, having eigenvalues $(A_1, \ldots, A_{n_r})$. We also use the notation $\det(f(i,j))$ for the determinant of the matrix with the (i,j)-th element given by $f(i,j)$ for any function f, and $\Delta(\mathbf{A}) = \det(A_i^{j-1})$. Expansion of the determinants (using the notation $\mathbf{a} = [a_1, \ldots, a_N]$, and with $\mathcal{P}_N$ the set of all permutations of $[1, \ldots, N]$) yields

$$P_{\mathbf{H}}(\mathbf{H}) = \frac{1}{\pi^N \Delta(\mathbf{A})} \left(\frac{n_r}{n_t E_0}\right)^{n_r^2} \prod_{n=1}^{n_r} \frac{1}{n!} \sum_{\mathbf{a},\mathbf{b} \in \mathcal{P}_{n_r}^2} (-1)^{\mathbf{a}+\mathbf{b}} \prod_{n=1}^{n_r} \int_{\lambda_n \in \mathbb{R}^+} \lambda_n^{n_r - n_t + a_n - 2} e^{-\frac{A_{b_n}}{\lambda_n}} e^{-\frac{n_r}{n_t E_0}\lambda_n}.$$

$$(22)$$

Letting $f_i(x) = \int_{\mathbb{R}^+} t^{n_r-n_t+i-2} e^{-x/t} e^{-\frac{n_r}{n_t E_0}t} dt$, we obtain finally the joint pdf of $\mathbf{H}$:

$$P_{\mathbf{H}}(\mathbf{H}) = \frac{n_r!}{\pi^N} \left(\frac{n_r}{n_t E_0}\right)^{n_r^2} \prod_{n=1}^{n_r} \frac{1}{n!} \frac{\det(f_i(A_j))}{\Delta(\mathbf{A})}. \tag{23}$$

CONCLUSION

We presented several analytical models for wireless MIMO flat-fading channels, based on various degrees of knowledge about the environment, using a modeling approach based on the maximum entropy principle. Our approach consists in first modeling the spatial covariance properties, both for the case of full correlation, and for the case of separable (transmit- and receive-side) correlation constraints. We showed that different Wishart distributions of the covariance matrix are obtained for the mean-constrained case and without constraints. In a second step, the covariance is marginalized over to obtain directly the channel probability density function. Analytical expressions for the channel pdf have been provided when possible.

ACKNOWLEDGMENTS

This work was funded by the I0 project of the Forschungszentrum Telekommunikation Wien (ftw.). ftw. is supported by the Kplus competence center program of the Austrian government.

REFERENCES

1. G. J. Foschini, and M. J. Gans, *Wireless Personal Communications* **6**, 311–335 (1998).
2. P. Almers, E. Bonek, A. Burr, N. Czink, M. Debbah, V. Degli-Espoti, et al., *EURASIP Journal on Wireless Communications and Networking* **2007** (2007).
3. A. M. Tulino, and S. Verdú, *Random Matrix Theory and Wireless Communications*, now Publishers, Delft, The Netherlands, 2004.
4. D.-S. Shiu, G. Foschini, M. Gans, and J. Kahn, *IEEE Transactions on Communications* **48**, 502–513 (2000).
5. B. Clerckx, L. Vandendorpe, D. Vanhoenacker-Janvier, and A. Paulraj, "Robust space-time codes for spatially correlated MIMO channels," in *Proc. IEEE International Conference on Communications (ICC)*, 2004, vol. 1, pp. 453–457.
6. E. T. Jaynes, *Probability Theory*, Cambridge University Press, 2003.
7. M. Guillaud, and M. Debbah, "Maximum Entropy MIMO Wireless Channel Models with Limited Information," in *Proc. MATHMOD Conference on Mathematical Modeling*, Wien, Austria, 2006.
8. M. Debbah, and R. R. Müller, *IEEE Transactions on Information Theory* **51**, 1667–1690 (2005).
9. M. Guillaud, M. Debbah, and A. L. Moustakas, "A Maximum Entropy Characterization of Spatially Correlated MIMO Wireless Channels," in *Proc. IEEE Wireless Communications and Networking Conference (WCNC)*, Hong Kong, 2007.
10. A. Edelman, *Eigenvalues and Condition Numbers of Random Matrices*, Ph.D. thesis, Massachusetts Institute of Technology, Cambridge, MA, USA (1989).
11. B. Maharaj, L. Linde, and J. Wallace, "MIMO Channel Modelling: the Kronecker model and maximum entropy," in *Proc. IEEE Wireless Communications and Networking Conference (WCNC)*, Hong Kong, 2007.
12. A. B. Balantekin, *Phys.Rev. D* **62** (2000).

Analysis of Intrusion Detection and Attack Proliferation in Computer Networks

Prahalad Rangan[a] and Kevin H. Knuth[a,b]

[a] University at Albany, Dept. of Informatics, Albany NY USA
[b] University at Albany, Dept. of Physics, Albany NY USA

Abstract. One of the popular models to describe computer worm propagation is the Susceptible-Infected (SI) model [1]. This model of worm propagation has been implemented on the simulation toolkit Network Simulator v2 (ns-2) [2]. The ns-2 toolkit has the capability to simulate networks of different topologies. The topology studied in this work, however, is that of a simple star-topology. This work introduces our initial efforts to learn the relevant quantities describing an infection given synthetic data obtained from running the ns-2 worm model. We aim to use Bayesian methods to gain a predictive understanding of how computer infections spread in real world network topologies. This understanding would greatly reinforce dissemination of targeted immunization strategies, which may prevent real-world epidemics.

The data consist of reports of infection from a subset of nodes in a large network during an attack. The infection equation obtained from [1] enables us to derive a likelihood function for the infection reports. This prior information can be used in the Bayesian framework to obtain the posterior probabilities for network properties of interest, such as the rate at which nodes contact one another (also referred to as contact rate or scan rate). Our preliminary analyses indicate an effective spread rate of only 1/5th the actual scan rate used for a star-type of topology. This implies that as the population becomes saturated with infected nodes the actual spread rate will become much less than the scan rate used in the simulation.

Keywords: internet, network, infection, attack, nodes, virus, security, epidemic, immunization
PACS: 07.05.Bx

NETWORK MODELS

Modern Internet worms have the capacity to proliferate among susceptible hosts at incredible rates. The line of research outlined in this paper aims to evaluate the impact played by network topology on the rate of spread of an epidemic in real-world networks of computers. The study of epidemic spreading is based upon the idea that infection is transmitted by contact between and infected individual and an uninfected (but susceptible) individual. Moore, et al. [1] outline the classic Susceptible-Infected (SI) epidemic model that describes the spread of a malicious attack through homogenous random contacts between susceptible and infected hosts. This model prescribes that the number of new infections is the product of the number of infected hosts (infectives), the fraction of susceptible hosts (susceptibles) and an average contact rate β. In the simulation β is the time interval between successive attempts made by an infected host to contact and infect a randomly chosen host from the vulnerable population.

To simulate an infection, we use Network Simulator v2 (ns-2) [2], which is a discrete event network simulator targeted at networking research. It is popular for its extensibility and vast online documentation. The simulator is written in C++; it uses OTcl (Object oriented Tool Command Language) as a command and configuration language. The ns-2 toolkit provides substantial support for simulation of TCP, routing, and multicast protocols over wired and wireless (local and satellite) networks. In ns-2 physical activities are translated to events. The events are queued and processed in the order of their scheduled occurrences, and time progresses as the events are processed. The worm propagation model, available in the ns-2 software package, models the internet with two parts: detailed and abstract part. For this reason, it is named the Detailed-Network and Abstract-Network (DN-AN) model. An example of a detailed network is an enterprise Local Area Network (LAN); whereas the rest of the Internet is abstracted into the abstract network. Network connectivity and packet transmission are also simulated in the ns-2 model. More details are available in the ns manuals [2].

In the SI model any infected individual can contact and infect any other susceptible individual. The infection, therefore, spreads at a particular contact rate. The contact rate is the rate at which an infected host attempts to contact and infect a susceptible host. At any given instant of time during the process of infection we can distinguish between the population of infectives and susceptibles. The infective population is the group of hosts who have been infected and the susceptible population is the group of vulnerable hosts, which have not yet been infected. The fraction of infectives at any given instant is the ratio of the infected population and the total vulnerable population. The fraction of susceptibles is the ratio of susceptible population and the total vulnerable population. The parameters in the SI model can be summarized as:

N – Size of the total vulnerable population of hosts in the network of interest
$S(t)$ – Susceptibles at time t
$I(t)$ – Infectives at time t
β – Contact Rate
$s(t)$ – Fraction of Susceptibles
$i(t)$ – Fraction of Infectives

The rate of change of infectives is given by,

$$\frac{dI}{dt} = \frac{\beta I(t) S(t)}{N}. \tag{1}$$

The rate of change of susceptibles is given by,

$$\frac{dS}{dt} = -\frac{\beta I(t) S(t)}{N}. \tag{2}$$

If $i = I/N$ and $S + I = N$, then (1) can be written as,

$$\frac{di}{dt} = \beta i(t)(1 - i(t)). \tag{3}$$

Integrating this differential equation from the time when the infective population is $i = 0$ to time T when the infective population is half the total vulnerable population $i = 1/2$, gives the proportion of infected individuals at time t

$$i(t) = \frac{e^{\beta(t-T)}}{1 + e^{\beta(t-T)}}. \tag{4}$$

This allows us to predict the number of infected nodes at time t given the model parameters.

BAYESIAN ANALYSIS

The connectivity of computer networks can be extremely complex and, in general, is unknown. Here we consider the problem of inferring the characteristics of the network by monitoring the propagation of an infection. In this initial exploration, we aim to infer the contact rate, β, which describes the rate at which an infected host attempts to contact and infect randomly chosen hosts from the vulnerable population. The contact rate is a function of the network topology and carries with it information about the connectivity. Equation (4) above describes how the proportion of infected computers at time t depends on β. We imagine having a system that can monitor the state (as it changes from susceptible to infected due to contact with another infected host) of a small fraction of the nodes in the population that reports to us when they have become infected. Our data will consist of the fact that node n_i reported an infection at time t_i. Our goal is to estimate β from the number of infected nodes $I(t)$ observed at time t using the infection information reported by a subset of nodes out of the entire network population.

In our example, the solution for the infected fraction $i(t) = I(t)/N$ is given by (4) above. By setting $t = T$ in the above equation we see that $i(t)$ reduces to 1/2, so that T is the time when half the population is infected. Applying Bayes' theorem, the posterior probability of β and T given the times at which the nodes report infection is proportional to the product of the prior probability of β and T and the conditional probability of the times of infection given β and T, which is also called likelihood. This can be written as

$$P(\beta, T \mid \{t\}, PI) = \frac{P(\beta, T \mid PI) P(\{t\} \mid \beta, T, PI)}{Z}, \tag{5}$$

where PI is the prior information, the prior probability is $P(\beta, T \mid PI)$, the likelihood is $P(\{t\} \mid \beta, T, PI)$, and the evidence is Z. In this calculation, we assign a joint uniform prior for β and T over a reasonable range of values

$$P(\beta, T \mid PI) = \frac{1}{(\beta_{max} - \beta_{min})} \frac{1}{(T_{max} - T_{min})}. \tag{6}$$

The likelihood is proportional to the rate of change of infectives as we can infer the times of infection from this quantity.

$$P(t \mid \beta, T, PI) \propto \frac{dI(t, \beta, T)}{dt}. \tag{7}$$

Since $I(t, \beta, T) = Ni(t, \beta, T) \Rightarrow N\, di(t, \beta, T)/dt = dI(t, \beta, T)/dt$, and (7) can be written as

$$P(t \mid \beta, T, PI) \propto \frac{di(t, \beta, T)}{dt} = \frac{1}{K} \frac{\beta e^{\beta(t-T)}}{\left(1 + e^{\beta(t-T)}\right)^2}, \tag{8}$$

where K^{-1} is the constant of proportionality. In general, K will depend on β and T, so to find K, we marginalize over t. The extreme values of T are taken to be the times when the first and last node report infection in our data. This normalization factor K is found to be

$$K = \frac{e^{\beta(t_f - T)}}{1 + e^{\beta(t_f - T)}} - \frac{e^{\beta(t_o - T)}}{1 + e^{\beta(t_o - T)}}. \tag{9}$$

If $t_o = 0$, then this reduces to

$$K = \frac{e^{\beta(t_f - T)}}{1 + e^{\beta(t_f - T)}} - \frac{e^{-\beta T}}{1 + e^{-\beta T}}. \tag{10}$$

Thus the likelihood for an infection at time t is

$$P(t \mid \beta, T, PI) = \frac{\left(1 + e^{\beta(t_f - T)}\right)\left(1 + e^{-\beta T}\right)}{e^{\beta(t_f - T)}(1 + e^{-\beta T}) - e^{-\beta T}\left(1 + e^{\beta(t_f - T)}\right)} \frac{\beta e^{\beta(t-T)}}{\left(1 + e^{\beta(t-T)}\right)^2}, \tag{11}$$

and the joint likelihood for a set of I_{tot} infections is

$$P(\{t\} \mid \beta, T, PI) = \frac{\left(1 + e^{\beta(t_f - T)}\right)\left(1 + e^{-\beta T}\right)}{e^{\beta(t_f - T)}(1 + e^{-\beta T}) - e^{-\beta T}\left(1 + e^{\beta(t_f - T)}\right)} \prod_{i=1}^{I_{tot}} \frac{\beta e^{\beta(t_i - T)}}{\left(1 + e^{\beta(t_i - T)}\right)^2}. \tag{12}$$

The non-normalized posterior probability from (5) becomes (13)

$$P(\beta, T \mid \{t\}, PI) \propto \frac{1}{(\beta_{max} - \beta_{min})} \frac{1}{(T_{max} - T_{min})} \frac{\left(1 + e^{\beta(t_f - T)}\right)\left(1 + e^{-\beta T}\right)}{e^{\beta(t_f - T)}(1 + e^{-\beta T}) - e^{-\beta T}\left(1 + e^{\beta(t_f - T)}\right)} \prod_{i=1}^{I_{tot}} \frac{\beta e^{\beta(t_i - T)}}{\left(1 + e^{\beta(t_i - T)}\right)^2},$$

which can be reduced to

$$P(\beta, T \mid \{t\}, PI) \propto \frac{\left(1 + e^{\beta(t_f - T)}\right)\left(1 + e^{-\beta T}\right)}{e^{\beta(t_f - T)}(1 + e^{-\beta T}) - e^{-\beta T}\left(1 + e^{\beta(t_f - T)}\right)} \prod_{i=1}^{I_{tot}} \frac{\beta e^{\beta(t_i - T)}}{\left(1 + e^{\beta(t_i - T)}\right)^2} \tag{14}$$

by absorbing the uniform prior into the implicit proportionality constant.

RESULTS

The worm propagation model starts with the infection of a node in the abstract network, which contacts and infects a node in the detailed network. Beyond this point the spread of the infection is contained within the detailed network. The scan rate β is set to 1000, such that an infected node in the detailed network attempts to contact and infect a randomly chosen host from within the detailed network every 1/1000 of a time step. As each susceptible (uninfected) node is contacted and infected we note the time of infection. So our data is made up of the times of infection of the nodes in the detailed network. Using the Matlab function fminsearch, β and T were estimated to be 156.4 ± 18.9 and 0.0417 ± 0.0015, respectively. Figure 1 shows the posterior probability as a function of both β and T. The estimated values of the model parameters were plugged into the original infection equation (4) and the results were compared against the data. Figure 2 shows a comparison of the plots of fraction infected versus time as predicted by (4) (red curve) and the fraction infected obtained as the nodes report infection during the simulation (blue circles). This indicates that the estimated β and T are in concordance with the times reported by the nodes during infection.

CONCLUSION AND FUTURE WORK

Even though the scan rate β is 1000 in the simulation, the estimated beta from the data produced works out to only 156.4 ± 18.9. This experiment has been performed in a closed star-type of network configuration where each node after getting infected sends out a probing packet to another randomly selected node every 1/β of a time step. As the population becomes saturated with infected nodes the actual spread rate will become much less than the scan rate used in the simulation. We are thus able to get a fair idea of how the infection progresses in the network by just having few nodes reporting the infection during its course. This is confirmed by the comparison of plots of the infection in Figure 2, as predicted by (4) and the actual times at which infection is reported by a subset of nodes in the network.

We have tried creating topologies using the GT-ITM topology generator package that comes with the current ns-2 release (http://www.isi.edu/nsnam/ns/ns-topogen.html). The topologies follow the Transit-Stub model as outlined in [3], where each domain in the internet is represented as a Transit or Stub domain. A stub domain carries only traffic that originates or terminates in the domain. The transit domains interconnect the stub domains. Our efforts focus on running the worm model against these topologies and study the variation in effective scan rate for such topologies. We hope this study will grant us insights as to how an epidemic spreads in computer networks with different topologies. This knowledge will provide us information that could be used to change the topology during an active epidemic to see if the spread of the infection can be controlled in real-time.

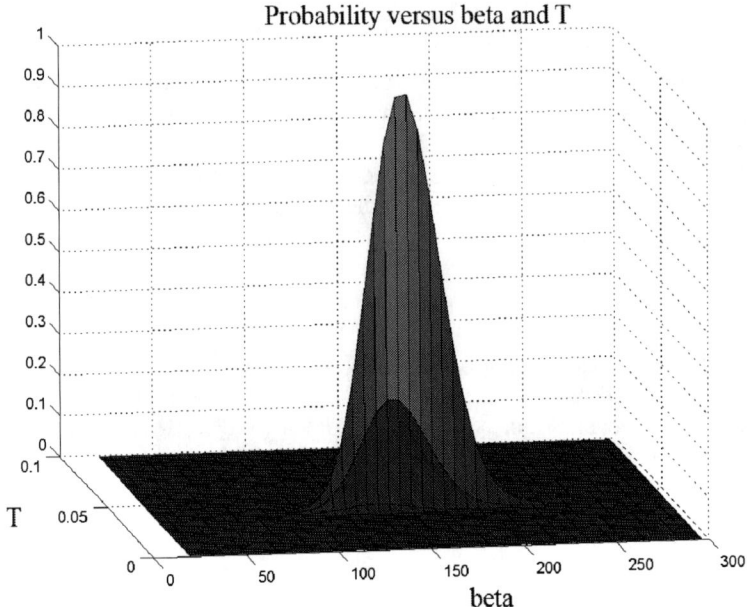

Figure 1. Plot of the two-dimensional posterior probability (non-normalized) of β and T for 50 nodes reporting with a scan rate of 1000.

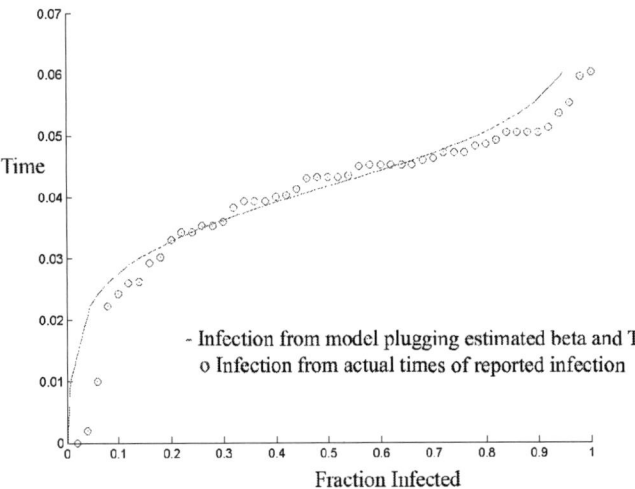

Figure 2. Comparison between the fraction of infected nodes predicted by the infection model using the estimated values of β and T and the original data obtained from simulation.

REFERENCES

1. D. Moore, C. Shannon, G. M. Voelker, and S. Savage, 2003. "Internet Quarantine: Requirements for Containing Self-Propagating Code," in *INFOCOM 2003*, April, 2003.
2. M. Greis, Tutorial and open source documentation for the network simulator.
   http://www.isi.edu/nsnam/ns/tutorial/index.html
   http://www.isi.edu/nsnam/ns/ns-topogen.html
3. K. Calvert, M. Doar, and E.W. Zegura, 1997. Modeling Internet topology, *IEEE Communications Magazine*, **35**(6), 160-163.

Bayesian Extraction of Deep UV Resonance Raman Signature of Fibrillar Cross-β Sheet Core based on H-D Exchange Data

V.A. Shashilov, I.K. Lednev

University at Albany, State University of New York, 1400 Washington Avenue Albany NY 12222
(e-mail: lednev@albany.edu)

Abstract. Amyloid fibrils are associated with many neurodegenerative diseases. The application of conventional biophysical techniques including solution NMR and X-ray crystallography for structural characterization of fibrils is limited because they are neither crystalline nor soluble. The Bayesian approach[1] was utilized for extracting the deep UV resonance Raman (DUVRR) spectrum of the lysozyme fibrillar β-sheet based on the hydrogen-deuterium exchange spectral data. The problem was shown to be unsolvable when using blind source separation or conventional chemometrics methods because of the 100% correlation of the concentration profiles of the species under study. Information about the mixing process was incorporated by forcing the columns of the concentration matrix to be proportional to the expected concentration profiles. The ill-conditioning of the matrix was removed by concatenating it to the diagonal matrix with entries corresponding to the known pure spectra (sources). Prior information about the spectral features and characteristic bands of the spectra was taken into account using the Bayesian signal dictionary approach[2]. The extracted DUVRR spectrum of the cross-β sheet core exhibited sharp bands indicating the highly ordered structure. Well resolved sub-bands in Amide I and Amide III regions enabled us to assign the fibril core structure to anti-parallel β-sheet and estimate the amide group facial angle Ψ in the cross-β structure. The elaborated Bayesian approach was demonstrated to be applicable for studying correlated biochemical processes.

Keywords: Bayesian source separation, deep UV resonance Raman, amyloid fibril.
PACS: 87.15.Cc, 87.64.Je, 02.60.–x.

INTRODUCTION

Amyloid fibrils are associated with many neurodegenerative diseases. Studying the structure of amyloid fibril is important for detailed understanding of fibrillogenesis at a molecular level. Amyloid fibrils are non-crystalline and insoluble, and thus are not amenable to conventional X-ray crystallography and solution NMR, the classical tools of structural biology[3]. Despite the recent advances in solid state NMR[4] and high resolution X-ray[5] the routine application of these techniques is yet to be established. Electron diffraction and scanning probe microscopy are not sensitive to fibrillar secondary structures. The application of optical techniques such fluorescence and circular dichroism, on the other hand, is difficult because of intense light scattering from fibril material. These limitations of conventional methods have made deep UV resonance Raman spectroscopy (DUVRR) a unique tool for structural characterization of amyloid fibrils[6]. The spectrum of the fibrillar core cross-β sheet is, however, not directly observable in the Raman experiment because of significant interference with Raman signals from β-turns and unordered protein structures. Here we repot on the elaboration and application of the Bayesian source separation algorithm for the extracting of the DUVRR signature of the cross-β structure on the basis of hydrogen-deuterium exchange (HX) data.

MATERIALS AND METHODS

Experimental DUVRR Spectroscopic Data

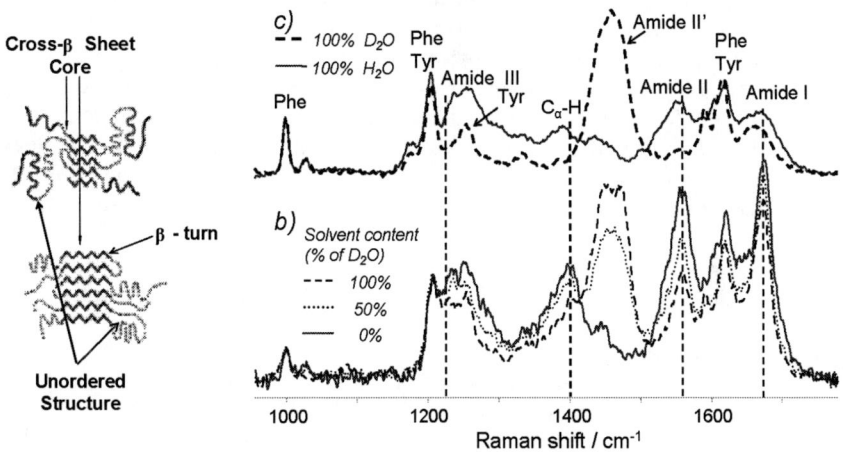

FIGURE 1. (a) Schematic representation of a growing amyloid fibril.[7] Cross-β sheet core strands are shown with saw-shaped solid lines. (b) DUVRR spectra of lysozyme fibrils in H_2O (solid), D_2O (dashed), and 50/50 % H_2O/D_2O mixture (dotted). (c) DUVRR spectrum of unordered lysozyme in H_2O (solid) and D_2O (dashed).

Amyloid fibrils (Fig. 1a) are composed of the highly ordered β-sheet core surrounded by unordered structure fragments. The main chain NH group protons in proteins show facile exchange when exposed to D_2O environment.[8] In the hydrophobic cross-β sheet core hidden from solvent, the rates of deuterium exchanges are strongly reduced. Accordingly, amide N-H protons in β-turns and unordered fragments of amyloid fibrils should be readily exchanged while those buried in the cross-β structure will remain protonated. Spectroscopically, HX will cause down-shift of Amide II DUVRR band from ~1555 cm^{-1} to ~1450 cm^{-1} (Amide II', Fig. 1b) and virtual disappearance of Amide III band[9] (Fig. 1c) originating from unordered protein while DUVRR amide bands of the protonated β-sheet core will remain unchanged.

The experimental spectra of fibril material suspended in H_2O/D_2O mixtures with the D_2O fraction ranging from 0 to 100 % at 2% increment were recorded to produce a dataset of 51 DUVRR spectra. Thus obtained dataset should be a linear combination of three individual components, *i.e.* Raman spectra of (i) the fibrillar cross-β sheet core, (ii) protonated β-turns and protonated unordered lysozyme structure (together referred to as unordered lysozyme below), and (iii) deuterated unordered lysozyme.

Development of the Bayesian Source Separation Approach

Why use the Bayesian Approach?

Earlier we have demonstrated the application of various source separation methods for extracting the pure DUVRR spectra of individual protein conformers including soft-constrained multivariate curve resolution (MCR),[10] non-negative independent component analysis (ICA) and pure variable methods[11], JADE and FSOBI ICA combined with 2D correlation spectroscopy[6].

The blind source separation algorithms such as ICA and pure variable methods do not easily allow for making use of prior information about the spectral features of individual components and concentration matrix, while the latter can be readily anticipated in our study. Namely, the fractions of protonated and deuterated unordered structures changed proportionally to the fractions of proton and deuterium ions in solution respectively, while the fraction of core β-sheet remained constant (Fig. 2). As can be further noticed from Fig. 2, the concentration profiles of cross-β sheet and deuterated and protonated unordered structures are linearly dependent. As a result, the inversion of their concentration matrix during the MCR search would yield three linearly-dependent, and therefore meaningless individual components. The inspection of the eigenvalues of the data matrix (Table 1) verified the assumption about the rank-deficiency of our data.

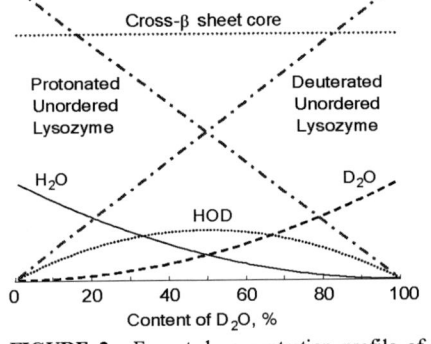

TABLE (1). Eigenvalues of the DUVRR data matrix.

Order # of Component	Eigenvalue
1	2.46E+2
2	1.28E+2
3	3.35E+1
4	2.54E+1
5	1.59E+1
6	1.53E+1

FIGURE 2. Expected concentration profile of major components vs. the fraction of D_2O: protonated unordered lysozyme (chain), cross-β sheet core (dotted), deuterated unordered lysozyme (chain), H_2O (solid), HOD (dotted), and D_2O (dashed)

As seen from Table 1, two rather that three largest eigenvalues (246 and 128) account for lysozyme spectral contribution, while the remaining eigenvalues represent the contribution of H_2O, D_2O, HOD, quartz, molecular oxygen (discussed below) and all types of error in the data. Thus, the problem under study seems to be prior-dominated and ill-posed, so we turned to the Bayesian statistics, a powerful tool for resolving such complicated cases.

Bayesian Formulation of the Source Separation Problem

The source separation problem is as follows

$$Data = C \cdot S + E \qquad (1)$$

where C is the concentration matrix, S is the matrix of pure component spectra, and E is error (random or systematic). The matrix $Data$ is assumed to be known while the matrices C and S are to be estimated. The Bayes theorem for problem (1) takes the form:

$$P(C,S \mid Data, I) \sim P(Data \mid C,S,I) \cdot P(C \mid I) \cdot P(S \mid I) \quad (2)$$

where $P(Data \mid C,S,I)$ is the likelihood measuring the quality of data fitting and $P(S \mid I)$ and $P(C \mid I)$ are prior probabilities for individual component spectra and concentrations respectively. Because finding either matrix C or S alone is enough for solving problem (1) the concentration matrix C is normally sought since it contains by far fewer elements. It was shown[1] that in the case of uniform prior for the concentration matrix and independent sources, the probability of the concentration matrix is given by

$$P(C \mid Data, I) \sim \int ds \cdot \prod_i \delta(Data_i - C_{ik} \cdot S_k) \cdot \prod_l p_l(s_l) \quad (3)$$

In the case of noise-free data (3) reduces to the logarithmic probability

$$P(C \mid Data, I) = \log(\det(W)) + \sum_l \log(p_l(s_l)) \quad (4)$$

where W is the separation matrix such that $S = W \cdot Data$.

Incorporating Prior Information about Concentrations and Pure Component Spectra

The actual concentration matrix was sought to have columns proportional to the concentration profiles sketched in Fig. 2. Hydrogen and deuterium ions of H_2O and D_2O molecules easily exchange to form mixed HOD molecules. The fractions of H_2O, D_2O, and HOD vs. the content of D_2O were analytically calculated taking into account that the probability of formation of a R_iOR_j molecule is:

$$P(R_iOR_j \mid [R_i],[R_j],I) \sim (2 - \delta_{ij}) \cdot [R_i] \cdot [R_j] \quad (5)$$

where $R_i=H$, $R_j=D$, brackets [] mark concentration, and δ_{ij} equals 1 if $i=j$ and zero otherwise. The DUVRR spectra of H_2O, D_2O, and 50-50 % mixture of H_2O/D_2O were recorded to calculate the spectrum of HOD. Furthermore, we hypothesized that the DUVRR spectrum of unfolded lysozyme from the supernatant of fibril samples may be close to the spectroscopic signature of unordered parts of the fibril. Table 2 summarizes the prior information about the spectra and concentrations of all individual components.

The contribution of the quartz signal to the experimental data was eliminated by the careful numerical subtraction from all DUVRR spectra, while the contribution of molecular oxygen was disregarded in a zero-order approximation.

TABLE 2. Prior information about the individual component spectra and concentrations.

Individual Component	Spectrum	Concentration Profile
H_2O	Known	Shape is known
HOD	Known	Shape is known
D_2O	Known	Shape is known
Protonated Unordered Lysozyme	*Unknown*	Shape is known
Deuterated Unordered Lysozyme	*Initial guess available*	Shape is known
Cross-β Sheet Core	*Unknown*	Shape is known
Quartz (from NMR tube)	Known	Small random contribution
Molecular Oxygen (mainly from water)	Known	Small random contribution

Constructing the concentration matrix. As seen from Table 2, problem (1) reduces to finding the concentration matrix C_{ij} where $i=1:N$ and $j=1:n$, with $N=51$ and $n=6$ representing the number of experimental spectra and the number of pure components, respectively. First, the template matrix T_{ij} was constructed, so that each column $T(:,j)$ is equal to the expected concentration profiles of a particular component shown in Fig. 2. The elements of each row of thus constructed T_{ij} matrix are proportional to the contributions of the individual components to the given spectrum of the data matrix *Data*. The heterogeneity of fibril suspensions did not allow for controlling the concentration of fibrils in samples and as a result, for the proper normalization of the DUVRR spectra. To compensate for the lack in data normalization, each row of the template matrix was multiplied by the scalar α_i yielding the matrix $C_{ij} = a_{ii} \cdot T_{ij}$, where a_{ii} is the $N \times N$ diagonal matrix of α_i elements. In addition to the normalization issue, fluctuation of fibril concentrations was taken into account by multiplying the contributions of three fibril components by scalars β_i to produce the concentration matrix:

$$C_{ij} = a_{ii} \cdot T_{ij} + \beta_{ii} \cdot T^0_{ij} \qquad (6)$$

where T^0_{ij} has zero columns for H_2O, D_2O and HOD elements and equals the T_{ij} matrix otherwise.

The matrix C_{ij} constructed based on eq. 6 remains ill-conditioned and therefore is not amenable to inverting. To deal with ill-conditioning, the concentration matrix C_{ij} and the data matrix D were replaced by the enhanced matrices C^+_{ij} and D^+ as follows:

$$C^+_{ij} = \begin{pmatrix} C_{ij} \\ 1\,0\,0\,0\,0\,0 \\ 0\,1\,0\,0\,0\,0 \\ 0\,0\,1\,0\,0\,0 \\ 0\,0\,0\,1\,0\,0 \end{pmatrix} \qquad D^+ = \begin{pmatrix} Data \\ S(H_2O) \\ S(HOD) \\ S(D_2O) \\ \lambda \cdot S(DUL) \end{pmatrix} \qquad (7)$$

where *DUL* stands for deuterated unordered lysozyme and λ is a small regularization parameter. Addition of four orthogonal rows to the concentration matrix makes it invertible. Moreover, enhancement of the data matrix *Data* with three known spectra of H_2O, D_2O and HOD constrains the sought spectra of water components to the experimental ones. The DUVRR spectrum of deuterated unordered lysozyme multiplied by the small parameter λ allows for the soft incorporation of the initial guess about this individual component.

Dictionary approach for modeling pure component spectra. A priori information about characteristic bands in the individual component spectra was incorporated via the Bayesian signal dictionary approach.[2] In this approach, individual components are presented as a linear combination of reference spectral bands. In our study, the reference band library was composed of mixed Gaussian / Lorentzian shapes with certain widths and spectral positions. The reference bands were obtained by fitting the DUVRR spectra of fibrils by a series of Gaussian / Lorentzian peaks having a pre-defined width range. The Bayesian choice[12] as to the number of peaks contributing to some DUVRR bands was made if no prior information about the structure of those bands was available. Some informative spectral features of the fibrillar cross-β sheet core were also extracted by analyzing fibril samples with different random coil content using

SMAC and SIMPLISMA methods.[13] The dictionary-based individual component spectrum was then calculated as follows:

$$S(v) = \sum_{i=1}^{M} H_i \cdot \frac{m_i}{4 \cdot \frac{(v-x_i)^2}{w_i^2}+1} + (1-m_i) \cdot H_i \cdot \exp(-4 \cdot \log(2) \cdot 4 \cdot \frac{(v-x_i)^2}{w_i^2}) \qquad (8)$$

where H_i, w_i and x_i are respectively the height, width, and the spectral position of the ith band, m_i is the contribution of the Lorentzian shape to the ith the reference band, and M is the total number of reference bands. Both reference band widths w and positions x were confined in a range of $w,x \pm 2\text{-}3$ cm^{-1} while the heights H were allowed to freely change. The individual component of cross-β sheet core was sought as a linear combination of the reference bands (8). In addition, the probabilities p_l (s_l) in (4) were assigned to the reciprocal elements of the individual component spectra, i.e.

$$p_l(s_l) \sim \frac{1}{\sum_k s_{lk}} \quad \text{or} \quad \log(p_l(s_l)) \sim -\sum_k s_{lk} \qquad (9)$$

where l is the order number of the individual component and k runs over all wavenumbers in the l-th pure component spectrum. Such assignment for the probabilities of sources makes sparse individual components more favorable, thus partly circumventing the common problem of under-resolved spectra (those with extraneous bands appearing as admixture from the other pure spectra) in MCR methods.

The posterior probability for the concentration matrix. The high signal-to-noise ratio of the experimental spectra enabled us to use eq. 4 elaborated for noise-free data as a reasonable approximation. Furthermore, the sought concentration matrix should be close to the model C_{ij} matrix (6) yet providing good data fitting. The sparse constraint on individual components should also be enforced. With a Student-t marginal likelihood the posterior becomes:

$$P(C \mid Data, C, S, I) = P(C \mid Data, I) + P(C \mid C, S, I) \qquad (10)$$

$$P(C \mid Data, I) = \log(\det(W)) - \frac{N \cdot k}{2} \cdot \log\{\|(Data - D_{ij})\|\} \qquad (10\text{-}a)$$

$$P(C \mid C, S, I) - \frac{N \cdot n}{2} \cdot \log\{\|(C - C_{ij})\|\} - \sum_k \sum_{l=1}^{n} \log(s_{lk}) \qquad (10\text{-}b)$$

where N is the number of experimental spectra, n-number of individual components, k – length of the individual component vector, C_{ij} and D_{ij} are model concentration matrix and data predicted by the model, respectively, C and $Data$ are actual concentration and data matrices, W – is a separation matrix, '$\| \ \|$' stands for Frobenius matrix norm. The problem now reduces to finding parameter set of α_i and β_i, $i=1{:}N$ and m_j, w_j, h_j and m_j, $j=1{:}M$ used in (8), which maximizes the posterior (10). It is worth noting that the prior information about the individual spectra of unordered deuterated lysozyme (available from experiment) and cross-β sheet core (composed of reference bands, eq. 8) missing in (10-b) was incorporated through the augmented C^+_{ij} and D^+, which now take the form:

$$C^+_{ij} = \begin{pmatrix} C_{ij} \\ 1\ 0\ 0\ 0\ 0\ 0 \\ 0\ 1\ 0\ 0\ 0\ 0 \\ 0\ 0\ 1\ 0\ 0\ 0 \\ 0\ 0\ 0\ 1\ 0\ 0 \\ 0\ 0\ 0\ 0\ 1\ 0 \end{pmatrix} \qquad D^+ = \begin{pmatrix} Data \\ S(H_2O) \\ S(HOD) \\ S(D_2O) \\ \lambda_1 \cdot S(DUL) \\ \lambda_2 \cdot S(\beta\text{-sheet}) \end{pmatrix} \qquad (11)$$

Algorithm Outline

(1) Initialize parameters α_i, β_i ($i=1:N$) and m_j, w_j, h_j and m_j, ($j=1:M$) by random values from the expected range. Fix parameters λ_1 and λ_2.

(2) Reduce dimension of the *Data* matrix from N to n spectra by SVD to produce the matrix $\overline{D}$.

(3) For *iter* = 1: *itmax* do
Sample parameters α_i, β_i, m_j, w_j, h_j and m_j.
  (a) Calculate the trial C_{ij} matrix as $C_{ij} = \alpha_{ii} \cdot T_{ij} + \beta_{ii} \cdot T^0_{ij}$ (eq. 6)
  (b) Calculate the dictionary-based spectrum $S(\beta\text{-sheet})$ (eq. 8)
  (c) Construct the C^+_{ij} and D^+ matrices (eq. 11).
  (d) Calculate individual components using the left pseudo-inverse of C^+_{ij}
$$s = (C_{ij}^{+T} C^+_{ij})^{-1} \cdot C_{ij}^{+T} \cdot D_{ij}^+$$
  (e) Set negative entries of the s matrix to zeros.
  (f) Calculate the concentration matrix C using the right pseudo-inverse of s.
$$C = D_{ij}^+ \cdot (s^T \cdot (s \cdot s^T)^{-1})$$
  (g) Calculate the matrices $D_{ij} = C(1:N,:) \cdot s$ and $W = s \cdot (\overline{D}^T \cdot (\overline{D} \cdot \overline{D}^T)^{-1})$.
  (h) Calculate the posterior $P(C \mid Data, C, S, I)$ (eq. 10).
  (i) Sample a new set of parameters α_i, β_i, m_j, w_j, h_j and m_j and go to (a).

(4) Stop when the maximum of $P(C \mid Data, C, S, I)$ (eq. 10) is found or *itmax* is reached.

The floating point genetic algorithm (GA)[10,14] was used as a sampling method. Previously,[10] we have demonstrated the excellent performance of GA in finding the global minimum, which is made possible by thorough exploration of the parameter space.

RESULTS AND DISCUSSION

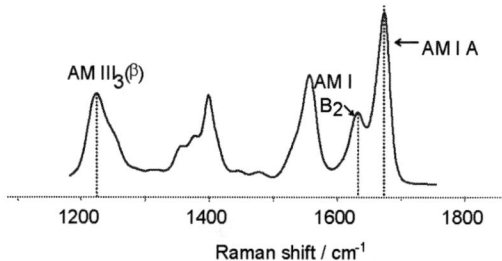

FIGURE 3. Extracted DUVRR spectrum of cross-β sheet fibrillar core of lysozyme. Amide I A (AM I A), Amide I B$_2$ (AM I B$_2$) and Amide III$_3$ (β) (AM III$_3$ (β)) are marked.

The DUVRR spectrum of the cross-β sheet core of lysozyme fibrils resolved by the proposed approach is shown in Fig. 3. The Bayesian analysis enabled us to find the exact position of the Amide III$_3$ (β) band related to the facial Ψ angle, which was found to be 133°. The isolating of the minor Amide I B$_2$ that is obscured in the experimental DUVRR spectrum of fibrils allowed for confident assignment of the lysozyme fibrillar core to anti-parallel β-sheet.

ACKNOWLEDGMENTS

We gratefully acknowledge Professor Kevin Knuth for valuable advice and discussion.

REFERENCES

1. K.H. Knuth, "Bayesian Source Separation and Localization" in SPIE'98 Proceedings: Bayesian Inference for Inverse Problems, San Diego, pp. 1998, 147-158.
2. M. Zibulevsky and B.A. Pearlmutter, B. A. *Neural Comp.* **13**, 863-882 (2001).
3. F. Chiti and C.M. Dobson, *Annu. Rev. Biochem.* **75**, 333-366 (2006).
4. R. Tycko, *Quart. Rev. Biophys.* **39**, 1-55 (2006).
5. R. Nelson, M.R. Sawaya, M. Balbirnie, A.O. Madsen, C. Riekel, R. Grothe, D. Eisenberg, *Nature (London)* **435**, 773-778 (2005).
6. V. Shashilov, M. Xu, V.V. Ermolenkov, L. Fredriksen, I.K. Lednev, *J. Am. Chem. Soc.* **129**, 6972-6973 (2007).
7. D.R. Booth, M. Sunde, V. Bellotti, C.V. Robinson, W.L. Hutchinson, P.E. Fraser, P.N. Hawkins, C.M. Dobson, S.E. Radford, C.C.F. Blake, M.B. Pepys, *Nature (London)* **385**, 787-793 (1997).
8. S.W. Englander, T.R. Sosnickt, J.J. Englandert, L. Maynet, *Curr. Opin. Struct. Biol.* **6**, 18-23 (1996).
9. A.V. Mikhonin, S.A. Asher, *J. Phys. Chem.* **109**, 3047-3052 (2005).
10. M. Xu, V. Shashilov, V.V. Ermolenkov, L. Fredriksen, D. Zagorevski, I. Lednev, *Protein. Sci.*, **16**, 815-532 (2007).
11. V.A. Shashilov, M. Xu, V.V. Ermolenkov, I.K. Lednev, *J. Quant. Spectrosc.& Radiative Transfer*, **102**, 46–61 (2006).
12. D.S. Sivia, C.J. Carlile, *J. Chem. Phys.* **96**, 170-178 (1992).
13. W. Windig, N.B. Gallagher, J.M. Shaver, B.M. Wise, *Chemom. Intell. Lab. Syst.* **77**, 85-96 (2005).
14. L. Elliott, D.B. Ingham, A.G. Kyneb, N.S. Mera, M. Pourkashanian, C.W. Wilson, *Progress in Energy and Combustion Sci.* **30**, 297–328 (2004).

Lessons about likelihood functions from nuclear physics

Kenneth M. Hanson

T-16, Nuclear Physics, Los Alamos National Laboratory,
Los Alamos, New Mexico, USA 087545
kmh@lanl.gov

Abstract. Least-squares data analysis is based on the assumption that the normal (Gaussian) distribution appropriately characterizes the likelihood, that is, the conditional probability of each measurement d, given a measured quantity y, $p(d|y)$. On the other hand, there is ample evidence in nuclear physics of significant disagreements among measurements, which are inconsistent with the normal distribution, given their stated uncertainties. In this study the histories of 99 measurements of the lifetimes of five elementary particles are examined to determine what can be inferred about the distribution of their values relative to their stated uncertainties. Taken as a whole, the variations in the data are somewhat larger than their quoted uncertainties would indicate. These data strongly support using a Student t distribution for the likelihood function instead of a normal. The most probable value for the order of the t distribution is 2.6 ± 0.9. It is shown that analyses based on long-tailed t-distribution likelihoods gracefully cope with outlying data.

Key Words: likelihood, Student t distribution, long-tailed likelihood functions, systematic uncertainty, inconsistent data, outliers, robust analysis, least-squares analysis

INTRODUCTION

The likelihood plays a central role in any inference process. The likelihood is the conditional probability $p(d|y)$, where d the measurement of a physical quantity y. In Bayesian analysis, the posterior, which fully describes the outcome of the analysis, is the product of the likelihood and the prior. The information brought to the analysis by the data comes from the likelihood. Least-squares (LS) analysis is a consequence of using the normal (Gaussian) distribution for the likelihood. The more general likelihood analysis, from which LS analysis is derived, is not so restricted; other forms for the likelihood are permissible.

Nuclear physics experiments offer substantial evidence of disagreement among repeated measurements. One indication is that the minimum χ^2 in data analyses is often significantly larger than the number of data points. In that situation, the analyst typically inflates the final uncertainty to cover the range of dispersion of the data. The large value of χ^2 indicates that either the data do not match the fitted model and/or the stated standard errors are too small and/or the assumption of a normal distribution is incorrect.

The fundamental question is, how well do experimentalists estimate the uncertainties in their results? The approach adopted here to answer that question is to examine a collection of repeated measurements and see what can be inferred about the distribution of measurements relative to their estimated uncertainties.

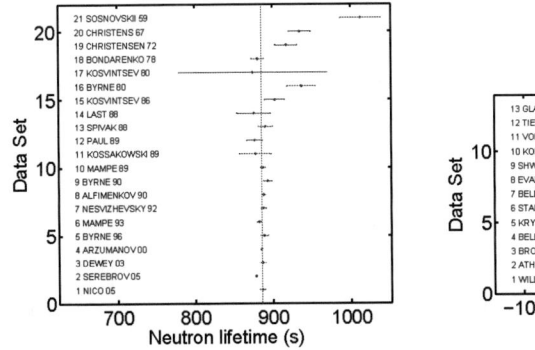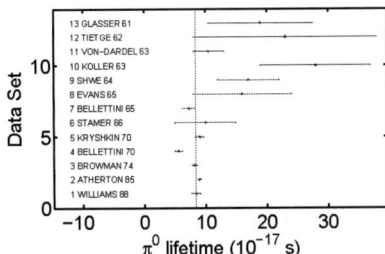

FIGURE 1. Plots of all measurements, in chronological order, of the lifetimes of the neutron and π^0 particles.

Lifetime measurements of elementary particles are useful for this purpose because there are approximately twenty repeat measurements for each of the longest known particles. I will show that these data strongly support using a Student t distribution for the shape of the likelihood function instead of a normal distribution. Furthermore, experimenters tend to underestimate the uncertainties in their results. It is shown that analyses based on long-tailed likelihoods, like the t distribution, gracefully cope with outlying data.

PARTICLE LIFETIME DATA

In 1957 Gell-Mann and Rosenfeld [1] published an authoritative review of the properties of elementary particles. Their work quickly led to the formation of the Particle Data Group, which now summarizes the known properties of elementary particles on an annual basis. For each particle property, the committee: (a) lists all relevant experimental data, (b) decides which data to include in its final analysis (outliers often rejected), and (c) states the best current value and its estimated standard error. The final results are typically obtained using the least-squares average of the accepted data. The standard error is often magnified by $\sqrt{\chi^2/(n-1)}$ to take into account the dispersion of the data.

Incidentally, in half of all 64 PDG tables involving three or more entries the standard error is adjusted, and when it is, the average scale factor is 2.0. These numbers indicate the frequent occurrence of significant disagreements among particle-physics measurements, relative to their quoted uncertainties. These observations are even more remarkable because the data involved have been carefully selected by the PDG.

The PDG reports [2] are an excellent source of information about measurements of unambiguous physical quantities. They are available online, and provide insight into how physicists interpret data.

Figure 1a shows all measurements of the lifetimes of the neutron compiled from Ref. [2] and earlier PDG reports. The vertical line is the PDG value, which includes

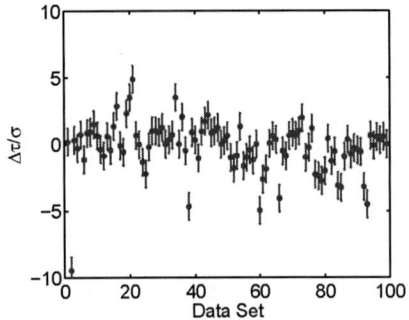 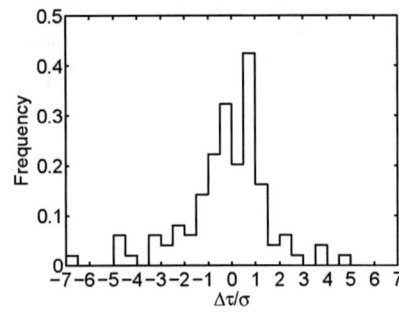

FIGURE 2. (a, left) Composite of lifetime measurements of five elementary particles, n, $\mu^{\pm}$, π^0, K_s^0, and Λ. The discrepancy of each measurement from the recent PDG value is divided by the quoted standard error. (b, right) Histogram of the normalized discrepancies.

the seven most recent data sets, except for #2 [3], because it disagrees with the rest by 9.5 standard errors. For all 21 points, χ^2 (relative to PDG) = 149. It is clear from the plot that the data set contains several outliers, that is, measurements that disagree with the PDG value by more than three times their stated standard errors.

Figure 1b shows all measurements of the lifetimes of the π^0 meson. The PDF value, indicated by the vertical line, is based on the four most recent values, excluding the latest one (#1). Included in the PDG average is #4, which is 4.7 standard errors away from the PDG average. For the 13 data points, χ^2 (relative to PDG) = 40.

Figure 2a shows the discrepancies of 99 lifetime measurements for five particles from PDG values, divided by their standard errors, $\Delta\tau/\sigma$. These five particles are chosen because they were among the first discovered of the unstable particles. Figure 2b shows the histogram of these normalized discrepancies. The objective of the present study is to characterize the distribution of discrepancies relative to their estimated uncertainties, $\Delta\tau/\sigma$. For these 99 data points, $\chi^2 = 367$, indicating their rms fluctuation is twice their stated standard errors.

For a first cut at analyzing any data set, the analysis suggested by John Tukey[4] is useful. The steps are: (a) find the quartile positions in the data set, Q1, Q2, Q3; (b) calculate the inter-quartile range, IQR = Q3 - Q1; (c) determine the fraction of data in the intervals $y <$ Q1 - 1.5 IQR and $y >$ Q3 + 1.5 IQR, called the suspected outlier fraction (SOF). For the normal distribution, IQR = 1.35 σ and SOF = 0.7%. The IQR is a measure of the width of the core of the distribution and the SOF the extent of its tail, relative to the width of the core. Q2 is the median, of course, which is a good estimate of the measurand.

For the composite lifetime data shown in Fig. 2a, IQR is 1.83. Thus, the width of the core of this distribution is 1.36 times larger than the value 1.35 that would obtain if the distribution were normal and the standard errors were correctly estimated. Furthermore, SOF = 6.6%, indicating this distribution has about ten times as many data in its tails than expected for a normal, denoting a distribution with long (fat) tails.

UNCERTAINTIES IN PHYSICS EXPERIMENTS

When an experimenter states his/her measurement of a physical quantity y as $y = d \pm \sigma$, the standard error σ represents experimenter's estimated uncertainty in d. This statement is interpreted probabilistically as a likelihood function, the conditional probability of d, given the measurand value y and the stated standard error σ, $p(d|y\sigma I)$, where I represents any relevant background information, for example, how the experiment is performed. The likelihood is a probability density function (PDF) in the variable d, so it is properly normalized to unit area with respect to d. However, the likelihood is usually viewed as function of y, and is not necessarily normalized wrt. to y. The likelihood is usually taken to be a normal distribution (Gaussian) with standard deviation σ.

Experimental uncertainties are usually thought of as consisting of two types. The first type is statistical uncertainty, which often arises from noise in the measured signals or events being counted. In the latter case, the uncertainty is usually estimated using the Poisson distribution. In the former, the rms fluctuations in the signals can be measured. These sources of uncertainty are usually Type A uncertainties, that is, they can be quantified by repeated measurements and estimated using frequentist statistical methods. Statistical uncertainties are likely to be estimated reliably.

The second type is called systematic uncertainty because it often affects many or all of the experimental results from an experiment. These arise from uncertainties in equipment calibration, experimental procedure, or corrections to the data. In nuclear physics, typical systematic uncertainties arise from detector efficiencies and deadtimes, target densities and thickness, and integrated beam fluence. Systematic uncertainties are usually Type B uncertainties; they are often determined by nonfrequentist methods, and may be based on the experimenter's judgment. Hence, these uncertainties may be subjective, difficult to assess, and possibly not well known.

Statistical and systematic uncertainties are usually added in quadrature (rms sum).

Uncertainty in the uncertainty

Suppose there is uncertainty in the stated standard error σ_0 for measurement d. Dose and von der Linden [5] presented the following plausible derivation of a suitable likelihood function. They assume the likelihood has an underlying normal distribution,

$$p(d|y\sigma I) \propto \exp\left[-\frac{1}{2}\left(\frac{d-y}{\sigma}\right)^2\right], \quad (1)$$

where σ is the standard deviation of the distribution. Because the experimenter's stated standard error σ_0 is uncertain, σ is assigned a probability density function (PDF). Rather than working directly with σ, Dose and van der Linden consider the variable ω, where σ is scaled by $\sigma = \sigma_0/\sqrt{\omega}$, for which they assign a Gamma distribution:

$$p(\omega|I) \propto \omega^{a-1}\exp(-a\omega). \quad (2)$$

The mean of this distribution is $\omega = 1$ and its variance is $1/a$. The value of a should be based on how uncertain one is in the uncertainty quoted by the experimenter. As a

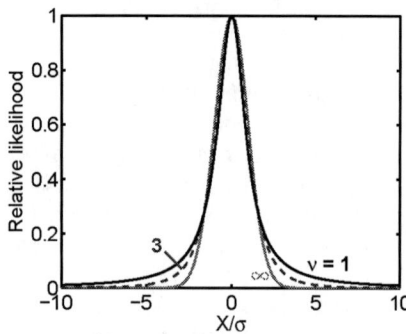
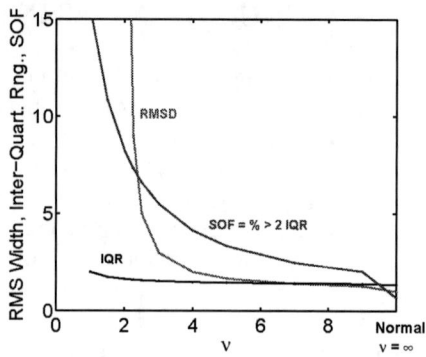

FIGURE 3. (a, left) Plots of t distributions for orders ν = 1, 3, and ∞ (normal). (b, right) Properties of t distributions as a function of ν, the rms deviation, the IQR, and the SOF. For ν < 9, SOF > 2%. The rms deviation diverges at ν of 2 and below.

approaches infinity, $p(\omega|I)$ approaches a delta function at $\omega = 1$. When viewed as a PDF in s, $p(s|I)$ has a plausible shape, with means of 1.59, 1.32, and 1.09, and rms deviations of 1.06, 0.69, and 0.30 at ν =1, 3, and 9, respectively.

The advantage of the above parameterization is that the integration over ω can be done analytically, resulting in the likelihood

$$p(d|y\sigma_0 I) \propto \left[1 + \frac{1}{\nu}\left(\frac{d-y}{\sigma_0}\right)^2\right]^{-\frac{\nu+1}{2}} \propto t_\nu\left(\frac{d-y}{\sigma_0}\right), \quad (3)$$

which is a Student t distribution of order $\nu = 2a$.

See Refs. [6–13] for other contributions to the outlier discussion.

Properties of Student t distributions

Figure 3a shows the Student[1] t_ν distribution for three ν values, ν = 1, 3, and ∞. For ν = ∞, the t distribution is the same as the normal distribution. For ν = 3, it is the Cauchy distribution, also called the Breit-Wigner or Lorentzian in physics. The tails of $t_\nu(x)$ asymptotically fall as $|x|^{-(\nu+1)}$. Because of that, the mean of the t distribution does not exist for $\nu \leq 1$, and its variance is infinite for $\nu \leq 2$.

Figure 3b displays the strong dependence of the rms width of $t_\nu(x)$ as a function of ν. The Tukey quantity, Inter-Quartile Range (IQR), does not change much with ν. However, the Suspected Outlier Fraction (SOF) is already 2% at ν = 9, and increases sharply as ν drops below 4. A good first guess as to what ν matches a given data set can be obtained from this curve and the SOF for the data set.

---

[1] Student (1908) was pseudonym for W.S. Gossett, who was not allowed to publish under his own name by his employer, Guinness brewery

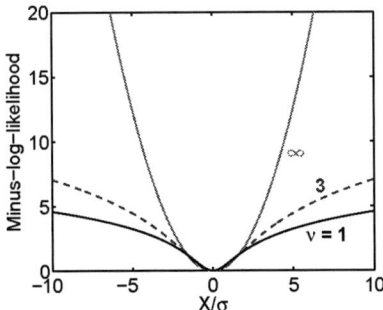 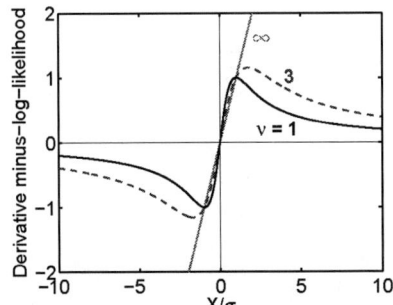

FIGURE 4. (a, left) Minus-log-likelihood dependence on the normalized residual for t distributions of order $\nu = 1, 3$, and ∞, the latter being identical to the normal distribution. (b, right) Derivatives of the same functions. For small ν values, the restoring force of t-distribution likelihood functions drops off at large residuals, as opposed to the linearly increasing force of the normal distribution.

Coping with outliers

The general Bayesian approach for coping with outliers is to use for the likelihood function a long-tailed, sometimes called a fat-tailed, distribution. Experience shows us that the exact form of the tail is not very important for ameliorating the effect of outliers. For examples of long-tailed likelihoods and their response to outliers, see [5, 8–10, 14].

The easiest way to understand how long-tailed likelihood functions deal with outliers is to employ the useful analogy between $\varphi(\Delta x)$ = minus-log-likelihood and a physical potential. Since the gradient of a potential is a force, $-\nabla \varphi$ is interpreted as the force with which the datum pulls on the model. Figure 4b shows the behavior of the derivative of the minus-log-t distribution. The slopes of the three curves at $x = 0$ are different, but that is not an issue here because of the scaling factor s used in the present analysis.

For likelihoods to be tolerant of outliers, the restoring force eventually decreases, or at least saturates, for increasing residuals. The extent to which a likelihood function accommodates outliers can be deduced from how fast the derivative of its logarithm falls off for large residuals. The normal distribution is not outlier tolerant because it pulls ever more strongly on the solution as its residual increases.

RESULTS

Analysis of composite data set

This section summarizes the results of analyzing of the full data set composed of the lifetime measurements for the five particles, shown in Fig. 2. The goal is to determine whether a t distribution appropriately describes the distribution of measurements.

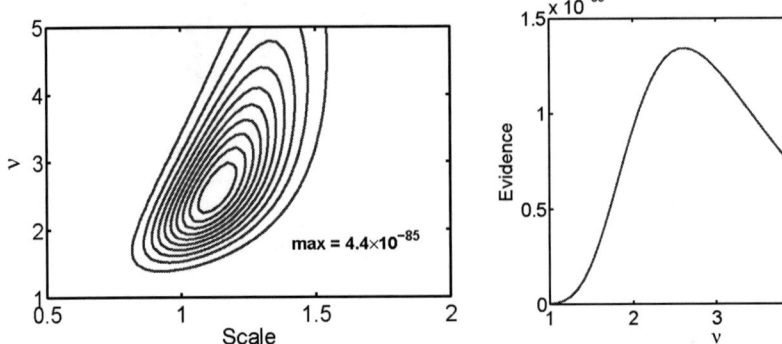

FIGURE 5. (a, left) Contour plot of the evidence as a function of order ν and the scale, s. (b, right) The distribution in ν has a broad peak with its maximum at ν = 2.6.

From the derivation above, the likelihood for each datum is given by the t distribution

$$p(d_i|\tau s \sigma_i I) \propto \left[1 + \frac{1}{\nu}\left(\frac{d_i - \tau}{s\sigma_i}\right)^2\right]^{-\frac{\nu+1}{2}}, \quad (4)$$

where the variable s has been introduced to scale all the uncertainties in the data set. The likelihood for a data set with n data points is the product of n such factors, $p(\mathbf{d}|\tau s \boldsymbol{\sigma}) \propto \prod_{i=1}^{n} p(d_i|\tau s \sigma_i)$, assuming the measurements are statistically independent[2]. The posterior for the lifetime is obtained by marginalizing the joint distribution for τ and s over the nuisance parameter s

$$p(\tau|\mathbf{d}\boldsymbol{\sigma}) = \int p(\tau s|\mathbf{d}\boldsymbol{\sigma})\,ds = \int p(\mathbf{d}|\tau s \boldsymbol{\sigma})p(\tau s)\,ds. \quad (5)$$

The dispersion of the data points is incorporated in this posterior, much in the same way as with the χ^2 scaling often applied in LS analysis.

To select between the two models, the t or normal distribution (T or N), Bayes rule [15] gives the odds ratio as

$$\frac{p(T|\mathbf{d}\boldsymbol{\sigma} I)}{p(N|\mathbf{d}\boldsymbol{\sigma} I)} = \frac{p(\mathbf{d}|T\boldsymbol{\sigma} I)}{p(\mathbf{d}|N\boldsymbol{\sigma} I)}\frac{p(T|I)}{p(N|I)}, \quad (6)$$

where $p(T|I)/p(N|I)$ is ratio of the priors for the two models, and $p(\mathbf{d}|T\boldsymbol{\sigma} I)$ is the evidence. The evidence is evaluated as the integral over the joint distribution in τ and s

$$p(\mathbf{d}|T\boldsymbol{\sigma} I) = \int p(\mathbf{d}|\tau s T\boldsymbol{\sigma} I)\,p(\tau s|T\boldsymbol{\sigma} I)\,d\tau ds, \quad (7)$$

---

[2] An alternate approach is to infer the properties of the likelihood function is to directly fit the histogram, as in Fig. 2b, using a multinomial distribution for the likelihood of each bin count. However, in the limit of infinitely narrow histogram bins, it is easy to show that this approach is equivalent to the one used here.

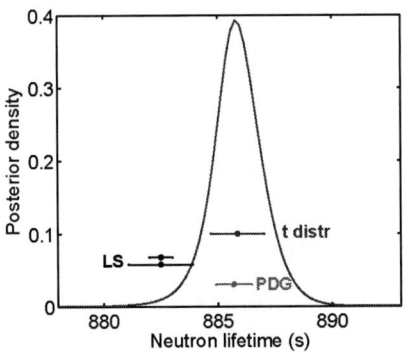
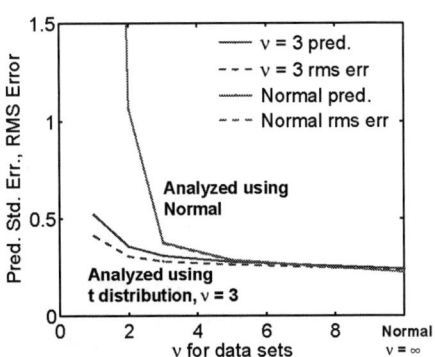

FIGURE 6. Posterior distribution for the t-distribution analysis of the neutron-lifetime data. The estimated lifetime and its standard error are shown for the t analysis and LS, based on the full set of 21 data. The PDG estimate is based on combining the seven most recent measurements, excluding that of Serebrov et al.

FIGURE 7. Results of tests comparing the performance of analyses based on t-distribution likelihoods to those based on using the normal distribution. The t-distribution analysis does much better than the normal for data sets from small ν, although the predicted or estimated uncertainties are slightly larger than the actual rms error.

shown in Fig. 5a. In this study, the evidence integral is estimated from the integrand values evaluated on a uniform grid. The maximum in the evidence in Fig. 5b occurs at $\nu \approx 2.6 \pm 0.9$. The odds ratios of the t distribution (at $\nu = 2.6$) to the normal is $1.3 \times 10^{-85}/2.2 \times 10^{-90} = 5.5 \times 10^4$, assuming the odds ratio of the priors on the models is unity and the priors on the parameters are the same for both models. The priors on τ, s, and ν are constant. Thus, the t distribution is strongly preferred by the data to the normal distribution. For the normal distribution, $\bar{s} = 1.95 \pm 0.14$. ; $s = 1$ is definitely rejected.

Excluding the data point with the largest discrepancy (9.5 sigma) yields an odds ratio of $2.1 \times 10^{-82}/5.0 \times 10^{-84} = 42$, which is considerably less than above, but the t distribution is still preferred over the normal. In this case, the scaling for the normal distribution is $\bar{s} = 1.71 \pm 0.12$.

Analysis of the neutron lifetime measurements

Figure 6 shows the result of analyzing all 21 measurements of the neutron lifetime shown in Fig. 1a using for the likelihood t-distributions with $\nu = 2.6$. The posterior in s is obtained by marginalizing the joint distribution in τ and s over s, as mentioned above. The posterior distribution for s has a mean value of $\bar{s} = 1.16$, indicating a fairly small scaling factor is required when using a t-distribution likelihood. By contrast, the least-squares result, shown in Fig. 6 with χ^2 scaling, applies a scaling factor of $s = 2.73$. This large scaling of the uncertainty is required because the data are more disperse than a normal distribution with the quoted standard errors.

The PDG estimate, based on the seven most recent measurements, excluding the one by Serebrov et al. [3] is reasonably consistent with the present result, although its standard error is somewhat smaller because of the relative good consistency of the seven chosen data. The standard error from the t-distribution analysis lies between that of the LS and PDG results, while deftly coping with the Serebrov outlier.

Tests of performance of t-distribution analysis

How well does the analysis based on a t distribution of a given order work for data drawn from different distributions? To answer that question, a series of Monte Carlo tests are performed. In these tests, 10000 data sets are created, each with 20 data values drawn from a specific t distribution. Each data set is analyzed with a likelihood function, either the t distribution of order $\nu = 3$ or the normal, which is equivalent to a least-squares (LS) analysis. The results from all data sets are summarized in terms of the mean estimate and its rms error. The rms estimated standard error is also calculated.

Figure 7 shows the results of these tests. The performance of the LS (normal) analysis on data drawn from t distributions with $\nu \leq 3$ is poor to very bad. This result is expected because these data tend to have a number of outliers. The analysis based on the t distribution with $\nu = 3$ excels when the data contain a significant fraction of outliers, that is, when they are drawn from t distributions with $\nu \leq 3$. For normally distributed data ($\nu = \infty$), however, the LS analysis has slightly smaller rms error than the t-distribution analysis by about 4%.

To summarize these results, LS analysis exposes the analyst to dire consequences in the presence of outliers in the data. On the other hand, using likelihoods based on t distributions with $\nu \approx 3$ gracefully deals with outliers while achieving very close to the same accuracy as the LS analysis when the data are normally distributed.

DISCUSSION

The present study demonstrates that particle lifetime data are much better described by a likelihood function based on the Student t distribution with $\nu \approx 2.6$ to 3.0 than by a normal distribution. Furthermore, likelihood or Bayesian analyses based on the t distribution cope well with outliers, while treating each datum in the same way. There is no need to identify outliers and specially deal with them. Furthermore, t-distribution analysis produces stable results when outliers exist in data sets, whereas the normal distribution does not. These results for particle lifetimes do not represent all physical measurements, but are worth keeping in mind.

A useful conclusion of this study is that repeat experiments are worthwhile to gain confidence and mitigate against outliers, even though they might not substantially improve on the accuracy of earlier experiments. While the use of t distributions reduces the influence of outliers, when an outlier is detected, every effort should be made to try to understand the details of the experiments and possible causes for the disagreements.

As a word of caution, long-tailed likelihood functions may result in posteriors with multiple maxima, which may complicate the analysis. While the posterior mean is the best estimator, it can be computationally costly to evaluate, especially for nonlinear models with many parameters.

It is expected that the experimental uncertainty in a given data set may contain statistical components that follow normal (or Poisson) distribution and systematic uncertainties that potentially follow t distributions. In that case, the likelihood is a convolution of normal and t distributions, which can not easily be represented analytically.

Some outlier models [5, 9] adopt the notion that the data set contains both good data and bad data. The likelihood is written as a mixture of normal and t distributions (or other long-tailed function), for example, $(1-\beta)N + \beta T$, where N stands for the normal and T the t distribution. This treatment allows for either T (with probability β) or N (with probability $(1-\beta)$), which form may satisfactorily approximate the convolution of normal and t distributions suggested in the previous paragraph.

ACKNOWLEDGMENTS

The author is grateful to Anthony Hill, Frederik Tovesson, Robert Haight, John Ullmann, Toshihiko Kawano, Patrick Talou, Gerald Hale, and Ludovic Bonneau for many useful discussions. This work was done under U.S. DOE Contract DE-AC52-06NA25396.

REFERENCES

1. M. Gell-Mann and A. H. Rosenfeld, "Hyperons and heavy mesons (systematics and decay)," *Ann. Rev. Nucl. Sci.* **7**, pp. 407–478, 1957.
2. W.-M. Yao et al., "Review of particle physics," *J. Phys. G* **33**, p. 1, 2006. Available at http://pdg.lbl.gov.
3. A. Serebrov et al., "Measurement of the neutron lifetime using a gravitational trap and a low-temperature Fomblin coating," *Phys. Let. B* **605**, pp. 72–78, 2005.
4. J. W. Tukey, *Exploratory Data Analysis*, Addison-Wesley, Reading, 1977.
5. V. Dose and W. von der Linden, "Outlier-tolerant parameter estimation," in *Maximum Entropy and Bayesian Methods*, W. von der Linden et al., ed., pp. 47–56, Kluwer Academic, (Dordrecht), 1999.
6. A. O'Hagan, "On outlier rejection phenomena in bayes inference," *J. Roy. Statist. Soc. B* **41**, pp. 358–367, 1979.
7. F. H. Fröhner, "Bayesian evaluation of discrepant experimental data," in *Maximum Entropy and Bayesian Methods*, J. Skilling, ed., pp. 467–474, Kluwer Academic, (Dordrecht), 1989.
8. K. M. Hanson and D. R. Wolf, "Estimators for the Cauchy distribution," in *Maximum Entropy and Bayesian Methods*, G. R. Heidbreder, ed., pp. 255–263, Kluwer Academic, 1996.
9. D. S. Sivia, "Dealing with duff data," in *Maximum Entropy and Bayesian Methods*, M. Sears et al., ed., pp. 131–137, N.M.B. Printers, (Port Elizabeth), 1996.
10. W. H. Press, "Understanding data better with Bayesian and global statistical methods," in *Unsolved Problems in Astrophysics*, J. N. Bahcall and J. P. Ostriker, eds., pp. 49–60, Princeton University, (Princeton), 1997.
11. K. M. Hanson, "Bayesian analysis of inconsistent measurements of neutron cross sections," in *Bayesian Inference and Maximum Entropy Methods in Science and Engineering*, K. H. Knuth et al., ed., *AIP Conf. Proc.* **803**, pp. 431–439, AIP, Melville, 2005.
12. G. E. P. Box and G. C. Tiao, "A Bayesian approach to some outlier problems," *Biometrika* **55**, pp. 119–129, 1968.
13. A. P. Dawid, "Posterior expectations for large observations," *Biometrika* **60**, pp. 664–667, 1973.
14. K. M. Hanson, "Bayesian analysis in nuclear physics (LANSCE tutorial)," Tech. Rep. LA-UR-05-5680, Los Alamos National Lab., 2005. Web: http://public.lanl.gov/kmh/talks/lansce05vgr.pdf.
15. D. S. Sivia and J. Skilling, *Data Analysis - A Bayesian Tutorial: Second Edition*, Clarendon, Oxford, 2006.

Assessment of Electron Energy Distributions in Discharges by Optical Emission Spectroscopy

Dirk Dodt*, Andreas Dinklage*, Rainer Fischer[†] and Roland Preuss[†]

*Max-Planck-Institut für Plasmaphysik, EURATOM Association
Wendelsteinstraße 1, 17491 Greifswald, Germany
[†]Boltzmannstraße 2, 85748 Garching, Germany

Abstract. A procedure for the reconstruction of electron energy distribution functions (EEDF) of low-temperature neon plasmas from optical emission spectroscopy data is presented. A data descriptive model, including the physics of the plasma discharge and the spectroscopic measurement is developed. Particular refinements of the model regarding the apparatus function of the spectrometer and the optical depth of the plasma for light emission with metastable final states are discussed. The effect of uncertainties in the atomic data entering the model is assessed. Maxwellian and Druyvestein parameterizations for the EEDF are employed. Discrepancies of the reconstructed EEDF with results from independent modelling are not within the error bounds of the data descriptive model.

Keywords: Electron Energy Distribution Function, Optical Emission Spectroscopy, Neon Plasma, EEDF
PACS: 52.25.Os

INTRODUCTION

Low-temperature plasmas are widely used nowadays, e.g. in industrial production processes or for lighting purposes [1]. The physics of these discharges is determined by the free-electron gas in the plasma, even with ionization degrees as small as $10^{-6}...10^{-5}$ ions per gas atom. The energy dissipation in inelastic collisions and heating processes requires a kinetic description of the electrons that takes into account the strong deviations from thermodynamic equilibrium. This non-equilibrium behaviour manifests in observed electron energy distribution functions (EEDF) that may substantially deviate from the Maxwellian distribution. A description with two-temperature distribution functions, like the Druyvestein distribution, is more appropriate.

Experimentally, EEDFs are usually determined by electrical probe measurements. This approach suffers from the substantial contact of the electrically biased probe with the plasma. Moreover, the spatial resolution of probe measurements is limited because of the formation of sheaths close to the probe. Therefore, a non-invasive assessment of EEDFs is attractive both for the validation of probe measurements and for physical modelling and process control.

The approach presented here aims at an assessment of the EEDF from emission spectroscopy in the optical regime. The approach employs the light emitted from gas atoms (line emission) that are excited by electron collisions. Since the discharges can be observed with appropriately designed imaging optics, emission spectroscopy may attain high spatial and temporal resolutions [2]. As the deviations from equilibrium distributions occur in the energy range of the inelastic processes, the light emission can be expected to directly reflect this phenomenon.

Therefore, the idea to use emission spectroscopy for EEDF assessments is long standing, see e.g. [3]. First attempts to use this approach were based on line-ratio techniques, mapping the intensities of different spectral lines onto temperatures. For the plasmas under consideration here, however, the line intensity ratios are affected by too many processes to infer the EEDF directly. Therefore, we extend the approach described in Ref. [4]. The data descriptive model is changed to directly model the raw data, rather than the analysis of pre-analyzed line intensities.

The basic idea is to fit the full physical model to a large number of spectral lines rather than inferring information from a few spectral lines as done in previous approaches. The benefit of this method is expected to result from the consistent use of correlations in the data, which are contained in the physical model of the measurement.

DATA MODEL

The data model maps the quantity of interest, the respective parameterization of the EEDF $f_e(E)$ onto a simulation of the measured data (spectrometer pixels). It consists of a chain of different elements, which are summarized in (1), more details about the data model can be found in [5].

$$\underbrace{\underbrace{f_e(E) \;\rightarrow\; n_i}_{\text{kinetic theory}} \;\rightarrow\; I_{ij} \rightarrow \underbrace{\int_{\text{l.o.s.}} I_{ij} dV}_{\text{radiation transport}} \rightarrow L(\lambda) \;\rightarrow\; \underbrace{D(\text{Pixel\#})}_{\text{measurement}}}_{\text{collisional radiative model}} \quad (1)$$

The EEDF determines the electron collision rates for the collisional-radiative model (CRM). Representations of EEDFs, derived from hybrid modelling of neon discharges accounting for a kinetic treatment of the electrons [6], [7] are considered for benchmarking purposes. The CRM yields the population densities n_i of the excited states of neon. The amount of emitted radiation is described by the *locally emitted power* I_{ij} measured in $[\text{W}/(\text{m}^3 \cdot \text{sr})]$. It is obtained by multiplication with the inverse lifetime of the excited states (Einstein coefficient for spontaneous emission), the photon energy and division by the full solid angle (4π). The radiation has to pass through the plasma before it leaves the discharge device. The apparent lifetime of the excited states is affected by the transport of photons if the absorber density is high, i.e. for transitions to the ground state of the atom. The description of this opacity gives together with the integration along the line-of-sight of the spectrometer, the *spectral radiance* $L(\lambda)$ as a function of the wavelength λ. The modelling of the actual measurement comprises the translation of $L(\lambda)$ into the detected signals and the mapping of wavelengths to pixel numbers. This requires details on the detector response, which were measured with a standard light source (sensitivity calibration). The calibration of the wavelength mapping is fitted to the data within the reconstruction.

Parameters of the model. In addition to the parameters f_e of the EEDF, a number of parameters $\vec{\eta}$ is used in the data model, which describe one of the following aspects:

- external properties of the discharge experiment and the spectrometric setup
- atomic data needed for the plasma model (e.g. cross section and life times of excited states)

Some of these parameters are not known a priori, or subject to significant uncertainty and are nuisance parameters for the reconstruction of the EEDF. The formalism of Bayesian data analysis is used to treat these parameters in a probabilistic way, allowing to quantify the amount of uncertainty introduced to the resulting EEDF. A list of nuisance parameters and the functional form of their prior information is given in table 1.

In the following subsections, specific examples for nuisance parameters are discussed. First, as a set of experimental nuisance parameters, the apparatus function of the spectrometer is described. Second, the incorporation of radiation trapping by metastable atoms is considered, a detail of the collisional-radiative model, which is not assessed in other publications known to the authors.

TABLE 1. Summary of the Parameters $\vec{\eta}$ used in the forward model of the spectroscopic measurement

Symbol	Parameter Description	remarks, prior
σ_{ij}	Electron impact excitation cross sections	scale varied, Gaussian prior
A_{ij}	Einstein coefficients	value varied, Gaussian prior
D_m	Diffusion constant of metastables	constant
Θ_{ij}	Escape factors of transitions to ground state	constants
Θ_{ij}	Escape factors of transitions to the metastable levels	flat prior for eff. densities
p_{Ne}	Gas pressure	constant
T_{Ne}	gas temperature	constant
r	Diameter of the discharge tube	constant
n_{Duran}	Refractive index of glass	constant
d	thickness of glass	constant
$n_{5si,4di}$	Populations of unmodelled atomic levels	Jeffreys prior
$n_{3si,3p,3d,..}$	radial profile of the excited state densities	constant
$\lambda_0, \lambda', \lambda''$	Wavelength calibration	flat priors
$C(\lambda)$	Intensity calibration	Gaussian prior

Apparatus Function

The data D_k as function of detector pixel number k are given by

$$D_k = C_k \cdot \sum_{ij} \int L_{ij}(\lambda') f(\lambda - \lambda') d\lambda' + \varepsilon_k \qquad (2)$$

where C_k is the sensitivity calibration of each pixel and the spectral radiance of the natural line width has to be convolved with the apparatus function f. The error ε of the data is discussed below. Since the width of the apparatus function is large compared to the natural line broadening, the latter may be neglected. In this case, the contribution to the spectral radiance of a specific line is given by the apparatus function multiplied with the radiance of the respective transition $\overline{L_{ij}}$. The radiance is obtained by the line-of-sight integration of the locally emitted power $I_{ij}(r)$ according to

$$\overline{L_{ij}} = \frac{1}{4\pi} A'_{ij} \frac{\hbar c}{\lambda_{ij}} \int_{-R(h)}^{R(h)} n_i(r) dr \qquad (3)$$

where R denotes the plasma boundary being a function of the distance h of the line-of-sight from the center of the plasma in cylindrical symmetry.

In first approximation the apparatus function of a spectrometer using a grating as its dispersive element can be described by a Gaussian function (see Fig. 1). A spline is fitted to the measured line profiles to model the deviance from the Gaussian behavior. For this purpose well separated spectral lines from different wavelengths are shifted and rescaled on top of each other and a smoothing spline is fitted to the sum of all data points.

To take into account the spectral variation of the apparatus function an error is determined from the residuals of the smoothing spline fit.

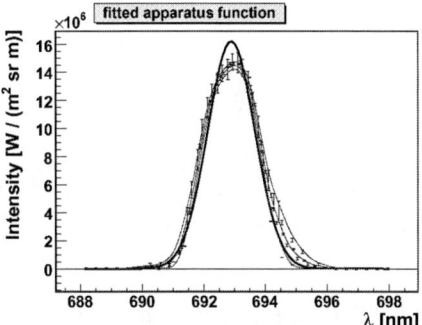
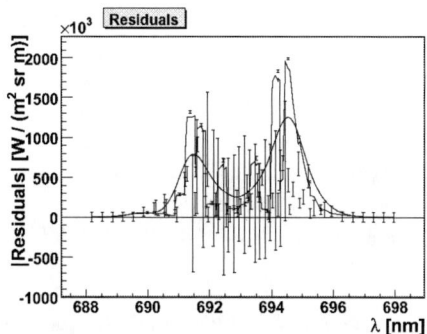

FIGURE 1. Generalized line profile of the spectrometer obtained by shifting and scaling different lines (left, see text). The smoothing spline with the assigned error band is shown together with a Gaussian fit for comparison. The error of the profile is determined using the residuals from the smoothing spline weighted with their respective statistical errors (right)

Optical Depth due to of Metastable Atoms

The light emitted by the excited atoms has to pass through the plasma before it leaves the discharge tube and enters the spectrometer. Atoms in the same atomic state as the emitting atom after its decay may absorb photons. This process leads to an opacity, which is measured by the optical depth for a specific wavelength. The mechanism affects primarily transitions to the ground state causing a higher population density e.g. of the so called resonant states 1s4 and 1s2 (Paschen's notation). In the collisional-radiative model the effect of optical depth is taken into account by considering effective Einstein coefficients $A'_{ij} = \Theta_{ij} A_{ij}$, which contain the *escape factors* $0 < \Theta_{ij} \leq 1$ ($\Theta_{ij} = 1$ meaning no effect of optical depth). Lawler and Curry [8] have published analytic formulas to obtain approximate escape factors in cylindrical geometry for a constant absorber density. These are used for the transitions to the ground state in our model.

The next highest populated states in the plasma are the metastable levels 1s5 and 1s3 of Neon. Their density is a function of the radial position in the cylindrical discharge tube and it vanishes at the glass walls. The result of Lawler and Curry's approximation which assumes constant absorber density gives a lower limit for the real escape factors for these absorbers. (e.g. transitions to the 1s5 level: Escape Factors have to be greater than ≈ 0.3, which lead to a variation of the modelled spectrum which is greater than the assigned statistical errors.)

To account for the optical depth of the metastable states effective densities are introduced, which are inserted to Lawler and Curry's approximation, instead of the actual population densities from the CRM. These densities are fitted to the spectral data and are expected to be somewhat lower than the population densities at the axis, which are the result of the collisional-radiative model.

DATA ANALYSIS

A Bayesian data analysis is used to find EEDFs leading to results of the data model consistent with the measured data. It is based on a Gaussian likelihood for the modelled intensities and incorporates prior information for nuisance parameters. The error statistics of the spectral measurement, which enter the likelihood described below is dominated by the electronic noise and is considered in Gaussian approximation. Its width is estimated by observing the fluctuations of the spectrometer response with a closed input aperture.

Likelihood. In general the likelihood $P(D|f_e,\vec{\eta},I)$ is the probability to obtain a certain outcome of the measurement D given a set of model parameters describing the EEDF f_e, a set of nuisance parameters $\vec{\eta}$ and other background assumptions I.

Here the probability to measure a pixel intensity D_k given the modelled intensity $D_{k,\text{sim}}(f_e,\vec{\eta})$ is given by the error statistics of the spectroscopic measurement and by the uncertainty of the apparatus function. According to [9] these two sources of uncertainty can be jointly described using a Gaussian likelihood with the effective width of $\sigma_k = \sqrt{\sigma_{k,\text{meas}}^2 + \sigma_{k,\text{app}}^2}$:

$$P(D|f_e,\vec{\eta},I) = \frac{1}{\prod_{k=1}^{N_d}\sqrt{2\pi}\sigma_k} \exp\left\{-\frac{1}{2}\sum_{k=1}^{N_d}\left(\frac{D_k - D_{k,\text{sim}}(f_e,\vec{\eta})}{\sigma_k}\right)^2\right\} \quad (4)$$

Bayesian Data Analysis. Bayes' rule is used to incorporate information which is not contained in the data of the spectroscopic measurement. At the current status of implementation this is used for the nuisance parameters listed in the data model section.

$$P(f_e|D,I) = \int P(D|f_e,\vec{\eta},I) \times \frac{P(f_e,\vec{\eta}|I)}{P(D|I)} d\vec{\eta} \quad (5)$$

The Posterior distribution $P(f_e,\vec{\eta}|D,I)$ is a product of the likelihood and the priors $P(f_e,\vec{\eta}|I)$. The $\vec{\eta}$ dependence is marginalized out by integration over $d\vec{\eta}$. The evidence $P(D|I)$ follows from the normalization constraint of the posterior distribution.

Expectation values and variance estimates for the parameters of interest are obtained by sampling the posterior with an implementation of the Metropolis Hastings Monte Carlo algorithm [11]

RESULTS

Figure 2 shows the result of the data model and measured data. The three rows show the spectrum of visible light, from green to near infrared, on a logarithmic scale for the spectral radiance. Almost all spectral features in the considered range are described by the model. For the EEDF reconstruction, a Druyvestein distribution function [10] was used. Figure 3 shows the corresponding result of the EEDF reconstruction. The variance of the reconstruction as obtained by the Monte Carlo method is also displayed. (green error bars, the size of the errors is similar to the width line in the plot and therefore hard to identify.) It incorporates the uncertainties of the spectral measurement, apparatus function and the uncertainty of the wavelength and intensity calibration. Obviously, the variance does not cover the discrepancy to the result from [6], which is shown for comparison. The normalized χ^2/N of the spectrum, that is modelled using a Druyvestein distribution, is $\chi^2/N \approx 7.6$, whereas the best fit Maxwellian distribution gives $\chi^2/N = 10.8$. The resulting electron densities are $n_e^D = (2.1\pm0.1) \times 10^{15}\,\mathrm{m}^{-3}$ and $n_e^M = (7\pm0.1) \times 10^{15}\,\mathrm{m}^{-3}$ for the Druyvestein and the Maxwellian distribution function, respectively. The reference value is $n_e = 2.93 \times 10^{15}\,\mathrm{m}^{-3}$ [6].

Both the large χ^2/N and the discrepancy of the electron density indicate shortcomings of the model. Nevertheless the Druyvestein distribution, which is closer to the reference distribution, gives a better description of the data and the electron density from [6]. The disagreement within errors with the well validated reference EEDF is taken as an indication, that the uncertainty of the reconstruction is not understood yet. To study this further the uncertainties of the underlying atomic data are addressed.

Influence of Atomic Data

Data for the electron impact excitation are available for Neon from theoretical calculations, e.g. [12]. The overall accuracy of the cross sections in comparison to experimental data is high for the excitation to the lowest excited levels [13] but differs close to threshold energies. The rate coefficient, however, is less affected because of the integration over the energy distribution function.

For a first assessment of the influence of uncertainties in the basis of atomic data, 36 nuisance parameters were introduced varying the scale of the excitation cross sections σ_{ij} of the more prominent processes. Gaussian priors where chosen, with an expectation value of one and a width of 0.25, assigning a 25% relative scale error to the cross sections. With this set of nuisance parameters it is possible to describe the data with an normalized $\chi^2/N \approx 0.8$. The variance propagating onto the reconstruction result (not

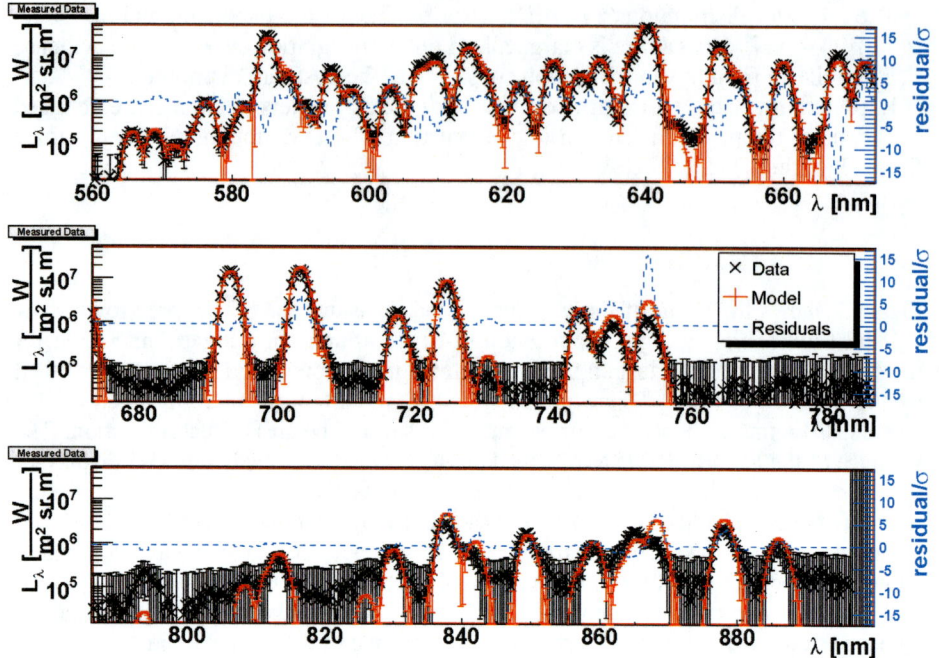

FIGURE 2. Spectral measurement (black crosses) together with the data model (grey/red curve). The parameters of the EEDF are set to their mean expectation values, the uncertainties of atomic data are not taken into account. The error bars correspond to the uncertainties of the measurement and the model (apparatus function). The dashed curve at half height is to be interpreted with the axis on the right hand side and shows the residuals between measurement and model in units of standard deviations. negative residuals correspond to the model having smaller values than the measurement.

shown), however, is comparable to the relative scale error introduced for the cross sections (i.e. 25%). In the plot the error band is still of similar width as the drawn lines. We conclude that the uncertainties considered so far are not able to explain the discrepancy to the EEDF shown for reference.

SUMMARY AND OUTLOOK

A data model for the reconstruction of electron energy distribution functions was set up. The fit of the apparatus function to the data is described. Systematic effects of the optical depth of transitions to the metastable levels were incorporated in the model. First studies of the uncertainties of atomic data do not reveal a satisfactory description of the error band of the reconstruction. The identification of further nuisance parameters and improvements of the model are the next steps. Especially the systematic effects caused by the choice of radial profiles of the excited state densities are to be studied.

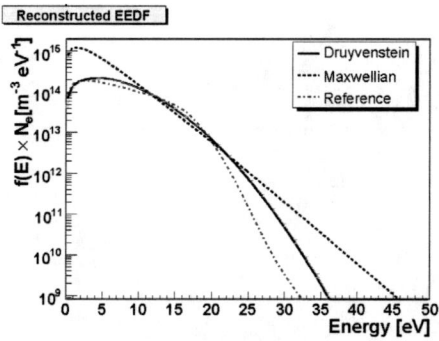

FIGURE 3. The result of the reconstruction is shown for Maxwellian and Druyvestein parameterizations of the EEDF. The dash dotted curve is the well validated result of hybrid modelling of a neon discharge [6], [7] and is shown for comparison. (The discharge parameters are chosen to match the ones of the modelled discharge.)

ACKNOWLEDGMENTS

This work was funded by Deutsche Forschungsgemeinschaft through Sonderforschungsbereich Trans Regio 24 (Fundamentals of Complex Plasmas).
 The authors are indebted to V. Dose, H. Dreier, M. Krychowiak, A. Werner, B. Pompe, D. Loffhagen ,F. Sigeneger and D. Uhrlandt for helpful discussions.

REFERENCES

1. J. Meichsner: *Low Temperature Plasmas*. In: *Plasma Physics: Confinement, Transport and Collective Effects*, Springer Lecture Notes Vol. 670, ed. by A. Dinklage et al (Springer, Berlin, 2005)
2. J. Röpcke, P.B. Davies, M. Käning, B.P. Lavrov:: *Diagnostics of non-equilibrium molecular plasmas using emission and absorption spectroscopy*. In: *Low Temperature Plasma Physics*, ed. by R. Hippler et al (VCH-Wiley, Berlin, 2001)
3. N. Brenning, J. Phys. D: Appl. Phys. **15** (1): L1 1982
4. R. Fischer, V. Dose, Plasma Phys. Control. Fusion **41** 1109 (1999)
5. D.Dodt, A. Dinklage, R.Fischer and D. Loffhagen, AIP CP **872** 264 (2006)
6. D. Uhrlandt, St. Franke, J. Phys. D: Appl. Phys. **35**, 680 (2002)
7. S. Franke, diploma thesis *Bilanzgleichungen zur Modellierung der Dynamik einer Neon-Glimmentladung*, Greifswald 1996
8. J. E. Lawler, J. J. Curry, J. Phys. D: Appl. Phys. **31**, 3235 (1998)
9. V. Dose, R. Fischer, and W. von der Linden, edited by G. Erickson, Maximum Entropy and Bayesian Methods, Kluwer Academic, Dordrecht, 147-152 (1998).
10. A. Rutscher *Characteristics of low-temperature plasmas under non-thermal conditions - a short summary*. In: *Low Temperature Plasma Physics*, ed. by R. Hippler et al (VCH-Wiley, Berlin, 2001)
11. Markov chain Monte Carlo in practice, edited by W.R. Gilks, S. Richardson and D.J. Spiegelhalter, Chapmann & Hall, 1996
12. O. Zatsarinny K. Bartschat, J. Phys. B: At. Mol. Opt. Phys. **37**, 2173 (2004)
13. M. Allan, K. Franz, H. Hotop, O. Zatsarinny and K, Bartschat, J. Phys. B: At. Mol. Opt. Phys. **39**, L139 (2006)

MAXENT 2007
GROUP PHOTO

1. Christian Anderson
2. Peter Sunehag
3. Karen Marutyan
4. Michel Chapron
5. Peter Martin
6. Mattias Nyberg
7. Ken Hanson
8. Prahalad Rangan
9. Udo von Toussaint
10. Tilman Neumann
11. Ramón Astudillo
12. Adom Giffin
13. Koki Kyo
14. Victor Shashilov
15. Timothy Sears
16. Man Kit Tse
17. Mark Ebden
18. Tom Loredo
19. Michael Osborne
20. Sam Tam
21. Angela Foudray
22. Stephen Roberts
23. Marcio Diniz
24. Fábio Mendes
25. Devinder Sivia
26. John Skilling
27. Rainer Fischer
28. Carlos Rodriguez
29. Shun-ichi Amari
30. Shahid Nawaz
31. Rafael Gutiérrez
32. Tilman Birnstiel
33. Andre Jalobeanu
34. Muhammad Mubeen
35. David Blower
36. Maxime Guillaud
37. Dirk Dodt
38. Robert Niven
39. Thomas Jefferys
40. William Jefferys
41. Danai Laksameethanasan
42. Chih-Yuan Tseng
43. Anna Pernestål
44. Carlo Cafaro
45. Jack Dodd
46. James Benson
47. Julio Stern
48. Ariel Caticha
49. Philip Erner
50. Julian Center
51. Phil Gregory
52. John Lee
53. Joshua Shimony
54. Larry Bretthorst
55. Kevin Knuth
56. Deniz Gençağa
57. Nabin Malakar
58. Vesselin Dimitrov
59. Sofia Fekih-Salem
60. Shona Schwab
61. Haley Maunu
62. Saleha Habibullah
63. Newshaw Bahreyni

MAXENT 2007 YEARBOOK

| Kevin | Koki | Danai | John | Thomas | Nabin |
| Knuth | Kyo | Laksameethanasan| Lee | Loredo | Malakar |

| Pete | Karen | Haley | Fábio | Robin | Muhammad |
| Martin | Marutyan | Maunu | Mendes | Morris | Mubeen |

| Shahid | Tilman | Robert | Mattias | Michael | John |
| Nawaz | Neumann | Niven | Nyberg | Osborne | Owens |

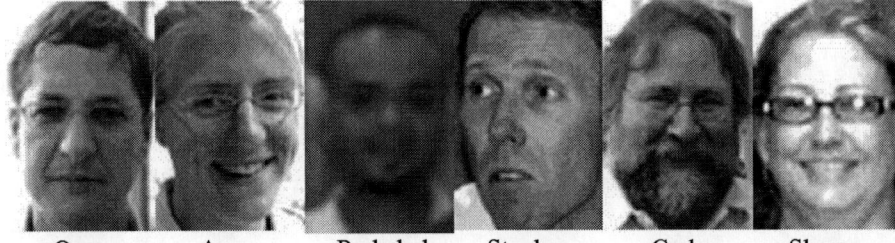

| Omer | Anna | Prahalad | Stephen | Carlos | Shona |
| Pelled | Pernestål | Rangan | Roberts | Rodríquez | Schwab |

Timothy Sears   Victor Shashilov   Joshua Shimony   Devinderjit Sivia   John Skilling   Julio Stern

  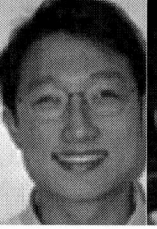 

Peter Sunehag   Sam Tam   Udo von Toussaint   Man Kit Tse   Chih-Yuan Tseng   Philip Tushscherer

Timothy Wallstrom   Ning Xiang

PHOTOGRAPHS FROM MAXENT 2007

Professor Shun-ichi Amari

Ramón Astudillo

José Bernardo and Keith Earle

Joshua Shimony

Angela Foudray

Geometry discussion with Ariel Caticha, Rainer Fischer,
Carlos Rodríguez, and Shun-ichi Amari

Tony Bell   Anna Pernestål

Philip Goyal gives his lecture to MaxEnt 2007

Shun-ichi Amari and Carlos Rodríguez

David Blower

Ariel Caticha explains his ideas to Philip Goyal

Adom Giffin

Saleha Haibibullah talks with Jack Dodd

The Musicians: Shun-ichi Amari, Philip Erner, Tony Bell

Devinder Sivia and Peter Sunehag

Deniz Gençağa

Adom Giffin and Carlo Cafaro enjoying the banquet

Man Kit Tse

A glimpse of the banquet

The Sagamore on Lake George

Carlos Rodríguez

Keith Earle and Rainer Fisher

Timothy Wallstrom and Robert Niven

Deniz Gençağa and Kevin Knuth work behind the scenes.

Udo von Toussaint and Seth Chaiken

Seth Chaiken and Ning Xiang

Nabin Malakar at the registration desk

Phil Erner and Kevin Knuth test their robot

John Skilling and Shun-ichi Amari

Chih-Yuan Tseng and Dirk Dodt

Jeannene and Ken Hanson

Julian Center

Julio Stern preparing for
MaxEnt 2008 in Brazil

Lake George

MaxEnt 2007 photographs provided by Ken Hanson, Kevin Knuth and Julian Center

Author Index

A

Anderson, C. C., 329, 337
Assemat, L., 402
Astudillo, R. F., 245

B

Ballard, J. A., 427
Barat, E., 381
Barnes, T. G., 315
Bell, A. J., 56
Blower, D. J., 109
Brandt, S. S., 354
Bretthorst, G. L., 329, 337, 346
Briggs, W. M., 101
Buitrago, E., 410

C

Cafaro, C., 165, 175
Camin, H. J., 427
Carbon, D. F., 322
Caticha, A., 11, 74, 165
Center, Jr., J. L., 229
Chan, C.-Y., 276
Chapron, M., 402
Choinsky, J., 322
Culver, R. L., 427

D

Dambis, A., 315
Dautremer, T., 381
Debbah, M., 435
de Faria, Jr., S. R., 268
de Souza Lauretto, M., 268
Dimitrov, V. I., 143
Diniz, M., 260
Dinklage, A., 195, 468
Dodt, D., 468

E

Epstein, A. A., 346
Erner, P. M., 203

F

Fekih Salem, S., 372
Finn, J. M., 293
Fischer, R., 195, 468
Foudray, A. M. K., 362
Frasso, S., 203

G

Giffin, A., 74
Goggans, P. M., 276
Gori, S., 212
Goyal, P., 153
Gregory, P. C., 307
Guillaud, M., 435
Gutiérrez, J. A., 237
Gutiérrez, R. M., 410

H

Hanson, K. M., 458
Hengartner, N., 293
Holland, M. R., 329, 337

J

Jalobeanu, A., 237
Jefferys, T. R., 315
Jefferys, W. H., 85, 315
Jemmott, C. W., 427
Jones, C. S., 293

K

Knockaert, L., 253
Knuth, K. H., 23, 203, 322, 443

Kolossa, D., 245
Kyo, K., 394

L

Laksameethanasan, D., 354
Lednev, I. K., 450
Lee, H. C., 386
Levin, C. S., 362

M

Martin, P., 185
Marutyan, K. R., 329, 337
Miller, J. G., 329, 337
Mohammad-Djafari, A., 372
Moustakas, A. L., 435

N

Neumann, T., 283
Niven, R. K., 133
Nyberg, M., 418

O

Oprea, A., 402
Orglmeister, R., 245

P

Pereira, B. B., 268
Pereira, C. A. B., 260, 268
Pernestål, A., 418
Preuss, R., 221, 468

R

Rangan, P., 443
Rodríguez, C. C., 47

S

Sears, T. D., 117, 125
Seghouane, A.-K., 253
Shashilov, V. K., 450
Shimony, J. S., 346
Sibul, L. H., 427
Skilling, J., 3, 39
Stern, J. M., 260, 268
Sultana, B., 402
Sunehag, P., 125

T

Tse, M. K., 322
Tseng, C.-Y., 386

U

Useche, G. M., 410

V

v. Toussaint, U., 212, 221
Vabre, A., 372

W

Wallstrom, T. C., 93
Wear, K. A., 329, 337